FROM EMBRYOLOGY TO EVO-DEVO

DIBNER
INSTITUTE
FOR THE HISTORY
OF SCIENCE AND
TECHNOLOGY

Dibner Institute Studies in the History of Science and Technology
George Smith, general editor

Jed Z. Buchwald and I. Bernard Cohen, editors, *Isaac Newton's Natural Philosophy*

Jed Z. Buchwald and Andrew Warwick, editors, *Histories of the Electron: The Birth of Microphysics*

Geoffrey Cantor and Sally Shuttleworth, editors, *Science Serialized: Representations of the Sciences in Nineteenth-Century Periodicals*

Michael Friedman and Alfred Nordmann, editors, *The Kantian Legacy in Nineteenth-Century Science*

Anthony Grafton and Nancy Siraisi, editors, *Natural Particulars: Nature and the Disciplines in Renaissance Europe*

J. P. Hogendijk and A. I. Sabra, editors, *The Enterprise of Science in Islam: New Perspectives*

Frederic L. Holmes and Trevor H. Levere, editors, *Instruments and Experimentation in the History of Chemistry*

Agatha C. Hughes and Thomas P. Hughes, editors, *Systems, Experts, and Computers: The Systems Approach in Management and Engineering, World War II and After*

Manfred D. Laubichler and Jane Maienschein, editors, *From Embryology to Evo-Devo: A History of Developmental Evolution*

Brett D. Steele and Tamera Dorland, editors, *The Heirs of Archimedes: Science and the Art of War Through the Age of Enlightenment*

N. L. Swerdlow, editor, *Ancient Astronomy and Celestial Divination*

FROM EMBRYOLOGY TO EVO-DEVO: A HISTORY OF DEVELOPMENTAL EVOLUTION

———

edited by Manfred D. Laubichler and Jane Maienschein

The MIT Press
Cambridge, Massachusetts
London, England

MIT Press books may be purchased at special quantity discounts for business or sales promotional use. For information, please e-mail special_sales@mitpress.mit.edu or write to Special Sales Department, The MIT Press, 55 Hayward Street, Cambridge, MA 02142.

This book was set in Bembo by SNP Best-set Typesetter Ltd., Hong Kong and was printed and bound in the United States of America.

Library of Congress Cataloging-in-Publication Data

From embryology to Evo-Devo / Manfred D. Laubichler and Jane Maienschein, editors.
    p. cm.
  Includes bibliographical references (p.)
  ISBN-13: 978-0-262-12283-2 (alk. paper)
  1. Developmental biology—History. 2. Evolution (Biology)—History. 3. Comparative embryology—History. I. Laubichler, Manfred. II. Maienschein, Jane.
  QH491.F76   2007
  571.809—dc22

                                                            2006047208

10 9 8 7 6 5 4 3 2

Contents

1

INTRODUCTION

Manfred D. Laubichler and Jane Maienschein

ONTOGENY AND PHYLOGENY OF THE VOLUME

Each summer, starting in 1989, the Dibner Institute has offered a seminar in the history of biology at the Marine Biological Laboratory (MBL) in Woods Hole, Massachusetts. Each year the seminar, founded by Garland Allen and Jane Maienschein, and currently coordinated by John Beatty, James Collins, and Jane Maienschein, is devoted to a different topic in the history of biology. As a rule, the Dibner Seminars in the History of Biology are organized by the coordinators in collaboration with experts in the respective topics and bring together a group of students and faculty for a week of intense discussions. In 2001 the Dibner Seminar "From Embryology to Evo-Devo (Evolutionary Developmental Biology)," organized by Manfred Laubichler and Jane Maienschein, focused on the history of the relations between embryology and evolution. Among the participants exploring the rich and diverse history of the subject were noted historians, philosophers, sociologists, and biologists, including John Tyler Bonner, Evelyn Fox Keller, Rudolf Raff, Sahorta Sarkar, and most of the authors in this volume.

The results of the week-long discussions at the Marine Biological Laboratory encouraged the organizers to convene another meeting with the specific goal of producing a tightly integrated edited volume. The Dibner Institute and its director, George Smith, agreed to continue to fund this project, and in October 2002, Manfred Laubichler and Jane Maienschein convened the Dibner Institute workshop "From Embryology to Evo-Devo." Most of the workshop participants had attended the Woods Hole seminar; a few of the seminar participants were unable to come, and others, such as Everett Mendelsohn and George Smith, joined the group.

The goal of the Dibner Institute workshop was not only further discussion of the work of the participants at the Woods Hole seminar in preparation for the planned volume, but also to have present three of the leading scientists in the current field of evolutionary developmental biology—Brian Hall, Gerd Müller, and Günter Wagner, all of whom have expressed an interest in the history of their discipline. The rationale for inviting scientists to a workshop devoted to the history of science was to break down disciplinary boundaries and to take advantage of diverse perspectives. This was the first time that scientists had participated in a Dibner Institute workshop other than as the subject for inquiry, and it was, by all accounts, a tremendously successful experience.

Discussions at the workshop benefited greatly from these new perspectives, and when Brian Hall, Gerd Müller, Günter Wagner, and Everett Mendelsohn presented their reflections on the papers and discussions, new avenues for thinking about the history of the relations between development and evolution emerged. The authors subsequently revised their papers in light of these ideas and suggestions, and our commentators graciously agreed to produce additional chapters, which accurately reflect the discussion at the workshop.

The initial seminar's venue at the MBL was a fortuitous match for the topic "From Embryology to Evo-Devo." After its founding in 1888, the MBL had been one of the premier research sites for embryology, physiology, *Entwicklungsmechanik,* and comparative biology and evolutionary biology, as well as a key meeting place for many of the leading scientists of those days. Many discussions about the relations between individual development (ontogeny) and evolutionary transformations (phylogeny) took place in this "marketplace of ideas," where people gathered at the Friday evening lectures to hear about the latest discoveries or theories, and continued their discussion in their labs, during collecting trips, and at (frequent) social events. From the beginning of the MBL, embryology has been one of the core areas of research and education.[1]

The seminar's topic, "From Embryology to Evo-Devo," proved to be extremely timely. The major events in biology during the 1990s, such as the announcement of Dolly, the first cloned mammal; the emerging debates about the therapeutic potential of stem cells (and the resulting regulatory and policy confusions); and the completion of the first sequence of the human genome marked the beginning of a new era in biological research. Development (embryology) clearly was to become a major focus in this "postgenomic" period in the history of biology. Focusing the attention of historians of biology on the largely neglected history of twentieth-century embryology therefore was appropriate, even more so because the seminar emphasized one particular aspect of this history, the discussions about the relations between ontogeny and phylogeny.

For many (historians as well as biologists), Ernst Haeckel's biogenetic law, "Ontogeny recapitulates phylogeny," still represents the canonical formulation of this relationship. The fact that even though it has long been disproven, at least in its radical form, the biogenetic law still discussed in textbooks is, at the very least, a testament to its intuitive appeal, if not to a more fundamental recognition that these two temporal processes in biology are in some way linked.

Exploring the various connections between ontogeny and phylogeny is also at the heart of a newly emerging discipline, evolutionary developmental biology. The field is often heralded as the new synthesis of developmental and evolutionary biology, especially since developmental biology (embryology) was not prominently featured in the last evolutionary synthesis, which centered more on genetics, systematics, and paleontology.[2] By the late 1990s evolutionary developmental biology had all the markings of a new scientific discipline. Two new journals were specifically devoted to the field: *Evolution & Development* and an independent section of the *Journal of Experimental Zoology, titled Molecular and Developmental Evolution* (now *JEZ Part B: Molecular and Developmental Biology*). In addition, the oldest journal in the field of experimental embryology (*Entwicklungsmechanik*)—*Roux's Archiv für Entwicklungsmechanik der Organismen,* founded in 1890—has been renamed *Genes, Development, and Evolution,* another reflection of the changed focus in developmental biology. Among granting agencies, the National Science Foundation has established a specific panel devoted to the "evolution of developmental mechanisms"; and professional societies, such as the Society for Integrative and Comparative Biology (formerly American Society of Zoologists), now have specific sections for evolutionary developmental biology. In short, by the time of the 2001 Dibner Seminar in the History of Biology, evolutionary developmental biology (Evo-Devo) had arrived, at least institutionally. As a consequence, the focus of developmental biology has been broadened substantially.

This did not happen overnight. Rather, the relations between development and evolution have been the subject of renewed, intense, and controversial discussions since the 1970s. Recognizing that these developments are an interesting episode in the recent history of biology, one that also had the potential to reconfigure the interpretation of the history of twentieth-century biology, the organizers decided to devote a Dibner Seminar in the History of Biology to this topic. It soon became obvious that while quite a bit was known about the late nineteenth-century discussions on the biogenetic law, comparative embryology,

Entwicklungsmechanik, and morphogenesis—and several participants in the recent emergence of evolutionary developmental biology had begun to document these events—next to nothing was known of what happened to these questions in the period between the 1920s and the 1970s. This historiographical vacuum provided the stimulus for expanding the topic of the seminar and focusing on a longer period, taking us "From Embryology to Evo-Devo."

Some readers misread the message and jumped to the conclusion that "from" implied direct causal change. For them, "From Embryology to Evo-Devo" seemed to imply a direct lineage of problems, ignoring the inherent complexity of the history. For others, the title evoked the image of a linear and gradual development, as if embryology had become Evo-Devo. Neither interpretation reflects either the organizers' intentions or the actual discussions at the seminar and the workshop. In these contexts embryology and Evo-Devo are merely historical markers that stand for the late nineteenth and the early twenty-first century, respectively. Exploring what happened in between was the goal of both events. Taken together, the contributions in this volume provide a first map of this extremely rich and fascinating part of the history of twentieth-century biology.

Several common themes emerged in the context of these discussions.

1. Development, as one of the central processes as well as one of the theoretical concepts of biology, has continuously been the focus of both empirical and conceptual attention. This in itself is not surprising. However, throughout the historical period covered here development has been studied and interpreted from different experimental angles and the results of these studies have been incorporated into often radically different conceptual systems, ranging from traditional comparative studies of embryology to molecular genetics and computational analysis. Studying the history "From Embryology to Evo-Devo" thus leads one to appreciate the diversity of conceptual interpretations and experimental strategies that characterizes twentieth century biology.

2. Throughout the twentieth century, multiple traditions of developmental biology coexisted; some of them are defined by their experimental strategies, others by their explanatory reference frames. In addition, different local and national traditions have persisted until the present day. These traditions are also reflected in different emphases within current evolutionary developmental biology.

3. Technology played an important role in the history of twentieth-century developmental biology, as it did in the emergence of *Entwicklungsmechanik* in the late nineteenth century (see especially Hall and Gilbert, this volume). In particular with respect to the question of ontogeny and phylogeny, the lack

of an adequate experimental approach is notable during several decades of the twentieth century.

4. Development, even more than physiology, has provided the foundation for an organism-centered perspective in biology. Even the most successful and in a sense most radical, molecular explanations of development, such as reaction diffusion models, lead to a concept, such as positional information, that implicitly refers to the three- and four-dimensional properties of organisms.

5. While the explanatory reference frames for development and evolution are different, roughly reflecting what Ernst Mayr calls proximate and ultimate explanations, at several times since the mid-1850s development has been considered essential to explanations of the patterns of phenotypic evolution. For example, in the context of evolutionary morphology, embryology provided a possible inference about genealogical connections between species (phylogeny); and in the context of developmental physiology and physiological genetics, the developmental and cellular context was considered essential to any mechanistic explanation of the evolution of phenotypes; and currently developmental mechanisms are implicated in discussions about the genotype-phenotype map, the limits of adaptation, and the origin of evolutionary novelties, among other things (see also chapters 15 and 16 in this volume).

6. All these observations contribute to a growing skepticism about what the current emphasis on a "synthesis" of evolutionary and developmental biology actually entails. On the one hand, current evolutionary developmental biology includes more than just developmental and evolutionary biology; on the other hand, it is still unclear whether the Evo-Devo focus can succeed in providing new perspectives that go beyond what would be possible within other explanatory schemata (see especially chapters 15 and 16 in this volume).

Of late, writing the history of science with a long-range perspective in order to get at the underlying persisting traditions, and thereby to be able to recognize and interpret significant patterns of change, has become unfashionable. Recent historiography of science has focused more on the local, immediate, contingent, and particular aspects of the scientific enterprise. The history of science has also matured as an academic discipline in its own right, and no longer sees its primary function as contributing to commemorative occasions and providing a "grand narrative of scientific progress" suitable for the first few pages of introductory textbooks. This new orientation in the historiography of science has led to important new insights, but it has also contributed to a growing alienation between the communities of historians of science and scientists.

In this volume with its long perspectives, we hope to bridge this unfortunate gap by including biologists as well as philosophers and sociologists. Our focus on a period of almost 150 years and our emphasis on a specific scientific problem—the relationship between ontogeny and phylogeny—allow us to analyze simultaneously continuities and transformations as well as discontinuities and novelties. Indeed, one would not be possible without the other, especially in areas of the history of biology where we do not have a commonly known, detailed historical narrative of their development. The chapters in this volume thus emphasize both the continuity of the general problem of defining the relationship between ontogeny and phylogeny, and the diversity of approaches, technologies, and concepts as well as the continuous transformation of the both the question and the scientific disciplines.

Finally, we want to mention one (tongue-in-cheek) observation that emerged during the workshop discussions: that the history of embryology and developmental biology has certain structural affinities with the individual and the phylogenetic history of organisms, and maybe this explains, in part, why among biologists, evolutionary and developmental biologists show the greatest interest in the history of their profession.

STRUCTURE OF THIS VOLUME

This volume collects the papers that were discussed at the 2002 Dibner Institute workshop. In addition, it includes a paper by Scott Gilbert that was presented at the 2001 Dibner Seminar in the History of Biology but not discussed at the workshop and a paper by Stuart Newman that the editors solicited after hearing a version presented at an international meeting.

Following the conceptual and epistemological introduction by Manfred Laubichler, the chapters are organized in three sections: part I, Ontogeny and Phylogeny in Early Twentieth-Century Biology; part II, Roots and Problems of Evolutionary Developmental Biology; and part III, Reflections. This structure reflects the course of the discussions at the Dibner Institute workshop and also broadly represents the chronological sequence of events "From Embryology to Evo-Devo." Everett Mendelsohn's observations were helpful in integrating the discussions at the Dibner Institute workshop and are reflected in several of the papers.

In chapter 2, Manfred Laubichler briefly discusses some of the conceptual and historiographic problems associated with writing the history of evolutionary developmental biology. He argues that the many transformations and discontinuities in that history are best understood if they are seen in the context of a specific scientific problem: defining the relations and connections between ontogeny and phylogeny. As a historical "object" of analysis, or an epistemic thing, this scientific problem has enough continuity that it can serve as a narrative anchor for telling the history "From Embryology to Evo-Devo." The conceptual changes associated with this scientific problem are then used to reconstruct the transformations of the epistemic space associated with the history of evolutionary developmental biology.

The chapters in part I of the volume provide a fresh perspective on events in the history of early twentieth-century embryology and developmental genetics. Fred Churchill discusses the fate of Haeckel's biogenetic law and analyzes several reformulations in major textbooks of embryology during the late nineteenth and early twentieth centuries. The picture that emerges is less black-and-white than many previous characterizations of the biogenetic law, and Churchill opens up new venues for analyzing the "internal critique" and reformulations of the biogenetic law, thus allowing for a careful reconstruction of the epistemic space of early debates about ontogeny and phylogeny. Stuart Newman's chapter focuses on William Bateson and his ideas about the physical determination of organic forms. Newman discusses Bateson's "vibratory theory," an attempt to incorporate physical principles (so-called Chladni patterns) into explanations of segmentation and repetitive pattern formation. Newman situates Bateson's ideas both within turn-of-the-century discussions about the nature of variation and inheritance and within the early twenty-first-century context of system biology. Analyzing the conceptual repertoire of both Bateson and modern system biologists, Newman argues for a broadening of the conceptual (epistemic) framework of explanations of development.

Jane Maienschein's contribution introduces cells and the process of morphogenesis into the discussions about the history of Evo-Devo. She emphasizes the role that cells played as an object of study and, in the form of slime molds in the work of John Tyler Bonner, also as model system for morphogenesis. Maienschein's chapter continues Newman's argument (via Bateson) that the physical and cellular characteristics of developing organisms are an important part of explanations of development.

Where Newman and Maienschein focus on specific objects and morphogenetic processes, Garland Allen takes a more conceptual and dynamic perspective. He presents a detailed account of the dialectics between analytic and synthetic explanations of development. In the interplay between these opposite yet complementary explanations of organic processes, Allen sees the organizing epistemological theme for the history of evolutionary developmental biology. Far from being abstract, his proposal provides a framework for the inclusion of many of the more detailed historical case studies.

Marsha Richmond's study of Richard Goldschmidt's role in uniting development and evolution concludes part I. During the 1920 and 1930s Goldschmidt, who remains an enigma to many even today, developed his idiosyncratic synthesis of ontogeny and phylogeny. Working with a different model organism (*Lymantria*) and beginning with a physiological and genetic account of sex determination, he developed a theory of the gene as a physiological agent. While during his lifetime many details of gene action remained beyond experimental reach, Goldschmidt nevertheless developed a conceptual framework that connected development, genetics, and evolution during a period when the prevailing attitude in science was to keep these domains apart.

Even though the chapters in part I revisit previously studied territory, each author has discovered new interpretative angles as well as new materials. In light of the theme of this volume, the history of the relations between ontogeny and phylogeny, they provide a fresh perspective on the history of embryology, genetics, and evolutionary biology in the first decades of the twentieth century.

The chapters in part II are more diverse. The topics covered here range from the history of comparative embryology in America and morphological and paleontological perspectives in the history of evolutionary developmental biology to a study of how developmental processes have been visually represented, a philosophical-historical analysis of research styles in embryology and genetics, a discussion of recent attempts to integrate development and evolution and the conceptual problems associated with these issues, and, finally, a philosophical-sociological analysis of research styles and conceptual change in biological research during the last few decades of the twentieth century. Most of these chapters focus on events during the period (1920s to the 1970s) that has not received much attention from historians of biology, and all of them provide perspectives that add dimensions to the problem of defining the relations between ontology and phylogeny.

John Wourms's study of comparative embryology in the American context brings to light an almost unknown chapter in the history of biology, especially with regard to the period that he includes. Wourms reminds us that even the most sophisticated conceptual schemata need to be grounded in empirical data, and that in the case of ontogeny and phylogeny, comparative embryology, in both a morphological and a systematic context, provides many of these data. Since during most periods in the history of biology, comparative embryology was not part of main-stream research, we have to look at different institutional settings (such as fisheries) to find the continuity of the work.

Alan Love's chapter similarly focuses on mostly neglected parts of the history of twentieth-century biology. His discussion of morphological and paleontological research by Dwight Davis and William Gregory is informed by the recognition that present-day evolutionary developmental biology is more than just a "synthesis" of development and evolution; rather, it involves the integration of various research traditions, including morphology and paleontology. As Love argues, it is in these fields which have not been part of mainstream twentieth-century biology that we find a closer continuity of the late nineteenth-century problem of ontogeny and phylogeny. His chapter thus opens an important new perspective on the history of midtwentieth-century biology.

William Wimsatt places attention on areas in the history of biology that have not received much attention from historians of biology, even though they are now becoming part of the canonical history of present-day evolutionary developmental biology. He discusses the problem of so-called internal factors of evolution, the idea that the details of the developing organismal system can have a major impact on the course of evolution. Internal factors of evolution are the prime candidates for a mechanistic integration of developmental processes into a theory of phe-notypic evolution, and have received considerable attention since the mid-1970s. Wimsatt focuses specifically on the work of Rupert Riedl and Wallace Arthur, and concludes his historical analysis with a brief discus-sion of his own work on generative entrenchment, which is one of several attempts to model internal factors of evolution.

Scott Gilbert discusses a current problem in developmental biology that has a long history. Since development is a complex process, visual rep-resentations of developmental transformations have always been a major conceptual as well as pedagogical tool. Gilbert argues that there is a concep-tual continuity between late nineteenth- and early twentieth-century attempts to trace the fate of cells in the developing embryo and modern approaches designed to establish gene expression patterns. He also states that one important aspect of the mapping program in the context of evolutionary developmental biology is to connect new (molecular) evidence with old (traditional) knowledge in embryology. In ways similar to Goethe's *Faust*, a major problem of current molecular approaches in developmental genetics is to reintegrate new molecular data into an organismal whole.

History, as a repository of knowledge, thus becomes an integral part of cutting-edge research in evolutionary developmental biology.

In his chapter James Griesemer uses a well-documented historical case study, the split between embryology and Morgan-style transmission genetics at the beginning of the twentieth century, to develop a philosophical argument about an important aspect of scientific representations and explanations of complex processes. He argues that in all explanations of complex processes, certain elements will necessarily be foregrounded while other will be relegated to the background. His analysis of the consequences of this dynamic contains several important lessons for all those— historians, philosophers, and biologists—who emphasize the synthetic character of modern evolutionary developmental biology.

The final chapter in part II presents an analytic perspective complementary to the ones proposed by Laubichler and Griesemer. Elihu Gerson focuses on the long-term pattern of relationships among lines of research in comparative biology by providing a framework for the inclusion of institutional and technological factors, what he calls a style of research. These factors cannot be separated from the epistemological concerns raised by Laubichler and Griesemer, so that Gerson's chapter (which includes a discussion of the effects of rationalization of work in both science and society) suggests further explorations of the interplay of contextual and contingent factors with epistemological factors in the history of evolutionary developmental biology.

The chapters in part III collect the reflective comments by Brian Hall, Gerd Müller, and Günter Wagner. These papers are expanded versions of the commentaries that these scientists and scholars gave after listening to (and participating in) two days of discussions at the workshop. In his chapter, Brian Hall, author of the first modern textbook of evolutionary developmental biology, who has a "second" career as a historian of developmental biology, reminds us about the many elements that contribute to the Modern Synthesis of Evo-Devo. In Hall's view, Evo-Devo is, and always has been, a model of an interdisciplinary science. To illustrate his point, Hall provides a few historical case studies that demonstrate how the problem of ontogeny and phylogeny has always been approached from a variety of conceptual perspectives and how specific institutions, such as the Naples Zoological Station, and instruments and their associated experimental practices—specifically the Cambridge Instrument Company and the automatic microtome—helped to establish an interdisciplinary environment. Hall's lessons are clear: any reconstruction of the history of evolutionary developmental biology will have to connect questions of conceptual integration with local issues of institutional and technological changes.

In his chapter, Gerd Müller takes the historical lessons of this workshop and volume and applies them to his analysis of the future of the field. His chapter is also a reflection of what he identifies as a phenomenon peculiar to the Evo-Devo discourse: the emphasis on metatheoretical reflections within Evo-Devo and the close collaboration of philosophers, historians, and biologists in shaping the future agenda of the field. He presents six memos that characterize the distinctiveness the Evo-Devo discipline. These memos, which capture the breadth of the Evo-Devo research agenda, are a fitting conclusion as well as a new beginning for the historical work presented in this volume. They show what happened "From Embryology to Evo-Devo," and they also invite the reader, as they did the workshop participants, to look back at the fascinating complexities of this history and ask: How did we get there? And what does it all mean?

Günter Wagner's comments move us forward by addressing the present state of evolutionary developmental biology, or developmental evolution (Devo-Evo), as he refers to it. This nomenclature reflects the internal disunity of present-day evolutionary developmental biology. Currently, several different questions are pursued in the context of this overarching synthesis. Wagner uses an episode in the history of biology—the decline of evolutionary morphology at the beginning of the twentieth century—to warn of the dire consequences for a field that fails to agree on standards of evidence to evaluate its results and interpretations. He goes on to suggest that if Evo-Devo (Devo-Evo) is currently entering its "academic phase," it will have to establish such evidentiary standards. Wagner then briefly sketches how such standards can be developed for evolutionary innovations.

ACKNOWLEDGMENTS

This project would not have come to fruition without the support of the Dibner Institute. Not only did the project take place through the auspices of the Dibner History of Biology Seminar at the MBL, and through a Dibner workshop at the wonderful Institute building in Cambridge, Massachusetts. In addition, the staff—including Jed Buchwald, Evelyn Simha, George Smith, and Carla Chrisfeld—provided vital support at critical points. The participants in both the seminar and the workshop provided the intellectual substance for the project. And the voices of those who had been at the seminar but could not join us at the workshop were heard as each participant responded to the substance of the seminar. This was truly a collaborative project, and each participant has an intellectual property interest in the product. Finally, George Smith encouraged us to pursue the publication of the collection and patiently waited for revision after revision during the preparation. His enthusiastic encouragement makes this, as well as other recent Dibner Institute volumes, possible.

NOTES

1. Jane Maienschein and Ruth Davis, *100 Years Exploring Life, 1888–1988. The Marine Biological Laboratory at Woods Hole* (Boston: Jones and Bartlett, 1989); Jane Maienschein, ed., *Defining Biology: Lectures from the 1890s* (Cambridge, Mass.: Harvard University Press, 1986).

2. Ernst Mayr and William B. Provine, eds., *The Evolutionary Synthesis: Perspectives on the Unification of Biology* (Cambridge, Mass.: Harvard University Press, 1980).

2

Does History Recapitulate Itself? Epistemological Reflections on the Origins of Evolutionary Developmental Biology

Manfred D. Laubichler

In recent years evolutionary developmental biology, or Evo-Devo for short, has established itself as one of the most dynamic research areas within the organismic branches of the life sciences. It represents a modern resynthesis of evolutionary and developmental biology that during the nineteenth century was associated with Ernst Haeckel, Carl Gegenbaur, Francis Balfour, and Alexander Kowalevsky.[1] Now, in the early twenty-first century, after roughly thirty years of gestation, evolutionary developmental biology is solidly entrenched within the conceptual framework of modern biology and has all the markings of a new discipline, such as representation in professional societies, scientific journals devoted to the field, academic programs and job searches, panels at grant agencies, and textbooks.[2]

However, today's evolutionary developmental biology is far from being a uniform scientific discipline with a single research agenda and a well-defined methodology. Despite the popularity of unifying images, including the iconic "conserved Homeobox genes" and the "genetic tool kit for development," many different research programs are currently pursued under the umbrella of evolutionary developmental biology. These include the search for the developmental and genetic basis of evolutionary change, the analysis of the evolution of developmental mechanisms, questions about the origin of evolutionary novelties and the reasons for conserved patterns of morphology, and the problem of the origin and diversification of animal and plant forms. All these questions can be grouped under several organizing categories that are reflected in the three most widely used terms for the field: "Evo-Devo," "Devo-Evo," and "developmental evolution."[3]

These divisions in today's field of evolutionary developmental biology reflect different research agendas as well as different conceptual and explanatory reference frames. "Evo-Devo" questions are those which look at developmental phenomena from an evolutionary perspective. These include the origin of developmental systems, the evolution of the developmental repertoire, and the modification of developmental processes.

In all these cases the object of study is development, while the explanation of these phenomena is given in the context of evolutionary biology. "Devo-Evo" questions, on the other hand, are those in which the scientific problem is one of phenotypic evolution, and developmental phenomena and processes are part of its explanation. Examples of such questions are the direction of phenotypic evolution, the origin of morphological novelties, the patterns of phenotypic variation, and the hierarchical organization of body plans.[4] This characterization is still only a preliminary approximation of the actual diversity of research questions currently pursued in the context of evolutionary developmental biology, but it is sufficiently accurate for the purpose of these brief epistemological reflections about the history of evolutionary developmental biology.

The diversity of research agendas in evolutionary developmental biology also raises several issues with regard to its supposed character as a synthesis. What, exactly, is the new and genuine product of this synthesis? What new questions and explanations provided by evolutionary developmental biology are not already part of either developmental or evolutionary biology (or any of the other disciplines and research agendas that contribute to the new focus)? Primarily these are conceptual, methodological, and philosophical questions of evolutionary developmental biology, and thus they are the subjects of intense discussions in the field, but these issues also have historiographical consequences.[5] If it is still not settled what the future contours of evolutionary developmental biology will be—let alone whether the field will actually be successful as an independent research program within the biological sciences—then how can we propose to write its (emerging) history? To be sure, we can follow the programmatic pronouncements, analyze the actual research done and its associated methodologies, study the politics of institutionalization, record the ideas, actions, and beliefs of its practitioners—and I would argue that this is what we also should be doing, and in a way began to do at the Dibner Institute workshop—but still this leaves open the question of what the historical "object" of this proposed history "From Embryology to Evo-Devo" actually is. What, in other words, has enough continuity throughout this long period that it can be recognized as an entity worth historical study?

Epistemological Questions in the History of Evolutionary Developmental Biology

If we wanted to write the history of biology as a whole, it would be comparatively easy to say what is legitimately part of the story and what is not. There is widespread agreement about what objects, questions, concepts, methods, and researches are properly biological. That, of course, leaves the extremely difficult question of selecting the most illuminating and comprehensible historical narrative. But works such as Francois Jacob's *The Logic of Life*, Robert Olby's *Origins of Mendelism* and *The Path to the Double Helix*, Michel Morange's *A History of Molecular Biology*, and Jan Sapp's *Genesis: The Evolution of Biology*, to name only a few, show that this difficult selection can be done and done well. In all these books as well as in the older classics, such as William Coleman's *Biology in the Nineteenth Century*, Garland Allen's *Life Science in the Twentieth Century*, and the monumental collection *Geschichte der Biologie*, edited by Ilse Jahn and E. M. Krausse, a history of ideas perspective is paired with a focus on the history of biological disciplines (or of biology itself).[6]

When we deal with subareas of biology, the question of what is legitimately within their history is often defined by a specific object of analysis. Recent historical studies of the gene, such as *The Concept of the Gene in Development and Evolution*, edited by Peter Beurton, Raphael Falk, and Hans-Jörg Rheinberger, and Evelyn Fox Keller's *The Century of the Gene*, are organized around the (contested) object of the "gene" as a the focus of analysis.[7] A similar object-based approach can be found in the historical parts of Jane Maienschein's recent book *Whose View of Life?*, which analyzes the history of embryo research.[8] In these cases, the object of study is well-defined, even when it is controversial.

When we turn to evolutionary developmental biology, however, we do not seem to be dealing with a single set of questions and methods; its distinctness from other areas of biology is still under debate; and we do not know how the story will turn out. This makes the historian's question of what is legitimately part of the story almost impossible to answer. The intellectual and scientific promise of Evo-Devo is so exciting, however, that we cannot *not* tell the story—even while knowing that it will have to be revised again and again in the light of accumulating knowledge.

As the workshop title, "From Embryology to Evo-Devo," indicates, this young field of modern evolutionary developmental biology is actually a member of a rather old cluster of scientific problems of embryos, development, and evolution. In the broadest sense, therefore, the problems that provide continuity through a long period of time are the scientific questions of the relationship between ontogeny and phylogeny. These problems represent what can be called the epistemic object we want to study.[9] Like all epistemic objects, this one derives its meanings and its epistemological function from its connections to various empirical and theoretical contexts. These contexts have included evolutionary morphology, phylogeny reconstruction, mechanistic explanations of morphological

change, and morphogenesis, as well as all the questions of current evolutionary developmental biology listed above. But an epistemic object is also
a historical object in its own right and is not defined solely by its function within a larger context. Thus far the category of an epistemic object
has been applied mostly to concrete material objects, such as the brain,
the TMV virus, and *Drosophila*. However, there is no logical reason why
the "object" has to refer to a material object and cannot be a specific
model, a technique, or, as in our case, a cluster of scientific problems.

If we take the problems related to "ontogeny and phylogeny" as the
epistemic object of our historical analysis, then we have to consider it in
relation to two kinds of contexts. First is its relation to the material objects
(such as embryos), experimental practices, and social organizations of scientific activities, and second is the set of theoretical contexts, such as the
idea of recapitulation or of constraints on phenotypic variation, that
provide the interpretative context for the research pursued at any given
time. Focusing on the continuity of the scientific problems of ontogeny
and phylogeny through history means that for each period, we have to
analyze the interactions among these three domains, each with its own set
of historical transformations. We thus have to study (1) changes in the
material object (embryos) and the research methodologies; (2) transformation of the questions themselves (the history of our epistemic object);
and (3) transformations in the theoretical contexts that ascribe the epistemological significance to our epistemic object. The continuity and transformations of the epistemic object (the scientific problems related to
ontogeny and phylogeny) then structure the rather diverse, and often contingent, changes in both the material objects and the methodologies, as
well as the developments in theoretical contexts, into a coherent narrative
"From Embryology to Evo-Devo." The result can eventually be a history
(in the sense of biography) of a scientific problem embedded within
accounts of the material, technological, institutional, social, and theoretical history of relevant areas of science. This is a history that transcends the
traditional dichotomies of theory and experiment, basic and applied, internal and external, material and ideal that have preoccupied historians and
sociologists of science for the last few decades.

To describe the history of the epistemic object "ontogeny and phylogeny" during the period "From Embryology to Evo-Devo," we have to
account for the transformations of the material conditions (objects, technology, institutions, etc.) as well as the theoretical contexts. For material
conditions this is rather straightforward; recent historiography of biology
has revealed many details of the changes in research methodology within

the biological sciences, such as those induced by the techniques of molecular biology or by increased computational power and new forms of data management and visualization.[10] The details of the theoretical contexts, on the other hand, have received far less attention from historians of science, although several biologists and historians and philosophers of biology have explored elements of this history.[11]

Therefore, for the remaining part of these epistemological reflections, I will focus on these transformations, which I will call the epistemic space surrounding "ontogeny and phylogeny." It was first established in the context of theories of generation and attempts to use embryological data to establish the relations between different types of organisms. In this early conceptualization we find, strictly speaking, no reference to "phylogeny," but rather to "system" or "order," even though after Lamarck, Erasmus Darwin, Goethe, and Buffon the potential historical dimension of this system was frequently discussed. With von Baer's laws we then have the first canonical formulation of the relation between ontogenetic sequences and systematic order. Von Baer concluded that during embryonic development the more general features appear first, and therefore the early stages of development of different organisms tend to be more similar than the later stages, but definitely are not a strict recapitulation of more primitive stages of adult organization. He put a damper on earlier attempts to see a closer temporal link between developmental stages and the complexity of organisms in the sense that the ontogeny of complex organisms recapitulates the stages of less complicated ones.

Despite this harsh pronouncement against simple recapitulation, the work of von Baer and others on *Entwicklungsgeschichte* of organisms (including vertebrates) contributed to the conceptual links among embryology, comparative anatomy, and systematics. It established that embryology was an important source of information about the relationships among different organismal types, and, after Darwin proposed a genealogical explanation for the relationships among species (and by implication also for higher systematic groups), the epistemic space for the problem of "ontogeny and phylogeny" had been constituted.

The Biogenetic Law, Evolutionary Morphology, and the Initial Synthesis of Ontogeny and Phylogeny: Deepening the Temporal Dimension of Organisms

The recognition of individual (ontogenetic) development of organisms had, in theories of generation, preformism, epigenesis, and life history, always introduced a temporal dimension into explanations of life and its philosophical and scientific interpretation. The story of how, through developments in natural history, geology, comparative anatomy, and cosmology, the temporal dimension of life on Earth had gradually been deepened has often been told, and need not be rehearsed here.[12] Suffice it to say that with the acceptance of the evolutionary history of organisms, the conceptual configuration for interpreting the *Entwicklung der Organismen* took on a new meaning.[13]

After Darwin, the topology of this newly emerging epistemic space was defined by the following problems and observations:

1. Both organisms and species vary.

2. Variation of both organisms and species is not arbitrary, but clustered. In other words, parents and offspring resemble each other closely, but not totally, and different species can be grouped into higher systematic groups that are united by common features.

3. The specific causes for the similarity and differences between parents and offspring were unknown, but it was clear that these causes had to act in the course of individual development (ontogeny).

4. The principles for grouping individual species into higher systematic groups were based on comparison and the establishment of homologies. Homologies—the same organs in different individuals, irrespective of form and function—were considered the basis for the hierarchical system of classification; it was, however, not always clear how homologies could be established.

5. Embryological observations (*Entwicklungsgeschichte*) had revealed that earlier (less complex) developmental stages of different species more closely resembled each other than did adult stages.

6. The genealogical perspective and the geological record suggested that less complex forms of life would have emerged earlier in the phylogenetic history (*Stammesgeschichte*) than more complex forms (an argument that had already been made in the context of cultural stages in the history of mankind).

7. It was clear that all explanations, as well as the specific details, of this history needed to be inferred—often, as with the geological record, from incomplete data sets—because direct observation was not an option.

Darwin himself, honed by detailed observations of developmental stages of barnacles, carefully suggested a connection between ontogeny and phylogeny: "Thus community in the embryonic structure reveals community of descent."[14] Others soon followed suit, exploring in what way the observable patterns of ontogeny could help to reveal the hidden patterns of phylogeny. Fritz Müller, studying the embryology of crustaceans, was rather skeptical. *Entwicklungsgeschichte* might be a record of phylogeny; however, since embryological stages are also subjected to the struggle for existence, this record tends to be erased through time as development becomes more direct in each case. Generally, Müller argued, a parallelism between ontogeny and phylogeny depends on whether new variations will be introduced early in development or as terminal additions to the developmental sequence.[15]

Haeckel, in all his writings on ontogeny and phylogeny, accepted a similar distinction. The biogenetic law, first called that in 1872, applied only to palingenetic forms of development. Should, on the other hand, the developmental sequence itself have changed, due to natural selection or any other factor, then ontogeny would not be a strict recapitulation of development.[16] However, the (sometimes aggressive) campaign by Haeckel and his followers for the heuristic value of the biogenetic law and the principle of recapitulation dominated the discussions. Furthermore, the developmental perspective implicit in the biogenetic law also seemed to offer insights into one of the major problems of comparative biology: the establishment of homology relations. A shared developmental history, Haeckel asserted, should be a solid basis for the assessment of homologies. This later assumption became one of the foundations of evolutionary morphology, a program initiated by Haeckel and his close friend Carl Gegenbaur that was focused on the establishment of a phylogenetic system as the goal of comparative anatomy and embryology. Gegenbaur, whose research school was the heart of the program, soon developed his own critical views on the relationship between ontogeny and phylogeny that were based on his conviction of the primacy of the methods of comparative anatomy.[17]

This new orientation, brought about by an emphasis on the relations between ontogeny and phylogeny, led to some spectacular insights. In the late 1860s the Russian embryologist Alexander Kowalevsky discovered similarities between the ontogenetic sequences of *Amphioxus* and those of vertebrates, as well as the existence of a *chorda dorsalis* in the larvae of ascidians.[18] These discoveries suggested that vertebrates were derived from the larvae of ascidians, a theory that was soon challenged by, among others, Anton Dohrn, the founder of the zoological station in Naples, who proposed an annelid ancestry for vertebrates.[19]

However, all these examples illustrate how, in the decades after the publication of the *Origin of Species*, the scientific problems of ontogeny and phylogeny (our epistemic object) were interpreted within a specific conceptual configuration that emphasized the importance of developmental data for problems of phylogeny and systematics, as well as the role of developmental processes in generating phenotypic variation, either caenogenetically or by means of terminal addition (palingenesis). The epistemic value of "ontogeny and phylogeny" was determined primarily by the context of evolutionary morphology and phylogeny, while it was at the same time embedded within the research methodology of comparative embryology and realized by comparative sequences of embryonic stages (and their illustrations).

THE SEARCH FOR CAUSAL MECHANISMS IN DEVELOPMENT AND EVOLUTION: FRAGMENTATION OF THE EPISTEMIC SPACE

Historians of biology have described in great detail the decline of the research program of evolutionary morphology and the simultaneous transformations of its associated epistemic space as due to the emergence of several new experimental approaches.[20] These accounts show that despite the revolutionary fervor of some proponents of the new biology, the epistemic space connected to the problem of ontogeny and phylogeny was not radically transformed, as a superficial reading of Garland Allen's phrase "revolt from morphology" might imply, but rather that this particular epistemic space had become fragmented. An increasing number of different theoretical and explanatory contexts became connected with the epistemic object "ontogeny and phylogeny," and its epistemic values multiplied. In some cases the epistemic value also shrank, as in the contexts of transmission genetics and *Entwicklungsmechanik*. Initially the latter research program emphasized the role of mechanical and physical-chemical causes in explanations of development; only later, in the context of a more physiological orientation, with Driesch, Goldschmidt, and Kühn, did evolutionary questions return, even though these had in the meantime been radically transformed and were quite different from the initial program of Darwin and Haeckel.[21]

In transmission genetics as well as in the emerging disciplines of population and quantitative genetics, the problem of development was excluded. These disciplines were grounded in several fundamental conceptual abstractions—the concept of the gene as unit of transmission and the genotype as the sum of all these genes—as well as in the additional assumption of the existence of an isomorphic mapping between genotype and phenotype, all of which were essential for the establishment of a coherent mathematical theory.[22] But, as might be expected, several researchers soon discovered problems with that approach, and in the context of some of these alternative proposals, including ideas about a "norm of reaction" and of *Dauermodifikationen*, aspects of development entered the genetic discourse.[23] Furthermore, transmission and population genetics were not the only research agendas within genetics. Especially in the context of physiological and developmental genetics, such as the work of Richard Goldschmidt and Alfred Kühn, problems related to "ontogeny and phylogeny" were frequently discussed.[24]

Comparative embryology and morphology, the two disciplines that had been associated most closely with "ontogeny and phylogeny," continued to be practiced as well. These disciplines were no longer in the spotlight, as they had been during the late nineteenth century, but they continued to flourish: students were still taught these fields, textbooks were written and updated, and new discoveries were made, especially in the context of deep-sea explorations that greatly expanded our knowledge of marine organisms.

All these activities continued in their respective scientific niches, which quite often were conceptually isolated from each other. The two unifying proposals of mid-to-late twentieth-century biology were the Modern Synthesis and the emerging molecular biology.[25] The former was based on the conceptual integration of the mathematical theory of population genetics with several aspects of biology, such as systematics, biogeography, population biology and ecology, paleontology, and behavioral biology, while the latter was based more on a unifying level of analysis and a shared repertoire of experimental approaches than on an integrative conceptual framework. Nevertheless, these two approaches and their sometimes acrimonious relations largely shaped the epistemic space of mid-twentieth-century biology. However, as even these brief remarks illustrate, the expansion of biology, both topically and even more institutionally, provided ample space for alternative niches. The problem of "ontogeny and phylogeny" survived in those niches and transformed itself in response to both conceptual and technological developments.

For our project of a history of the epistemic object "ontogeny and phylogeny," the period between the 1920s and the 1970s represents a historical vacuum. As I have suggested, the epistemic object continued to be important in several contexts, even though it had nowhere nearly as much intellectual appeal (or publicity) as during the late nineteenth century. Nevertheless, as soon as the issue surfaced again within several different contexts in the early 1970s, researchers could draw on newly accumulated knowledge as well as on earlier questions. One of the challenges for a history "From Embryology to Evo-Devo" thus is to uncover the many pathways through which the epistemic object "ontogeny and phylogeny" continued into the more recent period, as well as to reconstruct the associated fragmentation of its epistemic space.

THE EMERGENCE OF MODERN EVOLUTIONARY DEVELOPMENTAL BIOLOGY: OPENING OF A NEW EPISTEMIC SPACE

It has been the guiding hypothesis of this project that the problems related to "ontogeny and phylogeny" have continued to be a central question of biology since the idea of organic evolution was established beyond doubt by Darwin, Haeckel, and others in the middle of the nineteenth century. Nonetheless, we can identify a renewed confluence of activities that, since the mid-1970s, has led to the emergence of a new scientific focus and the establishment of the new discipline of evolutionary developmental biology. There is still no consensus about how one should interpret these recent events; many of the (still living) participants have their own interpretations, some of them collected in this volume.[26]

One recent programmatic paper was first presented at the inaugural symposium of the new division of Evolutionary Developmental Biology of the Society for Integrative and Comparative Biology by Wagner, Chiu, and Laubichler. They distinguished, adapting Gunter Stent's terminology, among romantic, enthusiastic, and academic phases in recent evolutionary developmental biology.[27] And in this volume Wagner has expanded the romantic phase to include most of the earlier history of the problem of ontogeny and phylogeny (from the middle of the nineteenth century to the mid-1980s).[28] This extension of the romantic phase is problematic, since the contributions to this volume show that the conceptualizations of the relations between ontogeny and phylogeny have always been extremely diverse. Therefore a single category of "romantic" does not do justice to this diversity. However, the main point of this "three-phase" model of the recent history of evolutionary developmental biology is not necessarily to provide an accurate history, but rather to emphasize the internal shifts in the theoretical structure of the field during the last decades.

Seen from this perspective, what has happened since the mid-1970s, then, has not been solely the gradual emergence of a new synthesis of developmental and evolutionary biology—and any number of other disciplines and traditions, such as comparative embryology, paleontology, cladistics, and molecular biology—but also a major reconfiguration of the epistemic space of what is now referred to as evolutionary developmental biology, within which both empirical and theoretical contributions are evaluated and interpreted.

Many of the "romantics" (i.e., the early contributors to the gradually emerging evolutionary developmental biology) focused on the problem of development and evolution because they were dissatisfied with the explanatory framework of the Modern Synthesis. Three related issues in particular were at the heart of their critique. For one, the assumption of the Modern Synthesis that macroevolutionary patterns can be explained by a simple extension of microevolutionary processes was questioned. Paleontologists, who led this line of attack, focused on two phenomena that they thought could not be reconciled within this framework: (1) the observation that the fossil record of several clades shows periods of rapid evolutionary change followed by extended periods of stasis led to the hypothesis of punctuated equilibrium; and (2) the observation that most conceivable morphological patterns are not realized (i.e., that the morphospace is largely empty) and that realized morphologies are clustered in certain domains of the morphospace seemed to support ideas that morphological change is primarily a consequence of developmental processes, such as heterochrony, rather than strictly determined by (incremental) genetic factors.[29]

The second set of concerns was related to more general trends in post–World War II biology, and can best be summarized as a growing dissatisfaction with the loss of an organismal perspective in biology. In this context the research program of "ontogeny and phylogeny" led to challenges to the prevailing epistemological focus of molecular reduction. As an alternative, proponents of an organismal perspective emphasized an organism- or system-centered approach to problems of morphogenesis and pattern formation, which they saw as complementary to the paradigm of genetic determination and causation that was present both in molecular biology (including developmental genetics) and in evolutionary biology.[30]

The third major source of dissatisfaction with the Modern Synthesis was the privileged role of adaptation in dominant theories of evolution.[31] The adaptationist paradigm was challenged by molecular biologists (e.g., with the neutral theory of evolution) as well as by evolutionary developmental biologists, who emphasized the role of internal factors in evolution. The most prominent early concepts in this context were the notion of developmental constraints as a limitation on possible phenotypic variation, and therefore on adaptation, and Riedl's idea of "burden," which postulates an internal, in addition to the external and environmentally induced, selection pressure.[32] The internal conditions, which act as a second (or rather first) selective environment, are those of the developing system. In the context of Riedl's theory, these internal factors, or system conditions, as he calls them, can explain both the evolvability and the hierarchical organization of organismal forms.[33]

At the end of the romantic period in the mid-1980s, a consensus had been reached among incipient evolutionary developmental biologists that the prevailing version of evolutionary theory needed to be reformed and that "development" needed to be integrated into the general framework of evolutionary theory. In other words, the epistemic space of evolutionary theory was in need of transformation. Consequently, the problems related to ontogeny and phylogeny again gained in epistemic value. However, even though there was general agreement that development needed to be part of evolutionary theory, there was far less agreement about the actual research agendas or about the best strategies to accomplish this desired new synthesis. The epistemic space had been transformed, but this transformation has resulted a multitude of different interpretative contexts for the problems of ontogeny and phylogeny. This trend grew stronger after the discovery of the conserved sequences of Homeobox genes initiated a transformation of developmental genetics and started the enthusiastic phase in the recent history of evolutionary developmental biology.

This plurality of approaches to the problems of ontogeny and phylogeny, which continues to this day despite a growing attempt to establish a genuine new "synthesis" of evolutionary developmental biology, is also reflected in the fragmented structure of the emerging epistemic space surrounding evolutionary developmental biology, as well as in a similar plurality of objects and methodologies within which these questions are embedded. For example, an increasing number of biological disciplines have recently discovered the importance of "Evo-Devo" questions. But in many cases these references to "Evo-Devo" only incorporate specific elements into a well-defined epistemic space of an established biological discipline. Examples are the emphasis on "Evo-Devo" elements within life history theory, cognitive and developmental psychology, and even evolutionary medicine. An integrative epistemic space that would unite all these different "Evo-Devo" applications within a genuine new "synthesis" has not yet emerged. What we can observe, however, is the already described clustering of research questions around two or three main emphases, such as "Evo-Devo," "Devo-Evo," and developmental evolution.

Conclusion: Does History Recapitulate Itself?

What lessons can we draw from these brief epistemological reflections on the history of evolutionary developmental biology? The study of evolutionary developmental biology, as an emerging scientific discipline based on a much older scientific problem focus, needs exploration of both the continuity and the transformations of its core scientific problems, along with the associated changes in the epistemic space and scientific practices (materials, technologies, and institutions). This historical approach gives a unique perspective on, and informs, current developments; in that way, historical awareness can actually improve scientific practice. Practitioners of evolutionary developmental biology have long recognized this fact, and there is probably no other group of scientists that is currently as deeply engaged in discussions about the history of their field.[34] The ambiguous status of evolutionary developmental biology as a new "synthesis" and the diversity of its multiple research agendas certainly contribute to this interest in history, as does the deeply historical character of its main scientific problems.[35]

Writing the history of evolutionary developmental biology and other similarly complex fields, such as theoretical biology, has its own set of historiographical challenges that can be met within the proposed focus on epistemology. In the case of evolutionary developmental biology, the first challenge is to identify the appropriate historical object—the specific research problems of "ontogeny and phylogeny"—as the relevant epistemic object that allows us to connect the research agendas of current evolutionary developmental biology with previous attempts. In the case of theoretical biology, which for most of its history had no disciplinary identity of its own, the challenge is to find a connecting thread that unites seemingly diverse activities. In both cases, we are dealing with a threefold relationship among epistemic object, epistemic space, and material manifestation that enables us to highlight conceptual connections through time and link those developments with the material, institutional, and social history of science. Even though the epistemological model presented here starts with the identification of a scientific problem as an epistemic object, which seems to privilege conceptual issues, these epistemic objects are, of course, embedded within their associated scientific practices as well as within the larger theoretical structure of the relevant scientific contexts. The epistemological focus advocated here thus suspends the history of a scientific problem within a web of relations (to material objects as well as ideas) that support each other. Or, to put it differently, scientific problems as epistemic objects act in a way similar to radioactive markers in molecular biology; they highlight synchronic scientific connections (similar to metabolic pathways) as well as diachronic modes of transmission.

Emphasizing the continuity of the scientific problems related to "ontogeny and phylogeny" and describing the transformations of these problems "From Embryology to Evo-Devo" as part of a history of evolutionary developmental biology also anchors our analysis of the associated transformations of the epistemic space and the conceptual configurations of relevant areas of biology. The resulting narrative is thus not a linear history of ideas, although ideas, concepts, and theories will be at its center. Rather, our reconstructions emphasize the relationships among the triad of scientific problems as the epistemic object, the theoretical context as the source of epistemic values and power, and the material objects as the embodiment of the scientific problem. Our historiographic model bears a close resemblance to the triadic nature of the sign in Peirce's semiotics. And here, as in Peirce's semiotics, this triadic relation is transitive. No matter what the initial focus is—scientific problems, experimental practices, or theoretical contexts—none can ultimately be privileged over any of the others. As we see in all the chapters in this volume, it is simply not possible to discuss one element of this triadic relation without references to the others.

Finally, I want to address the question of the audience(s) for such a project of writing the history of evolutionary developmental biology. For what purpose and for whom are we writing this history? Historians of science, philosophers, and biologists attended the Dibner Institute workshop and the Dibner seminar. They were united by their interest in the topic, but they also brought different interests and agendas to the table that need either to be reconciled or to be made explicit in the further process of writing the history of evolutionary developmental biology.

For historians of biology looking back at twentieth-century biology, the perspective "From Embryology to Evo-Devo" allows them to reevaluate and challenge received views about this history, especially in the post-World War II period. The current conception of late twentieth-century biology is dominated by the narratives of molecularization, biomedicalization, increased specialization, expansion (both institutional and financial), quantification, and industrialization. This perspective also affects which scientific disciplines and research problems historians of biology currently consider. Molecular biology, biomedicine, biotechnology, genetics, and neurobiology receive most of the attention. Immunology, ecology, evolutionary biology, paleontology, and, more recently, the behavioral sciences also are studied, at least by some. In contrast, until very recently developmental biology hardly appeared in the historical reconstruction of late twentieth-century biology.[36] The history of evolutionary developmental biology and the observation that the research agendas of "ontogeny and phylogeny" had a continuous history throughout the twentieth century will thus contribute to a reevaluation of the current interpretation of the recent history of biology.

Such a new focus will also have methodological consequences, because the history of evolutionary developmental biology will reinforce a growing concern with theoretical and conceptual questions within the history of biology at large. Historians have devoted a good deal of their attention to problems of instrumentation and representation, as well as to experimental practices (which quite appropriately followed the major developments within the life sciences in the twentieth century). But as theoretical and conceptual questions become more prominent throughout the life sciences, and especially in fields such as evolutionary developmental biology, the historiographical approach suggested here can serve as a model for other areas in the history of life sciences, such as genomics, ecology, neuroscience, or theoretical biology, where we also observe an increasing emergence of theoretical issues.

For philosophers of biology, the history of evolutionary developmental biology will be an important resource for the project of developing a historically informed, critical epistemology of biology. Among philosophers of science, philosophers of biology have always paid the most attention to the history of science, since it provided exemplary cases for the main problems in the philosophy of biology, such as the reductionism problem; the questions of theory formation, integration, and explanation; and the analysis of central concepts of biology such as the gene and species concepts. Some, like David Hall, have even extended a "biological" methodology to the history of biology.[37] Therefore, for philosophers of biology, the proposed reconstruction of the transformations of the epistemic space surrounding the problems of ontogeny and phylogeny will be of special interest, since questions of synthesis, theory integration, and explanatory pluralism, as well as the emerging focus on organisms as an integrative biological concept, are widely discussed within the field.

For biologists, being aware of the history of evolutionary developmental biology also has practical implications. The way this history is reconstructed by practitioners reflects a lot about their assumptions concerning the current status of the field. Several themes stand out in recent discussions of evolutionary developmental biology and its history. One is the emphasis that "Evo-Devo" represents a new synthesis or that it actually completes the Modern Synthesis, which had largely ignored development. Another theme, championed by, among others, Brian Hall, is the view that evolutionary developmental biology brings together many more disciplines and approaches than just developmental and evolutionary genetics. In a similar vein, Love and Raff have argued for the inclusion of a tradition of comparative embryology, but so far have focused mostly on the British and American literature. And Scott Gilbert has argued that ecology needs to be incorporated more forcefully into evolutionary developmental biology, and has presented several historical arguments to support his claim.[38] This list could go on, but it is obvious that having even the rudiments of a more accurate history of evolutionary developmental biology that is connected to the history of biology at large will have a positive impact on discussions about its current state and its future. An integrative history written within the framework proposed here will reveal a lot about conceptual constraints, patterns of variation, and the problems of (conceptual) innovation within evolutionary developmental biology. In that sense, history does recapitulate itself in the context of current discussions in evolutionary developmental biology.

NOTES

1. Balfour and Gegenbaur have been the subjects of vignettes in the history of evolutionary developmental biology. See Hall, 2003; Laubichler 2003. The Alexander Kowalevsky Medal, awarded by the St. Petersburg Society of Naturalists, is the premier recognition in the field of evolutionary developmental biology; its recent winners include Walter Gehring, Brian Hall, Rudy Raff, and Rupert Riedl. See also the special issue of the *Journal of Experimental Zoology, Part B: Molecular and Developmental Evolution* 302, no. 1 (2004).

2. The journals are *Evolution and Development*, edited by Rudy Raff; *Journal of Experimental Zoology, Part B: Molecular and Developmental Evolution*, edited by Günter Wagner; and *Genes, Development and Evolution*, edited by Diethard Tautz (the latter is the continuation of *Roux's Archiv für Entwicklungsmechanik*, founded in 1890 as the first journal devoted to experimental embryology). The main textbooks are *Evolutionary Developmental Biology* by Brian Hall (1992; 2nd ed., 1998) and Sean Carroll, Jennifer K Grenier, and Scott D. Weatherbee's *From DNA to Diversity: Molecular Genetics and the Evolution of Animal Design* (2001; 2nd ed., 2005).

3. See Hall, 2000; Wagner, Chiu, and Laubichler, 2000; and chapters 15 and 16 in this volume.

4. See also chapter 16 in this volume; and Wagner, Chiu, and Laubichler, 2000.

5. For discussions about the conceptual and philosophical problems of evolutionary developmental biology, see the editorials and perspective pieces in *Evolution and Development* and *Journal of Experimental Zoology, Part B: Molecular and Developmental Evolution,* and the special issue of *Biology & Philosophy* 18, no. 2 (2003) devoted to these topics.

6. Jacob, 1993; Olby, 1974, 1985; Morange, 1998; Sapp, 2003; Coleman, 1971; Allen, 1975; Jahn and Krausse, 1998.

7. Beurton, Falk, and Rheinberger, 2000; Keller, 2000.

8. Maienschein, 2003.

9. On the role of epistemic objects in the history of science, see Hagner, 1997; Rheinberger, 1997; and Hagner and Rheinberger, 2003.

10. E.g., Kohler, 1994; Rheinberger, 1997; Creager, 2002; De Chadarevian, 2002.

11. E.g., Oppenheimer, 1967; Gould, 1977; Hall, 1992; Gilbert, 1994; Keller, 2002; Maienschein, 2003.

12. See, e.g., Zimmermann, 1953; Bowler, 1989; Jahn and Krausse, 1998.

13. The German term *Entwicklung* refers to both development and evolution.

14. Darwin, 1859, p. 449.

15. Müller, 1864.

16. Haeckel, 1866.

17. See Nyhart, 1995, 2002; Laubichler, 2003; Laubichler and Maienschein, 2003.

18. Kowalevsky, 1866, 1867, 1871, 1877.

19. Dohrn, 1875.

20. E.g., Allen, 1975; Nyhart, 1995, 2002.

21. See, e.g., Mocek, 1998; Laubichler and Rheinberger, 2004; and chapters 3, 5, and 7 in this volume.

22. E.g., Provine, 1971; Beurton, Falk, and Rheinberger, 2000.

23. See, e.g., Sarkar, 1999.

24. See, e.g., Laubichler and Rheinberger, 2004; see also chapter 7 in this volume.

25. E.g., Mayr and Provine, 1980; Mayr, 1982; Morange, 1998.

26. See especially chapters 14–16, but also chapters 5 and 9, in this volume.

27. Wagner, Chiu, and Laubichler, 2000; Stent, 1968.

28. Chapter 16 in this volume.

29. E.g., Eldredge and Gould, 1972; Gould, 1977.

30. E.g., Bonner, 1974; Sarkar, 1998.

31. See Gould and Lewontin, 1979, as the locus classicus for this critique.

32. Riedl 1975; Maynard-Smith et al., 1985.

33. Riedl, 1975; Wagner and Laubichler, 2004.

34. This is shown not only in the willingness of several leaders in the field to participate in both the seminar and the workshop but also in the amount and quality of historical work done by active (rather than retired) scientists in the field.

35. See Laubichler and Wagner, 2003.

36. See Keller, 2002; and Maienschein, 2003 for some recent exceptions.

37. Hall, 1988.

38. Hall, 1992; Love and Raff, 2003; Gilbert, 2001; Gilbert and Bolker, 2003.

REFERENCES

Allen, G. E. (1975). *Life Science in the Twentieth Century*. New York: Wiley.

Beurton, P., R. Falk, and H.-J. Rheinberger. (2000). *The Concept of the Gene in Development and Evolution*. Cambridge: Cambridge University Press.

Bonner, J. T. (1974). *On Development. The Biology of Form*. Cambridge: Harvard University Press.

Bowler, P. (1989). *Evolution: The History of an Idea*. Berkeley: University of California Press.

Carroll, S. B., J. K. Grenier, and S. D. Weatherbee. (2001). *From DNA to Diversity: Molecular Genetics and the Evolution of Animal Design*. Oxford: Blackwell Science.

Coleman, W. (1971). *Biology in the Nineteenth Century: Problems of Form, Function, and Transformation*. New York: Wiley.

Creager, A. (2002). *The Life of a Virus: Tobacco Mosaic Virus as an Experimental Model, 1930–1965*. Chicago: University of Chicago Press.

Darwin, C. (1859). *The Origin of Species*. London: Murray.

Daston, L., ed. (2000). *Biographies of Scientific Objects*. Chicago: University of Chicago Press.

De Chadarevian, S. (2002). *Designs for Life: Molecular Biology After World War II*. Cambridge: Cambridge University Press.

Dohrn, A. (1875). *Der Ursprung der Wirbelthiere und das Princip des Functionswechsels: Genealogische Skizzen*. Leipzig: Engelmann.

Eldredge, N., and Gould, S. J. (1972). Punctuated equilibria: an alternative to phyletic gradualism. In T. J. M. Schopf, ed., *Models in Paleobiology*. San Francisco: Freeman.

Gilbert, S. (2001). "Ecological Developmental Biology: Developmental Biology Meets the Real World." *Developmental Biology* 233: 1–12.

———. ed. (1994). *A Conceptual History of Modern Embryology*. Baltimore: Johns Hopkins University Press.

Gilbert, S., and J. A. Bolker. (2003). "Ecological Developmental Biology: Preface to the Symposium." *Evolution and Development* 5: 3–8.

Gould, S. J. (1977). *Ontogeny and Phylogeny*. Cambridge, Mass.: Belknap Press of Harvard University Press.

Gould, S. J., and R. Lewontin. (1979). "The Spandrels of San Marco and the Panglossian Paradigm: A Critique of the Adaptationist Programme." *Proceedings of the Royal Society of London* B205: 581–598.

Haeckel, E. (1866). *Generelle Morphologie der Organismen*. Berlin: Reimer.

Hagner, M. (1997). *Homo Cerebralis*. Berlin: Berlin Verlag.

Hagner, M., and H.-J. Rheinberger. (2003). "Prolepsis: Considerations for Histories of Science After 2000." In B. Joerges and H. Nowotny, eds., *Social Studies of Science and Technology: Looking Back, Ahead*, pp. 211–228. Dordrecht: Kluwer.

Hall, B. K. (1992). *Evolutionary Developmental Biology*. London, Chapman & Hall.

———. (1998). *Evolutionary Developmental Biology*. 2nd ed. London: Chapman & Hall.

———. (2000). "Evo-Devo or Devo-Evo—Does It Matter?" *Evolution and Development* 2: 177–178.

———. (2003). "Francis Maitland Balfour (1851–1882): A Founder of Evolutionary Embryology." *Journal of Experimental Zoology, Part B: Molecular and Developmental Evolution* 299: 3–8.

Jacob, F. (1993). *The Logic of Life: A History of Heredity*, B. E. Spillman, trans. Princeton, N. J.: Princeton University Press.

Jahn, I., and E. M. Krausse, eds. (1998). *Geschichte der Biologie: Theorien, Methoden, Institutionen, Kurzbiographien*, 3rd ed. Jena: Fischer.

Keller, E. F. (2000). *The Century of the Gene*. Cambridge, Mass.: Harvard University Press.

———. (2002). *Making Sense of Life: Explaining Biological Development with Models, Metaphors, and Machines*. Cambridge, Mass.: Harvard University Press.

Kohler, R. E. (1994). *Lords of the Fly. Drosophila Genetics and the Experimental Life*. Chicago: University of Chicago Press.

Kowalevsky, A. (1866). "Entwicklungsgeschichte der einfachen Ascidien." *Memoires of the Academy of Science, St. Petersburg* 7: 1–19.

————. (1867). "Entwicklungsgeschichte des *Amphioxus lanceolatus.*" *Mem. Acad. Sci. St. Petersburg* 11: 1–17.

————. (1871). "Weitere Studien über die Entwicklungsgeschichte der einfachen Ascidien." *Archiv für mikroskopische Anatomie* 7: 101–130.

————. (1877). "Weitere Studien über die Entwicklungsgeschichte des *Amphioxus lanceolatus.*" *Archiv für mikroskopische Anatomie* 13: 181–204.

Laubichler, M. D. (2003). "Carl Gegenbaur (1826–1903): Integrating Comparative Anatomy and Embryology." *Journal of Experimental Zoology, Part B: Molecular and Developmental Evolution* 300: 23–31.

Laubichler, M. D., and J. Maienschein. (2003). "Ontogeny, Anatomy, and the Problem of Homology: Carl Gegenbaur and the American Tradition of Cell Lineage Studies." *Theory in Biosciences* 122: 194–203.

Laubichler, M. D., and H.-J. Rheinberger. (2004). "Alfred Kühn (1885–1968) and Developmental Evolution." *Journal of Experimental Zoology, Part B: Molecular and Developmental Evolution* 302: 103–110.

Laubichler, M. D., and G. P. Wagner. (2003). "Editorial: A New Series of Vignettes on the History of Evolutionary Developmental Biology." *Journal of Experimental Zoology, Part B: Molecular and Developmental Evolution* 299: 1–2.

Love, A. C., and R. A. Raff. (2003). "Knowing Your Ancestors: Themes in the History of evo-devo." *Evolution and Development* 5: 327–330.

Maienschein, J. (2003). *Whose View of Life?* Cambridge, Mass.: Harvard University Press.

Maynard-Smith, J., et al. (1985). "Developmental Constraints and Evolution." *Quarterly Review of Biology* 60: 265–287.

Mayr, E. (1982). *The Growth of Biological Thought: Diversity, Evolution, and Inheritance.* Cambridge, Mass.: Belknap Press.

Mayr, E., and W. B. Provine, eds. (1980). *The Evolutionary Synthesis: Perspectives on the Unification of Biology.* Cambridge, Mass.: Harvard University Press.

Mocek, R. (1998). *Die werdende Form: Eine Geschichte der kausalen Morphologie.* Marburg: Basilisken Presse.

Morange, M. (1998). *A History of Molecular Biology*, M. Cobb, trans. Cambridge, Mass.: Harvard University Press.

Müller, F. (1864). *Für Darwin.* Leipzig: Engelmann.

Nyhart, L. K. (1995). *Biology Takes Form: Animal Morphology and the German Universities, 1800–1900.* Chicago: University of Chicago Press.

———. (2002). "Learning from History: Morphology's Challenges in Germany Circa 1900." *Journal of Morphology* 252: 2–14.

Olby, R. (1974). *The Path to the Double Helix*. Seattle: University of Washington Press. First published 1966.

———. (1985). *Origins of Mendelism*, 2nd ed. Chicago: University of Chicago Press.

Oppenheimer, J. M. (1967). *Essays in the History of Embryology and Biology*. Cambridge, Mass.: MIT Press.

Provine, W. B., comp. (1971). *The Origins of Theoretical Population Genetics*. Chicago: University of Chicago Press.

Rheinberger, H.-J. (1997). *Toward a History of Epistemic Things: Synthesizing Proteins in the Test Tube*. Stanford, Calif.: Stanford University Press.

Riedl, R. (1975). *Die Ordnung des Lebendigen*. Hamburg: Parey.

Sapp, J. (2003). *Genesis: The Evolution of Biology*. Oxford: Oxford University Press.

Sarkar, S. (1998). *Genetics and Reductionism*. Cambridge: Cambridge University Press.

———. (1999). "From *Reaktionsnorm* to the Adaptive Norm: The Norm of Reaction, 1909–1960." *Biology and Philosophy* 14: 235–252.

Stent, G. (1968). "That Was the Molecular Biology That Was." *Science* 160: 390–395.

Wagner, G. P., C.-H. Chiu, and M. D. Laubichler. (2000). "Developmental Evolution as a Mechanistic Science: The Inference from Developmental Mechanisms to Evolutionary Processes." *American Zoologist* 40: 819–831.

Wagner, G. P., and M. D. Laubichler. (2004). "Rupert Riedl and the Re-synthesis of Evolutionary and Developmental Biology: Body Plans and Evolvability." *Journal of Experimental Zoology, Part B: Molecular and Developmental Evolution* 302: 92–102.

Zimmermann, W. (1953). *Evolution: Die Geschichte ihrer Probleme und Erkenntnisse*. Freiburg: Alber.

I

ONTOGENY AND PHYLOGENY IN EARLY TWENTIETH-CENTURY BIOLOGY

3

LIVING WITH THE BIOGENETIC LAW: A REAPPRAISAL
Frederick B. Churchill

There is a well-merited tradition in the scholarly literature of the nineteenth and twentieth centuries of surveying the histories of the past before revising them. This continues to be true in most fields of history, including the history of science. I support this practice, for it not only acknowledges the work of predecessors, but it forces us to keep our own modest contributions in honest perspective and impels all participants to stand back and look at the limitations of interpreting the past. To present a paper on the biogenetic law from Haeckel to World War I as a prelude to a conference on the history and philosophy of Evo-Devo is, however, to enter three levels of virtual reality. The first is a metahistory to acknowledge and revise; the second is a utilitarian history to frame the past in order to provide a lesson for today; the third is a biological history embedded in evolution and development themselves. All three present challenges to our understanding of past reality. I start with the metahistory of acknowledgment and revision. The utilitarian and biological histories will gradually emerge.

HISTORIANS AND THE BIOGENETIC LAW

The nineteenth-century outlines of Haeckel's biogenetic law are well known to historians of science.[1] Foreshadowed in Darwin's *Origin of Species* and Fritz Müller's strong endorsement *Für Darwin*; promulgated and broadcast afar by the inventive zoology and popular texts of Ernst Haeckel; promoted, modified, and internalized by at least the first two generations of post-Darwinian zoologists; and criticized, even condemned, by some contemporaries and later generations who possessed different touchstones, the biogenetic law left an enduring legacy worthy of continued historical interest.

Of the twentieth-century Anglo-American historians of the biogenetic law, Edward Stuart Russell stands out as a careful and wide-ranging scholar well read in the primary sources. As a biologist he also engaged in scientific debates, used his history to show the weaknesses of the Haeckelian tradition of morphology, and, in so doing, furthered his own zoological beliefs in the greater value of the more recent experimental tradition.[2] He spent much effort deciphering the empirical details and agenda of Haeckel's program, but his cardinal point was to expose Haeckel's depiction of individual development as a dialectic between an epitome or recapitulation of the phylogenetic past (i.e., a "palingenesis") and a falsification or vitiation of that past (i.e., a "cenogenesis"). In short, Russell pointed out, as others had before him, that there was no way of empirically refuting the biogenetic law because of its circular reasoning—any deviation from a true recapitulation of the embryo's ancestors could always be passed off as a falsification caused by cenogenesis. Despite the impressive morphological work in constructing hypothetical lineages by Haeckel's generation, and despite the fact that Haeckel and many of his contemporaries believed that paleontology would eventually offer an independent confirmation of conclusions drawn from comparative morphology, Russell inferred that the results of "classical morphology" remained questionable. Experimental embryology and functional morphology, not the pursuit of missing links, Russell argued, would lead to an understanding of the causal connection between evolution and development.

Writing fifty years after Russell, Stephen Jay Gould turned his attention to Haeckel and the biogenetic law as part of a more general historical and contemporary study of "The evolutionary importance of *heterochrony*."[3] Thus, he, too, saw his science and its history as intimately connected. Where Russell had focused more on the construction of phylogenies in the wake of Haeckel's programmatic statements, Gould was more interested in explaining the details of the biogenetic law. Since "Evolution implied a true, physical continuity of forms through time,"[4] Gould explained, Haeckel and his generation responded by suggesting physical mechanisms for development, heredity, and recapitulation.

At times I have trouble with Gould's chronological condensation of history, but he was certainly right that Darwinians, neo-Darwinians, neo-Lamarckians, and orthogeneticists alike found the biogenetic law to be reflective of a past reality and a useful generalization for contemporary research. Furthermore, Gould correctly pointed out that neither the comparative morphology nor the paleontology of those days could resolve the widely differing opinions about whether any given feature resulted from palingenesis or cenogenesis.[5] Whereas Russell had stopped his story about the decline in acceptance of the biogenetic law at the end of the nineteenth century, Gould concluded that the demise came later, at the hands of classical genetics.

In 1991, Nicolas Rasmussen, a young historian of biology, wrote a provocative piece in which he endorsed Gould's claim that the rapidly multiplying empirical findings of cenogenesis could not by themselves overthrow the biogenetic law. Instead of seeing classical genetics as its definitive refutation, however, Rasmussen envisioned a grander disciplinary confrontation between the Haeckelian worldview centered on the biogenetic law as a causal explanation of development and what Rasmussen called "the new world order." In the hands of Thomas Hunt Morgan, this "new world order" consisted of an expansion of experimental morphology and genetics in a way that co-opted from the Haeckelian worldview the mantle of providing a causal explanation for evolution. The historical shift between two contrasting "world orders" was not simply a rejection of a particular biological theory—that is, the biogenetic law—but entailed a redefinition of what was central to the disciplinary practice of biology.[6]

Finally, there are two more recent accounts that, among many other things, address the historical question of the rise and fall of the biogenetic law. Peter J. Bowler's *Life's Splendid Drama* forces us to reconsider the empirical morphological practices so derisively dismissed by Russell, Gould, and Rasmussen.[7] Bowler's objective, to be sure, is broader than simply a reexamination of the rise and fall of the biogenetic law. He purposefully refrains from reciting the standard story about the contrast between descriptive and experimental biology. Instead, he provides historical case studies of morphological efforts to establish the phylogenies of arthropods, vertebrates, fish/amphibians, and birds/mammals. Throughout a sixty-year period, roughly centered on 1900, comparative anatomists, embryologists, paleontologists, and biogeographers, according to Bowler, added their unique perspectives to the search for human ancestors. They not only contributed a wealth of new information but unwittingly and inescapably drifted into a "Darwinian" frame of mind by ignoring the formalism inherent in palingenesis and refocusing on the functional nature of adaptation inherent in cenogenesis. By the 1920s and 1930s, Bowler's argument continues, these zoologists, recognizing the impasse in solving phylogenetic questions through morphology, helped set the stage for the evolutionary synthesis through their newly found support of natural selection as the primary mechanism of adaptation. In so doing, they forced a reevaluation of many morphological assumptions, including the biogenetic law.[8]

Explicit in Bowler's account is the conscious effort to understand comparative morphology, paleontology, and biogeography as ongoing, developing traditions of research that provided their own internal critique of the methods and assumptions which led to the biogenetic law. One of the central themes of Lynn Nyhart's historical studies may be interpreted in the same way. In her innovative *Biology Takes Form*,[9] Nyhart examines in particular the comparative anatomy school led by Carl Gegenbaur and his students, who challenged the effective use of embryology as a tool for ascertaining phylogeny. Since there was so much confusion in determining what were palingenetic and what were cenogenetic structures, Gegenbaur insisted that comparative anatomy, rather than comparative embryology, must be the final arbiter. This *Kompetenzkonflikt*, as it was called at the time, meant that the functionalism implied in determining the purpose of adult structures, rather than the morphological formalism embedded in the biogenetic law, must guide the process of phylogeny construction.

It is important to point out that, like Bowler, Nyhart finds that important biological issues in the last third of the nineteenth century provided an effective internal critique of the biogenetic law. More important for her overall story, Nyhart places the debate between anatomists and embryologists within the broader framework of institutional and professional changes that took place in the German universities as generations of anatomy and zoology professors promoted their own careers and programs. As generational cohorts changed, nineteenth-century morphology, and implicitly the biogenetic law with it, gave way to twentieth-century issues and research.

For the purposes of this chapter I wish to emphasize that an unstated disagreement exists between Russell, Gould, and Rasmussen, on the one hand, and Bowler, on the other, in evaluating the ability of evolutionary morphology to work its way through the circular reasoning in Haeckel's version of the biogenetic law. For Bowler the technical dialogue of constructing phylogenies (that is, the internal critique) contributed to the downfall of the biogenetic law. For Russell, Gould, and Rasmussen an external critique, be it cognitive, disciplinary, or cultural, was essential for the displacement of Haeckel. These rival claims, however, need not be mutually exclusive, as Nyhart has shown by integrating them. In what follows, I (1) examine the "hardening" of Haeckel's biogenetic law,[10] (2) detail four elements of the internal critique that responded to Haeckel's claims, (3) delineate a contrast between what I call the "strong" and "weak" versions of the biogenetic law, (4) review three texts in comparative embryology that reflect differing attitudes toward Haeckel's strong version, and (5) come to some summary conclusions and general reflections. An analysis on the metalevel of history will continue, but in a less focused way.

THE "HARDENING" OF HAECKEL'S BIOGENETIC LAW

Given this historiographical introduction, let us look more closely at Haeckel's biogenetic law. How did it shape an understanding about phylogeny and ontogeny between the 1860s, when Haeckel (Figure 3.1) began promoting their causal linkage, and the third decade of the twentieth century, after embryology, biogeography, and taxonomy had substantially changed their focus and the new sciences of genetics and endocrinology had reoriented the science of form?[11] First, I provide a few words about terminology.

Haeckel did not use the expression "biogenetic law" in his *Generelle Morphologie*, as is so often claimed in the secondary literature, but at that time he laid down five "theses" which dictated patterns of recapitulation. Briefly paraphrased, these claimed:

1. Ontogeny is directly "caused" (*bedingt*) by phylogeny.

2. Ontogeny is "the short and rapid recapitulation" of phylogeny "caused" (*bedingt*) by heredity and modified by adaptation.

3. Ontogeny of the individual repeats the most important changes in form of its ancestors.

4. ". . . the complete and accurate repetition of phyletic by ontogenetic development is obliterated and abbreviated" as subsequent ontogenies take an ever shorter path toward the adult stage.

5. As successive individual ontogenies become adapted to new circumstances, the record of phylogeny becomes "falsified and altered" by secondary adaptation. Therefore, the truer the repetition, the more similar the conditions of existence in which the embryo and ancestors have developed.[12]

These theses are recognizable formulation of what in 1870 Haeckel was to designate "das biogenetische Grundgesetz," and yet there are subtle differences dictated by an evolving cognitive context. To anticipate, the anatomist Franz Keibel pointed out at the end of the century that these theses of 1866, despite being identified by Haeckel as "Thesen von dem Causalnexus der biotischen und der phyletischen Entwickelung," were really more descriptive and wishful in intent than mechanistically causal.[13]

In 1866 Haeckel was responding to the challenge of unifying biology through Darwin's mechanistic theory of evolution and Rudolf Virchow's

Figure 3.1
Portrait of Ernst Haeckel by Marie Rosenthal-Hatschek, a highly accomplished artist
who had painted portraits of members of the Habsburg family. She had married
Haeckel's former student, Berthold Hatschek, who at the time was professor of zoology,
director of the second Zoological Institute in Vienna, and an important contributor to
evolution and development. In the spring of 1911 the artist, accompanied by her
husband, visited Jena to paint Haeckel's portrait. Shortly before their visit, however,
Haeckel had fallen and broken the neck of his left femur. Nevertheless, according to
the recollections of the artist's daughter, he cheerfully sat for his portrait. One should
note Haeckel's extended left leg, the gorilla skull, the bookcase in the background, and
particularly the sparkle in face of the 70-year old evolutionist. The artist's signature is
visible in the lower right hand corner of the original, which is owned by Indiana University and hangs prominently in its Lilly Library. Further details have been collected
in Armin Geus, "Der achtzigjährige Ernst Haeckel—ein Altersporträt von Marie
Rosenthal-Hatschek," *Medizin Historisches Journal*, 15. (1980), 172–176.

materialistic vision of cells, both of which were substantial advances in biology made at the end of the 1850s.[14] Direct cell division provided Haeckel with a mechanical model of how a complex organism might arise from a bud, a single-celled zygote, or a spore, and how the next generation of buds, gametes, and spores arose in turn. At the end of the cycle these reproductive cells had to conserve what the cycle embraced up to that point in order to start the process over again and allow for evolutionary change. In other words, reproduction was the material extension or cellular overgrowth of the sexually mature individual. Darwin's mechanism of natural selection and his overwhelming documentation of the reality of evolution provided Haeckel with a template for lineage construction, both diverging and direct, upon which to map hypothesized physical phylogenies.[15] Thus the recapitulation of phylogeny in subsequent ontogenies was mediated through direct cell division. The phylogeny, being antecedent, literally caused the course of ontogeny—that is, up to a point.

Abbreviations and secondary adaptations also played a role. These, however, were not the pivotal issue of the *Generelle Morphologie*. The "causal nexus" of phylogeny and ontogeny, paraphrased above, was epitomized in just five of forty-five formal, ontogenetic theses that were concerned, among other things, with the mechanical and physiological nature of development, the formative forces in organic material, the three basic stages of development, and the cycle of three levels of genealogical individuality. Unable to repress his penchant for formalization, Haeckel presented, a hundred pages later, thirty-eight "phylogenetic theses," none of which had anything to do with recapitulation.[16] He thus unwittingly revealed how marginal "recapitulation" was in his two-volume sketch of general morphology.

Four years later Haeckel elevated, in passing, those five descriptive theses to a fundamental causal process with the phrase "das höchst wichtige biogenetische Grundgesetz."[17] In 1872, in a two-volume work on the classification of calcareous sponges he devoted a full section to "Das Biogenetische Grundgesetz,"[18] and in 1874, combining the logic of this biogenetic law and his conviction that invagination was the true primitive mechanism of germ layer formation, he detailed his daring and notorious extrapolation of the purported ancestor of the metazoans in his gastraea theory.[19] Finally, in the same year he coined the terms "palingenesis" and "cenogenesis."[20]

This chronology may seem antiquated, but I believe there is some merit therein. As scientists, politicians, and ideologues know, to name a process or idea is to crystallize the ephemeral and stake out one's ownership. For Haeckel to turn the five theses of recapitulation into "das biogenetische Grundgesetz" was to do even more. In nineteenth-century science, Grundgesetz identified a fundamental, true law of nature. Besides exhibiting a certain epistemic bravado, there was a strategic reason for rendering recapitulation into a concise expression. Haeckel's specification of phylogeny as the cause of ontogeny had been under attack by a card-carrying reductionist embryologist, Wilhelm His.[21] To announce a "biogenetische Grundgesetz" was to parry the physical laws used by His with an organic law.

Haeckel's neologisms "palingenesis" and "cenogenesis" were also significant, for they emphasized the antithesis between recapitulation of phylogeny and its falsification. What had been descriptive words in the two final theses of 1866 became antithetical causal expressions in 1874. Above all, I find in Haeckel's 1872 discussion of the biogenetic law his most interesting statement: "In my General Morphology," Haeckel explained in an English abstract of his work, "I sought to demonstrate synthetically that all the phenomena of the organic world of forms can be explained and understood only by the monistic philosophy; and now this demonstration is furnished *analytically* by the morphology of the Calcispongiae."[22] The analytic component of his argument then was contained in the morphological details of his classification of those sponges and in the central role that the germ layers played in his taxonomic efforts (Figure 3.2). A belief in the homologous nature of the germ layers throughout the animal kingdom had recently emerged as a result of the extraordinary empirical achievements in comparative embryology by Alexander Kowalevsky, Elie

Figure 3.2
A composite of two plates from Haeckel's *Kalkschwämme* (1872) showing gastrulation of the calcareous sponges (a) *Leuculmis echinus* (from *Tafel* 30, nos. 8 and 9) and (b) *Sycyssa huxleyi* (from *Tafel* 44, nos. 14 and 15). Clearly shown, but not explained by Haeckel, is the contrast between the palingenetic gastrulation of the *Leuculmis* and the cenogenetic (i.e., "falsified") gastrulation of *Sycyssa* amphiblastula larvae. For Haeckel the contrast represented the phylogenetic branching of the Leuconen/Syconen lineage of calcareous sponges from the more primitive Asconen calcareous sponges. He did not understand the inversion process of the blastulae, which brings the flagellated cells into the interior, but he did illustrate in these two species (e.g., *Tafel* 44, no. 16) the contrasting ectodermal and endodermal cells following gastrulation. Otherwise, spicule morphology (e.g., *Tafel* 30, nos. 1 and 11, *Dyssycus echinus*) rather than germ layer formation and larval development played the important role in his elaborate descriptions and classification of the calcareous sponges.

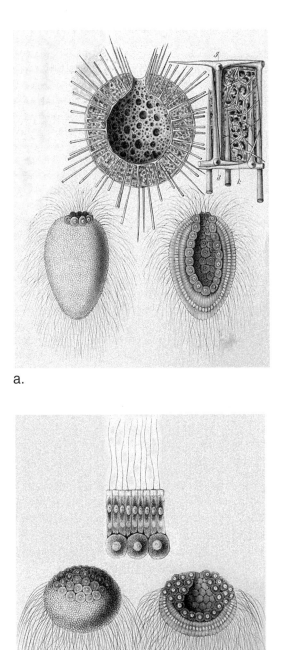

a.

b.

Metchnikoff, and Nicolai Kleinenberg from 1867 on. Haeckel's version of recapitulation helped validate the germ layers as the first morphological elements in which phylogenetic lineages of metazoans could be easily detected, and, as circular as the argument might seem, the germ layers in turn provided Haeckel with his analytic justification of recapitulation. There were both synthetic and analytic grounds for invoking a causal *Grundgesetz*.

Finally, the illustration of the imaginary ontogeny of the hypothetical gastraea became the totem demonstration of the *Grundgesetz*. Haeckel referred to it repeatedly throughout his career. It is curious, though, that the gastraea theory was mentioned in his monographic work on calcareous sponges, for Haeckel had all but ignored cleavage, gastrulation, and the early ontogeny of sponge larvae. In 1872 he was still uncertain whether the sponges might not be an evolutionary lineage separate from the Metazoa. Three years later, however, a closer examination of cleavage, inversion, gastrulation, and larva formation of a calcareous sponge by Franz Eilhard Schulze convinced Haeckel that early sponge development, like coelenterate development, evinced a loose "mesodermal-like" layer and so should be considered the base of a grand monophyletic evolutionary tree of all Metazoa.[23]

THE INTERNAL CRITIQUE

Given the preceding discussion, it is time to examine, if only in brief vignettes, the nature of what I have alluded to as the "internal critique." Not intended to be exhaustive, the following catalog is nonetheless illustrative of the extent of self-evaluation that comparative embryologists underwent as they took full measure of the biogenetic law in their research. Many of these areas of internal critique have been described in the secondary literature, but to my knowledge they have not been systematically gathered together. I gratefully draw upon the historians who have studied these issues, and present these vignettes in the order in which embryologists might consider them as they follow development from egg to adult. I begin with discussions of egg segmentation and cell lineage studies, move on to the conception of the primitive germ layers and accounts of larval forms, and finish with a dialogue about the meaning of individual variations. In the end I argue that these areas of critique presented empirical challenges to the biogenetic law, for they asserted the role of cenogenesis over palingenesis; that is, they emphasized innovation and adaptation over conservation and recapitulation.

Cleavage Patterns

The most obvious step in early development, as seen through the optical microscope, is the segmentation of the fertilized egg into blastomeres. A year after he wrote the extended version of his gastraea theory, Haeckel added a sequel to his treatise, in which he examined the gastrula in the light of early cleavage of the egg. (Figure 3.3)[24] In brief, he recognized four basic patterns of cleavage, and his segmentation taxonomy bears a rough resemblance to today's distinctions between the complete cleavage of holoblastic eggs and the incomplete cleavage of meroblastic and centrolicithal eggs. As became characteristic of Haeckel's writings, much of this text consisted of renewed assertions of what he had written before—in this case, reminders of the difference between palingenesis and cenogenesis.

Nevertheless, a number of idiosyncratic arguments stand out. Haeckel organized his taxonomy of cleavage around his gastraea theory. He characterized the primordial form of segmentation as a radially symmetrical cleavage leading to a radial gastrula—that original ancestor of all metazoans. He decided an unequal complete cleavage reflected a future amphiblastula, which possessed a few larger blastomeres at the vegetal pole and many smaller ones at the animal pole. He further reasoned that discoidal and superficial segmentation patterns were determined by the further augmentation and placement of nutritive yolk. Haeckel's basic assumptions driving these distinctions were that the protoplasm of the egg (i.e., the active formative yolk, *Bildungsdotter*), must be primitive (that is, the essence of cell division and palingenesis). Nutritive yolk (*Nahrungsdotter*), on the other hand, was inert, the mere storehouse of the egg, which "in all cases . . . is a secondary cenogenetic product."[25] Finally, despite his relatively rich store of information, he continued to focus on complete homologies of whole organisms in his quest for phylogenies, rather than pushing the homologies of his cleavage taxonomy.

Jane Maienschein[26] has examined the American cell lineage tradition, which explored radial, particularly spiral, cleavage patterns and which made determinations about the fates of different blastomeres. What is fascinating about this group of investigators is that they were zealously focused on the cellular details of cleavage and development in a few organisms, especially annelids and mollusks, with spiral cleavage. Furthermore, Maienschein detected in this tradition a progressive critique of Haeckel's biogenetic law. Charles Otis Whitman diverted attention from Haeckel's overwhelming concern for the fate of primitive germ layers to the blastomeres. Edmund Beecher Wilson disavowed Haeckel's stress on the

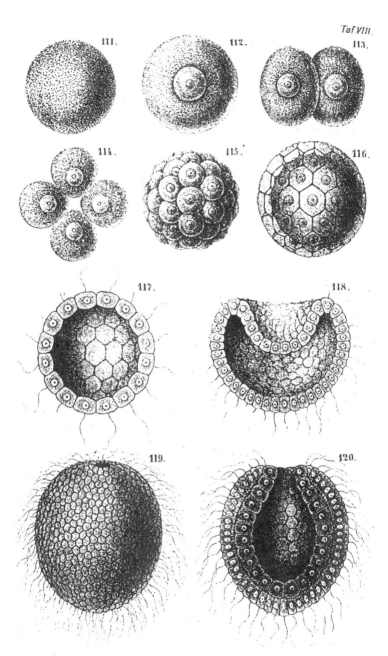

biogenetic law and established an intermediate position of recognizing a combination of "ancestral reminiscences, hereditary material, and external factors" determining early development.[27] Wilson, Davis Mead, and Aaron Louis Treadwell increasingly stressed adaptations of larvae and adults to living conditions. Edwin Grant Conklin appears to have been more concerned about the relationship between ontogeny and phylogeny. Nevertheless, his detailed lineage studies, in which he distinguished determinate from indeterminate cleavage and focused on the importance of intrinsic protoplasmic structure, led him to "reject Haeckel's recapitulation" as a causal explanation of development (Figure 3.4). Finally, according to Maienschein, Frank Rattray Lillie exhibited "a modern understanding of individual development" by describing various cleavage patterns as expressions of diverging adaptations of the whole organism, larvae and adults. Homologies between blastomeres might be found in determinate, but not indeterminate, cleavage. She concluded that "This cell-lineage work provided a transition away from Haeckel's naïve causal biogenetic law and his preoccupation with tracing phylogenetic trees."[28]

More recently, Robert Guralnick has traversed much of the same territory, including an examination of the work of students of the first generation of American cell "lineagists," and has reached a different conclusion about the general trend to modernity.[29] He finds that as the research program continued into the twentieth century, the mechanical explanations of cleavage patterns urged by Wilson became less important than discussions of prospective adaptations at the larval and adult stages. Investigators perceived that regional, even cellular, homologies were a better way of demonstrating the conservation of the ancestral past. By breaking down the field for homologies to a finer morphological level and by emphasizing the process of precocious segregation of adult and larval

Figure 3.3
From Haeckel, "Studien zur Gastraea-Theorie," part 2, "Die Gastrula und die Eifurchung der Thiere" (1875). Haeckel illustrated what he considered the five basic stages in the palingenetic sequence of earliest development in the living *Gastrophysema*, variously classified in Haeckel's day among the lowest coelenterates or the rhizopodean protozoans. Stage 1 (no. 111): the archimonerula, the anucleated egg cell prior to fertilization; stage 2 (no. 112): the archicytula, the fertilized cell, and (nos. 113 and 114) the first two cleavages of the archicytula; stage 3 (no. 115): the archimorula; stage 4 (nos. 116 and 117): an external view and a cross section of the hollow archiblastula; stage 5 (nos. 118, 119, and 120): the invagination, external view, and cross section of the gastrula.

Pl. II.

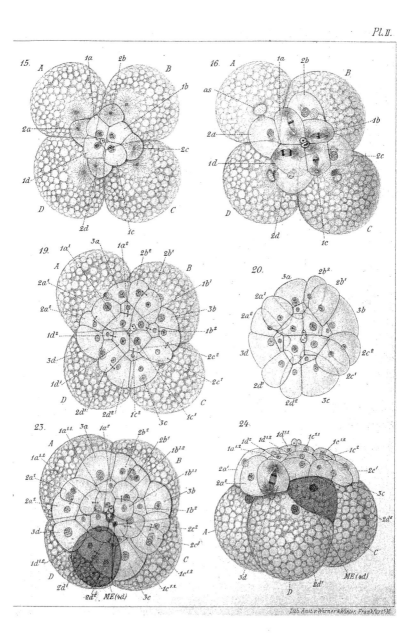

structures to earlier stages, they were in essence muting the "strong" Haeckelian search for complete homologies. According to Guralnick, the cell lineage program came to an end after 1907 "because of too much conservation *and* variation in the dataset."[30] There was little future for younger zoologists to detail yet one more variant in the range of cleavage patterns.

Both Maienschein's and Guralnick's works serve my purposes well. What seems clear is that cleavage studies hardly remained slavishly bound to Haeckel's biogenetic law. They may have resorted to recapitulation as part of the armory of analysis, but they also discussed adaptations and variations as well as ancestral reminiscences. They fixed on the homologies of cells rather than organs and whole organisms, and accepted a perplexing range of variations among the early cleavage cells and their fates.

Germ Layer Formation

As suggested above, the identification of the primitive germ layers was at the heart of Haeckel's biogenetic law and gastraea theory. It is surprising, then, that the history of these structures played such a small role in twentieth-century secondary literature.[31] For our purposes we can bypass the earliest history of the germ layer doctrine in the first half of the nineteenth century, a story that stretches from Pander, Von Baer, and Rathke through Huxley, Remak, and Allman.

In the post-Darwinian period the primitive germ layers acquired an evolutionary meaning, but as Jane Oppenheimer has pointed out, Haeckel did not even refer to germ layers in his *Generelle Morphologie*. It took a decade and the innovative descriptive work of Kowalevsky, Metchnikoff,

Figure 3.4
From Edwin Grant Conklin, "The Embryology of Crepidula: A Contribution to the Cell Lineage and Early Development of Some Marine Gasteropods" *Journal of Morphology* 13: 1–226 (1897). Nos. 15, 16, 19, 20, 23, and 24, which appear on the right hand side of plate II, offer a representative sample of the cell lineage method of systematic blastomere identification and of determinative spiral cleavage from the second quartet to a twenty-nine-cell structure found in Crepidula. Note the lettered macromeres, the shaded mesoentoblast, ME(4d), and the polar bodies atop the side view in no. 24. Conklin was clear about his conclusions: (1) a recognition of mechanical conditions in cleavage; (2) the identification of homologies of cleavage cells between large groups of organisms; (3) "The principal significance of any determinate form of cleavage is *prospective* rather than *retrospective*." This final conclusion undercut the retrospective nature of Haeckel's method and emphasized the functional value of cleavage for the formation of the adult organism (pp. 202–203).

and Kleinenberg to develop the evolutionary application of these primitive structures.[32] As described above, Haeckel recognized the phylogenetic importance of the endoderm and ectoderm in the diploblastic calcareous sponges; in his gastraea theory he assumed not only their primitive nature but also that the blastopore, formed through invagination, was the primitive mouth—gastrulation produced the gastraea (Figure 3.5).

Earl Ray Lankester, exposed to the same body of research, simultaneously exploited the primitive germ layers in a different way. He challenged Haeckel's claim that invagination of the blastula was the archetypal pathway, on the grounds that the planula larva of coelenterates often formed from the blastula through circumcircular cell divisions of the blastula cells and a delamination of the inner daughter cells so as to achieve a diploblastic organism. He found it easier to consider invagination as an adaptation to the delamination process than the other way around. Besides—and this was the clincher for him—the blastopore could not be considered the primitive mouth, as Haeckel claimed, because the mouths of many major groups of organisms arose as a secondary break in the diploblastic structure. In opposition to Haeckel's gastraea theory, Lankester proposed his planula theory, which appeared simpler and more primitive.[33]

After working with Haeckel in Jena and after jointly researching the embryology of sea anemones, comb jellies, and arrowworms, Oscar and Richard Hertwig devoted a four-part work to the germ layer doctrine. They came to the conclusion that zoologists were inclined to confuse the morphological status of the initial germ layers with the organ-forming tissues of later development. They recommended employing the terms "epiblast," "endoblast," and "mesoblast" for the primitive germ layers and restricting the familiar terms "ectoderm," "endoderm," and "mesoderm" to the tissues of organ formation.[34] The parallel terminology reflected the conclusion that whatever homologies they might be prepared to accept within the diploblastic sponges and coelenterates would be irrelevant in higher organisms. Moreover, the mesoderm of triploblastic organisms, in their minds, became an even more ambiguous structure because "an homology between the mesoderm [i.e., interstitial structures] of the coelenterates and the mesoblast of the remaining animals completely falls to the ground."[35]

The objections and revisions of Lankester and the Hertwig brothers represent merely the tip of the iceberg of discontent with Haeckel's gastraea theory. Although they did not reject Haeckel's morphological approach, they were actively engaged in reinterpreting its embryological

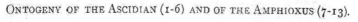

ONTOGENY OF THE ASCIDIAN (1-6) AND OF THE AMPHIOXUS (7-13).

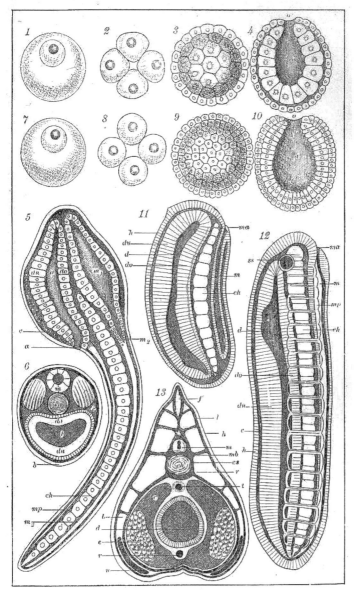

Figure 3.5
From Haeckel, *Anthropogenie oder Entwickelungsgeschichte des Menschen* (1874a; 3rd ed., 1877). Plate X, nos. 1–6, show early development of an ascidian with a lacuna between early development and the larval stages. Nos. 7–13 show the same development of *Amphioxus*. In both cases there is a palingenetic development through the gastrula stage. Together the two revealed to Haeckel that protovertebrates formed a direct lineage back to the gastraea of the Laurentide epoch. The letters are abbreviations of the German names of anatomical parts. The designations are the same in the English editions from which this plate was taken.

evidence. Their disagreements were in principle disagreements about what was palingenetic and what was cenogenetic. In a critical essay review at the end of the century, F. Braem struck a more hostile note. He argued that twenty years of descriptive embryology had documented that invagination, delamination, and epiboly were all ancestral pathways for producing the gastrula. "How are these layers to be judged from a morphological standpoint?" Braem asked, and immediately answered his own question: "Everything that has been taught about the origin of germ layers, fits like a square peg in a round hole."[36]

By then the privileged data upon which Haeckel's gastraea theory and the conjectured specified fates of the primitive germ layers rested, and upon which the "analytical" side of the biogenetic law depended, appeared to have been overwhelmed by a mass of newly acquired embryological details. None of it would refute the "synthetic" side of the biogenetic law, but in the aggregate the wealth of new information diminished its dogmatic rendering by demonstrating that development was far more complex in structures and processes than Haeckel and his program had envisioned.

Larvae

Animal development presented the post-Darwinian zoologists with a bewildering diversity of larval forms that often complicated the effort to unify the phylogenetic picture. On the surface nothing might seem more recalcitrant, yet more interesting for such a program, than the ubiquity of larvae and the cycles of alternating generations. The marvelous forms of life cycles of coelenterates, the nauplius and zoea larvae of crustaceans, the pluteus larvae of echinoderms, the trochophore and veliger larvae of annelids and mollusks, the variety of insect larvae and asexual generations, and—foremost—the tadpole larvae or appendicularia of ascidians provided both confusion and a key to many a taxonomic endeavor. But what was palingenetic and the result of deep inheritance? What was cenogenetic and attributable to variations and adaptations? From the 1830s on, despite these difficulties, larval and alternating forms had been essential in the overhaul of taxonomy—witness John Vaughn Thompson's discovery of barnacle larvae and Darwin's extended study of them.[37]

Johannes P. Müller, who had worked out the pluteus larvae among echinoderms, familiarized Haeckel with the diversity of life cycles of marine organisms.[38] The latter's early work on the causal union between phylogeny and ontogeny shows the importance of larval forms in his constructions. Interestingly, when Haeckel discussed the larvae (pluteus) of what he considered the four basic types of echinoderms, he returned to

Müller's earlier contention that they were not larvae at all but the initial, nurse organisms (i.e., *Ammen*) in an alternation of generations.[39] He probably agreed with Müller's interpretation because the adult form appears to arise, echinoderm-like, from a bud on the pluteus. Such budding emphasized the larva's prior independent status as an organism. An independent ur-form was clearly on his mind when Haeckel illustrated and discussed the nauplius and zoea larvae of crustaceans: "We can therefore conclude with certainty as to the common descent of all those orders from a common *Urkrebse*, which was in reality like today's nauplius." Haeckel failed to discuss the larval forms of insects or the trochophore larvae of worms and mollusks, but he triumphantly paraded forth the appendicularia of ascidians, recently discovered by Kowalevsky, for they allowed him to link vertebrates with lower invertebrates—perhaps even to connect man with worms.[40]

Bowler's first-rate presentation of Haeckel's "ascidian theory" and its rivals is surpassed, in my mind, by his chapter on the post-Darwinian attempts to unravel the phylogeny of arthropods.[41] Modern evolutionary biologists recognize this taxon as "the most diverse phylum" with "some of the most complex and interesting phylogenetic problems of any animals."[42] Its nineteenth-century history provides a fine example supporting Bowler's (and this chapter's) contention that normal descriptive embryology between the time of Haeckel's early pronouncements and the first decade of the twentieth century trenchantly, though not conclusively, challenged the biogenetic law.

Bowler's chapter focuses on three substantive debates. The first involves the relationship between the trachea-breathing arthropods (i.e., Tracheata or arachnids, insects, and myriapods) and the gill-breathing arthropods, (i.e., the Branchiata or crustaceans). The canonical recapitulationist view, developed by Fritz Müller and seconded by Haeckel, considered the nauplius and zoea larvae as ancestral forms of all crustaceans.[43] This phylogeny, however, presented a significant problem. From annelids to nauplius and zoea to Tracheata and Crustacea implied an evolution from a segmented organism to two unsegmented ones and then back to segmented forms in crustaceans only. Such an improbable sequence could be addressed only if one assumed that the tracheates and crustaceans were independently evolved from annelids. This strategy would render arthropods an unnatural group. Berthold Hatschek and Anton Dohrn considered nauplius and zoea to be cenogenetic, and thus explained their phylogenetic appearance on functional rather than biogenetic grounds. In 1874 Henry N. Moseley, zoologist on the *Challenger* expedition, provided

convincing morphological evidence for the independent origins of tracheates through a detailed study of the onychophoran *Peripatus*. This mul-
tisegmented, primitive organism with tracheae appeared to be the natural
link between the annelids and tracheates but reinforced the polyphyletic
nature of the arthropods.[44]

The second area of debate deals with Lankester's 1881 claim that
Limulus (i.e., the horseshoe crab) should be classified as a primitive arachnid rather than a crustacean. This view generated a bitter, personal controversy between Lankester and the Austrian zoologist Carl Claus.
Eventually both sides became reconciled to the rending of the Treacheata
into two independently evolved lineages. By the end of the century, others
had entered the quarrel with arguments about their common compound
eyes and segmentation, which indicated to some that the Treacheata and
Branchiata had converged for functional reasons. Bowler's third area of
debate concerns the entrance of paleontology into the discourse about the
phylogenetic nature of the arthropods, also with conflicting results: "The
new century [thus] began," Bowler argues, "with no agreement among
experts on whether or not the arthropods were polyphyletic."[45]

Bowler's chapter, and indeed his entire book, offers strong evidence
that the increasing study of larval forms challenged the monophyletic
imperative of Haeckel's biogenetic law. Moreover, the story demonstrates
that during the last two decades of the nineteenth century, comparative
embryology was pushing the limits of historical explanations. Instead of
simply insisting that phylogeny caused ontogeny, as Haeckel had done,
instead of agonizing over which structures were palingenetic and which
were cenogenetic, morphologists became increasingly interested in the
functional and environmental causes behind adaptations.

Individual Variations

Haeckel acknowledged small differences between individuals of the same
species, and he incorporated his discussion of what we commonly call
"variations" into a broader picture of the adaptation of the organism to
the surrounding world. In his *Generelle Morphologie*, he codified variations
as direct and indirect adaptations. The former were those changes brought
about by the environment, nutrition, and use and disuse of mental activities during the individual's lifetime; the latter consisted of the same kinds
of influences on the parent(s), but the changes induced by them became
manifest only in the offspring.[46] With his articulation of the biogenetic
law and gastraea theory in the early 1870s, Haeckel made a clear distinc-

tion between the processes of heredity and adaptation, and identified them with palingenesis and cenogenesis, respectively.[47] He presented an idiosyncratic theory to explain heredity, but I am unaware that he examined in detail the production and range of variations in related species or individuals of the same species. The subject became the focus of research in the 1890s for a group of German comparative morphologists.[48]

Russell recognized the group, which included A. Oppel, Ernst Mehnert, Franz Keibel, and A. Fischel, all of whose work in comparative embryology and anatomy explored heterochronic and heterotopic variations in organs of closely related species and individuals within a single species.[49] The group forced morphologists to question the biogenetic law on the basis of the functional requirements of such variations for evolution. Gould is more dismissive of their accomplishments, but his conclusion appears to me to be somewhat colored by his opinion of similar research done in the 1930s.[50] Both of these historical accounts were in part derivative from Franz Keibel's lengthy essay review, which included extensive passages from many of the original texts.[51]

Each of these morphologists pursued detailed descriptive studies of vertebrate development. Oppel fashioned a series of comparative tables in which he recorded measurements of organs, the precise embryological ages, the structural form, and the temporal appearance and juxtaposition of embryonic organs, such as the primitive streak, somites, and the notochord. He insisted that varying organs, not whole embryos, held the key to an understanding of the link between ontogeny and phylogeny. Their temporal advances or regressions were independent of one another. In toto they provided heterochronic and heterotopic mechanisms for phylogenetic change. His research still lay within the spirit of Haeckel's phylogenetic program, but in reviewing his own work, Oppel could conclude that "Ontogeny is not a recapitulation of phylogeny."[52] Keibel elaborated the approach into an ambitious series of *Normentafeln*, which enlisted many authors and portrayed in table form and magnificent lithographs the temporal stages of a developing vertebrate species.[53] These became standard for the profession and reinforced Oppel's conclusions without adopting certain of their Haeckelian features, such as a belief in the inheritance of acquired characters and a stress on adult forms. Mehnert and Fischel provided detailed comparative examinations of variations in embryonic organs of the same species and their temporal dislocations. Mehnert concluded that the multitude of recorded variations found cannot be considered cenogenesis "in the sense of the older authors or as falsifications of ontogenies."[54] One feature that emerges from Keibel's review is the extent to which these detailed morphological studies appear to be a reaction to Haeckel's negative understanding of cenogenesis as *Fälschungsgeschichte*.

This line of research entailed an examination of what "cenogenesis" really implied. As indicated above, Haeckel had been very clear about its antithesis to palingenesis when he first coined the terms: "die *Palingenesis* oder Auszugsgeschichte und *Cenogenesis* oder Fälschungsgeschichte."[55] His words were immediately translated into English as "Palingenesis or extended development and Cenogenesis or false development." His definition with its emphasis on "false" development set the tone for a constricted version of the biogenetic law.[56] At the end of the century, in a lengthy monograph on cenogenesis, Mehnert insisted in bold font that from his "point of view the acceleration and retardation of the unfolding are nothing other than expression of embryonic cenogenesis." He immediately added that "Insofar as the embryonic cenogenesis is only a direct regular result of phylogenetic developmental energy, it may also be wholly unwarranted, when one—as has happened up to now—has designated the same as falsifications." As suggested by this highly speculative conclusion, Mehnert continued to support the biogenetic law, but Keibel's retort elucidates my point: "Mehnert once again stands up decisively for the biogenetic law. Yet it is perfectly obvious that Mehnert's biogenetic law is something quite different from Haeckel's."[57]

Further Studies

Neither space nor time allows me to present other focused studies of the internal critique. For example, the debate over the origin of the vertebrate limbs, so nicely presented by Nyhart, reminds us that embryologists and anatomists began to disagree as to the comparative value of embryonic and adult forms for adjudicating what was palingenetic and what was cenogenetic.[58] Keibel, Russell, and Paul Weindling span a century of extolling Oscar Hertwig's role in providing a cellular perspective, which rejected the biogenetic law and promoted a polygenesis throughout the phylogenetic tree.[59] Bowler discusses at length how by the end of the century, paleontologists became the preferred judges of the phylogenetic past and indirectly, often unwittingly, substituted a functional mechanism for Haeckel's causal arrow leading from phylogeny to ontogeny.[60] My central message remains the same: between 1870 and circa 1900, internal critiques within morphology brought into stark focus the inadequacies of a strict rendering of Haeckel's biogenetic law.

STRONGER AND WEAKER VERSIONS OF THE BIOGENETIC LAW

When examining the ultimate fate of Haeckel's program, there is some merit in distinguishing between a stronger and a weaker version of the biogenetic law.[61] The differences may be associated with the following indicators. A belief in any single indicator or combination of indicators, however, should not be construed as necessary or sufficient for membership in either club.

The stronger version of the biogenetic law is by design reflective of Haeckel's 1872–1874 analytic formulation.

1. Haeckel's "palingenesis" and "cenogenesis," coined in his *Anthropogenie* of 1874, emphasized that evolution consisted of an unequal dialogue between two fundamental processes: "extended development" and "falsification" of the phylogenetic record. Paligenesis was primary because it produced the true phylogeny; cenogenesis was secondary because it created a "corrupted" text.[62]

2. Haeckel's biology conjured up a physical, mechanistically derived pattern in phylogeny.[63] It repeatedly insisted that phylogeny caused ontogeny. With the exception of the Protista and Protozoa, the animal kingdom must be understood as branching and overall monophyletic. Unlike later zoologists, however, Haeckel did not dwell on the process of terminal additions.[64]

3. Haeckel considered the ancestral segmentation pattern to be holoblastic, and the gastrula stage with its archenteron to be the recapitulation of the invaginated ancestor of all Metazoa. The blastopore corresponded to the adult *Urmund*.[65]

4. Haeckel specified the fate of the primitive germ layers through phylogeny.[66]

5. Haeckel considered larval forms, such as nauplius, zoea, and appendicularia, as fair representations of ancestral forms.[67]

6. Haeckel appears to have envisioned development as sequentially specified and hierarchical.

7. Variations in embryos and adults of the same species were not considered significant.

8. Throughout his career Haeckel defended the neo-Lamarckian belief in the inheritance of acquired characters.

The weaker version reflects a composite of positions maintained by comparative embryologists as they worked within the framework of the biogenetic law and the program to establish phylogenetic lineages. Many such versions may have descended from views of recapitulation that predate Haeckel's coining of his *Grundgesetz*. Collectively, they seem to emphasize the complexity of life and offer more flexibility for the study of phylogenies to move in new research directions.

1. There was a clear reluctance on the part of adherents to use Haeckel's vocabulary. "Biogenetic law," "palingenesis," and "cenogenesis," for whatever reason, were deemed not to be useful terms.

2. Morphologists increasingly entertained the possibility of the polymorphic origins of major complex groups, such as the Arthropoda.

3. No single segmentation or primitive germ layer formation pattern seemed to be primary. The uncertainty encouraged thinking about functional rather than hereditary formative factors. A variety of gastrula-producing patterns appeared to be equally primitive. The adult mouth and anus rarely corresponded to a blastopore.

4. The primitive germ layers did not necessarily correspond to organ-producing tissues. Major changes in form were often explained in functional scenarios.

5. Larval stages were considered in adaptive rather than hereditary terms. Selection processes became a common way of considering these adaptations.

6. Sequentially specific and hierarchical development was not a significant concern.

7. Morphologists recognized that embryos and adults of the same species differed widely in size and shape, and thus began to question the underlying type concept of homologous embryonic stages.

8. Proponents did not consider the inheritance of acquired characters to be a foregone conclusion, and they may even have rejected this Lamarckian form of inheritance.

When we consider the rise and fall of the biogenetic law as historical events and when we measure the impact of the internal critique, it is important to ask whether we are dealing with a strong or a weak version of the biogenetic law. My instincts tell me that the strong version was repeatedly found wanting within morphology itself, and was easily ignored by most zoologists except Haeckel and some of his close followers. The strong version may, in fact, have prevailed in popular literature, of which Haeckel was a master craftsman, but it was less characteristic of scholarly research. The weak version may have become even milder as the century drew to a close, and focused on patterns of functional adaptation throughout ontogeny. The phenomenon of recapitulation became recognized as simply a backdrop of conserved development that needed to be understood against the foreground of somatic variations, adaptations to living conditions, and a testable theory of heredity.

TEXTBOOK VERSIONS

When intellectual historians confront the issue of the reception or rejection of a theory, they necessarily confront quantitative questions about representation. Who counts? To what extent? For how long? I find a practical way of gauging the shift in perceptions of the biogenetic law to be to look at certain basic textbooks in comparative embryology. Depending on their level of sophistication, they reflect the status of shared knowledge of the day.[68] If one compares two classical embryology texts—Francis Maitland Balfour's *Treatise on Comparative Embryology* (1880–1881) and Eugen Korschelt and Karl Heider's *Entwicklungsgeschichte der wirbellosen Thiere* (1890–1893 and 1902)—I believe that one finds a progressive erosion of the formal details of Haeckel's program, yet, at the same time, a continued, resolute conviction that comparative embryology has as one of its missions the establishment of phylogenetic links. A glance at E. W. MacBride's *Text-book of Embryology*, volume 1, *Invertebrata*, of 1914, however, will readily persuade us that Haeckel's strong version was alive and well in the second decade of the twentieth century.[69]

Francis Maitland Balfour

There is no question that Balfour pursued his embryology in Haeckel's shadow, but never slavishly so. He had mastered the refined details of German morphology at the zoological station in Naples, where he interacted with Dohrn and his coworkers. He had also established a morphology department at Cambridge, which became one of the centers of comparative embryology prior to his untimely death in 1882. In his classic monograph on elasmobranchs, Balfour tangled with Haeckel over the meaning of partial segmentation in the meroblastic eggs of cartilaginous fish, osseous fish, and birds, and he argued that invagination was too variable a process to be accommodated easily by Haeckel's gastraea theory. Nevertheless, Balfour insisted that "I wish it to be clearly understood that my disagreement from his [Haeckel's] opinions concerns matters of detail only, and that I quite accept the Gastraea theory in its general bearings."[70]

Only a few years later this acquiescent tone had changed. In the introduction to his *Treatise on Comparative Embryology*, Balfour recognized two "antagonistic principles which have rendered possible the present order of the organic world," but he described them as Darwin's "laws of heredity and variation" rather than the processes of palingenesis and cenogenesis." In his opinion, there were two aims of comparative embryology:

"(1) to form a basis of Phylogeny and (2) to form a basis for Organogeny or the origin and evolution of organs."[71] The thrust of the second contravened Haeckel's stress on adult forms in the first. Occasionally in the body of the text Balfour referred to Haeckel—but almost exclusively to his specialty publications; when he declined to choose between invagination and delamination as the primary process of gastrula formation, he acknowledged, only indirectly, Haeckel's phylogenetic program. As far as I can ascertain, Balfour never referred explicitly to the biogenetic law or even to "recapitulation" in his influential text.

A more interesting aspect of Balfour's comparative embryology was the functionalism embedded in his interpretation of form. This meant emphasizing mechanical accommodations and adaptations at all stages of development. For example, Balfour repeatedly interpreted the variations in segmentation and gastrulation in terms of the quantity of yolk in an egg. "It may be laid down as a general law," he urged at one point, "which holds very accurately for the Vertebrata, that in eggs in which the distribution of food-yolk is not uniform, the size of the cells resulting from segmentation is proportional to the quantity of food-material they contain."[72] At another point, after recounting the great diversity in the location and fate of the blastopore, he insisted that many of these differences "can only be accounted for as secondary adaptations for the convenience of development."[73] Balfour clearly had a larger agenda than simply establishing phylogenetic ancestors. He was a sensitive, judicious commentator on the descriptive embryology of others; he was cautious in his own conclusions, and besides organizing the descriptive details of development across the animal spectrum, he had a real sense of the adaptive effects of the conditions of life and of the body in directing animal form. Finally, he was open to the possibility of polyphyly rather than monophyly of the Metazoa. Balfour's comparative embryology represents a move from an attachment to the strong version to a weaker version of the biogenetic law.

Eugen Korschelt and Karl Heider

Barely a decade later, Eugen Korschelt, professor of zoology and comparative anatomy at Marburg, and Karl Heider, professor of zoology at Berlin, assembled another comprehensive survey of comparative embryology. So rapidly had the basic information of the discipline grown, that they restricted their examination to invertebrates, tunicates, and *Amphioxus*—and even then it took the two of them and three volumes to cover the field.[74]

Again there is no entry on the "biogenetic law," "recapitulation," "cenogenesis," or "palingenesis" in the subject indices. Having read a number of key chapters carefully, I feel certain that these subjects do not come up in an easily recognizable manner. There is no focused discussion about the opposing forces of evolution (i.e., heredity versus adaptation)—though both are assumed to operate. The strong Haeckelian program with its emphasis on a recapitulation of evolution and its falsification is completely absent. This does not mean the authors ignored Haeckel's monographic work, particularly that with Porifera, Cnidaria and Ctenophora. Only at the end of the last volume, when discussing the development of Tunicata and *Amphioxus*, do Korschelt and Heider refer to Haeckel's *Anthropogenie* and criticize his and Gegenbaur's advocacy of the tunicates as the direct phylogenetic link between the invertebrates and Chordata.

There is no question, however, that Korschelt and Heider were interested in how comparative embryology elucidated phylogeny. They constantly reflected on how individual embryonic and adult structures differed from class or order within the phylum under discussion. In a section titled "General Considerations," which appeared at the end of nearly every chapter, they hypothesized about how the accumulated information reflected on the phylogenetic relationship between the phylum and neighboring phyla. They were keenly aware of the assumptions that went into the task of lineage construction. Decisions had to be made as to what was primitive and what was advanced, and to what extent these could be made by assuming a recapitulation of individual structures during development and to what extent by other assumptions, such as the simplicity and complexity of structures. Furthermore, decisions had to be made between progressive and degenerative evolution, between intercalations and heterochronies, and between examples of monophyly, polyphyly, and evolution through convergence.

Above all, Korschelt and Heider went far beyond Balfour in being attentive to functional changes at all stages of development: segmentation, germ layer formation, gastrulation, organ formation, and adult requirements. Following in Balfour's footsteps, they made much of the presence and location of yolk and patterns of segmentation; they saw in larval forms—for example, in the nauplius and zoea—the dynamic interaction of postembryonic development and the conditions of life. They concluded with respect to all arthropods that "Zoologists were for a long time inclined to ascribe to the larvae of the Arthropoda an important phylogenetic significance. But when it was recognized that these larvae often represented secondarily modified (adapted) forms. . . . , the comparison of

TABLE TO FACILITATE A COMPARISON OF THE EXTREMITIES FOUND IN THE PRINCIPAL GROUPS OF THE ARTHROPODA.

| Crustacea. | Xiphosura. | Arachnida. | Onychophora. | Myriopoda. | | Hexapoda. |
				Chilopoda.	Diplopoda.	
First antennae	—	—	Antennae*	Antennae	Antennae	Antennae
Second antennae	Chelicerae	Chelicerae	Jaws	—	—	—
Mandibles	First pair of legs	Pedipalps	Oral papillae	Mandibles	Mandibles	Mandibles
First maxillae	Second pair of legs	First pair of legs	First pair of legs	First pair of maxillae	Maxillae	First maxillae
Second maxillae	Third pair of legs	Second pair of legs	Second pair of legs	Second pair of maxillae	First pair of legs	Second maxillae (lower lip)
First pair of thoracic limbs	Fourth pair of legs	Third pair of legs	Third pair of legs	Maxillipedes	Second pair of legs	First pair of legs
Second pair of thoracic limbs	Fifth pair of legs	Fourth pair of legs	Fourth pair of legs	First pair of legs	Third pair of legs	Second pair of legs
Third pair of thoracic limbs	First pair of† abdominal limbs	First pair of‡ abdominal limbs	Fifth pair of legs	Second pair of legs	Fourth pair of legs	Third pair of legs

* [Goodrich (No. II.) regards the antennae of *Peripatus* as belonging to the peristomial segment, while those of the Crustacea, Myriopoda, and Insecta he traces to the second post-oral segment.—Ed.]
† [According to Kishinouye and Packard (No. VII.), this would be the chilaria. See Vol. ii., p. 345, footnote.—Ed.]
‡ [This would be the transitory pre-genital segment of the Scorpion, according to Brauer, and a pre-genital segment which forms the waist in the Araneae (p. 57).—Ed.]

Figure 3.6
From E. Korschelt and K. Heider, *Text-book of the Embryology of Invertebrates*, vol. 3, pp. 428–430. At the end of their chapter on development in the arthropods, the authors felt confident that the varied embryological evidence from many principal groups indicated a common origin from the annelids. Given the widespread appearance of larval forms, they nevertheless considered many of the accepted homologies to be questionable. Of this table of the appendages of arthropods they concluded, "that the different regions of the body (the head, the thorax, and the abdomen) are not, in the various divisions of the Arthropoda, precisely homologous, for they are not formed of the same number of segments in all cases, nor are the same segments in all cases included in the similarly named body-region. . . . Although we thus see that . . . the consecutive segments develop heteromorphously, we shall still be inclined to explain this fact by the requirements of the different functions, and shall not homologise the regions bearing similarly formed appendages."

the adult forms received more attention, a far higher value being set upon this branch of inquiry."[75] This, too, had its dangers, and they discouraged homologizing appendages on the basis of the sequence in the body segments of six subgroups of arthropods. Presenting a table showing the diverse origins of appendages (Figure 3.6), Korschelt and Heider concluded that "we shall still be inclined to explain this fact by the requirements of the different functions, and shall not homologize the regions bearing similarly formed appendages."[76] They represent adherents of the weak version of the biogenetic law.

The greatest difference between Balfour, on the one hand, and Korschelt and Heider, on the other, can be seen in the latter's final, general volume, the *Allgemeiner Teil*.[77] Here we find extensive accounts of the results of two subdisciplines that emerged out of descriptive embryology in the 1880s. Half of the 750-page volume treats advances in experimental embryology and the physical dynamics of cell growth and differentiation; the other half deals with microscopical studies of cellular, nuclear, and chromosomal movements as they informed the understanding of the origins, maturation, and fertilization of gametes. These were subdisciplines that had come into their own after Balfour had written, and were exerting a powerful influence on the nature of biology as it entered the twentieth century.

Since experimental embryology has been seen by twentieth-century historians as breaking radically from morphology and contributing to the overthrow of the biogenetic law, it is worth noting that the discipline was subsumed into a textbook on comparative embryology. The authors saw no opposition between descriptive and experimental embryology. According to them, each, in its own fashion, provided causal explanations that supplemented the other.[78] As for the second half of the *Allgemeiner Theil*, that is the kinematics of cells, nuclei, and chromosomes, it represents a subdiscipline that does not even appear on the radar screen of the historical discussions of the rise and fall of the biogenetic law.[79] Korschelt and Heider at the very beginning of the twentieth century, however, had a different perspective as they provided lengthy discussions of many aspects of cytology.

Finally, the linchpin that appears to justify the inclusion of the *Allgemeiner Teil* as part of their *Text-book of the Embryology of Invertebrates* is a single paragraph that the authors use as a transition from experimental embryology to cytology. It is so revealing that I present it here in full (Figure 3.7). It is the one place I have found where Korschelt and Heider explicitly mention the biogenetic law and its opposing components of heredity and variation (note: not palingenesis and cenogenesis):

> Even if it were possible to survey causally the entire ontogeny of any given form, so as to render understandable the relationship of cause and effect in every single developmental process, we still would not have gained an understanding of why that form is produced with all the traits that possess so completely the character of purposefulness in relation to its environment. The historical approach offers us a certain causal explanation in a general sense, in that we view the entire mechanism of the ontogeny as the result of a prior developmental process of the sequence of ancestors of the relevant form, and we try to reconstruct by means of the so-called *biogenetic law*, in light of the facts of comparative

III. Capitel. Ermittlungen der im Innern wirkenden Entwicklungsfactoren. 247

L i t t e r a t u r.

Driesch, H. *Entwicklungsmech. Studien. No. X. Ueber einige allgemeinere ent-wicklungsmechanische Ergebnisse. Neapl. Mittheilungen. 11. Bd. 1895. Auf-stellung des Begriffes der Position.*

Driesch, H. *Analytische Theorie der organischen Entwicklung. Leipzig. 1894.*

Driesch, H. *Die Localisation morphogenetischer Vorgänge. Ein Beweis vitalistischen Geschehens. Arch. f. Entw.-Mech. 8. Bd. 1899.*

Driesch, H. *Resultate und Probleme der Entwicklungsphysiologie der Thiere, in: Merkel-Bonnet, Ergebnisse. 8. Bd. 1898.*

v. Hanstein. *Referat über Driesch: Die Localisation etc. Naturw. Rundschau. 15. Jahrg. 1900.*

Morgan, T. H. *Some Problems of Regeneration. Biol. Lect. Woods Holl. (1897 bis 1898.) Boston. 1899.*

H. Die Phylogenie als erklärender Factor für das Specifische der Gestaltungen.

Wenn es möglich wäre, die gesammte Ontogenese irgend einer Form derart causal zu überblicken, dass uns jeder einzelne Ent-wicklungsprocess in seinem Verhältnisse von Ursache und Wirkung verständlich wäre, so hätten wir doch noch kein Verständniss dafür gewonnen, warum gerade die vorliegende Form mit allen Merkmalen, die mit Rücksicht auf die Lebensbedingungen derselben so sehr den Character der Zweckmässigkeit an sich tragen, erzeugt wird. Eine gewisse causale Erklärung im weiteren Sinne eröffnet sich uns nach dieser Richtung durch die historische Betrachtungsweise, indem wir den ganzen Mechanismus der Ontogenie als das Resultat eines in der Reihe der Vorfahren der betreffenden Form vor sich gegangenen Entwicklungsprocesses betrachten und an der Hand des sog. bio-genetischen Grundgesetzes unter Berücksichtigung der That-sachen der vergleichenden Anatomie und Entwicklungsgeschichte diesen in der Stammesgeschichte abgelaufenen Entwicklungsprocess zu reconstruiren versuchen. Eine causale Erklärung des letzteren im strengen Sinne würde sich nur ermöglichen lassen, wenn wir den beiden uns hier entgegentretenden complexen Componenten der V e r-e r b u n g und V a r i a t i o n in exacterer Weise näher zu treten in der Lage wären. Es eröffnet sich hier ein neues und weiteres Gebiet der Entwicklungsphysiologie, als dessen Zukunftsprogramm eine auf exacter Grundlage beruhende T h e o r i e d e r A r t u m w a n d l u n g zu betrachten wäre. Die hier in den Vordergrund tretende Frage, auf welchen Bedingungen die Entstehung vererbungsfähiger Abänderungen und ihre weitere Erhaltung beruht, ist bereits vielfach und nach verschiedenen Richtungen in Angriff genommen worden. Die Einzel-ergebnisse der vergleichenden Morphologie hinsichtlich der Stammes-geschichte der Organismen würden dann nur als speciellere An-wendung dieser allgemeinen Theorie zu betrachten sein. Naturgemäss wird diesen speci[e]leren stammesgeschichtlichen Forschungen mit Rück-sicht auf die geringe Kenntniss, die wir von den Lebensbedingungen der Vorzeit besitzen, stets in hohem Grade der Character des Hypo-thetischen anhaften. Nichtsdestoweniger wird diesen Versuchen mit Rücksicht auf die Nöthigung, den Zusammenhang der Formenreihen systematisch zu überblicken, immer eine gewisse Bedeutung zu-erkannt werden müssen.

Figure 3.7

The text of the original German in E. Korschelt and K. Heider, *Lehrbuch der vergle-ichenden Entwicklungsgeschichte der wirbellosen Thiere. Allgemeiner Theil* (1902). An English translation and a discussion of the context of the paragraph appear in the text.

anatomy and embryology, the process by which the family tree developed. Strictly speaking, a causal explanation of this process would only be possible if we were in a position to approach in a more exact way both of the complex components of *heredity* and *variation* that are involved. A new and broader domain of developmental physiology presents itself here, in which one hopes future research will establish a rigorous foundation for a *theory of the transformation of species*. The main question here, as to what kinds of conditions occasion the rise of hereditary changes and their further preservation, has already been confronted often and from different angles. Individual results of comparative morphology with respect to the lineage of organisms could be seen only as a special application of this general theory. Naturally these specialized genealogical researches will inherently possess in high degree the character of the hypothetical, as a consequence of the slight knowledge that we have about conditions of life in pre-history. Nevertheless a certain significance must be attributed to these efforts because of the need to survey systematically the continuity in the sequence of forms.[80]

There are many messages wrapped into this transition paragraph, as ponderous as the text may seem to our modern ears. First, note the implicit recognition of the distinction between proximal and ultimate causation (i.e., the physiology of development and the causes of evolution). These were distinct domains of biology, and there was no strategic ground for considering phylogeny as the cause of ontogeny. This aspect of the above quotation reinforces Nyhart's point that many morphologists, particularly Hermann Braus, were at the turn of the century deeply engaged in debating the meaning of causation.[81] Second, the reference to the "biogenetic law" may seem to be straight out of Haeckel, but I see it differently. It is now the "so-called" law. It is definitely not a *Grundgesetz*, for the modifier denotes an element of skepticism. Third, the law's two components, "heredity" and "variation," have a different connotation in 1902 than in 1872. Whereas Haeckel's formulation of palingenesis and cenogenesis literally meant "recapitulated development" and "false development," Korschelt and Heider wrote "heredity" and "variation" within the context of the twenty years of experimental and cytological research they were reviewing: that is, twenty years of demonstrating the flexibility of the embryo through experimentation and microscopy and twenty years of nuclear theories of fertilization and gamete maturation division; of August Weismann's germ plasm theory, which they discussed at length at the end of the volume; and of Theodor Boveri's recent demonstration of the individuality of the chromosomes.

Fourth, Korschelt and Heider were nevertheless prepared to consider their descriptive embryology as only a special application of a more general program. It would have been surprising if they had mentioned Mendel and his laws, but classical genetics of course had its cytological as well as its hybridization dimension. It is the cytology of gamete maturation and fertilization that made it clear how and where evolutionary changes might take place. The discoveries to come in classical genetics of chromosomal and genic mutations would show that the biogenetic law in the strong sense would be false.

In the meantime, if comparative embryologists who read the *Allgemeiner Theil* had gotten up to speed on both the experimental and the cytological sides of their discipline, the internal critique would have done its job of substantially eviscerating the biogenetic law—but not for the believers in the strong version.

E. W. MacBride

In 1914, E. W. MacBride published the first volume of a new survey of comparative embryology.[82] His assignment was to deal with the development of the Invertebrata and Protochordata, but he sandwiched his 600-page filling of technical descriptions between an introduction and a summary that were programmatic justifications of the "so-called fundamental law of biogenetics." This time there was little skepticism shading the modifier. "If this law can be substantiated," MacBride confidently opined, "the interest in embryology becomes immense, it binds all the innumerable phenomena of development into one coherent scheme, and opens the door to the hope that we may yet be able to sketch the main history of life on the earth."[83]

Substantiation of the biogenetic law served as the goal of this volume, and was to be found in the details of larval development and parasitic life. Where other embryologists—he referred specifically to Balfour's successor and his former teacher, Adam Sedgwick—had stressed that larval forms were indicators of secondary adaptations, MacBride emphasized that they were ancestors adapted to ancestral conditions. In addition, he stated that the "embryonic phase is the remnants of a former larval phase."[84] The pattern of terminal addition, condensation, and recapitulation was emphasized by MacBride in italicis:

> *If, then, the last stages in developmental history are, so to speak, the record of the last habits assumed by the species, the main framework of all developmental history must be the condensed record of ancestral experience; for each stage*

in the development of an animal bears the same relationship to the one which immediately precedes it as the adult stage does to the last larval stage.[85]

MacBride of course allowed room for secondary and "superficial" adaptations within the chain of embryonic and larval forms. He rationalized that nutrition of larvae and adults became the physical link between changing habits and adaptations. The chemicals of nutrition might over the course of many generations effect parallel changes in the germinal cells; thus adaptations could become imprinted on the phylogenetic lineage.

The long summary at the end of the volume reasserted the claims and mechanisms MacBride had outlined in his introduction, and it added one further suggestion. The recent discovery and isolation of hormones, particularly by Ernest Starling and his associates, offered a theoretical mechanism for chemically affecting the shapes of organs and possibly for explaining how acquired changes could be inherited.[86] MacBride's examinations of the marvelous and even fantastic larval and parasitic forms within the main body of the text reinforced his stated objective—to demonstrate the legitimacy of the biogenetic law—and gave his text a veneer of modernity.[87]

The challenge for comparative embryologists, as MacBride saw it, was in distinguishing, in the fine details of complex life cycles, the intercalated secondary changes from the original evolutionary progression of terminal additions. This, however, took patience and time—which, he complained, modern experimental embryologists did not possess.[88] MacBride paid token recognition to the advances being made in experimental embryology, cytology, and Mendelian genetics—disciplines that, to his dismay, were seducing younger investigators away from laboring in the vineyards of comparative embryology. His conclusions appear uncompromising and single-minded when compared with Balfour's and Korschelt and Heider's presentations.

One might wonder whether there are national differences between the German and English embryologists just reviewed. After all, they wrestled with the meaning and importance of the biogenetic law when they looked at the same biological material—often in the same great marine laboratories; they contributed to the same journals, read and translated one another's books, and were inspired by the same leaders of the field. Leaving the contrast in university traditions aside, the biggest difference between the two nations with regard to accepting or softening the biogenetic law may have been in the highly contingent, personal persuasion of MacBride himself. By 1914 MacBride appears to have been the major spokesman among comparative embryologists for the strong version of the biogenetic law—at a time when Haeckel, now in his eighties, was more involved in the propagation of his monistic philosophy than in concentrated embryological research. After World War I, MacBride's obsession with Haeckel's biogenetic law became more a defense of Lamarck's inheritance of acquired characters, but I suspect that his intransigence on both counts was central to arousing the well-known critiques of his fellow countrymen Walter Garstang and Gavin de Beer.[89] From their hostile comments, however, it might appear that Haeckel's biogenetic law continued to thrive in comparative embryology into the early 1930s.

At the same time, in Germany, two outstanding general textbooks, Richard Hertwig's *Lehrbuch der Zoologie* (1900, 1922) and Claus and Grobben's *Lehrbuch der Zoologie* (originally *Grundzüge der Zoologie* [1876]), briefly instructed students about the biogenetic law, but their commentary hardly promoted the law as the keystone to the understanding of phylogeny. Over the years, both these works became increasingly Darwinian and functional in outlook.[90]

SUMMARY AND REFLECTIONS

The biogenetic law in its strong version had an appeal in the 1870s, for it was completely understandable in the simple mechanical terms of the reigning positivistic and deterministic science. From Haeckel's arguments of a genetic continuity through generations of organic individuals at each level, and from his belief in the specificity of the germ layers and gastraea theory to MacBride's discussions of nutrition, habits, and hormones, there appears to be a primacy placed on organic or chemical mechanisms that might lock ontogeny and phylogeny into a unified and transparent process. The motivation for Haeckel seems to have been the need to demonstrate the reality of organic evolution and the ascent of mankind in a monistic universe. For MacBride the bottom line is less clear, but he certainly felt a need to validate his life's research in comparative embryology and to defend his work against emerging fields of biology. Unfortunately, he appears in retrospect more like King Canute "the great" standing at Westminster against the incoming tide.

I interpret the texts of Balfour and of Korschelt and Heider as providing weakening versions of the biogenetic law. The appeal of a mechanical reading of development was very real for these embryologists, too, but more often this was applied simply to ontogeny rather than to the node articulating ontogeny with phylogeny. Reference to the biogenetic law, if made at all, was slight and cautious, and Korschelt and Heider, in particular, seemed ready to embrace whatever new biology appeared for the better understanding of variations and heredity. Given the subjects covered and the structure of their *Allgemeiner Theil*, they appeared, above all, to be sympathetic to the contrast between the germinal constitution and somatic development. This distinction made it easier for them to disengage from an immediate causal connection between ontogeny and phylogeny, and to focus on function, chromosomes, and the dynamic between recapitulation and innovation.

A final metahistorical word on the biogenetic law is in order. I continue to find value in each of the secondary sources I reviewed at the outset, but I find their take often too coarse to depict the great differences within the community of comparative embryologists as they faced the claims and promises of the biogenetic law. Both the strong and the weak versions of the law—as well as intermediate versions—might have been swept away in time by some combination of experimental embryology, classical genetics, and a "new world order." Nonetheless, I remain persuaded by Bowler's demonstration that the internal critique supplied ample incentive and opportunity for comparative morphologists to move beyond Haeckel. To my mind, this was the principal lesson of the lengthy passage from Korschelt and Heider discussed above.

This leads me to a challenge laid down by Everett Mendelsohn to consider continuities and discontinuities in the history of Evo-Devo. I find an irony in reviewing the rise and fall of the biogenetic law. As general historians, we should be used to the daunting task of integrating the continuous and the unique in history's flow. As historians of embryology and evolution, we should recognize the same problem within the scientific study of development and evolution. We should also recognize certain biological models that promote the presentation of both together, such as the continuous and discontinuous changes that occur separately on the somatic and germinal levels of the same individual, or the continuity and discontinuity in isolated populations mirrored in Niles Eldredge and Stephen Jay Gould's theory of punctuated equilibrium. Until now, however, when we have presented our histories of the same biology, we have tended to frame our narratives in terms of either continuities or discontinuities or one succeeding the other. Our instincts should tell us that both are necessarily involved and intertwined. So the challenge for historians of biology might be much the same as for the evolutionary and developmental biologists we study. How are we to develop a biological model for capturing and clarifying both the continuous and the discontinuous in the history of biology? Nyhart's discussion of stable academic cohorts at evolving German university programs provides us with an important example of how to embrace both as part of the historical reality.

Now, did the biogenetic law meet a natural death—that is, did it suffer the ultimate discontinuity, as all the histories mentioned imply? The strong version certainly did. As for a weaker version, which balanced recapitulations and adaptations, found value in the Gaussian spread of individual variations, and turned from formal to functional explanations, I am not so sure. So here is my utilitarian message: When I hear developmental biologists today talking about molecular homologies, conservative phylotypes, genetic constraints, and Hox genes, and find therewith ontogenetic generalities that help explain phylogeny, I feel confident that Balfour, Korschelt, Heider, and many other comparative embryologists mentioned in this chapter would feel very much at home with the new Evo-Devo.

ACKNOWLEDGMENTS

I wish to thank Rasmus Winther, Alan Love, and Sandy Gliboff for helpful comments during the process of revision. The challenging work by Ron Amundson (2005) came to my attention too late to be integrated into this chapter.

NOTES

1. For discussions of the pre-Darwinian law of parallelism and evolution, see Kohlbrugge, 1911; Russell, 1916; Lenoir, 1982; and Müller, 1998.

2. Russell, 1916, pp. v–vi, is clear about this: "In the course of this book I have not hidden my own sympathy with the functional attitude."

3. Gould, 1977, p. 2.

4. Ibid., p. 74.

5. Ibid., pp. 177–184.

6. Rasmussen, 1991.

7. Bowler, 1996.

8. Ibid., pp. 442–443.

9. Nyhart, 1995.

10. This expression is a purposeful echo of Gould's well-known paper "The Hardening of the Modern Synthesis" (1983).

11. Uschmann, 1953. This is one of the more sophisticated accounts of Haeckel's biogenetic law. It accords in many ways with my analysis, which I made prior to being aware of Uschmann's paper.

12. Haeckel, 1866, vol. 2, p. 300. Note that Haeckel uses *Wiederholung* more often than *Rekapitulation*.

13. Keibel, 1897. See also Peters, 1980.

14. Haeckel expressed his gratitude in *Generelle Morphologie*, vol. 1, "Vorwort," to Virchow, under whom he worked as an assistant. The third *Buch* of the *Generelle Morphologie*, in which he deals with the notion of individuality in organisms, indicates how dependent he had become on the Virchowian theory of direct cell division.

15. Rinard, 1981, p. 275, claims that "the aggressively mechanistic self-image he [Haeckel] presented when talking about ontogeny and phylogeny" has been overstressed.

16. Haeckel, 1866, vol. 2, pp. 295–300, 418–422.

17. Haeckel, 2nd ed., 1870, pp. 361, 482. In the first edition (1868), Haeckel described ontogeny as a short and quick *Wiederholung* (recapitulation) of phylogeny but made no mention of a *biogenetisches Grundgesetz*.

18. Haeckel, 1872, vol. 1, pp. 471–473. A translation of two chapters of this work appeared in English (Haeckel, 1873).

19. Haeckel, 1874b. In following the lead of Kowalevsky, Haeckel had briefly mentioned the existence of the gastraea ancestor in his monograph on calcareous sponges. See Haeckel (1872), vol. 1, p. 467.

20. Haeckel, 1874a, p. 10.

21. Haeckel, 1872, vol. 1, fn. 3, pp. 471–472. Other Continental zoologists besides His reacted negatively to Haeckel's conclusions; for Metchnikoff's reaction, see Tauber and Chernyak, 1991, chap. 2.

22. Quotation in Haeckel, 1872, vol. 1, pp. 483–484, and Haeckel, 1873, p. 430. Contrasting synthetic and analytic knowledge and drawing upon the empirical knowledge gained from calcareous sponges, Haeckel explicitly acknowledges J. S. Mill's *Inductive Logic* rather than Kant. See Haeckel, 1872, vol. 1, p. 66.

23. Haeckel, 1875, pp. 499–501; Schulze (1875).

24. Haeckel, 1875.

25. Ibid., p. 417.

26. Maienschein, 1978. Beginning with Whitman's study of the annelid *Clepsine* (1878, 1888), this tradition extended to Wilson's work on the annelids *Polygordius* and

Nereis (1890–1898), Mead's on gastropods and annelids (1897, 1898), Treadwell's on the annelid *Podarke* (1898, 1900), Conklin's on the gastropod *Crepidula* (1896, 1897), and Frank Rattray Lillie's on the freshwater bivalve *Unio* (1895). As Maienschein points out, this cluster of zoologists interacted with one another and often performed their work in close proximity at the Marine Biological Laboratory at Woods Hole.

27. Maienschein, 1978, p. 146.

28. Ibid., p. 157.

29. Guralnick, 2002.

30. Ibid., p. 564.

31. Exceptions are Russell, 1916, and Oppenheimer, 1940.

32. Oppenheimer, 1940. Oppenheimer points out that Kowalevsky and Kleinenberg also recognized the evolutionary significance of the germ layers, but not in the grand manner of Haeckel and Lankester.

33. Lankester, 1877.

34. The terminological refinement, in fact, reversed A. Thomson's original suggestion, which only added to the confusion. Braem, 1895, p. 432, fn.

35. Hertwig and Hertwig, vol. 4, 1881, p. 205. Their coelome theory, which flowed naturally from their conclusions about the connection between diploblastic and triploblastic organisms, nevertheless appeared to some to be a reinstatement of Haeckel's search for an ur-form, but at a higher level in phylogeny.

36. Braem, 1895, p. 440: "Wie sind diese Schichte vom morphologischen Standpunkte aus zu beurteilen? Alles was über die Entstehung der Keimblätter gelehrt wird, passt hier wie die Faust aufs Auge."

37. Winsor, 1976.

38. Haeckel studied with Müller for only a year, during which time Müller introduced him to marine life in Helgoland and the Mediterranean.

39. Winsor, 1976, pp. 106–127.

40. Haeckel, 1870, chaps. 18–20; quotation on p. 487.

41. Bowler, 1996, chaps. 3–4.

42. Raff, 1996, pp. 123ff.

43. See Uschmann (1953), pp. 132–134, for the important differences between Darwin and Müller, on the one hand, and Haeckel, on the other.

44. Bowler, 1996, pp. 103–123.

45. Bowler, 1996, p. 131.

46. Haeckel, 1866, vol. 2, pp. 191–223.

47. Haeckel, 1879, vol. 1, chap. 1.

48. I have benefited greatly from a lecture by and discussion with Nick Hopwood on this group of anatomists. As this paper went to press, I received Hopwood (2005), which discusses this material.

49. Russell, 1916, pp. 348–352.

50. Gould, 1977, pp. 174–175.

51. Russell, 1916, has translated many of these quotations at length.

52. Quoted in Keibel, 1897, p. 744.

53. Keibel wrote a number of articles describing his series before the first volume appeared as his *Normentafel zur Entwickelungsgeschichte des Schweines* (Jena: Gustar Fischer, 1897).

54. Quoted in Keibel, 1897, p. 776.

55. Haeckel, 1874a, vol. 1, p. 10.

56. It is hard to pin down an "official" translation of these terms, for both Haeckel and his translators varied their texts. Haeckel, 1879, vol. 2, p. 460, fns 8 and 9 read: "Palingenesis (Gk.) = original evolution, from palingenesia (Gk.) = new-birth, renewal of the former course of evolution. Therefore, Palingeny = inherited history (from Gk. = reproduced, and Gk. = history of evolution)." and "Kenogenesis (Gk.) = modified evolution, from kenos (Gk.) strange, meaningless; and genea (Gk.) = history of evolution. The modifications introduced into Palingenesis by Kenogenesis are vitiations, strange, meaningless additions to the original, true course of evolution. Kenogeny = vitiated history." Haeckel's greek orthography has been eliminated where indicated by "(Gk.)".

57. Mehnert, 1897, quotation appears on p. 106. Keibel, 1898, p. 790. All other quotations are in Gould, 1977, pp. 174–175. The combined quotations from Mehnert, p. 106, read as follows: "Von diesem Gesichtspunkte betrachtet, sind **die Acceleration und Retardation der Enfaltung auch nichts anderes als Ausdruck der embryonalen Kainogenese.**

"Insofern als die embryonale Kainogenese nur ein direkter gesetzmässiger Ausfluss phyletischer Entfaltungsenergien ist, dürfte es auch ganz ungerechtfertigt sein, wenn man—wie bisweilen geschehen ist—dieselbe als Fälschungen bezeichnet hat."

Note that Mehnert uses *Kainos* rather than *Kenos* as the root for kainogenesis. This implies "recent" or "new" rather than "strange," "vitiated," or *Fälschungen*.

58. Nyhart, 1995, chap. 8.

59. O. Hertwig, 1898, pp. 271–277; Keibel, 1898, pp. 791–792; Russell, 1916, pp. 354–357; Weindling, 1991, pp. 160–166.

60. Bowler, 1996, chaps. 5–7.

61. Guralnick, 2002, pp. 549, 554–555, has also written of "a weak form of the biogenetic law."

62. After introducing the two terms, Haeckel (1879, vol. 1, pp. 10–11), wrote: "This critical distinction between the primary palingenetic, and the secondary kenogenetic processes is of course of the greatest importance to scientific Phylogeny, which, from the available empiric material supplied by Ontogeny, by comparative Anatomy, and by Paleontology, seeks to infer the long extinct historical processes of tribal evolution." Further on in the same paragraph Haeckel draws the analogy between "corrupt and genuine passages in the text of an old writer."

63. There is an idealistic streak in Haeckel that can be seen particularly in his consideration of man. See Russell, 1916, pp. 256–258; Rinard, 1981; and Müller, 1998.

64. I thank Sanders Gliboff for pointing this out to me.

65. Haeckel's gastraea theory was first suggested in 1872 but expanded into a comprehensive theory in 1874.

66. Haeckel, 1872, vol. 1, pp. 464–467.

67. This seems clear in Haeckel, 1866, vol. 2, lxxxvi–lxxxvii; it is less clear in Haeckel, 1894–1895, vol. 2, pp. 656–657.

68. In the second and subsequent editions of his *Structures of Scientific Revolutions* (1970), Thomas Kuhn emphasized the importance of science textbooks in establishing and reinforcing paradigms. See postscript, pp. 174–210.

69. The first two of these textbooks, and perhaps the third, were not simply student primers as we think of textbooks today. They were intended and used as comprehensive reference sources, and were quickly translated into the opposite language (i.e., English into German and vice versa).

70. Balfour, 1878, pp. 64–70, quotation on p. 70. The last phrase strikes me as the most diplomatic hedge one could ask.

71. Balfour, 1880–1881, vol. 1, pp. 2–6. For a detailed discussion of Darwin's theory of variations, see Winther, 2000.

72. Balfour (1880–1881), vol. 2, pp. 231, 238. Balfour writes freely of "the mechanical effects of food-yolk"; see p. 298 and chap. 11.

73. Ibid., vol. 2, p. 282.

74. Korschelt and Heider, 1890–1892. In contrast to the German edition, the English translation is divided into four volumes. It is this version that I follow. Each chapter is attributed to a single author, but the chapters are uniform in structure, and the wording in the text implies that both authors assumed responsibility for what was written.

75. Ibid., vol. 3, p. 426.

76. Ibid., pp. 429–430.

77. After the turn of the century the same authors published the *Allgemeiner Theil* (1902). This volume has not been translated into English.

78. Ibid., pp. 5–6.

79. An exception to this is Oppenheimer, 1940.

80. Korschelt and Heider, *Allgemeiner Teil*, p. 247. I wish to thank Nancy Boerner for assisting me with this translation. Emphasis is in the original.

81. Nyhart, 2002, pp. 8–13.

82. MacBride, 1914, vol. 1.

83. Ibid., p. 20.

84. Ibid., p. 21.

85. Ibid., pp. 24–25.

86. Ibid., pp. 652–654. Besides coining the word "hormone," Ernest Henry Starling had suggested that an unidentified secretion of the thyroid influenced the growth and ultimate form of the body; this suggestion was soon picked up by Julian Huxley. Starling, 1905; Churchill, 1993.

87. For example, see his discussion of the parasitic crustacean *Portunion maenadis*—MacBride, 1914, pp. 219–220—or his discussion of larval stages of insects and arthropods, pp. 285–288.

88. Ibid., p. 29.

89. It is worth noting that MacBride opens his textbook with a discussion of contemporary work on the cytology of fertilization and reduction division, then discounts it as highly problematic. We need a detailed historical study of MacBride.

90. Compare the emphasis and expansions in R. Hertwig, 1900, pp. 20–46, with its counterpart of 1922, pp. 20–51.

References

Amundson, Ron. 2005. *The changing Role of the Embryo in Evolutionary Thought Roots of Evo-Devo.* Cambridge: Cambridge University Press.

Anonymous. 1840. "Prof. E. W. MacBride, F.R.S." *Nature*, 146: 831–832.

Balfour, Francis Maitland. 1878. *A Monograph on the Development of Elasmobranch Fishes.* London: Macmillan.

———. 1880–1881. *A Treatise on Comparative Embryology.* 2 vols. London: Macmillan.

Bowler, Peter J. 1996. *Life's Splendid Drama: Evolutionary Biology and the Reconstruction of Life's Ancestry, 1860–1940.* Chicago: University of Chicago Press.

Braem, F. 1895. "Was ist ein Keimblatt?" *Biologisches Centralblatt,* 15: 427–443, 466–476, 491–506.

Churchill, F. 1993. "On the Road to the *k* Constant: A Historical Introduction." In Julian S. Huxley, ed., *Problems of Relative Growth.* Facs. reprint. Baltimore: Johns Hopkins University Press.

Claus, Carl. 1905. *Lehrbuch der Zoologie,* begründet von C. Claus. neu bearbeitet von Dr. Karl Grobben. Marburg: N. G. Elwert.

Gould, Stephen Jay. 1977. *Ontogeny and Phylogeny.* Cambridge, Mass.: Belknap Press of Harvard University Press.

———. 1983. "The Hardening of the Modern Synthesis." In Marjorie Grene, ed., *Dimensions of Darwinism.* Cambridge: Cambridge University Press.

Guralnick, Robert. 2002. "A Recapitulation of the Rise and Fall of the Cell Lineage Research Program: The Evolutionary-Developmental Relationship of Cleavage to Homology, Body Plans and Life History." *Journal of the History of Biology,* 35: 537–567.

Haeckel, Ernst. 1866. *Generelle Morphologie der Organismen.* 2 vols. Berlin: Georg Reimer.

———. 2nd ed., 1870. *Natürliche Schöpfungsgeschichte: Gemeinverständliche wissenschaftliche Vorträge über die Entwickelungslehre im allgemeinen und diejenige von Darwin, Goethe und Larmarck im besonderen, über die Anwendung derselben auf den Ursprung des Menschen und andere damit zusammenhängende Grundfragen der Naturwissenschaft.* Berlin: Georg Reimer.

———. 1872. *Die Kalkschwämme. Eine Monographie.* 3 vols. Berlin: Georg Reimer.

———. 1873. "On the Calcispongiae, Their Position in the Animal Kingdom, and Their Relation to the Theory of Descendance." *Annals and Magazine of Naural History,* 4th ser., *11*: 241–262, 421–430. This consists of a translation by W. S. Dalles of the last two chapters of vol. 1 of Haeckel (1872).

———. 1874a. *Anthropogenie: Oder, Entwickelungsgeschichte des Menschen.* [*Keimes- und Stammesgeschichte.*] 2 vols. Leipzig: Wilhelm Engelmann.

———. 1874b. "Die Gastraea-Theorie, die phylogenetische Classification des Thierreichs und die Homologie der Keimblätter." *Jenaische Zeitschrift für Naturwissenschaft, herausgegeben von der Medicinisch-Naturwissenschaftlichen Gesellschaft zu Jena,* 8: 1–55.

———. 1875. ["Studien zur Gastraea-Theorie," part 2] "Die Gastrula und die Eifurchung der Thiere." *Jenaische Zeitschrift für Naturwissenschaft herausgegeben von der Medicinisch-Naturwissenschaftlichen Gesellschaft zu Jena,* 9: 402–508.

———. 1877. "Nachträge zur Gastraea-Theorie." *Jenaische Zeitschrift für Naturwissenschaft, herausgegeben von der Medicinisch-Naturwissenschaftlichen Gesellschaft zu Jena,* 11: 55–98.

————. 1879. *The Evolution of Man: A Popular Exposition of the Principal Points of Human Ontogeny and Phylogeny*, 2 vols. London: C. Kegan Paul. Translation of the German 3rd ed.

————. 1894–1895. *Systematische Phylogenie: Entwurf eines natürlichen Systems der Organismen auf Grund ihrer Stammesgeschichte*. 3 vols. Berlin: Georg Reimer.

————. 1900. *The Riddle of the Universe at the Close of the Nineteenth Century*, trans. Joseph McCabe. New York: Harper & Brothers.

Hertwig, Oscar. 1898. *Die Zelle und die Gewebe. Grundzüge der allgemeinen Anatomie und Physiologie*, 2 vols. Jena: Gustav Fischer.

Hertwig, Oscar, and Richard Hertwig. 1879–1881. *Studien zur Blättertheorie*. vol. 1, *Die Actinien, anatomisch und histologisch mit besonderer Berücksichtigung des Nervenmuskelsystems untersucht* (1879); vol. 2, *Die Chaetognathen: Ihre Anatomie, Systematik und Entwicklungsgeschichte* (1880); vol. 3, *Über den Bau der Ctenophoren* (1880); vol. 4, *Die Coelomtheorie: Versuch einer Erklärung des mittleren Keimblattes* (1881). Jena: Gustav Fischer.

Hertwig, Richard. 1900. *Lehrbuch der Zoologie*. Jena: Gustav Fischer. 5th edition.

————. 1922. *Lehrbuch der Zoologie*. Jena: Gustav Fischer. 13th edition.

Hopwood, Thick. 2005. "Visual Standards and Disciplinary Change: Normal Plates, Tables, and Stages in Embryology," *History of Science*, 63: 239–303.

Keibel, Franz. 1898. "Das biogenetische Grundgesetz und die Cenogenese." *Ergebnisse der Anatomie und Entwickelungsgeschichte*, 7: 726–792.

Kohlbrugge, J. H. F. 1911. "Das biogenetische Grundgesetz. Eine historische Studie." *Zoologischer Anzeiger*, 38: 447–453.

Korschelt, Eugen, and Karl Heider. 1890–1893. *Lehrbuch der vergleichenden Entwickelungsgeschichte der wirbellosen Thiere*. Jena: G. Fischer, 3 vols. I have referred to the English edition: 1895–1900. *Text-book of the Embryology of Invertebrates*, trans and ed, Edward L. Mark, W. McM. Woodworth, Matilda Bernard, and Martin F. Woodward. 4 vols. London: Swan Sonnenschein.

————. 1902. *Lehrbuch der vergleichenden Entwicklungsgeschichte der wirbellosen Thiere. Allgemeiner Theil*. Jena: Gustav Fischer.

Kuhn, Thomas. 1970. *The Structure of Scientific Revolutions*, 2nd ed. Chicago: University of Chicago Press.

Lankester, Earl Ray. 1877. "Notes on the Embryology and Classification of the Animal Kingdom: Comprising a Revision of Speculations relative to the Origin and Significance of the Germ-layers." *Quarterly Journal of Microscopical Sciences*, 17: 399–454.

Lenoir, Timothy. 1982. *The Strategy of Life: Teleology and Mechanics in Nineteenth Century German Biology*. Dordrecht: D. Reidel.

MacBride, E. W. 1914. *Text-book of Embryology*, vol. 1, *Invertebrata*, ed. Walter Heape. London: Macmillan.

Maienschein, Jane. 1978. "Cell Lineage, Ancestral Reminiscence, and the Biogenetic Law." *Journal of the History of Biology*, 11: 129–158.

Mehnert, Ernst. 1897. "Kainogenese eine gesetzmässige Abänderung der embryonalen Entfaltung in Folge von erblicher Uebertragung in der Phylogenese erworbener Eigenthümlichkeiten." *Morphologishe Arbeiten*, 7: 1–152.

Müller, Irmgard. 1998. "Historische Grundlagen des biogenetischen Grundgesetzes." In *Welträtsel und Lebenswunder: Ernst Haeckel—Werk, Wirkung und Folgen*. Staphia 56 [Kataloge des OÖ. Landesmuseums, Neu Folge 131], pp. 119–130.

Nyhart, Lynn K. 1995. *Biology Takes Form: Animal Morphology and the German Universities, 1800–1900*. Chicago: University of Chicago Press.

———. 2002. "Learning from History: Morphology's Challenges in Germany ca. 1900." *Journal of Morphology*, 252: 2–14.

Peters, D. Stephan. 1980. "Das biogenetische Grundgesetz—Vorgeschichte und Folgerungen." *Medizin historisches Journal*, 15: 57–69.

Oppenheimer, Jane M. 1940. "The Non-Specificity of the Germ-Layers." Reprinted in her *Essays in the History of Embryology and Biology*, pp. 256–294. Cambridge, Mass.: MIT Press, 1967.

Raff, Rudolf A. 1996. *The Shape of Life: Genes, Development, and the Evolution of Animal Form*. Chicago: University of Chicago Press.

Rasmussen, Nicolas. 1991. "The Decline of Recapitulationism in Early Twentieth-Century Biology: Disciplinary Conflict and Consensus on the Battleground of Theory." *Journal of the History of Biology*, 24: 51–89.

Rinard, Ruth G. 1981. "The Problem of the Organic Individual: Ernst Haeckel and the Development of the Biogenetic Law." *Journal of the History of Biology*, 14: 249–275.

Russell, E. S. 1916. *Form and Function: A Contribution to the History of Animal Morphology*. London: John Murray.

Salvini-Plawen, Luitfried. 1998. "Morphologie: Haeckels Gastraea-Theorie und ihre Folgen." In *Welträtsel und Lebenswunder: Ernst Haeckel—Werk, Wirkung und Folgen*. Staphia 56 [Kataloge des OÖ. Landesmuseums, Neu Folge 131] pp. 147–168.

Schulze, Franz Eilhard. 1875. "Über den Bau und die Entwicklung eines Kalkschwammes, *Sycandra raphanus* Haeckel." *Tageblatt der 48. Versammlung Deutscher Naturforscher und Ärzte in Graz* vom 18–24 September (graz: 1875), pp. 101–102.

Shumway, Waldo. 1932. "The Recapitulation Theory." *Quarterly Review of Biology*, 7: 93–99.

Starling, Ernest Henry. 1905. *Croonian Lectures on the Chemical Correlation of the Functions of the Body*. Delivered before the Royal College of Physicians of London on June 20, 22, 27, and 29, 1905.

Tauber, Alfred I, and Leon Chernyak. 1991. *Metchnikoff and the Origins of Immunology: From Metaphor to Theory.* New York: Oxford University Press.

Uschmann, Georg. 1953. "Einige Bemerkungen zu Haeckels biogenetischem Grundgesetz." *Urania,* 16: 131–158.

Weindling, Paul Julian. 1991. *Darwinism and Social Darwinism in Imperial Germany: The Contribution of the Cell Biologist Oscar Hertwig, 1849–1922.* Stuttgart and New York: Gustav Fischer.

Winsor, Mary P. 1976. *Starfish, Jellyfish, and the Order of Life: Issues in Nineteenth-Century Science.* New Haven, Conn.: Yale University Press.

Winther, Rasmus G. 2000. "Darwin on Variation and Heredity." *Journal of the History of Biology,* 33: 425–455.

4

WILLIAM BATESON'S PHYSICALIST IDEAS
Stuart A. Newman

William Bateson has occupied a peculiar place in the history of biology: lauded for his early recognition of Mendel's accomplishments and his role in introducing Mendel's ideas to the English-speaking world, but ridiculed for his reluctance to accept a notion of genes as spatially confined chromosomal particles, his claimed lack of appreciation of the implications of Mendelism for evolutionary theory, and his espousal of alternative notions of inheritance of form and of speciation, such as the "vibratory theory" and the "presence-and-absence" hypothesis. It is argued here that, contrary to the view which stamps him as a holdout against a revolution in biological understanding, Bateson was ahead of his time in advocating a dynamical systems approach to gene action, and a progenitor of strains of evolutionary developmental biology that began to emerge only a century after he began his work. The technological level of biological science during Bateson's lifetime was inadequate to support empirical implementation of his research program. Nonetheless, subsequent scientific work has shown his vibratory theory regarding segmentation and other repetitive pattern formation to have been essentially correct, and his presence-and-absence hypothesis and the related notion of evolution by loss of inhibitors, to have embodied a "systems" view of the relation of genotype to phenotype, with implied possibilities for abrupt morphological transitions in development and evolution—that is, increasingly replacing the "corpuscular," and gradualist, population biology-based views of his contemporary and later critics.

In September 1891, William Bateson wrote to his sister Anna about a new idea he had found to account for discontinuous variation, a problem that had been perplexing him since he was a student, and to which he was devoting a book in progress. Bateson wondered how novelties (e.g., eyes, limbs) arose with apparent abruptness in the course of evolution, a phenomenon he could not square with Darwin's theory, based as it was on the gradualist paradigm of animal domestication. He also wondered how series of discrete forms could arise during embryonic development—the segments of an earthworm, the somites that provide the primordia for the vertebrae, the digits of the hands and feet. By considering these two sets of phenomena as linked, he distinguished himself as a progenitor of the modern field of evolutionary developmental biology.

The new idea may have been suggested to Bateson by some passages in St. George Mivart's two-decade-old critique of Darwin's theory, *On the Genesis of Species* (Mivart, 1871), or by W. K. Brooks of Johns Hopkins University, who was also taken by the notion, which he credited to Mivart in a book he had just completed (Brooks, 1883) when Bateson was a postgraduate visitor in his laboratory during the summers of 1883 and 1884. Whether or not he actually thought of it himself, or was conscious of any prior influences, Bateson considered the idea his own. In his letter to Anna, he effused:

> Did I tell you anything about my new *VIBRATORY THEORY of REP-ETITION of PARTS* in animals and plants? I have been turning it over lately, and feel sure there is something in it. It is the best idea I have ever had or am likely to have—Do you see what I mean?—Divisions between segments, petals, etc. are *internodal lines* like those in sand figures made by sound, i.e. lines of maximum vibratory strain, while the mid-segmental lines and the petals, etc. are the *nodal* lines, or places of minimum movement. Hence all the *patterns* and *recurrence of patterns* in animals and plants—hence the perfection of symmetry—hence bilaterally symmetrical variation, and the *completeness* of repetition whether of a part repeated in a radial or linear series etc. etc. I am, as you see, in a great fluster. I have been talking to F. D. [Francis Darwin] about it—and he thinks it "Really is very neat, upon my word," also "Oh!" he says. (Bateson and Bateson, 1928, pp. 42–43)

And, in "a note scribbled on the envelope," according to his wife, Beatrice, in her published memoir, he pressed the point with Anna, herself a botanist: "You see, an eight-petalled form stands to a four-petalled form as a note does to the lower octave" (Bateson and Bateson, 1928, p. 43).

For Bateson, his vibratory theory was more than just a piece of the puzzle of the transmission of organic form that was increasingly occupying the efforts of "naturalists"[1] during this period between the emergence of Darwinism and the rise of Mendelism. In a postscript to the letter announcing the idea, he states, "Of course heredity becomes quite a simple phenomenon in light of this." And in another letter to Anna later that year, he states, "I am tremendously pleased with the *IDEA*. . . . You'll see—

it will be a common-place of education, like the multiplication table or Shakespeare, before long!" (Bateson and Bateson, 1928, p. 44).

Bateson, aged thirty at the time, had been searching for several years for a basis for transmission of organismal form and pattern, and changes thereof, in the dialectic between the inherent material properties of the organism and the external environment. Five years previously, he had written to his mother from Kazalinsk in western Central Asia, where he was on a research expedition: "By the way, whoever originated that ridiculous piece of bad logic about variations due to environmental change seeming not to be 'permanent'? . . . If iron in soil make[s] hydrangeas blue, why is this to be regarded as a false variation? Because the same hydrangea without iron is *not* blue?" (Bateson and Bateson, 1928, p. 20).

The objective of the Russian expedition was to test Darwin's notion that there should be consistent, stable phenotypic differences between similar organisms living in different environments. Examining the fauna in lakes of distinctly different salinity yielded inconsistent results and no general rules. But even the best example of covariation of phenotype with environment, that of the bivalve *Cardium edule*, left Bateson skeptical about whether natural selection had really been at work: "Upon this point I have no evidence; but that the animals would, if they lived and propagated, ultimately regain their former structure, appears probable; for since it can be shown that certain variations are constantly produced by water of certain constitution, it practically follows that maintenance of these variations depend [*sic*] on the same cause" (Bateson, 1889, p. 298).

This concern with what is now termed "phenotypic plasticity" (West-Eberhard, 2003) was to set Bateson against contemporary Darwinians, whose bane was the notion that evolution was guided by inherent propensities to change in certain directions—"orthogenesis." It also led him, once he became a convinced Mendelian a decade later, to marginalize himself further by rejecting static notions of the gene. Indeed, treating genetic determination on a par with environmental determination was to be a hallmark of Bateson's lifelong thought. In an 1888 letter to Anna (a dozen years before the "rediscovery" of Mendel's work) he laid out this framework, at the same time providing a neat (though apparently unwitting) synthesis of Cuvier's "correlation of parts" and Geoffroy's "laws of form" perspectives, the two contending worldviews of the previous generation of biologists:

> My brain boils with evolution. It is becoming a perfect nightmare to me. I believe now that it is an axiomatic truth that no variation, however

small, can occur in any part without other variation occurring in correlation to it in all other parts; or rather, that no system, in which a
variation of one part had occurred without such correlated variation in
all other parts, could continue to be a system. This follows from what
one knows of the nature of an "individual," whatever that may be. If
then, it is true that no variation could occur if it were not arranged
that other variations should occur in correlation with it, in all parts, all
these correlation variations are dictated by the initial variation acting as
an environmental change. Therefore the occurrences of any variation in
a system is a proof that all parts have the power of changing with environmental change and of necessity must do so. Further any variation
must always consist chiefly of the secondary correlated variations and to
an infinitely small degree of an original primary variation. You will
observe that if any variation occurring in one system is acting through
the mechanism of correlation as a cause of further variation, it would
then happen that on the occurrence of one variation, general variation
must be expected, for if all the parts are to work with the new variation, a long time must elapse before the organism is again a system.
. . . The accomodatory mechanism is the thing to go for. I don't believe
it is generally recognized as existing, though when stated, it seems
obvious. (Bateson and Bateson, 1928, pp. 38–39)

Standard narratives of Bateson's career have him hunting for a plausible mechanism for discontinuous inheritance until he was nearly forty,
and believing he had found it when he came upon Mendel's work (or a
reference to it in a paper by Hugo de Vries) in May 1900 (Darden, 1977).
Initially persuaded by Francis Galton's views, in which selection of ordinary biological variation alone was held to be incapable of generating
qualitatively new traits, Bateson threw in his lot with De Vries and the
"mutationists," who saw Mendelian heredity as a discontinuous complement to standard inheritance of quantitatively varying traits. Within a short
time, however, perhaps spurred by W. F. R. Weldon's critique of Mendel's
work (Weldon, 1902; discussed in Olby, 1987), Bateson recognized that
since all alternative alleles in the Mendelian framework were not constrained by the dominant-recessive relationship, blending inheritance could
occur without the factors losing their separate identities, and therefore
quantitative traits could be accommodated within Mendelism (Bateson and
Saunders, 1902; Olby, 1987). However, although he became a committed
Mendelian in all matters of standard inheritance, Bateson (like Galton)
continued to believe that *between*-species differences were somehow qualitatively distinct from *within*-species differences (Bateson, 1914; Cock,

1983). It was only at the end of his life that Bateson gave up his resistance to the idea that Mendelian factors were material entities which resided on visible structures (i.e., the chromosomes).

Thus Bateson, though applauded for his role in introducing Mendel's ideas to the English-speaking world (see Henig, 2000, for a popular account), is basically seen as a holdover from an earlier scientific tradition, never accommodating himself to the emerging programs of the neo-Darwinian synthesis and biochemical genetics/molecular biology. In fact, to some he is a decided villain of science, consigned to the same circle of Hell as Jean-Baptiste Lamarck and Richard Goldschmidt of "hopeful monster" fame. According to the evolutionary biologist and historian Ernst Mayr, "Bateson was pig-headed, intemperate, and intolerant . . . uncompromising . . . and . . . quite incapable of understanding the nature of natural populations" (Mayr, 1973, p. 146), and his "stubborn resistance to the chromosome theory resulted in much effort by members of the Morgan school that could have been devoted to new frontiers in genetics" (Mayr, 1973, p. 147). Indeed, in Mayr's view, Bateson's vibratory theory "simply retarded scientific progress" (Mayr, 1982, p. 42).

In the remainder of this chapter I will argue that Bateson's resistance to the coming syntheses, his rejection of gradualist notions of speciation and of the structural reification of Mendel's factors, and his search for physical mechanisms of morphogenesis were all of a piece. However, I suggest that rather than representing a reactionary failure to come to grips with new scientific findings and thought, they point to a coherent set of views (or, at least, intuitions—the means for scientific investigation of Bateson's notions were not available during his lifetime) which we can recognize as underlying modern evolutionary developmental biology ("Evo-Devo"), which began to take form a century after Bateson worked. I will focus my discussion initially on the vibratory theory, since Bateson returned to it again and again in his writings and addresses. Moreover, far from being "a view that resides comfortably in the world of transcendental morphology, a shovelful of paradigms beneath the chromosomal theory" (Hutchinson and Rachootin, 1979, p. xii), the vibratory theory has proved to be extraordinarily prescient and, if some allowance is made for the state of biological knowledge in the era in which it was propounded, even correct. I will then discuss the "presence-and-absence" hypothesis, which Bateson advanced to account for the phenomenon of genetic dominance, but which then became the basis for his remarkable ideas on the primordial phenotypic plasticity of living matter.

CHLADNI PATTERNS AND SPATIAL PERIODICITY

The German physicist Ernst Chladni (1756–1827), known as the founder of the field of acoustics, was the first to note that sand spread on the wooden back of a violin or on a metal or glass plate will become arranged into "nodal" patterns, or periodic ridges or spots, if the surface is set into motion by passing a violin bow across the strings of the violin or the edge of the plate (Chladni, 1787). The rosined bow variously slips and catches the material it is drawn across, and therefore imparts energy with a range of frequencies. A plate, depending on its composition, will resonate to one or more of these frequencies. Then, depending on the plate's shape and where it is clamped, certain patterns of vibratory motion will be sustained as long the energy is supplied, and others will die out. The "selected" patterns are the plate's stationary modes of vibration. Like a vibrating string, whose stationary modes are sine waves, the vibratory modes of a plate are realized as patterns of displacement from the quiescent, flat configuration. Sites of zero displacement are called "nodes." Any movable material on the plate, such as sand or talc, will accumulate at the nodes, since once it is displaced to those sites by the motion of the underlying plate, it will not be agitated any further and be caused to move away from them (see figure 4.1).

Bateson became fascinated with the pertinence of this and related physical processes to the establishment of repetitive morphological patterns.[2] In his first major work, *Materials for the Study of Variation* (1894), he defines "merism" as the "phenomenon of repetition of parts, generally occurring in such a way as to form a symmetry of pattern" (p. 20), and considers its basis:

> Looking at simple cases of meristic variation, such as that of the tulip or of *Aurelia*, or of the cockroach *Tarsus*, there is, I think, a fair suggestion that the definiteness of these variations is determined *mechanically*, and that the patterns into which the tissues of animals are divided represent positions in which the forces that effect the division are in equilibrium. On this view, the lines or planes of division would be regarded as lines or planes at right angles to the directions of the dividing forces; and in the lines of meristic division we are perhaps actually presented with a map of the lines of those forces of attraction and repulsion which determine the number and positions of the repeated parts, and from which symmetry results. If the symmetry of a living body were thus recognized as of the same nature as that of any symmetrical system of mechanical forces, the definiteness of the symmetry in meristic varia-

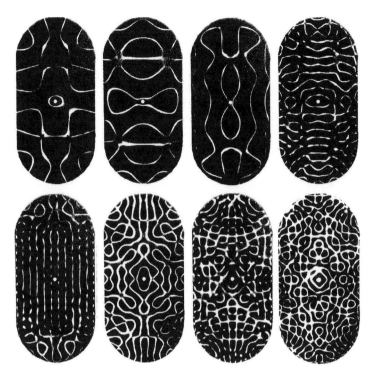

Figure 4.1
Chladni patterns formed by sand on the surfaces of "stadium-shaped" plates vibrated
at various frequencies. Upper row, from left: 387.8 Hz, 519.1 Hz, 649.6 Hz, 2667.3 Hz.
Lower row, from left: 2845.0 Hz, 3215.0 Hz, 4583.0 Hz, 6005.3 Hz. Courtesy of Dr.
Stephen Morris, Experimental Nonlinear Physics Group, Department of Physics, Uni-
versity of Toronto: http://www.physics.utoronto.ca/~nonlin/chladni.html.

tion would call for no special remark, and the perfection of the symmetry of a tulip with its parts divided into four, though occurring sud-
denly as a "sport," would be recognized as in nowise more singular than
the symmetry of the type. Both alike would then be seen to owe their
perfection to mechanical conditions and not to selection or to any other
gradual process. (Bateson, 1894, p. 70)

Bateson returns to the idea in *Problems of Genetics*:

If anyone will compare one of our animal patterns, say that of a zebra's
hide, with patterns known to be of purely mechanical production, he
will need no argument to convince him that there must be an essential similarity between the processes by which the two kinds of patterns
were made and that parts at least of the analysis applicable to the

mechanical patterns are applicable to the zebra stripes also. One of the most familiar examples, and one presenting some especially striking analogies to organic patterns, is that provided by the ripples of a mackerel sky, or that made in a flat sandy beach by the wind or the ebbing tide. With a little search we can find among the ripple-marks, and in other patterns produced by simple physical means, the closest parallels to all the phenomena of striping as we see them in our animals. We cannot tell what in the zebra corresponds to the wind or the flow of the current. . . . Our tissues therefore are like a beach composed of sands of different kinds, and different kinds of sand may show distinct and interpenetrating ripples. (Bateson, 1913, pp. 35–36)

In considering that segmentation and other periodic structures in tissues have mechanistic affinities to "patterns known to be of purely mechanical production," Bateson's instincts were on target physically. The rearrangement of a mobile material such as sand by a periodic driving force, as in Chladni's experiments, is, as noted above, physically straight-forward. The formation of ripples by wind or water currents has some similarities to this, but is a physically more subtle matter. Here the periodicity emerges only through the interaction of the driving force, which need not be periodic, with some inherent property of the mobile material. (For sand ripples this is the average distance individual grains saltate, or jump, after being randomly dislodged: Prigozhin, 1999; Scherer et al., 1999). Such phenomena are termed "self-organizing instabilities" (Nicolis and Prigogine, 1977). In a classic case devised by the mathematician A. M. Turing in the mid-twentieth century, spatial periodicity (i.e., stripe- and spotlike patterns) of the concentration of a diffusible chemical (which could be a morphogenetic molecule in an embryo) emerges from the interplay among the reactions which produce that chemical and a second chemical that inhibits its production, and the diffusion of the two chemicals (Turing, 1952).

Bateson's Dynamical Materialism

It is obvious that the patterns formed by the systems described above are both "physical" and "material." Even if it is unknown what substance, if any, in tissues assumes the role of the rippling sand or the diffusible chemical, or if "[w]e cannot tell what in the zebra corresponds to the wind or the flow of the current," it is a scientifically valid move to suggest that something in a developing embryo behaves according to such dynamical laws. Indeed, a "sand model" was used to provide insight into mechanisms of insect development as recently as the 1970s (Lawrence, 1971; Bryant, 1975). Strangely though, some commentators have interpreted the vibratory theory to mean that Bateson held a prescientific view of how characters were transmitted. Provine, for example, writes that "Bateson . . . did not believe that Mendel's 'differentiating elements' were material bodies. As early as 1893 Bateson had developed a 'vibratory theory of heredity' that did not fit with a materialist view of heredity, and he maintained this theory with some misgivings to the end of his life. It even caused him to reject the chromosome theory of heredity" (Provine, 2001, p. 61). Coleman (1970) makes a similar charge, as do Hutchinson and Rachootin (1979), as noted above. Allen (1974, p. 62) places Bateson among "the most recalcitrant idealists.

These views are eloquently dispatched by Cock, who warrants quoting at some length:

> Bateson was reluctant to accept any solution to the problem of transmission genetics which did not at the same time throw light on the developmental aspect, let alone one which seemed to make it more intractable. His difficulties were real and serious, for the chromosome content of all cells of the body (with the trivial exception of the gametes), was the same; nor was there any consistent difference of shape or structure between the chromosomes of different tissues. How then, if the genes were carried on the chromosomes, could they play any part in differentiation? Yet the way in which the effects of genes were often restricted to particular organs, such as wings or hairs, demanded that they play such a part. Bateson repeatedly pointed to the uniformity of the chromosomes over the whole body as a serious difficulty for chromosome theory. There was also the converse point: that the different Mendelian factors act in widely diverse ways, yet the chromatin particles which are supposed to represent them were to all appearances (as far as contemporary evidence went) uniform. (Cock, 1983, p. 45)

Bateson might reasonably stand accused of having stuck too long with a "nonparticulate" view of the inheritance of a character such as segmentation, for example, if it actually turned out to be the case that specific genes for segmentation reside on the chromosomes. Bateson rejected this notion on principle, however, and in this he would appear to have been affirmed when other early geneticists finally acknowledged the distinction between the Mendelian factor—a "differentiating element" for a character—and the character so differentiated. Mendel himself, in fact, recognized this distinction when he noted that "the distinguishing traits of two plants can, after all, be caused only by differences in the composition and grouping of the elements existing in *dynamical interaction* in their primordial cells" (quoted in Stern and Sherwood, 1966, p. 42 emphasis added).

Indeed, from the time Bateson took up Mendelism and became its most avid exponent in the English-speaking world, he refrained from identifying evidence for the existence of Mendelian factors influencing a character with the capacity of the factors to generate the character. In his book *Mendel's Principles of Heredity* he states:

> Let us recognise from the outset that as to the essential nature of these phenomena we still know absolutely nothing. We have no glimmering of an idea as to what constitutes the essential process by which the likeness of the parent is transmitted to the offspring. We can study the processes of fertilisation and development in the finest detail which the microscope manifests to us . . . but of the nature of the physical basis of heredity we have no conception at all. . . . We are in the state in which the students of physical science were, in the period when it was open to anyone to believe that heat was a material substance or not, as he chose. (Bateson, 1902, p. 3)

The vibratory theory was Bateson's most explicit attempt to understand how transmissible factors may generate a morphological character.[3] This effort led Mayr to castigate Bateson (along with other "physicalists" such as Johannsen) for refusing to accept a "corpuscular gene," wanting rather "to interpret everything in terms of forces" (Mayr, 1982, p. 737). Paradoxically, Mayr also places Bateson among the early geneticists who did not distinguish phenotypes from the corresponding genes (Mayr, 1982; see also Olby, 1987). Despite Bateson's loose use of the term "character" for the Mendelian factor in the 1902 book, this categorization is not compelling. In the same book Bateson note that "from the fact of the existence of the interchangeable characters we must, for purposes of treatment, and to complete the possibilities, necessarily form the conception of an *irresoluble base*" (Bateson, 1902, p. 28). The notion of a "base" (sometimes also referred to as a "residue"), which appears repeatedly in Bateson's writings, corresponds to what we would now identify as a "dynamical system": a context within which elements of a complex whole, such as the Mendelian factors, play out their differentiating roles.

Until the twentieth century, little was known about the properties of certain dynamical systems that allow them to exhibit self-organizing pattern formation. It is now recognized that unless a system is open to the flow of energy, matter, or both, it will simply descend to an equilibrium state in which there is no further macroscopic change—no new sand ripples, no chemical nonuniformities, certainly no life (Nicolis and Prigogine, 1977). Bateson's concept of living systems anticipated this understanding, and renders entirely reasonable his reluctance to sign on to a static notion of gene function:

> We commonly think of animals and plants as matter, but they are really systems through which matter is continually passing. The orderly relations of their parts are as much under geometrical control as the concentric waves in a pool. If we could in any real way identify or analyze the causation of growth, biology would become a branch of physics. Till then we are merely collecting diagrams which someday the physicist will interpret. He will I think work on the geometrical clue. (Bateson, 1917, in Bateson and Bateson, 1928, p. 209)

THE VIBRATORY THEORY AND MODERN DEVELOPMENTAL BIOLOGY

In the century since Bateson presented his vibratory theory, the role of wavelike phenomena in the generation of pattern and form during embryogenesis has moved from speculation to reality. Weiss (2002) has discussed the connection between Bateson's ideas and the "reaction-diffusion" mechanism proposed by Turing (1952), mentioned above (see also Newman, 1984). Turing's idea is particularly relevant to embryonic development since the movable material is postulated to be a diffusible molecular species, and the secretion and diffusion of growth and differentiation factors is a well-established and ubiquitous component of developmental mechanisms across all taxonomic groups (Tabata, 2001; Gurdon and Bourillot, 2001). For nearly forty years after the appearance of Turing's paper, which was audaciously titled "The Chemical Basis of Morphogenesis," the role of reaction-diffusion mechanisms in forming patterns of any kind, let alone biological patterns, was no less hypothetical than Bateson's vibratory theory. But by 1990 experimental conditions had been devised such that spatial patterns of chemical concentration in the form of spots, stripes, and even spirals, due unambiguously to Turing's mechanism, could be demonstrated in simple reaction-diffusion systems (Castets et al., 1990; Ouyang and Swinney, 1991; Vanag and Epstein, 2001).

Suggestions for developmental processes based on the Turing mechanism began to appear in the biological literature even before Turing's theoretical reasoning was confirmed in an actual chemical system, and afterward the pace of such proposals accelerated. Applications included the streaming patterns of the social amoeba *Dictyostelium discoideum* (Keller and Segel, 1970), budding in the polyp *Hydra* (Gierer and Meinhardt, 1972), skeletal pattern formation in the vertebrate limb (Newman and Frisch, 1979), formation of striped pigment patterns in the skin of fish (Kondo and Asai, 1995), and establishment of feather bud tracts in the skin of birds (Jiang et al., 1999). In some cases it has been possible to fairly confidently assign the role of Turing "morphogens" (the term was introduced in the 1952 paper) to actual molecular components of the systems in question (see, for example, Jiang et al., 1999; Miura and Shiota, 2000). One such case discussed in relation to the vibratory theory by Weiss (2002) is the formation of dental patterns (Jernvall and Jung, 2000; Salazar-Ciudad and Jernvall, 2002), a particular focus of Bateson in his treatise on discontinuous variation (Bateson, 1894).

However, Turing systems, which generate chemical "standing waves," are not the only manifestations of oscillatory phenomena in development. During vertebrate embryogenesis certain genes, such as the transcription factor *c-hairy1*, are expressed in cyclic waves sweeping across the paraxial mesoderm (Palmeirim et al., 1997) (figure 4.2). As this occurs, this region of the embryo becomes subdivided into the paired somites—blocks of tissue along the main axis that eventually give rise to the vertebrae and trunk and limb muscles. Each cell of the paraxial mesoderm experiences only temporal oscillations in the concentration of one or more of the relevant gene products, but the *phase* of the oscillation is different from point to point in space (i.e., across the tissue). Ultimately, the sweeping oscillation is converted into segments by a "gated" regulation of cell-cell adhesion (McGrew and Pourquié, 1998), but the details of how this occurs are still unclear.

The research group that discovered this phenomenon characterized it in the following fashion: "This gene [*c-hairy1*] is expressed in a cyclic fashion in the presomitic mesoderm with a periodicity corresponding to the formation of one somite. The periodic expression of *c-hairy1* mRNA appears as a wavefront travelling along the anteroposterior axis, and this scheduled expression constitutes an autonomous property of the paraxial mesoderm. We discuss these results in terms of a developmental clock linked to segmentation" (Palmeirim et al., 1997, p. 639). It is instructive to compare this with Bateson's speculation in his book *Problems of Genetics*, eighty-four years earlier: "the rhythms of segmentation may be the consequence of a single force definite in direction and continuously acting during the time of growth. The polarity of the organism would thus be

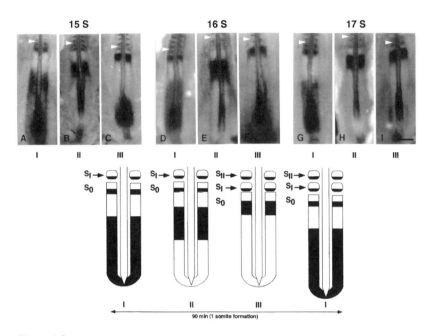

Figure 4.2
Formation of somites (segmented blocks of tissue along the main body axis) in chicken
embryos that are associated with traveling waves of expression of a regulatory protein
(*c-hairy1*). The protein is seen as dark, stained areas in the top photographic panel and
as black regions in the lower schematic drawing. *c-hairy1* is expressed in a temporally
periodic fashion in individual cells, but since the phase of the oscillator is different at
different points along the embryo's axis, the areas of maximal expression sweep along
the axis in a periodic fashion. Expression is confined to the caudal (toward the tail)
half of each somite, where it plays a functional role in causing separation from adjacent, presegmented tissue. Reprinted, with modifications, from I. Palmeirim, D. Henrique, D. Ish-Horowicz, and O. Pourquié, "Avian Hairy Gene Expression Identifies a
Molecular Clock Linked to Vertebrate Segmentation and Somitogenesis," *Cell*, *91*, 642.
Copyright 1997, with permission from Elsevier.

the expression of the fact that this meristic force is definitely directed once
it has been excited" (Bateson, 1913, p. 80).

The contemporary scientists quoted above can hardly be charged
with ignoring the existence of genes, not accepting the existence of
chromosomes, and so forth. Their findings exemplify the fact that genes
function by specifying components of dynamical systems, which in
turn produce pattern and form. The fact that individual genes reside at
particular sites on chromosomes, and are therefore "corpuscular" or

"particulate" (Mayr, 1982), could not be more irrelevant to Bateson's program of understanding the physical basis of the inheritance and generation of form in its continuous and discontinuous varieties, which is also the program of modern evolutionary developmental biology.

THE PRESENCE-AND-ABSENCE HYPOTHESIS

Both Mendel and the geneticists who took up his ideas in later decades were fascinated by the phenomenon of genetic dominance. In this case, as often happens, one version (allele) of a gene has a greater effect on the phenotypic outcome than the alternative version, such that when both are present, the phenotype of the heterozygote is like that of the dominant homozygote. Accumulating evidence suggested that this relationship applied in an asymmetric fashion to wild type characters and the alternative counterparts of those characters which resulted from mutations. Specifically, most mutated alleles are recessive to the corresponding wild type alleles under natural and experimental conditions (reviewed in Falk, 2001). The reasons for this were considered to bear importantly on the nature of the gene, and occupied experimental and theoretical geneticists for at least half a century, with competing ideas being put forward by many of the most important figures in the field, including Bateson, Morgan, Fisher, Wright, Haldane, Müller, and Dobzhansky (Provine, 2001; Falk, 2001). Astonishingly, given all the effort devoted to it, interest in the phenomenon eventually faded as more physiological concepts, informed by better knowledge of gene function, replaced models that proved to be bound to static notions of transmission genetics (Falk, 2001).

Close attention to Bateson's ideas concerning dominance suggests that, as with the vibratory theory of segmentation, he was thinking in dynamical terms akin to those which almost a century later provided means to address and solve these problems. His theory of dominance, which came to be known as the "presence-and-absence" hypothesis (Bateson, 1907; E. A. Carlson, 1966, pp. 5–65), asserted that the dominant version of a character is associated with the primordial (or default, as we might now put it) action of a Mendelian factor, while the recessive version results from the inhibition of this action. While Bateson (as noted above) used terminology which, like that of some of his contemporaries, did not distinguish clearly between Mendelian factors and the characters affected by such factors, for Bateson, as we have seen, the factors were altogether different from the hereditary particles conceived of by, for example, De Vries (see also Darden, 1977, pp. 100–103). Falk (2001, p. 293) notes in this same context that "Bateson emphasized vibration, motion, and waves rather than static material units. . . . Bateson referred to functions, not to structures." In a 1914 address by Bateson, the presence-and-absence hypothesis took the form of a rejection of the notion that addition of new Mendelian factors could produce new organismal qualities:

> We have to reverse our habitual modes of thought. At first it may seem rank absurdity to suppose that the primordial form or forms of proto-plasm could have contained complexity enough to produce the divers types of life. But is it easier to imagine that these could have been con-veyed by extrinsic additions? Of what nature could these additions be? Additions of material cannot surely be in question. We are told that salts of iron in the soil may turn a pink hydrangea blue. The iron cannot be passed on to the next generation. How can the iron multiply itself? The *power* to assimilate the iron is all that can be transmitted. (Bateson, 1914, in Bateson and Bateson, 1928, p. 292)[4]

Falk's insightful study of the evolution of the concept of dominance over the past century (Falk, 2001) shows Bateson's prescience as a proto-systems biologist. A key step in this transition was the emergence of complex systems models of gene function such as that of Kascer and Burns (1981) and Cornish-Bowden (1987). Such systems may contain feedback circuits and other buffering mechanisms, and frequently provide insight into why it is often the case (at least in systems selected for homeostatic behavior) that alterations (e.g., mutations) in individual system components will have minimal effect on the system (biochemical) phenotype. Although there continues to be debate on which organizational aspects of such complex systems are responsible for their resistance to perturbation (see, for example, Savageau, 1992; Forsdyke, 1994; J. M. Carlson and Doyle, 2000), there is little question that what has traditionally gone under the name of "dominance" in genetics can be understood only in systems terms, not in terms of the properties of individual gene products (Sarkar, 1998).

Throughout the period in which this change of perspective was gaining ground, Bateson's presence-and-absence hypothesis was criticized and rejected for various reasons (reviewed by Falk), essentially all of which missed the main point, which is that it was essentially a phenotype-oriented model pertaining to the presence and absence of *effects* in systems of interacting factors, rather than a genotype-based model postulating the presence or absence of the factors themselves (Falk, 2001). While Bateson cannot be said to have provided a definitive understanding of dominance (which would have been impossible for a scientist of his era), it is unquestionable that the *style* of his thinking on this question is more in tune with how modern biologists approach such phenomena than that of his contemporaries and their intellectual successors.

The most profound implications of the presence-and-absence hypothesis, which show signs of outlasting its application to the dominance question, pertain to speciation and the evolutionary process in general. As early as 1902 Bateson raised questions about whether the newfound Mendelian theory, about which he was so enthusiastic (particularly inasmuch as it seemed to provide insight into discontinuous variation; Darden, 1977), could really explain the origin of distinct species. He wondered:

> Has a given organism a fixed number of unit-characters? Can we rightly conceive of the whole organism as composed of such unit characters, or is there some residue—*a basis*—upon which the unit characters are imposed? . . . We are thus brought to face the further question of the bearing of Mendelian facts on the nature of species. The conception of species, however we may formulate it, can hardly be supposed to attach to allelomorphic or analytical varieties. We may be driven to conceive "species" as a phenomenon belonging to that "residue." (Bateson and Saunders, 1902, pp. 26–27)

Notwithstanding Bateson's dynamical conception of the operation of differentiating factors in the generation of alternative phenotypes, he was unconvinced that such factors, even acting dynamically, could move the organism beyond the confines of a particular species identity. Bateson returned to this theme on several occasions, notably in his address to the 1922 Toronto meeting of the American Association for the Advancement of Science, later published as "Evolutionary Faith and Modern Doubts":

> Analysis has revealed hosts of transferable characters. Their combinations suffice to supply an abundance of types which might pass for new species, and certainly would be so classed if they were met with in nature. Yet critically tested, we find that they are not distinct species and we have not reason to suppose that any accumulation of characters of the same order would cumulate in the production of distinct species. . . . Specific difference [i.e., pertaining to the generation of species] therefore must be regarded as probably attached to the base upon which these transferables are implanted, of which we know absolutely nothing at all. (Bateson, 1922, pp. 59–60)

Earlier, Bateson had tried to fathom the nature of the "base" or "residue" (as noted above, "dynamical system" would do just as well) that would both permit transferable characters operating within it to be alternatively expressed, and also be capable of undergoing more profound reorganization so as to give rise to "specific," or species, differences. In doing so, Bateson applied the same logic that, decades earlier, had led him to espouse the vibratory theory. Variation, whether of the "transferable" or the "specific" category, was held to be a complex and dynamic, rather than particulate and static, property of an organism's components:

> That which is conferred in variation must rather itself be a change, not of material, but of *arrangement*, or of motion. The invocation of additions extrinsic to the organism does not seriously help us to imagine how the power to change can be conferred. . . . By the re-arrangement of a very moderate number of things we soon reach a number of possibilities practically infinite. That primordial life may have been of small dimensions need not disturb us. Quantity is of no account in these considerations. Shakespeare once existed as a speck of protoplasm not so big as a small pin's head. To this nothing was added that would not equally well have served to build up a baboon or a rat. (Bateson, 1914, in Bateson and Bateson, 1928, pp. 292–293)

In this formulation, living matter, from the very start, had the propensity to take all the forms seen in nature, with Mendelian factors, acquired successively over the course of evolution, uncovering rather than originating morphological and other phenotypic possibilities:

> . . . we must begin seriously to consider whether the course of evolution can at all reasonably be represented as an unpacking of an original complex which contained within itself the whole range of diversity which living things present. . . . But as we have got to recognise that that there has been an evolution, that somehow or other the forms of life have arisen from fewer forms, we may as well see whether we are limited to the old view that evolutionary progress is from the simple to the complex, and that after all it is conceivable that the process was the other way about. (Bateson, 1914, in Bateson and Bateson, 1928, p. 292)

This was a radical notion that went well beyond the limited use Bateson had made of the presence-and-absence hypothesis in accounting for dominance effects in "transferable" Mendelian characters. This broader application of the notion presumed to establish a basis for a non-Darwinian approach to the generation of "specific" characters, and therefore for evolution as a whole. Thus, it became an easy target during the subsequent half-century, when neo-Darwinism reigned supreme. Richard Goldschmidt, himself hardly a partisan of those ideas when they were applied to the evolution of development, nonetheless wrote derisively of Bateson's "embarrassing idea of evolution by loss of inhibitors" (Goldschmidt, 1955, p. 478).

Bateson specifically disavowed the notion that he was formulating an alternative theory of evolution in this extension of the presence-and-absence hypothesis. (He precedes the passage quoted above with the comment "As I have said already, this is no time for devising theories of evolution, and I propound none."[5]) It is unquestionable, however, that in these passages he was pointing in a new direction, one which, given the science of the time, he was unable to navigate. As with the application of the presence-and-absence hypothesis to the understanding of dominance, its extension to the problem of the origin of species did not so much solve the problem as embody an approach that, with the availability of modern techniques of gene sequence and expression analysis, and of conceptual advances in the mathematical and computational modeling of complex systems, has the potential to do so. This approach, still more of a loose research agenda than a body of results, is often termed "systems biology" (Newman, 2003a). Patrick Bateson explicitly identifies William Bateson as a pioneer in this field (P. Bateson, 2002).

CONCLUSION: BATESON AS A FORERUNNER OF EVO-DEVO

Recent systems biological attempts, in light of detailed knowledge of genetic relationships among all the major taxonomic groups, to understand the evolutionary origins of body and organ morphologies, have concluded that the gradual acquisition of new genes by neo-Darwinian scenarios has not been the primary mechanism of diversification. For instance, many, if not most, of the components of the "genomic tool kit" that are active in morphogenesis were present well before the split between the protostomes (e.g., insects) and deuterostomes (e.g., vertebrates) (Erwin and Davidson, 2002; Newman, 2006). There is little evidence that the common ancestor of these modern forms had features—segments, limbs, eyes—that the modern forms themselves have in common. A reasonable conclusion is that during the course of evolution, the corresponding gene products, as elements of complex systems, were mobilized again and again to build similar structures. Bateson himself was famous for the suggestion that segmentation arose independently in more than one phylogenetic lineage (Bateson, 1886).

How this could have occurred is the subject of scenarios considered in the recent literature of Evo-Devo, which draw on several interconnected ideas first propounded by Bateson. Phenotypic plasticity—in Bateson's terms, the presence in individual organisms of the "range of diversity which living things present"—is now recognized as being a major factor in evolution, such that phenotypic change can precede, rather than proceed in lockstep with, genetic change (West-Eberhard, 2003). Genetic change accompanying evolution is dependent on changes in interactions among genes acting in networks rather than on the acquisition of new genes (Erwin and Davidson, 2002). (Bateson quoted on p. 99, above: "That which is conferred in variation must rather itself be a change, not of material, but of *arrangement*, or of motion.") Together, these insights make plausible Bateson's suggestion (see p. 97, above) that "the primordial form or forms of protoplasm could have contained complexity enough to produce the divers types of life," and that (see p. 99, above) "evolution can . . . reasonably be represented as an unpacking of an original complex."

But shifting focus away from a gene-centered, incrementalist, population-based approach to the evolution of form to a organism-centered, dynamical, development-based approach, raises in a particularly sharp fashion the question of how novel forms come to be produced by these developmental systems—that is, the question of origination (Müller and Newman, 2003, 2005). Here, Bateson's vibratory theory provides an apt paradigm, having been confirmed in its main elements for vertebrate segmentation (Pourquié, 2003). Indeed, the "generic" physical properties of tissues (i.e., those held in common with nonliving, chemically active viscoelastic materials), which include, but are not confined to, the propensity to form segments and other repeated structures, can take us a great distance in understanding the origination of body types and organ forms (Newman, 2003b). While the "unpacking of [the] original complex" during the course of evolution was not literally a matter of loss of inhibitors, in this contemporary picture the production of any given form from a delimited set of physically possible ones means that some latent system properties have been realized at the expense of others (Newman and Müller, 2005).

What might be considered Bateson's legacy to modern biology, however, is rarely treated as such, Bateson's reputation having suffered from charges of conservative thinking, aestheticism, and irrationalism (Coleman, 1970); anti-materialism (Provine, 2001, among others); and refusal to accept the "corpuscular gene" (Mayr, 1982; an accusation that rings particularly hollow in light of the foregoing discussion), views that remain current (Henig, 2000).[6] Although many of Bateson's insights have been rediscovered by others, and evolutionary developmental biology appears to be back on track, the record needs to reflect the accomplishments of this prescient scientist and the historical context of their temporary derailment.

ACKNOWLEDGMENTS

I benefited from discussions with Gerd Müller (University of Vienna) and Stephen Morris (University of Toronto) during the preparation of this paper. Support from the National Science Foundation is gratefully acknowledged.

NOTES

1. This is Beatrice Bateson's term for her husband in the title of her 1928 memoir, published two years after Bateson's death. By this time neither morphologists, evolutionary biologists, nor geneticists—Bateson was all three— were referring to themselves as naturalists. Mayr (1982, p. 540) identifies the "naturalists" of the later nineteenth and early twentieth centuries with the zoologists, botanists, and paleontologists who worked with whole organisms and were "particularly fascinated by diversity, its origin and meaning." This would seem to be an apt description of Bateson, and epitomizes the nonreductionist style of his thought, which led him, among other things, to resist a solely gene-based mechanism for speciation. Nonetheless, Mayr excludes Bateson from this esteemed company, placing him instead with De Vries, Johanssen, and Morgan among the "experimental evolutionists," former embryologists and incipient geneticists who either never accepted, or were late to accept, the relationship between Mendelism and Darwinism.

2. Cock (1983, p. 34) cites an unpublished manuscript by Bateson in the Cambridge University Library, dating from 1891, titled "A 'Vibratory' Theory of Linear and Radial Segmentation as Found in Living Bodies."

3. It should be noted that if the components of a dynamical system are transmissible, and the physical laws that govern such a system are also transmitted (as they must be), then the dynamical behavior itself (e.g., the capacity to form segments) is inherited.

4. We might note, from the hindsight of contemporary biology, that such "power" is typically due to the product of a specific gene (i.e., Mendelian factor), such as an iron transporter protein. But such factors, apart from rare horizontal transfers, are not acquired whole. In the case discussed by Bateson, the blue phenotype would arise from the action, in the appropriate environment, of a power *present* regardless of environment. In a constitutively pink mutant the power (but not the factor) would be *absent*.

5. Bateson never disavowed the efficacy of natural selection, but he saw a more limited role for it than many of his and our own contemporaries. His

views on this were in place early on: "In the view of the phenomena of variation here outlined, there is nothing which is in any way opposed to theory of the origin of species 'by means of natural selection, or the preservation of favoured races in the struggle for life.' But by a full and unwavering belief in the doctrine as originally expressed, we shall in no way be committed to representations of that doctrine made by those who have come after. . . . For the crude belief that living beings are plastic conglomerates of miscellaneous attributes and that order of form or symmetry have [sic] been impressed upon this medley by selection alone; and that by variation any of these attributes may be subtracted or any other attribute added in indefinite proportion, is a fancy which the study of variation does not support" (Bateson, 1894, p. 80).

6. See, however, Webster (1992), Forsdyke (2000), Falk (2001) and P. Bateson (2002) for more sympathetic views of Bateson. Each of these writers finds elements in Bateson that were missed by his contemporaries and successive generations of geneticists and developmental biologists. Webster explicitly sees him as having carried the nineteenth-century rational morphology program into the post-Darwin, post-Mendel era.

REFERENCES

Allen, G. E. (1974). Opposition to the Mendelian-chromosome theory: The physiological and developmental genetics of Richard Goldschmidt. *Journal of the History of Biology*, 7: 49–92.

Bateson, B., and Bateson, W. (1928). *William Bateson, F.R.S., Naturalist: His Essays & Addresses, Together with a Short Account of His Life*. Cambridge: Cambridge University Press.

Bateson, P. (2002). William Bateson: a biologist ahead of his time. *Journal of Genetetics*, 81: 49–58.

Bateson, W. (1886). The ancestry of the Chordata. *Quarterly Journal of Microscopical Science*, 26: 535–571.

———. (1889). On some variations of *Cardium edule* apparently correlated to the conditions of life. *Philosophical Transactions of the Royal Society of London*, B153: 297–330.

———. (1894). *Materials for the Study of Variation*. London: Macmillan. Reissued Baltimore: Johns Hopkins University Press, 1992.

———. (1902). *Mendel's Principles of Heredity: A Defence*. London: Cambridge University Press.

———. (1907). The progress of genetics since the rediscovery of Mendel's papers. In J. P. Lotsy (ed.), *Progressus Rei Botanicae*. Jena: G. Fischer.

———. (1913). *Problems of Genetics*. New Haven, Conn.: Yale University Press. Reissued 1979.

————. (1914). Address to British Association for the Advancement of Science, Melbourne, Australia. In Bateson and Bateson (1928).

————. (1917). "Gamete and zygote." In Bateson and Bateson (1928).

————. (1922). Evolutionary faith and modern doubts. *Science*, 55: 55–61.

Bateson, W., and Saunders, E. R. (1902). The facts of heredity in the light of Mendel's discovery. *Reports to the Evolution Committee of the Royal Society*, vol. 1, pp. 125–160.

Brooks, W. K. (1883). *The Law of Heredity*. Baltimore: John Murphy.

Bryant, P. J. (1975). Regeneration and duplication in imaginal discs. *Ciba Foundation Symp*, 71–93.

Carlson, E. A. (1966). *The Gene: A Critical History*. Philadelphia: Saunders.

Carlson, J. M., and Doyle, J. (2000). Highly optimized tolerance: Robustness and design in complex systems. *Physical Review Letters*, 84: 2529–2532.

Castets, V., Dulos, E., Boissonade, J., and DeKepper, P. (1990). Experimental evidence of a sustained standing Turing-type nonequilibrium chemical pattern. *Physical Review Letters*, 64: 2953–2956.

Chladni, E. F. F. (1787). *Entdeckungen über die Theorie des Klanges*. Leipzig: Weidmanns Erben und Reich.

Cock, A. G. (1983). William Bateson's rejection and ultimate acceptance of chromosome theory. *Annals of Science*, 40: 19–59.

Coleman, W. (1970). Bateson and chromosomes: Conservative thought in science. *Centaurus*, 15: 228–314.

Cornish-Bowden, A. (1987). Dominance is not inevitable. *Juornal of Theoretical Biology*, 125: 333–338.

Darden, L. (1977). William Bateson and the promise of Mendelism. *Journal of the History of Biology*, 10: 87–106.

Erwin, D. H., and Davidson, E. H. (2002). The last common bilaterian ancestor. *Development*, 129: 3021–3032.

Falk, R. (2001). The rise and fall of dominance. *Biology and Philosophy*, 16: 285–323.

Forsdyke, D. R. (1994). The heat-shock response and the molecular basis of genetic dominance. *Journal of Theoretical Biology*, 167: 1–5.

————. (2000). Bateson, William. In *Nature Encyclopedia of Life Sciences*. London: Nature Publishing Group. Also available at http://www.els.net/[doi:10.1038/npg.els.0002359].

Gierer, A., and Meinhardt, H. (1972). A theory of biological pattern formation. *Kybernetik*, 12: 30–39.

Goldschmidt, R. B. (1955). *Theoretical Genetics*. Berkeley: University of California Press.

Gurdon, J. B., and Bourillot, P. Y. (2001). Morphogen gradient interpretation. *Nature*, 413: 797–803.

Henig, R. M. (2000). *The Monk in the Garden: The Lost and Found Genius of Gregor Mendel, the Father of Genetics*. Boston: Houghton Mifflin.

Hutchinson, G. E., and Rachootin, S. (1979). Historical introduction. In William Bateson, *Problems of Genetics*. New Haven, Conn.: Yale University Press. Book first published in 1913.

Jernvall, J., and Jung, H. S. (2000). Genotype, phenotype, and developmental biology of molar tooth characters. *American Journal of Physical Anthropology*, suppl. 171–190.

Jiang, T., Jung, H., Widelitz, R. B., and Chuong, C.-M. (1999). Self-organization of periodic patterns by dissociated feather mesenchymal cells and the regulation of size, number and spacing of primordia. *Development*, 126: 4997–5009.

Kacser, H., and Burns, J. A. (1981). The molecular basis of dominance. *Genetics*, 97: 639–666.

Keller, E. F., and Segel, L. A. (1970). Initiation of slime mold aggregation viewed as an instability. *Journal of Theoretical Biology*, 26(3): 399–415.

Kondo, S., and Asai, R. (1995). A reaction-diffusion wave on the skin of the marine angelfish *Pomacanthus*. *Nature*, 376: 765–768.

Lawrence, P. A. (1971). The organization of the insect segment. *Symp Soc Exp Biol*, 25: 379–390.

Mayr, E. (1973). The recent historiography of genetics. *Journal of the History of Biology*, 6: 125–154.

———. (1982). *The Growth of Biological Thought: Diversity, Evolution, and Inheritance*. Cambridge, Mass.: Belknap Press.

McGrew, M. J., and Pourquié, O. (1998). Somitogenesis: Segmenting a vertebrate. *Curr Opin Genet Dev*, 8: 487–493.

Miura, T., and Shiota, K. (2000). TGFβ2 acts as an "activator" molecule in reaction-diffusion model and is involved in cell sorting phenomenon in mouse limb micromass culture. *Dev Dyn*, 217: 241–249.

Mivart, S. G. J. (1871). *On the Genesis of Species*. New York: Appleton.

Müller, G. B., and Newman, S. A. (2003). Origination of organismal form: The forgotten cause in evolutionary theory. In G. B. Müller and S. A. Newman (Eds.), *Origination of Organismal Form: Beyond the Gene in Developmental and Evolutionary Biology*, pp. 3–10. Cambridge, Mass.: MIT Press.

Müller, G. B., and Newman, S. A. (2005). The innovation triad: an EvoDevo agenda. *Journal of Experimental Zoology Part B: Molecular and Developmental Evolution*, 304: 487–503.

Newman, S. A. (1984). Vertebrate bones and violin tones: Music and the making of limbs. *The Sciences* (New York Academy of Sciences), 24: 38–43.

———. (2003a). The fall and rise of systems biology. *GeneWatch*, 16: 8–12.

———. (2003b). From physics to development: The evolution of morphogenetic mechanisms. In G. B. Müller and S. A. Newman (Eds.), *Origination of Organismal Form: Beyond the Gene in Developmental and Evolutionary Biology*, pp. 221–239. Cambridge, Mass.: MIT Press.

Newman, S. A. (2006). The developmental-genetic toolkit and the molecular homology-analogy paradox. *Biological Theory*, 1: 12–16.

Newman, S. A., and Frisch, H. L. (1979). Dynamics of skeletal pattern formation in developing chick limb. *Science*, 205: 662–668.

Newman, S. A., and Müller, G. B. (2005). Genes and form: inherency in the evolution of developmental mechanisms. In E. Neumann-Held and C. Rehmann-Sutter (Eds.), Genes in development: re-reading the molecular paradigm, Duke University Press, Durham, NC, pp. 38–73.

Nicolis, G., and Prigogine, I. (1977). *Self-organization in Nonequilibrium Systems: From Dissipative Structures to Order Through Fluctuations.* New York: Wiley.

Olby, R. C. (1987). William Bateson's introduction of Mendelism to England: A reassessment. *British Journal of the History of Science*, 20: 399–420.

Ouyang, Q., and Swinney, H. (1991). Transition from a uniform state to hexagonal and striped Turing patterns. *Nature*, 352: 610–612.

Palmeirim, I., Henrique, D., Ish-Horowicz, D., and Pourquié, O. (1997). Avian hairy gene expression identifies a molecular clock linked to vertebrate segmentation and somitogenesis. *Cell*, 91: 639–648.

Pourquié, O. (2003). A biochemical oscillator linked to vertebrate segmentation. In G. B. Müller and S. A. Newman (Eds.), *Origination of Organismal Form: Beyond the Gene in Developmental and Evolutionary Biology*, pp. 183–194. Cambridge, Mass.: MIT Press.

Prigozhin, L. (1999). Nonlinear dynamics of aeolian sand ripples. *Physical Review*, E60: 729–733.

Provine, W. B. (2001). *The Origins of Theoretical Population Genetics.* 2nd ed. Chicago: University of Chicago Press.

Salazar-Ciudad, I., and Jernvall, J. (2002). A gene network model accounting for development and evolution of mammalian teeth. *Proceedings of the National Academy of Sciences of the U.S.A.*, 99: 8116–8120.

Sarkar, S. (1998). *Genetics and Reductionism*. Cambridge: Cambridge University Press.

Savageau, M. A. (1992). Dominance according to metabolic control analysis: Major achievement or house of cards? *Journal of Theoretical Biology*, 154: 131–136.

Scherer, M. A., Melo, F., and Marder, M. (1999). Sand ripples in an oscillating annular sand–water cell. *Physics of Fluids*, 11: 58–67.

Stern, C., and Sherwood, E. R. (Eds.). (1966). *The Origin of Genetics: A Mendel Source Book*. San Francisco: W. H. Freeman.

Tabata, T. (2001). Genetics of morphogen gradients. *Nature Reviews Genetics*, 2: 620–630.

Turing, A. M. (1952). The chemical basis of morphogenesis. *Philosophical Transactions of the Royal Society of London*, B237: 37–72.

Vanag, V. K., and Epstein, I. R. (2001). Inwardly rotating spiral waves in a reaction-diffusion system. *Science*, 294: 835–837.

Webster, G. (1992). William Bateson and the science of form. In 1992 reissue of Bateson (1894), pp. xxix–lix.

Weiss, K. M. (2002). Good vibrations: The silent symphony of life. *Evolutionary Anthropology*, 11: 176–182.

Weldon, W. F. R. (1902). Mendel's laws of alternative inheritance in peas. *Biometrika*, 1: 228–254.

West-Eberhard, M. J. (2003). *Developmental Plasticity and Evolution*. Oxford: Oxford University Press.

_____

5

TO EVO-DEVO THROUGH CELLS, EMBRYOS, AND
MORPHOGENESIS

Jane Maienschein

Evo-Devo finally brings us a new synthesis, it is claimed, with evolution of development as the central focus. There is a sense of triumphalism in the declarations that this is a much better synthesis than the so-called evolutionary synthesis before it. Discussion has, naturally enough, centered on the question "Why not before, and why now?"

Fans of the evolutionary synthesis have invoked a variety of explanations for why development was left out of the synthesis of the 1940s and 1950s. Yet we can also ask the question rather differently: Why was evolutionary biology so foolishly distracted by philosophy and theology that it failed to do "real" science and missed the boat of experimental progress? We might, with Yale developmental biologist J. P. Trinkaus (known as Trink), hold that far from feeling left out of some important self-declared "synthesis," those concerned with development actually felt sorry for their evolutionary biology counterparts. As Trink put it, "Hell, no, we didn't feel left out of anything. They were just jealous because they couldn't figure how to get the NIH funding!"[1]

In 1884, Karl Nägeli had already noted the tendency of evolutionists to wander away from what he considered the core biological questions. He emphasized the importance of examining physical and mechanical understanding of organic nature. Nägeli complained that

> The theory of evolution touches also philosophy and theology in very sensitive spots and interests the intelligent general public partly for this reason and partly because human vanity has always attached much importance to origin and relationship.
>
> On this account we have seen philosophers, theologians and, in addition, literati of all sorts and conditions take possession of the problem. This too would have been quite in order, if every one had but utilized the established results of scientific investigation for his own field and rendered to his own circle a clarifying and instructive account of them; and if so many had not considered this field of difficult physiological problems to be a free-for-all arena for senseless argumentation.[2]

Citing Nägeli, Ross Harrison echoed this sentiment in 1936, in his address as retiring chairman of the Section on Zoological Science at the American Association for the Advancement of Science meeting in Atlantic City, New Jersey. Discussing "Embryology and Its Relations," Harrison saw evolution as having gone astray because of its engagement with fundamentalist theology and lamented that "The scientific investigation of evolution has suffered severely from this emotional conflict." In particular, he pointed to the failure to achieve scientific—by which he meant experimental—results. He acknowledged that the long time frame required for evolution makes experimental investigation more difficult, but it is nonetheless necessary, for "it can scarcely escape any one accustomed to scientific thinking that the processes of evolution can be elucidated only by painstaking experimental work carried on over a long period of years." Fortunately, Harrison saw "hopeful signs of the applicability of exact methods to the study of evolutionary processes" in the "development of modern genetics, the experimental study of the origin of mutations and the new mathematical theory of natural selection."[3]

Harrison was not alone in his thinking; he represented a community of researchers exploring embryology and its relations to other fields and other processes and patterns of biology. They felt that true science must be experimental and analytical, and must avoid the speculative distractions to which they felt evolution had succumbed.

In the 1920s Harrison had been less sanguine about the prospects for such science and for embryology. In his address as retiring president of the American Society of Zoologists in 1925, he had urged a "Return to Embryology."[4] After a period of quiescence and even depression within the field, he had pointed to what he hoped would soon become a resurgence of embryological study. The concept of the "organizer" had brought promise for analytical approaches to development, and there had been reason to hope that younger researchers would take up the study of embryos again. By 1936, Harrison was clearly pleased to announce, the resurrection had taken place. And even though the organizer theory had given way to concepts of "induction," and to closer and more careful analysis of internal regulation within organisms—and indeed within individual cells within the organisms—this was progress. Embryology had come much farther than scientific study of evolution. For Harrison and his likes, there could be a coming together of evolution and development only if those studying evolution, those studying heredity through genetics, and embryologists all relied on the same experimental scientific approaches.

It was not in the 1930s, or even during the next half-century, that evo and devo began to come together in the new and promising ways that have given rise to today's self-proclaimed "Evo-devo" movement. And that has come, as Harrison predicted, because of technical and experimental advances more than because of additional speculation and theory. Some chapters in this volume explore aspects of these recent advances, while others look at episodes of study of devo and earlier attempts at evo-devo syntheses. In this chapter, I concentrate instead on lines of research that lay within what would have been considered embryology and that largely ignored evolution as a factor to be addressed directly. Yet I contend that these studies that focused on exploring embryos and cells really did seek to address fundamental questions about development in the light of evolution.

I look at studies of morphogenesis. This field focused on the proximate or local and immediate causal mechanisms of the emergence of the parts that make up an individual organism's form. Yet attempts to make sense of morphogenesis also bring together the different time scales of individual development and evolutionary history because "morphogenesis" presumes the development of a particular "morph" that conforms to the form of its species. The question, then, was how an individual comes to acquire the particular form of its species, which is a product of the different long-term time scale of evolution.

Therefore, morphogenesis was at its heart one way of bringing together the devo and the evo, respecting the proximate mechanisms of individual development and also the "ultimate" factors brought by evolution and revealed through systematics. In the twentieth century, fundamental questions about morphogenesis found tractability in study of cells and of the whole, interacting, developing embryos of which the cells are the parts. I therefore concentrate on studies at the conjunction of cells, embryos, and morphogenesis. This allows us to get at one set of ways in which researchers thought they could meaningfully bring together development and evolution, and this historical perspective should illuminate current discussions.

Cells and Morphogenesis in the 1890s

In the 1890s, the development of morphological structures was labeled "morphogenesis." At the same time, because of tremendous advances in cytology during the last quarter of the nineteenth century, researchers had begun to focus on cells and on the ways that cells interact to generate structure. While some researchers moved toward hereditary accounts of development, pointing to the inherited material inside cells as determining what follows, those concentrating on development drew on epigenetic accounts. Cells, embryos, morphogenesis, and epigenesis converged in exciting new research programs.

Years later, the committed agnostic Harrison emphasized the epigenetic outlook that underlay this research by quoting a passage from the Biblical Psalm 139: "And in Thy book all my members were written, which in continuance were fashioned, when as yet there was none of them." Development is not seen as an unfolding of something preexistent, but as a coming into being. Obviously, Harrison agreed and wanted to emphasize that therefore the best approach for understanding the genesis of an individual's life was through embryology as the study of epigenetic emergence, and focused on "Cellular Differentiation and Internal Environment."[5]

The central question was what role cells play in development. Do cells serve as causal agents in morphogenesis, actually bringing about the generation of form and the function that comes with form through the actions and interactions of individual cells? If so, in what way? What is the relative importance of local, proximate, internal environmental factors, and how can the interplay of these factors in shaping each cell give rise to complex, multicellular forms? Or, alternatively, are cells just epiphenomenal results that come only after the real work of development has occurred through other forces? And if they are, what are the forces and how do they do the job of morphogenesis?

One view held that cells do carry significant causal force for development and differentiation. Edmund Beecher Wilson took this view in the first edition of his *The Cell in Development and Inheritance* (1896). He noted that cell theory and evolution provide the two foundations for biology. He asked what cell division does, and how we get from one fertilized egg cell to differentiated cells and eventually to a formed organism. How does development of the form, or morphogenesis, work?

Wilson answered that cell division brings differentiation, and the series of cell divisions leads to the gradual and epigenetic process of morphogenesis. Wilson noted that "for two reasons the cleavage of the egg possesses a higher interest than any other case of cell-division. First, the egg-cell gives rise by division not only to cells like itself, as is the case with most tissue-cells, but also to many other kinds of cells. The operation of cleavage is therefore immediately connected with the process of differentiation, which is the most fundamental phenomenon in development."[6]

Therefore, cleavage and differentiation are connected, with cleavage apparently causing or at least leading to differentiation. There was some "promorphological" arrangement in the segmented egg that brought a sort of "germinal localization," but there was no preformation or even predetermination in any meaningful sense. The form was not there already. Rather, for Wilson there was some "organization" or prelocalization in the segmented egg that established the starting point for and process that gives rise to an individual organism.[7] Then it was the cell division that brings differentiation, and with it morphogenesis. For Wilson, morphogenesis occurred one cell division at a time, against a background of cytoplasmic differentiation.

Wilhelm Roux took a much more extreme version of this cell-division-as-cause-of-developing-form. His mosaic interpretation involved parceling out differentiated inherited material to each cell, so that every cell division brought specialization and localization of differentiated cells and parts. Again, morphogenesis occurred one cell division at a time, very decidedly *because of* the cell divisions. Yet on this interpretation the cells were little more than containers for the hereditary units. In Roux's case, however, the form was effectively predelineated in the inherited units parceled out to each cell.

Alternatively, a second theory held that cells are epiphenomena that follow rather than cause cell division. In this case, some other causes drive differentiation and morphogenesis. For example, Thomas Henry Huxley saw them this way, for cells "are no more the producers of vital phenomena, than the shells scattered in orderly line[s] along the sea-beach are the instruments by which the gravitational force of the moon acts upon the ocean. Like these, the cells mark only where the vital tides have been, and how they have acted."[8] Instead of cells and cell division, properties of protoplasm and evolutionary factors drove development, according to Huxley; cells were simply secondary, or epiphenomenal.

Charles Otis Whitman agreed with Huxley's view in his essay "The Inadequacy of the Cell-Theory."[9] Whitman insisted that Wilson had it backward. Organization was *not* the product of cell-formation; rather, "organization precedes cell formatting and regulates it."[10] He said that "an organism is an organism from the egg onward," and that cleavage simply followed and divided up the material. He explained morphogenesis and differentiation in terms of a predelineation within the egg that provided the "organization" for the future "organism."

Charles Manning Child agreed with Whitman. It must be "the organism—the individual, which is the unit and not the cell."[11] For Child, morphogenesis was driven by internal gradients set up by inherited nuclear and cytoplasmic factors, and responding to external and internal environmental considerations.

If this interpretation that cells are secondary effects rather than causes were right, what causes differentiation and morphogenesis? It cannot simply be genes, since, as Thomas Hunt Morgan liked to remind his colleagues, all the genes seem to be the same in every cell. How, then, can we get difference from the inherited sameness? How does morphogenesis—and with it differentiation—occur? It was the failure to address these questions adequately that kept those most seriously committed to explaining development from seeing genetics or evolution as important for understanding embryology.

"Embryology and Its Relations"

Ross Harrison agreed with Morgan that the answer could not be in the genes, and that this was a central problem of biology. In his 1936 speech to the American Association for the Advancement of Science, Harrison noted: "The prestige of success enjoyed by the gene theory might easily become a hindrance to the understanding of development by directing our attention solely to the genom." Instead, Harrison insisted that "cell movements, differentiation, and in fact all developmental processes are actually effected by the cytoplasm."[12] They are effected through the cytoplasm—that is, through local action and particularly through local chemical action—and not through the action of some remote and distance-inherited material or purported information.

Harrison suggested that morphogenesis involves a sort of crystallization process that brings chemical compositions of parts and differentiation through the relations among them.[13] He lamented that excessive enthusiasm about Hans Spemann's "organizer" had distracted embryologists, as genetics had, and had led them to ignore other important factors in development, especially relations among differentiating parts. Yet embryologists were making great progress by looking at the microstructures of eggs, cells, and developing organic parts. Harrison acknowledged that he had questions with few answers, but that it was very important to work hard—and to continue working hard, using proper scientific experimental analysis—on the hard questions, and not to give in to temptations to unwarranted theorizing or guesswork.

Harrison quoted Max Planck to the effect that "We must never forget that ideas devoid of a clear meaning frequently gave the strongest impulse to further development of science . . . they [can] give rise to thought, for they show clearly that in science as elsewhere fortune favors the brave."[14] The brave included Joseph Needham, with his ideas about chemical morphogenesis through the internal chemical relations of parts and "morphogenetic hormones,"[15] or those offering the mathematical, mechanical models that Evelyn Fox Keller discusses in her book *Making Sense of Life*.[16]

As J. H. Woodger had pointed out earlier, it is especially important to have some brave theorizers or big thinkers when the data and details pile up and threaten to overwhelm our thinking: "The continual heaping up of data is worse than useless if interpretation does not keep pace with it. In biology, this is all the more deplorable because it leads us to slur over what is characteristically biological in order to reach hypothetical 'causes.'"[17]

MORPHOGENESIS

For Harrison, and for many others, biological form and the apparent "organization" of individuals were "characteristically biological" and ought not to be slurred over, no matter how difficult to address. Among the many making scientific sense of the emergence and establishment of form in later decades was another brave man, John Tyler Bonner. Bonner's *Morphogenesis: An Essay on Development* appeared in 1952, roughly a half-century after Wilson's *Cell* and roughly a half-century before our current enthusiasm for Evo-Devo.[18]

Yale biologist John Spangler Nicholas reviewed Bonner's book and wrote that "Bonner deserves our thanks. He makes no pretense of giving the answer to the problem of form. He has, however, placed it succinctly before us and has focused attention on what we do not know but need to know before a more definite answer can be given to the significant factors underlying the formative pattern of development which results in the specific form of the organism."[19]

We still owe Bonner our thanks for keeping a focus on form and pattern, on morphogenesis and internal relations, when so many were rushing to embrace the "modern synthesis" of the evolutionists, or the molecular interpretations brought by DNA and genetics. As they took up other methods and other questions, they set aside and often forgot about scientific efforts to understand these fundamental life processes. Bonner has continued to keep our eye on the "problem of form," especially the question of how form emerges, and, within that problem, on questions about the role of cells and their internal relations in development and differentiation. Bonner retained his focus on the mechanisms and proximate time frame of individual development, while also remaining mindful of the longer time scale and causal shaping of form by evolution. He sought to bring devo and evo together through the study of morphogenesis.

Bonner wrote his *Morphogenesis* in Woods Hole, Massachusetts, at the Marine Biological Laboratory. He worked in the library there, and in Edwin Grant Conklin's laboratory, writing the sort of general, problem-oriented, big-picture book that young biologists at places such as Princeton can no longer afford to write if they hope to remain on a normal track toward tenure. Bonner tackled the tough problems and tried to bring order to our thinking about them. He tried to weigh the range of theories, data, and worries without succumbing prematurely to any one interpretation or to giving up trying to address the big questions.

Bonner set out to consider the "problem of form" in terms of three things—growth, morphogenetic movements, and differentiation. He sought to avoid the "treacherously hypothetical" by including a full range of organisms—animals, plants, and microbes: slime molds (especially slime molds), *Hydra*, sponges, frogs, and ants.

First came growth and the patterns of growth. Growth is a basic process and does not bring about any morphogenesis or differentiation by itself, but it makes these processes possible. Next come the patterns of morphogenetic movements. These are like the actions of a sculptor who has already added the clay, through growth, and now shapes it. But it is the cells, rather than the sculptor, doing the shaping in the biological organism. So, yes, cells are inevitably involved, but the movement comes through groups of cells or cell interactions as they move. Morphogenetic movements lead to differentiation, by which Bonner means differences in parts because of the chemical composition and also because of the positions and needs of the whole organism.

Differentiation very clearly does *not* result from parceling out of genetic information, as Roux and his later genetic determinist followers had suggested. At least for all practical purposes this is true, Bonner was sure. Rather, differentiation can be caused by position in the organism. Bonner cited Henry van Peters Wilson's research on sponges, in which Wilson had separated (disaggregated) cells to discover whether they were all the same or were already differentiated at various points throughout development.[20] That research and other studies on *Hydra*, slime molds, sea urchins, and other organisms showed that cells and groups of cells might be differentiated, but could be redifferentiated by changing position. That is, the role of the individual cell could be influenced by the needs of the whole organism.

There seemed to be internal diffusion gradients and some organized microstructure that functioned as "ultrastructure" to guide the "regulation" of the whole. Bonner saw this regulation of the whole, with its responsiveness to changing conditions and to the internal and external environments, as an important driver for differentiation. Organisms can have gradients or fields that affect the "patterns of differentiation," and these play out in different ways in different organisms. The differentiation and morphogenetic processes of different organisms can be different not just because the organism started differently, or has different heredity, but also because of the particular interactions within each whole.

Bonner clearly saw differentiation as a problem of the whole, living organism and the way the parts interact. Understanding this is a challenge, however, and Bonner was brave enough to insist that we not ignore it. The question was how to make sense of the "wholeness" or "organization." As Bonner put it, we must not "forget the most important fact that the organism always differentiates as a whole, and that the particular cause of the differentiation of a part is determined by its position with respect to the whole."[21] There is indeed a wholeness that begins with "localization of differences in different parts of the egg or sperm."[22] That is, there is some initial structure that provides a starting point, as E. B. Wilson had suggested, though it is not a "promorphological" delineation in Wilson's sense or any sort of preformation or predetermination of form.

Nonetheless, Bonner cautioned that we should not be "overly impressed by the [special] significance of this wholeness."[23] The wholeness is not mysterious, as some would suggest, but is a product of interactions and microprocesses within the context of the initial germ cell that is itself influenced by natural selection. Organisms inherit the tendency for cells "to migrate and [to] respond to the substances given off by the other cells," through a sort of chemotaxis and interactions among the cells.

Bonner constantly tries to find a balance among microexplanations of chemical and mechanical causation, and also attempts to preserve wholeness and a sense of integrated life—whatever that might mean and even when that might seem vague and mushy. It is precisely this grappling with making sense of form and his unwillingness to fall into reductionistic geneticism or to succumb to mysterious organicism that is appealing. Bonner wrote, "There must be some factor which transcends the cell wall and unifies this cottony mass, but what this factor or factors might be is another matter. Already we have come to the deep-rooted sign of the least understood problem of this sort of development that makes us say that growth and development is a problem. Really it is many problems; but this one, the unification of great masses of protoplasm into a oneness, a wholeness, has us more mystified than others."[24]

Yet, "More than anything else, this . . . making of a perfect whole from a small bit of a previous whole, is what seems marvelous to us, so much so that we become, I think, psychologically affected and troubled, and cannot believe that a solution to such a problem would be anything but difficult, if not impossible."[25] Yet difficult—or even apparently impossible—as this "problem of development" might seem, Bonner tried, and felt it vitally important to continue trying.

Surely Bonner's drawing on evidence from diverse organisms and trying to bring together different kinds of evidence shaped his thinking in important ways. It kept his focus on the bigger problems, even when they were difficult to solve. Model, select organisms can work to solve particular problems, but would not have served Bonner's purposes unless he made the a priori assumption that all organisms differentiate in the same way. And he did not make that assumption, that the "evo" affects the "devo." This brings us back to cells.

Each germ cell has an internal arrangement, and every cell division brings new arrangements. The particularities of growth and morphogenetic movements, with regulatory responses to the conditions and needs of the organism, bring differentiation—and the material basis for the mechanisms of morphogenesis. But each set of opportunities and pattern of responses is shaped by evolution as well, and each represents a set of adaptations. For, as Bonner put it: "In each case, there is a . . . unity which comes with the structure, and this we have related to the advantage of functional wholes, for without being functionally cohesive they would either not live, or at least not withstand the rigors of natural selection. The very fact that they are wholes must be adaptively advantageous, and natural selection, by differential reproduction, would tend to keep them that way."[26] Evolution and development, long and short time scales: all there to be studied in cells through morphogenesis.

CONCLUSION

In sum, for Bonner cells are not the cause of morphogenesis (as Wilson and Roux had suggested in their very different ways). Nor are they mere epiphenomena (or shells on the shore, as Huxley and Whitman held). Rather, cells and cell interactions are primary players in the processes that shape organisms, but the processes are also influenced by genetics and by evolutionary adaptations. Bonner just was not sure how, though he felt it important to try to understand.

Bonner ended his book by declaring that if anyone comes up with a microstructural account of the causes of morphogenesis, "the world will acclaim his discovery as a most satisfying explanation and a great advancement in science."[27] But in 1952, it was clear that Bonner did not expect such an account anytime soon.

Compare Bonner's tone with Wilson's ending of his book in 1896, Wilson wrote:

> I can only express my conviction that the magnitude of the problem of development . . . has been underestimated. . . . Yet the splendid achievements of cell-research in the past twenty years stand as the promise of its possibilities for the future, and we need set no limits to its advance. . . . We cannot foretell its future triumphs, nor can we repress the hope that step by step the way may yet be opened to our understanding of inheritance and development.[28]

For Bonner a half-century later, the problem of development remained, and would not be solved through cell research alone or even primarily. Today, another half-century later, we are swimming in data, yet the fundamental biological problems of morphogenesis and development of form remain as challenging and exciting as ever—in new ways. It is important not to lose sight of the cells or of morphogenesis as we embrace Evo-Devo enthusiasm for other levels of analysis. Joining evo (and with it molecular genetics) with devo surely offers the greatest promise for achieving the greatest advances in understanding the problem of development that Wilson sought, and also in giving us the microstructural account of development that Bonner would still like to see.

Notes

1. J. P. Trinkaus, personal discussion at the Marine Biological Laboratory, Woods Hole, Massachusetts. He thought this was a silly question that missed the point that, as he saw it, developmental biology had been considered a success even when it was not as much advertised or heralded as genetics or evolution. There is, he noted, no Ernst Mayr for developmental biology, and he clearly did not think that was a bad thing.

2. Nägeli, 1884, translated and quoted by Harrison, 1937, p. 371.

3. Harrison, 1937, p. 371.

4. Harrison, 1925.

5. Harrison, 1940, p. 77.

6. E. B. Wilson, 1896, pp. 264–265.

7. Dröscher.

8. Huxley, 1853, p. 243. For discussion, see Richmond, 2000.

9. Whitman, 1893.

10. See Lillie, 1911.

11. Child, 1900, p. 265. Also see Child, 1915.

12. Harrison, 1937, p. 372.

13. Haraway, 1976.

14. Planck, 1936, p. 112, cited in Harrison, 1937, p. 374.

15. See Haraway, 1976, chap. 4, for further discussion.

16. Keller, 2002.

17. Woodger, 1929, p. 318.

18. Bonner, 1952.

19. Nicholas, 1952, p. 492.

20. H. v. P. Wilson, 1907.

21. Bonner, 1952, p. 260.

22. Ibid., p. 201.

23. Ibid., p. 268.

24. Ibid., p. 100.

25. Ibid., p. 241.

26. Ibid., p. 268.

27. Ibid., p. 276.

28. E. B. Wilson, 1896, p. 330.

R EFERENCES

Bonner, John Tyler. 1952. *Morphogenesis: An Essay on Development*. Princeton, N.J.: Princeton University Press.

Child, Charles Manning. 1900. "The Significance of the Spiral Type of Cleavage and Its Relation to the Process of Differentiation." *Biological Lectures of the Marine Biological Laboratory*, 1899: 232–265.

Child, Charles Manning. 1915. *Individuality in Organisms*. Chicago: University of Chicago Press.

Dröscher, Ariane. "Edmund B. Wilson's *The Cell* and Cell Theory Before 1896 and 1925." Unpublished Manuscript.

Haraway, Donna. 1976. *Crystals, Fabrics, and Fields: Metaphors of Organicism in Twentieth-Century Developmental Biology*. New Haven, Conn.: Yale University Press.

Harrison, Ross. 1925. "The Return to Embryology." Address as retiring president of American Society of Zoologists. Ross Granville Harrison Papers, Yale University Archives.

Harrison, Ross. 1937. "Embryology and Its Relations." *Science* 85: 369–374.

Harrison, Ross. 1940. "Cellular Differentiation and Internal Environment." Reprint issued as publication no. 14 of the American Association for the Advancement of Science, pp. 77–97.

Huxley, Thomas Henry. 1853. "The Cell Theory." *British and Foreign Medico-Chirurgical Review* 221–243.

Keller, Evelyn Fox. 2002. *Making Sense of Life: Explaining Biological Development of Models, Metaphors, and Machines*. Cambridge, Mass.: Harvard University Press.

Lillie, Frank Rattray. 1911. "Charles Otis Whitman." *Journal of Morphology* 22: xv–lxxvii.

Nägeli, Karl. 1894. *Mechanisch-physiologische Theorie der Abstammungslehre*. Munich: R. Oldenburg.

Nicholas, John Spangler. 1952. Review of Bonner. *Science* 116: 491–492.

Planck, Max. 1936. *The Philosophy of Physics*, trans. H. W. Johnson. New York: Norton.

Richmond, Marsha. 2000. "T. H. Huxley's Criticism of German Cell Theory: An Epigenetic and Physiological Interpretation of Cell Structure." *Journal of the History of Biology* 33: 247–289.

Trinkaus, John P. Personal discussions at the Marine Biological Laboratory, Woods Hole, Mass.

Whitman, Charles Otis. 1893. "The Inadequacy of the Cell-Theory." *Journal of Morphology* 3: 639–658.

Wilson, Edmund Beecher. 1896. *The Cell in Development and Inheritance*. New York: Macmillan.

Wilson, Henry van Peters. 1907. "On Some Phenomena of Coalescence and Regeneration in Sponges." *Journal of Experimental Zoology* 5: 245–258.

Woodger, J. H. 1929. *Biological Principles: A Critical Study*. London: Kegan Paul, Trench and Trubner.

6

A Century of Evo-Devo: The Dialectics of Analysis
and Synthesis in Twentieth-Century Life Science
Garland E. Allen

Background

In 1977, when Stephen Jay Gould published *Ontogeny and Phylogeny*, his
first major book-length excursion into the history of biology, he introduced both biologists and historians of science to the long and tortuous
history of the fascinating but elusive relationship between embryonic
development and organic evolution. Gould's intent, he informed the
reader, was twofold: (1) to elucidate the many varieties of views on recapitulation and its controversial interpretations over the years, and (2) to
bring the issue back into the scientific limelight, as a legitimate area for
biological study.[1] Although the whole topic had lain fallow since the early
decades of the century, all biologists knew it was no mere coincidence
that the embryos of related forms showed numerous parallels in their early
development. While it might be difficult to document, it is likely that
Ontogeny and Phylogeny did, in fact, contribute in an important way to the
resurgence of interest in "evo-devo" since its publication. And a clear
resurgence it has been. What was in 1991 a trifling subfield in the National
Science Foundation's (NSF) Developmental Mechanisms Program, with a
mere $250,000 budget, the "evolution of development" (as it was called
then), mushroomed by 1999 into a full-scale program within the Developmental Mechanisms Cluster, with a budget of $1.6 million.[2] If all other
NSF and National Institutes of Health (NIH) grants that touch on the
relationship between evolution, development, and genetics are included,
"evo-devo" is clearly taking center stage among modern biological
disciplines.

For clarity, I am defining "evo-devo" rather broadly, as a synthetic
paradigm or research program in which developmental biology is combined in some ways with both genetics and evolutionary theory. Some
versions of "evo-devo" in the past, as well as the present, emphasize the
genetic, while others stress the evolutionary, aspect, but the common
element is that in some way all three components are intertwined.

"Evo-devo" was in the late nineteenth century, and is becoming again today, a unifying research program, though different takes on it emphasize different "unifications."

OVERVIEW

The rapid rise of evo-devo in the biological sciences in recent years has raised a number of historical, sociological, and philosophical questions. In this paper I would like to explore one of the main claims about the fate of "evo-devo" studies over the past century: that historically, the late nineteenth-century study of evolution, development, and heredity constituted a single research problem that was split by the rise of Wilhelm Roux's *Entwickungsmechanik* developmental program. Out of that split arose two separate fields (embryology and genetics) that went their separate ways, largely ignoring, throughout the first half of the century, one another as well as the evolutionary questions that impinged on both. While I do not think the claim is wrong, I would like to look at it as a constantly shifting dialectic between a series of opposing theoretical interpretations and the philosophical, sociological, and methodological issues associated with them. These differences, especially when taken together by the 1920s, made the synthesis of embryology and either genetics or evolution seem a distant and a difficult prospect. At a more general level, these debates raised questions about the very nature of biology, and specifically *rigorous* biology of a sort that approximated physics and chemistry of the day.

In the first half of the century there were, of course, attempts to bridge some of the gaps between heredity, development, and evolution—from Hermann Braus to Richard Goldschmidt, Boris Ephrussi, C. H. Waddington, L. C. Dunn, Salome Glücksohn-Waelsch, and Ivan Schmalhausen.[3] However, none of these gave rise to a real and lasting synthetic program, even though they often produced novel and intriguing experiments or suggestive hypotheses along the way. As Lynn Nyhart has shown, Hermann Braus, as "intellectual grandson" of Carl Gegenbaur, attempted a synthetic *experimental morphology* program that showed great promise for resolving old debates about the origin of vertebrate limbs. Ingenious as some of the experiments Braus performed on spiny dogfish at Naples in 1905–1906 were, World War I and later institutional constraints prevented his establishing a permanent following. The physiological genetics of Goldschmidt could have made a greater impact had it not been for his disruptive exile from Germany to the United States in 1936, and for his subsequent unorthodox views on the gene.[4] The closest synthesis of embryology and genetics, which was also tied to a practical experimental program, came in the developmental genetics of L. C. Dunn, Salome Glücksohn-Waelsch (on the T-locus of the mouse), and Walter Landauer (on the creeper fowl); only Schmaulhausen in the Soviet Union (virtually unknown in the West) and Waddington in Britain kept the torch alive for evolution as well.[5]

Factors Contributing to the Decline of Evo-Devo in the Twentieth Century

I will examine five sets of dialectical processes that I believe led to, and perpetuated, the decline of interest in "evo-devo" questions in the early twentieth century. The first was the tension created by reaction to the speculative and hotly contested claims of Ernst Haeckel's biogenetic law. As Gould and others (for example, Churchill, in this volume) have documented, Haeckel's various claims about the telescoping of ontogeny, evolutionary advancement through terminal addition, and the realization that the developmental process itself could be modified by selection made his formulation of recapitulation theory highly controversial. The fact that Haeckel's views were only speculative, with no real way to test his conclusions or subsidiary claims (such as the role of hetereochrony) cast a long shadow over the whole recapitulationist program for generations to come.[6] It led, dialectically, to a strong movement toward experimentation and analytical approaches to biology that set the stage for a new generation of biologists in the early twentieth century.

A second factor derived directly from the critique of Haeckel's recapitulation theory, and focused on the distinction between historical and mechanical, or proximate, causation. Haeckel's recapitulation theory, in which he argued that phylogeny was the *cause* of ontogeny, is an example of historical causation, since the evolutionary history of the species determined, or caused, the specific sequence of stages that made up the embryonic development of each organism today. The alternative, or mechanical, view of causation focused on proximate causes, for example, the specific mechanisms of differentiation in embryogenesis. As we will see, historical causation, so well described by Lynn Nyhart, by 1900 had become an outmoded and suspect form of discourse. The ongoing dialectic between historical and mechanical causation played a major role in the split between embryology and genetics in the twentieth century.

A third factor grew out of the ongoing tension between synthetic and analytic methods of investigating biological systems. Synthetic theories, such as Haeckel's, promoted an integrative agenda that sought to bring together disparate biological fields, such as heredity, development, and evolution. By contrast, analytic theories, such as Mendelian genetics, tended to break down complex processes—for example, reproduction—into their specific components, such as genetic transmission and embryonic differentiation, studying each separately rather than as an integrated, whole system. (In a similar way, the *Naturphilosophie* movement of the late eighteenth and early nineteenth centuries has been described as a response to the mechanical view of the organism put forward by the mechanically minded philosophers of the Enlightenment.) I will argue that this dynamic was at play in the late nineteenth and early twentieth centuries, when the first "evo-devo" program—Haeckel's version of recapitulation theory—led to the call, especially among younger biologists, for a more analytical biology in which specific hypotheses could be tested by controlled experiments. With an envious eye cast toward the physical sciences, increasing numbers of biologists from the 1890s onward argued that for biology to establish itself as a hard science, on the same footing as physics and chemistry, it was essential to break complex organic processes down into their simpler components, in which specific questions could be answered by specific, empirical tests.[7]

The fourth factor is a spin-off of the synthetic-analytic tension: the emerging dispute, from the early 1900s through the 1930s, between mechanistic and holistic or organicist views of nature. Particularly prominent in Germany, where a holistic biology was championed by such figures as Baron Jakob von Uexküll, Ludwig von Bertalanffy, and Hans Driesch, it was argued that an organism functions as a whole, not as a mosaic of separate parts, and that this "whole" possesses emergent properties that are greater than the sum of the parts.[8] Partly because such claims (especially in the hands of writers such as von Uexküll and Driesch) drifted into various forms of mysticism and vitalism, the analytical view emerged even more clearly as the form in which biological investigations should proceed. Holism became associated with metaphysics, a throwback to an older style of German theorizing and speculation, and the antithesis of the way younger biologists thought their science should be practiced.

A final set of dialectical processes, particularly important in retaining and reinforcing the divergence between studies in heredity and development, arose from the increasing success of experimental methods in both embryology and genetics in the early twentieth century, and the corresponding lack of alternative methods for rigorously investigating "evo-devo" hypotheses. Thus, the extirpation and transplantation methods of embryology, and the correlation of breeding ratios with the cytological study of chromosome structure in classical (Mendelian) genetics, seemed distinctly different techniques and approaches that could not be interchanged between fields. At the cellular level, embryologists continued to claim that the real events of development occurred in the cytoplasm, while geneticists argued that the main basis for both transmission and development lay in the nucleus (i.e., with the genes). I refer to these factors collectively as *methodological constraints*, which would include not only experimental techniques but also intellectual commitments to methods of investigating a particular phenomenon: for example, the views of embryos as "harmonious equipotential systems" (Driesch) as opposed to mosaics of separate and qualitatively distinct cells (Roux).

A further factor that played a role in the divergence between embryology and genetics is the turf battle which emerged by the 1930s, in which the methods of the geneticists seemed not only irrelevant for embryologists but also imperialistic in their claim that problems of development could ultimately be resolved by genetic analysis. Geneticists portrayed the gene as the pivotal element in development: it was the turning on and/or off of genes that accounted for cell and tissue differentiation. Thus, development was to be "reduced" to genetics—that is, to the differential action of genes over time.

Before turning to a more detailed examination of each of these sets of interactions, it will be useful to examine what it means to cast the discussion in "dialectical " terms and how such a discussion is particularly useful in understanding the historical split between, and ultimate reintegration of, genetics and development.

DIALECTICS AND DIALECTICAL MATERIALISM

Dialectical thinking (the juxtaposition of opposing ideas or processes) has a long history, but its modern incarnation derives from the German philosopher G. W. F. Hegel. It was given a materialist formulation in the nineteenth century by Karl Marx and Friedrich Engels as the basis for their historical materialism, and sharpened into a systematic philosophy known as *dialectical materialism* in the twentieth century by Georgi Plekhanov (1891) and later by Vladimir Lenin and other Marxists. The materialist part of dialectical thinking has had a particularly prominent place in the sciences since the nineteenth century, where explanations have been increasingly grounded on the interaction of material components of systems: atoms or molecules, cells, tissues and organs/organ systems, or individual organisms, populations, and the abiotic environment.[9]

One of the most important aspects of the Hegelian dialectic is the interaction of opposing tendencies (in dialectical terms, thesis and antithesis) to produce a new state of development, or "synthesis" (not Hegel's term, but incorporated into his system by post-Hegelians such as Eduard Gens and Ludwig Feuerbach). The new state differs sometimes only quantitatively (that is, by small increments) and sometimes qualitatively (by large increments) from the state that existed before the synthesis took place. A prominent example is Darwin's mechanism of evolution by natural selection, in which two opposing genetic processes, faithful replication (heredity) and unfaithful replication (variation) are both necessary for evolution to occur. Their constant interaction in the face of environmental selection ensures that the species never stays in a static, unchangeable state. Thus, the opposition of heredity and variation (in a selective environment) produces continual syntheses (new variant populations) that alter the species makeup to various degrees in each successive generation. As long as the syntheses occur in small steps, the changes will be considered simply quantitative, but they will still represent a synthesis with respect to the previous state.

As a number of such small, quantitative changes accumulate, however, a new state may be reached that differs qualitatively from the old. The point at which a quantitative change gives rise to a qualitative change is known as the threshold (or threshold effect) and, once reached, leads to emergent properties—that is, properties of the new, postthreshold state which are not observed in the earlier state. In the heredity-variation dialectic, for example, a qualitative change would be represented by passing over the species threshold. This could occur in a single phyletic line, where the new species, if it could be brought back into contact with its original ancestor, would no longer be able to interbreed (or whatever criteria are used in specific cases to distinguish between species), or it could occur between two divergent lines that each, in its own right, becomes species distinct from the other and from its ancestor.

An important aspect of dialectical thinking is that it emphasizes the dynamic, ever-changing nature of a system i.e., its evolution. Dialectical thinking thus focuses the investigator's attention on the regular or repeatable factors that lead to the changes any system undergoes, and insists that the system's history is important in understanding how these factors produce the observed change. Except for systems that are truly programmed (such as embryonic development), dialectically evolving systems are *not* teleogical and have no predetermined goal or end point.

With this basic mode of analysis in mind, we turn now to a more detailed examination of the various dialectical factors that contributed to the split between embryology, genetics, and evolution between 1900 and 1950, and how other dialectical factors may be operating at present to bring the fields back together through "evo-devo."

FACTORS LEADING TO THE DISSOLUTION OF HAECKEL'S NINETEENTH-CENTURY EVO-DEVO PROGRAM

Almost immediately after the publication of Haeckel's recapitulation theory in his *Generelle Morphologie* (1866), objections were raised to the theory in general, and to the claims of the biogenetic law in particular. These objections have been summarized by Gould, and need not be repeated in detail here.[10] The main problems centered on assumptions behind the biogenetic law (the claims of terminal addition, condensation of stages, and the acknowledged process of embryonic evolution), Haeckel's claim that phylogeny is the *mechanical cause* of ontogeny (rather than the other way around), and that the whole scheme, including the hereditary particles that Haeckel called "plastidules," which supposedly explained variation, was purely speculative. There was simply no way to test any of the components of the theory, and little empirical evidence in its behalf.

Although Haeckel met these objections with various modifications or further elaborations of his original theory, it was becoming clear by the early 1900s that the biogenetic law could not be applied readily to answering questions about the relationship of development to evolution, or to unraveling the phylogeny of any specific group. (How could one tell, for example, whether a particular ontogenetic stage was an interpolated embryonic/larval adaptation or an accelerated/retarded ancestral stage?) Because of its breadth and the arbitrary nature of its many assumptions and postulated causalities, the biogenetic law was ultimately rendered incapable of providing a sound "theory to work by." In the meantime it did, however, provide a generation or two of investigators with the research program of *Morphologie*, which dominated much of late nineteenth- and early twentieth-century biology.[11]

Morphologists combined phylogenetic interests with excruciatingly detailed anatomical and microscopical studies of embryonic development—particularly the early cell divisions following cleavage (for

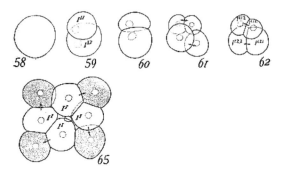

Figure 6.1
Example of cell lineage studies in the mollusk (clam) *Patella*, showing the first three cleavages (up to the eight-cell stage). 58 is the fertilized egg; 59, the first two blastomeres resulting from the first cleavage; 60, the same two blastomeres seen from the side; 61, the four-cell stage resulting from the second cleavage, seen from below; 62 shows the same four cells from above; 65 is the eight-cell stage resulting from the third cell division. The numbering system indicates the cell division products and their designation in the developing embryo. For example, in 62, the two larger cells on the bottom are daughter products of the first division, and are numbered I 1.2.1 and I 1.2.2. In this way the cells can be followed in exact order of their formation from divisions of each given blastomere. (After Wilson, 1904, p. 220.)

example, in cell lineage studies shown in figure 6.1)—in an attempt to resolve such controversial phylogenetic questions as how various taxa (for example, mollusks and annelids, as shown in figure 6.2) were related to each other, or determining how tetrapod limbs could have arisen from the pectoral fins of fish.[12] Thomas Hunt Morgan, not known for his love of morphological work (which he later took pride in contrasting negatively to physiology and biochemistry) earned his spurs as a biologist with a morphologically based Ph.D. dissertation titled "The Embryology and Phylogeny of the Pycnogonids [Sea Spiders]."[13] The fundamental question, which he posed in the opening paragraph of this work, lays out a classical morphological problem: "In the year 1767 Karl Linné . . . described . . . a Pycnogonid, and here for the first time is the question raised whether the group is to be ranged under the Arachnids or the Crustacea."[14] After detailed analyses of many aspects of early development of a variety of organ systems in the sea spiders, Morgan concluded with a discussion of the development of the eyes in arachnids and pycnogonids. Germ layer developments, folding of tissue layers, and optic cup formation all suggested that the pycnogonids are more closely allied to the arachnids than to the crustaceans. The reasoning was sound, but inferential; and there was no way to test it.

In addition to the speculative nature of this sort of analysis, morphology was also primarily a descriptive science. The morphologist observed embryos growing at various stages, and intervened only to the extent of sectioning and staining specimens for more detailed observation. For the most part, morphologists observed *normal* development step by step. They did not specifically alter embryonic development to determine what changes, if any, resulted. While most morphologists were excellent observers (Morgan's detailed descriptions of the pycnogonids are quite impressive), observation alone limited the degree to which they could test their various phylogenetic hypotheses. The *comparative method* of both embryos and adults was the way to buttress particular claims about phylogenetic relationships (the importance of comparative anatomy in the overall enterprise is exemplified by Haeckel's dedication of *Generelle Morphologie* in 1866 to Carl Gegenbaur, the "Grundzüge der allgemeinen Anatomie"), but because of the great modifications to which embryonic development had been subjected in the course of evolution, comparison could take the morphologist only so far. Valuable as it was, the comparative method had its limitations in terms of drawing rigorous conclusions.

In this light it is not surprising that even more serious objections to Haeckel's morphology focused on his basic method of doing science. Forgetting about his well-known transgressions in using the same woodcuts for eggs and early developmental stages of different species, the basic Haeckelian method was one of proposing likely, possible, or seemingly logical phylogenies, or mechanisms for developmental processes such as differentiation, with no way to distinguish between one hypothesis and another. It was the largely speculative nature of the conclusions resulting from Haeckel's method that seemed so problematic to many, especially younger generation, biologists. While any given scenario *might be right*, there was no way to distinguish between scenarios, and thus no way to reach any certain conclusions. As Jane Maienschein, Ron Rainger, and Keith Benson initially, and others—most recently Lynn Nyhart—have clearly pointed out, the true juxtaposition or dialectic here is not experiment versus observation, since after all any good experiment requires both prior observations and careful, detailed observations of the results, but speculative and synthetic versus experimental and analytical. The real complaint many younger investigators had against the morphological program was that it was an example—indeed, one of the most prominent—of reliance *only* on observational methods, with the consequent generation of only speculative, and largely untestable, hypotheses.

As a consequence of the dialectic between speculative/synthetic and experimental/analytic methodology, experimentation became an important

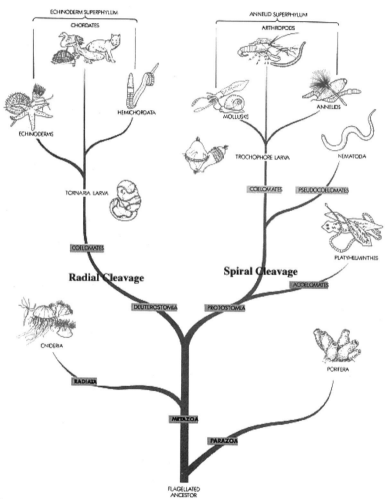

a

(a)

(b)

b

rallying point among younger workers as the only way to obtain rigorous answers. As Wilhelm Roux put it, "'*Certainty*' in causal deduction *can only come from experiment* . . ."[15] The experimenter presented the organism with a specific question and, by virtue of direct intervention, extracted a specific answer. It was active, it involved taking control. Experiment was the methodological arbiter between competing hypotheses. The inability of Haeckel's method to accomplish this sort of rigorous distinction between alternative hypotheses appeared to younger biologists to make the whole morphological enterprise futile.

There was yet another, albeit related, issue at stake. Younger biologists, following Roux, Driesch, and Chabry, among others, were interested in "how" questions—that is, in keeping with their emulation of the physical sciences, in the proximate, causal mechanism behind phenomena (for example, how differentiation of cells takes place during embryogenesis). Haeckel's passion was never so much focused on mechanisms as it was on the tracing of phylogenies.[16] Thus the Haeckelian paradigm ignored that most intriguing of all proximate questions for the younger generation of biologists: How do cells in the embryo become different? This is, of course, exactly what experimentation on separated blastomeres or dorsal lips was intended to find out. And to give Haeckel his due, perverse as it might seem, it is no accident that several of those who pioneered in the new experimental embryology—most notably Roux, Driesch and Curt Herbst—were Haeckel's students.

As Lynn Nyhart has suggested, however, Haeckel and morphologists in general were not uninterested in causation; their causation was of another type, what they saw as historical causation.[17] Taking their view from Karl Friedrich Burdach in 1817—"All knowledge is insight into the causal connection of phenomena"[18]—morphologists asked "What for?" and "Where from?" Phylogenetic explanations were a central form of

Figure 6.2
Spiral and radial cleavage patterns (a) and the relationship between the two largest subgroups within the animal kingdom, the Echinoderm and the Annelid superphyla (b). In radial cleavage the four blastomeres resulting from the third cleavage rest directly on top of the four cells below (the results of the first two cleavages); in spiral cleavage, the top four blastomeres sit in the spaces between the lower four. Features of embryonic development that were useful to morphologists in assigning phylogenetic relationships included not only cleavage patterns, but also patterns of mesoderm formation and similarity of larval types found at various stages in the life cycles of related forms. (Modified from Jeffrey J. W. Baker and Garland E. Allen, *The Study of Biology*, 4th ed. [Reading, Mass.: Addison-Wesley, 4th ed., 1982], pp. 878, 925.)

historical causation that seemed eminently satisfying to morphological thinkers. Even Roux recognized that historical causation was an essential part of understanding the development of any organism, but by itself it was not enough.[19] Haeckel's biogenetic law was the epitome of historical causation in that phylogeny was seen as the *cause* of ontogeny: it was the historical processes of acceleration, retardation, and elimination of adult ancestral stages that produced the particular sequence of developmental stages through which the embryos of each species passed. History did count; for the embryonic process it was determinative.

In summary, by the 1890s, the overall effect of the Haeckelian "evo-devo" program was to undermine for a whole new generation of biologists interest in, and a sense of the explanatory value of, descriptive, large-scale, and historical explanations. If biology were to attain the rigor and status of the physical sciences, then it had to abandon such enterprises and turn to questions that could be answered definitively and causally. Sometimes those answers could come from making further observations; mostly, though, they would come from intervention into the system by experiment. The interacting series of dialectics here—description versus experimentation, historical versus proximate causation, speculation versus rigorous testing of alternative theories, were all part of the historical change leading from morphology to experimental embryology in the 1880s and 1890s. In addition there was the related dialectic between synthetic and analytic theories.

ANALYTIC VERSUS SYNTHETIC

A variant of the dialectic between mechanical and historical causation was that between analytical and synthetic explanation. Synthetic theories bring together under one roof (in a formal consilience) evidence and arguments from a variety of disciplines or approaches, while analytical theories were based on the splitting apart of complex processes or theoretical constructs into their component parts, each of which could be investigated more rigorously in isolation. Following Darwin, the consummate synthesizer, Haeckel had brought together evolution, development, and heredity under one grand explanatory scheme. Although not interested in mechanisms in any detailed fashion, Haeckel's recapitulation theory offered an explanation of how new variations originated (inheritance of acquired characters; that is, it explained the origin of adaptations) by postulating that environmental factors affected the rate of vibration of the hypothetical hereditary particles, the plastidules; it also could account for how new and successful variations were introduced into the evolutionary process as terminal additions. These were the mechanism, if you will, by which environmental influences were incorporated into the organism's ontogeny, and could then be acted upon by selection. It was all very neat and comprehensive.

But Haeckel's "evo-devo" synthesis could not be tested. It made logical sense, but younger biologists argued that science was not about what could possibly be true, but what was actually true. If the pycnogonid eye could be said to resemble the arachnid eye only by assuming it telescoped a number of stages of embryonic folding into a single step, it could just as easily be argued that the pycnogonid eye was a wholly new invention in evolution, and thus could not be used to resolve the arachnid-crustacean debate. Nyhart has shown a similarly irreconcilable debate during the same period on the origin of the tetrapod limbs. How could the morphologist distinguish between limbs evolved from gill arches or from lateral folds? Synthetic theories, by their very definition, were not amenable to any single, rigorous test. As a result, many younger biologists, while not uninterested in evolution per se, increasingly came to feel that narrower questions, which could be answered analytically, were the important ones to pursue. For many, the questions of whether pycnogonids were more closely allied to arachnids or crustaceans, or whether vertebrate limbs originated from gill arches or lateral folds, lost a good deal of their interest.

Far more interesting problems lay hidden in ontogeny than the reconstruction of phylogenies. The *ultimate* questions of historical causation gave way to the proximate questions of mechanical causation, that is, how differentiation of cells came about. Precise mechanisms of "how does it work?" stimulated more interest than vague, all-encompassing speculations about how, historically, the various taxonomic groups came to be. Mechanisms of development, from Roux's and Driesch's testing of the mosaic theory, to Jacques Loeb's studies of artificial parthenogenesis, to Spemann's early constriction experiments, all suggested there was a lot to be learned about the proximate, mechanical causes involved in various developmental events. Analytical thinking provided a more firm and certain direction in which biology could move.

THE SPECIFIC ROLE OF "EXPERIMENTATION"

Along with analytical thinking came experimentation. Experiment meant to younger biologists the opportunity to formulate precise questions and then intervene and *force* the organism to yield a precise answer. Teratologists had previously used nature's own experiments (developmental processes that had gone awry and produced deviations in development, known at the time as "monsters") to try to answer proximate questions about mechanisms of development. The new generation of experimentalists did not wait for nature to make a mistake and thus provide them with a question and a potential answer—they formulated the questions themselves and actively extracted the answers from organisms. Perhaps more important, the value of experimentation lay in the possibility of accurate prediction and thus of *control*, that is, the manipulation of nature for human purposes. If one can demonstrate a causal connection between A and B— that is, the prediction that A will always lead to B—then it is possible to use that prediction to alter, or "control life," in Phil Pauly's terms, to meet various human needs.[20]

It may be too bold and speculative to bear out, but let me suggest that there may be a much broader economic and social context which gave an increasing interest in, and even urgency to, the experimental ethos. It derives from Pauly's concept of the introduction of the engineering ideal into biology. I would like to suggest that it may be no accident that much of the background for interest in embryology—especially experimental— from the 1870s onward in western Europe and North America, came from concerns about declining marine food resources. By the end of the nineteenth century there was a noticeable decline in the populations of many marine species, and developmental biology clearly had much to contribute to understanding the causes underlying this occurrence.

Descriptive studies were of course important—understanding life histories, ecological relationships, reproductive behavior, and the like. But also, there was the problem of how to manipulate and control these processes in nature to increase yields.[21] Problems related to spawning (temperature, currents, light), fertilization (pH, light-dark cycles, oxygen availability), early development (factors affecting cleavage), and food resources for larval and adult growth were all important variables for understanding how to manage commercially valuable marine resources. Experimentation helped elucidate proximal causes, which led to prediction and ultimately to control, a necessary step in managing marine life. It is likely no mere coincidence that marine laboratories were opened throughout the world between 1875 and 1895 in the very countries that were experiencing rapid industrialization and urbanization (and thus increased demands for food). Many of the early workers in developmental biology in the United States, for example, began their work with marine organisms at the U.S. Bureau of Fisheries Laboratory, established at Woods Hole, Massachusetts, in 1876 by Spencer Fullerton Baird, and later expanded into the Marine Biological Laboratory (MBL) in 1888. (The MBL was modeled on a major predecessor, the Stazione Zoologica in Naples, founded by Anton Dohrn in 1872.)[22]

In those seaside laboratories, descriptive work went on side by side with experimental work. Once biologists became familiar with the advantages of working with marine organisms—their visual transparency, easy manipulation in the laboratory, fecundity, and ready availability—these groups became model organisms for the study of all manner of general biological problems. Historically, we can recognize those very real attributes of marine organisms which attracted the interest and attention of so many biologists, while also understanding that the impetus for the expansion of research, including the institutional form it took in marine stations, may owe as much to economic needs as to the various intellectual and professional issues that were becoming prominent within the biological community at the time. The important point is that intellectual interests or theoretical considerations alone are seldom enough to explain how a particular set of claims and practices in science emerges into a prominent and widely based research program. In terms of the movement toward a more analytical and mechanistic biology at the end of the nineteenth and beginning of the twentieth centuries, I want to suggest that both sets of factors—intellectual/scientific and economic/social—were intimately and causally interconnected in expanding that ideology into real scientific practice.

Thus far, I have tried to suggest that rejection of the Haeckelian paradigm began the process of a social and intellectual reorientation of the field of embryology by turning it from a primarily descriptive, historical, and speculative field into an experimental, causal, mechanical science. The continued interaction of new sets of dialectical processes that perpetuated the division, especially from the embryological side, is the topic of the next section.

THE DIALECTICS OF EXPERIMENTATION: EMBRYOLOGY AND GENETICS, 1880–1940

The Early Development of Experimental Embryology

The introduction of experimentation into embryology—first with Chabry, His, Roux, Driesch, Herbst, Spemann, and others—has been well and thoroughly elucidated by a number of scholars.[23] Separation of blastomeres, egg constrictions, and ultimately transplantations provided means of asking precise questions of the embryo and getting precise answers. The experimental method also provided insights into the possible *mechanisms*—especially of differentiation—that might be at work in embryonic development. By "mechanism" I mean here a specific set of entities (ions, cells, membranes) and processes (cell-to-cell contact, cell movement, temperature effects), material conditions that can serve to trigger cells which are initially alike to eventually follow different lines of development.[24] All of these factors became parts of the different levels at which mechanisms of differentiation and embryonic organization could be investigated.

The rise of experimental embryology occurred in two phases: (1) the analytical, reductionist program of Roux's *Entwicklungsmechanik*, and (2) the holistic yet still causal-analytical work of the Spemann school at Freiburg. Although Reinhard Mocek, an astute scholar in this field, has argued differently,[25] I view Roux's program as primarily mechanistic and reductionist in character. It focused on the problem of how differentiation occurs by testing hypotheses such as the mosaic theory (the view that cells in the developing embryo become increasingly restricted, or determined, by qualitative parceling-out of hereditary units during cleavage). If Roux's theory were correct, so the argument went, with each successive cleavage, embryonic cells should become more and more restricted in the developmental paths they can follow (i.e., they become less and less totipotent, in today's terminology). Roux's experiments in 1888 supported (confirmed) the mosaic theory, while Driesch's sea urchin experiments three years later contradicted it.[26] Working with sea urchin embryos, Driesch separated the two blastomeres by vigorous shaking (rather than by killing one, as Roux had done). The result was two complete (though slightly smaller) embryos developing from each blastomere. For Driesch, this meant that the embryo was a self-regulating entity, what he would later refer to as a "harmonious equipotential system." Although for Driesch the ultimate result was despair of ever understanding the mechanism of development and an ultimate retreat to vitalism, his legacy to later workers, such as Spemann, was to emphasize the interactive nature of the parts of an embryo, and to deemphasize the purely mechanical, mosaic nature of organic processes.

This emerging dialectic—between a highly mechanistic and a more holistic treatment of development (and of biological processes in general)—was to become a major obstacle to integration (synthesis, if you will) of the new experimental embryology with Mendelian genetics in the

1920s and 1930s. The major claim I want to make here is that regardless of the actual results, the mosaic theory was not confined to the realm of mere speculation: it could be tested experimentally. This same experimental approach, albeit at the organismic rather than the cellular level, became the centerpiece of Spemann's causal-analytical program at Freiburg. This program became the cutting edge of experimental embryology contemporaneous with the expansion of Mendelian genetics.

The Spemann School, 1901–1935

Spemann began his career in experimental embryology by designing a new version of Roux's and Driesch's experiments. Borrowing a technique and model organism from Oskar Hertwig, Spemann took a fertilized salamander (*Triton*) egg (rather than the frog used by Roux) that was undergoing its first cleavage, and with a very fine baby hair, separated the first two blastomeres by tying a loop through the cleavage furrow (figure 6.3a). When the contents of the two cells remained in contact across the constriction bridge, Spemann was able to obtain partially double embryos

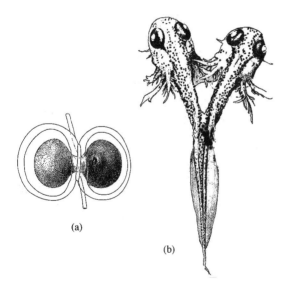

(a)

(b)

Figure 6.3
(a) Hans Spemann's "hair-loop" experiment, in which he found that constriction of the first two blastomeres with a fine baby's hair provided the basis for analyzing the causal factors in differentiation. (b) When the hair loop was tied in the plane of cell division, but allowed some cytoplasmic exchange between the two cells, a partial, or two-headed, larva resulted. (Modified from Spemann, 1903, pp. 577–579.)

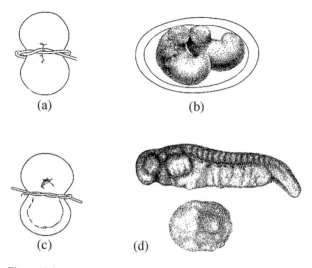

Figure 6.4

Spemann's constriction experiments led him to identify the region of the two-celled stage that would become the future dorsal lip of the blastopore (a). In experiments where the constriction bisected the future dorsal lip region, thus allowing each daughter cell to receive a portion of the dorsal lip tissue, two complete embryos developed (b). When the constriction was on one side of the hair loop (and thus restricted to one of the two developing blastomeres, as in (c), the results were one complete embryo and a mass of unorganized "belly tissue" (d). (From Spemann, 1924, pp. 69 and 1903, pp. 577–578.)

(figure 6.3b). By varying the amount, timing, or position of constriction—and thus allowing for various degrees of contact between the contents of the two cells—Spemann found he could produce different degrees of double-headed larvae and other "monstrosities." Like Driesch, Spemann concluded that embryonic cells had the ability to self-regulate or adjust to changed circumstances. The purely mosaic theory took a step backward. More important, the experimental procedure itself opened the door to additional questions, including what turned out to be Spemann's most significant discovery: when the hair loop constriction was made so that it crossed the median plane of the embryo, and bisected the blastopore, he obtained two complete embryos (figure 6.4a and b). If, however, he tied the loop so that only one cell contained the future blastopore region, that cell developed into a complete embryo, while the other formed an undifferentiated "belly mass" (*Bauchstück*) that eventually died (figure 6.4c and d). Spemann's constriction experiments thus had two profound results: (1) they showed that the cleavage furrow established the dorsoventral axis of the future body organization, and that the future blastopore region of the egg was essential for differentiation; (2) they emphasized the holistic nature of development: it was a process not limited to changes in individual cells, but a property of the entire organism. Thus, for Spemann, what was important about the constriction experiments was that they allowed him to analyze the formation of the entire dorsoventral axis of the organism.

This methodology and its attendant holistic philosophy were to remain the focus of Spemann's work throughout his career and to form the basis for much (if not all) of his students' work as well. The problem was always a dialectical one: to answer the questions of how separate parts (eyes, limb buds, nerve cells) differentiated in the context of the organism as a whole. Spemann had no abiding interest in the biochemical or cellular-level mechanism by which constriction produced its various effects. He wrote in 1901: "My experiments do not give any information concerning the kind of 'differentiation substance' which is lacking in the ventral blastomere. One can think of an unorganized substance which is necessary either for the release of the formation of [axial] organs; or else it could be organized embryonic material which has the capacity to differentiate into the respective organs and perhaps to incite other cells to differentiate."[27] Within the next several years Spemann was to refine his experimental techniques and coin the term "induction" to describe the effects of one tissue on another during differentiation.

Between 1901 and 1903, Spemann studied the differentiation of the lens of the eye (in salamanders) from the ectoderm overlying the developing optic cup as it grew out from the brain. When the optic cup makes contact with overlying ectoderm, it triggers differentiation of that tissue into a lens (figure 6.5). Lacking contact with the optic cup, the ectodermal tissue does not differentiate; conversely, if optic cup tissue is transplanted to some other part of the body, such as the flank, a lens develops from the overlying ectoderm in the new locality. Spemann coined the term "induction" to describe the process by which the optic cup triggers the ectoderm to differentiate. Inductions operated at various levels throughout embryonic development. The induction of the lens by the optic cup was a relatively specific example; other inductions were of a more general nature, leading to the formation of organ systems or, in the case of the dorsal lip of the blastopore, of the entire anterior-posterior axis of the organism (figure 6.6). This primary region of influence Spemann called the *organizer*. This hierarchy of inductions provided a basis for the analysis of development, for it was amenable to experimental extirpation and transplantation of developing tissues between one part of the embryo

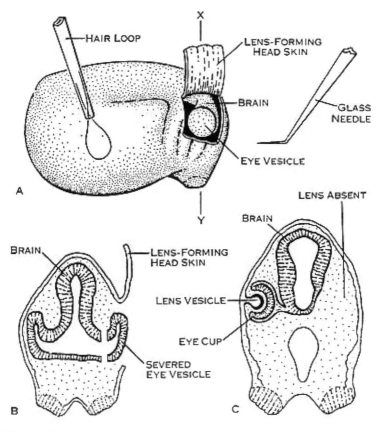

Figure 6.5

Spemann's induction experiments with the optic cup, 1901–1903. (a) A flap of ecto-derm at the anterior end of the developing tadpole is cut and lifted to expose the optic cup (eye vesicle), a structure developing as an outgrowth of the embryonic brain stem. (b) The optic cup is removed with a glass needle and hair loop, and the ecto-dermal flap is replaced. (c) Results of the experiment show that where optic cup tissue has been able to make contact with the overlying ectoderm, the latter is induced to form a lens (left); where the optic cup has been removed, however, no lens differen-tiation occurs (right). It was this work that originally led Spemann to the concept of induction and the "organizer." (From Hamburger, 1988, p. 19.)

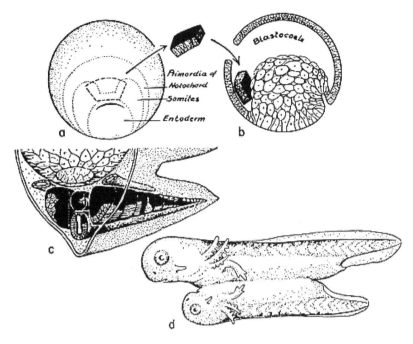

Figure 6.6
The organizer experiment carried out by Spemann's graduate student Hilde Proschelt Mangold and published in 1924. (a and b) A section of the upper lip of the blastopore of a salamander (*Triturus cristatus*) early gastrula is removed and transplanted into the blastocoel of another species of salamander (*Triturus taeniatus*). (c) The upper lip tissue "organizes" the development of the axial system of a whole second embryo, which develops into a double-tadpole stage with the two organisms joined at the ventral surface (d). Because the tissues of *T. cristatus* are naturally pigmented while those of *T. taeniatus* are not, Spemann and Mangold were able to observe that the secondary embryo consisted of cells from both the transplanted dorsal lip tissue and others recruited from the host tissue. (From Holtfreter and Hamburger, 1955; reprinted in Hamburger, 1988, p. 51.)

and another, or between embryos of different ages and species. The analytical method allowed the embryologist to unravel the chain of causal processes that occurred during development.

In all these experiments, however, Spemann remained focused on causal analysis at the organismic level. It was the *pattern* in development, from the earliest cleavages to the layout of the adult axial organs, that informed his view. In describing Spemann's basic approach, his student Viktor Hamburger wrote: "The individual organism is the most important organizational unit in the biological world. The embryologist is aware of the continuous identity of the individual, from the fertilized egg to the adult."[28] Hamburger, one of Spemann's most successful and productive students, and one of my longtime colleagues, told me that he imbibed his mentor's concern for the "wholeness" of biological processes, enunciated toward the end of his life in what he termed his *pact with the embryo*: "I promised the embryo that if it revealed to me some of its secrets, I would never grind it up in a Waring blender. I think we have both kept our promise." It was this same concern for "wholeness" that motivated Spemannn's work, and in some respects came to characterize the tradition that he created in experimental embryology.

When another of Spemann's students, Johannes Holtfreter, showed in a series of ingenious experiments that not only would heat-killed or alcohol-dried dorsal lip tissue (as well as a variety of foreign substances) still induce the formation of a secondary embryo, Spemann found the results surprising, perplexing, and alien to his own way of thinking. He recognized the validity of Holtfreter's empirical results, but they did not fit his paradigm of organic wholeness. Induction of the axial system reflected an organizational process that involved the embryo as a whole; it was not simply reducible to a single chemical trigger.

When others, such as Gavin de Beer and Joseph Needham, attempted to identify the molecular nature of the "organizer substance" (as a fatty acid or peptide), Spemann took little notice. As Hamburger has pointed out, it was not that Spemann believed that embryonic events had no chemical or molecular basis; rather, his own view of the process was that it was a problem of *organization*. Individual biochemical events were surely involved, but finding those cues, or triggers, would not answer the larger question: How does the anterior-posterior axis of the entire organism get laid out in such a precise way? This was the *biological* problem that interested Spemann and that he set out to solve. Finding a single molecular cause (or even multiple molecular causes) was not going to provide the answer to this larger question. Wholes and parts indeed formed a dialectic, just as the analytical and synthetic methods did. The problem was ultimately to bring them together in a true Hegelian synthesis. This prospect did not seem possible experimentally either to Spemann or to most biologists in 1930.

Understanding Spemann's organismic view is important in the present context for two reasons. The first is that Spemann was often accused by contemporaries of harboring a vitalistic view akin to Driesch's, which attributed the organizational properties of the embryo to a nonmaterial, nonanalyzable force. The second is that Spemann's view of how to pursue analytical, experimental work at the *organismic* level was quite at odds with the mechanical and reductionistic approach of Mendelian genetics at the time. Indeed, the two formed the opposite poles (in terms of methodology) , or contradictory approaches, to understanding the basic evo-devo process so well envisioned by Haeckel.

With regard to the first point—that Spemann was a vitalist— Viktor Hamburger has argued effectively that Spemann's interest in "vital processes" is not to be equated with the metaphysical vitalism of someone like Driesch.[29] Spemann did use the term "vital" in a number of instances, even in the context of "consciousness," but a careful reading suggests he was talking not about "vital*ism*," but about pluripotency of all cells. In the translation Hamburger made of portions of Spemann's *Autobiography*, this point becomes more clear when Spemann discusses his early encounter with August Pauly, a lecturer at the University of Munich in 1893:

> It [conversation with Pauly] strengthened my natural conviction that all parts of an organism possess an element of consciousness—not just the organ with which we are familiar, the brain. Today [1930s] I am more convinced than ever before of the fundamental affinity of all living processes: my own experimental work has shown that the same cell assembly which seemed to be destined to form skin can become brain, if it is transplanted in an early developmental stage (early gastrula) to the region of the later brain; that we stand and walk with parts of the body by which we might have been able to think, if they were developed in a different part of the whole.[30]

At first glance, it is understandable why many readers concluded Spemann had a strong vitalistic streak. Attributing consciousness to individual cells is certainly not the usual mechanistic point of view. It becomes clear in his overall context, however, that what Spemann is referring to is the *potentiality* of all cells to develop into associations that are capable of consciousness. An embryonic cell was itself not conscious, but it had the capability to produce consciousness under the proper circumstances, and therefore must possess the components out of which consciousness could emerge. But that emergence was not automatic, that is, it was not an unfolding or triggering of a specific locked-up potency. It was a result of context, or associations of cells in certain parts of the body, of complex organization for which there was no mechanical counterpart. For Spemann this provided an insight into the unity of life—again, not a metaphysical vision per se, but a recognition that pluripotency was a characteristic of all cells, and therefore a unifying feature of life itself.

Spemann's interests were in causal analysis at a different level of organization than that of the molecule or even the cell. It is clear that this orientation was at odds with the mechanistic approach characteristic of Mendelian genetics at the time. The causes of induction were material; the use of the term "vital" did not imply nonmaterial or nonchemical. "Vital" did mean a set of phenomena that somehow "emerged" from chemical and physical events by the nonlinear interaction of parts. The organization of the vertebrate axis was a process peculiar to vertebrate embryos, and the best way to unravel its complex causal pathways was to study in vivo vertebrate embryos, not in vitro chemical inducers. I see nothing in Spemann's writings to suggest he was a nonmaterialist. However, he was clearly not a *mechanistic materialist*, and in that sense was philosophically on a different wavelength than the Mendelian geneticists, especially those in the United States. It is to that point that I would now like to turn in order to understand more clearly why genetics and embryology in the 1930s had so little to say to one another, especially from the point of view of embryologists in general, and those of the Spemann school in particular.

The Development of Mendelian Genetics

Mendelism's Causal-Analytic Thought   Like embryology, genetics achieved its initial impetus because it was so preeminently amenable to experimental treatment and causal-analytical thinking. This approach was seen as clearly inherent in Mendel's own approach, which ignored both evolutionary and developmental considerations. Mendel's work was interpreted after 1900 as falling within the same mechanistic materialist framework as other theories of heredity in the late nineteenth century, such as those of August Weismann, Carl von Nägeli, Hugo de Vries, and others (with the difference that it was based on a far more extensive set of experimental results).[31] These theories had some features in common: they were particulate (atomistic), meaning that particles controlling specific traits were passed as material entities from parent to offspring; consequently, these theories viewed the adult organism as a mosaic of independently inherited traits. The view that the organism is the sum of its separate parts, and no more, was a clear example of mechanistic materialism, leading to a highly reductionist research program.

What distinguished Mendel's work from the others' was, among other things, its focus on transmission, to the exclusion of evolutionary or developmental concerns; its strong empirical (experimental) basis; and its quantitative reasoning. These characteristics were enough to recommend it to younger biologists frustrated at the inability to test existing synthetic theories that tried to combine all three concerns in one overarching explanation.[32] However, for many of those younger biologists brought up on Roux's program for causal analysis of development, Mendel's work provided an equally (and, for some, more) compelling approach to the other side of the picture, hereditary transmission. Mendel's work had all the desirable characteristics of the much-envied physical sciences at the time: it was mechanistic, experimental, mathematical, and predictive.

The Corpuscular (Atomistic) Gene Particularly prevalent in the early decades of twentieth-century genetics was the emphasis on corpuscularity of the hereditary material. The literature of genetics between 1901 and 1930 abounds with references to genes as the biologist's atoms, and genetic methodology as analogous to that of the chemist in the combination, dissociation, and recombination of atoms in strict mathematical proportions. Bateson was perhaps the first to see the comparison, referring to it as early as 1901:

> In so far as Mendel's law applies, the conclusion is forced upon us that a living organism is a complex of characters of which some, at least, are dissociable and are capable of being replaced by others. We thus reach the conception of *unit characters*, which may be rearranged in the formation of reproductive cells. It is hardly too much to say that the experiments which led to this advance in knowledge are worthy to rank with those that laid the foundation of the Atomic laws of Chemistry.[33]

A similar claim was made by W. E. Castle, who wrote that before the Mendelian scheme could be accepted unequivocally, it was necessary to know whether "all observed inheritance phenomena can be expressed satisfactorily in terms of genes, which are supposed to be to heredity what atoms are to chemistry, the ultimate, indivisible units, which constitute gametes much as atoms in combination constitute compounds."[34]

Castle's Harvard colleague E. M. East echoed the same view when, in discussing the notational nature of Mendelism, he remarked, "Mendelism is therefore just such a conceptual notation as is used in algebra or in chemistry."[35] The Danish plant breeder Wilhelm Johannsen, originally trained in pharmacology, used the comparison of heredity to chemistry quite overtly, though he continued to think of Mendelian theory as only a useful notational system even in the face of the chromosomal evidence. H. S. Jennings and C. B. Davenport expressed similar views about the analogy of genes to atoms, and of the whole Mendelian theory to Dalton's atomic-molecular theory.[36] And in his 1889 *Intracelluläre Pangenesis*, long before he encountered Mendel's work, De Vries compared pangenes to the atoms and molecules of the physicist and chemist.[37]

Early depictions of genes in the formal Mendelian notation showed discrete, independent units sorting and resorting in a classically kinetic model. After the wedding of Mendelian genetics to the chromosome theory (forming what has been referred to as the "Mendelian-chromosome theory of heredity"), primarily through the work of the Morgan school, even the more refined chromosome maps of the 1930s retained the element of genes as discrete units, or particles of definite molecular dimensions. H. J. Müller's target theory of mutation by bombardment with X-rays reinforced the same view of the corpuscularity of the gene: the more mutations within a gene created by a given X-ray dosage, the larger the gene (target) in question must be.

Another important aspect of the analogy of genes to atoms was their ability to combine and recombine in different ways to produce different outcomes, or phenotypic effects (i.e., analogous to atoms combining in different ways to produce different molecules). Furthermore, like atoms in chemical combinations, genes emerged from each association unchanged in their fundamental properties (the doctrine of the "purity of the gametes," as it was known in early Mendelian language). In other words, recessive and dominant genes were not altered (contaminated) by their association with each other in the same cell. A recessive allele came out of a heterozygous combination just as recessive as before. Although later work on such cases as epistasis, modifying factors, and position effect added elements to the Mendelian scheme for which there was no exact counterpart in chemistry, these nuances did not undermine the most basic property of the gene as a stable unit that could combine and recombine, always yielding predictable ratios, generation after generation.

A consequence of the ability of genes to emerge unchanged from various combinations was the view that their history was unimportant. Like atoms, whose history has always been considered irrelevant to their current combining properties, the combinations in which genes have existed in preexisting organisms have no bearing on their properties in the organism in which they now reside. Johannsen found this aspect of the gene particularly important in countering the (then) prevalent Galtonian law of ancestral inheritance (the idea that each organism is composed in some fraction of hereditary elements from all of its previous ancestors: half from each parent and, through them, one quarter from each grandparent, one eighth from each great-grandparent, and so on).[38] Galton's law was inconvenient for making predictions about future offspring because it did not say anything about individuals, only about groups; it thus had little heuristic value for breeders or for those, like Francis Galton himself, who were interested in eugenics.

More important, perhaps, Galton's law appeared to invoke a role for the entire history of a family line in influencing the traits of the present generation, a view that remained both impractical for making specific predictions and intellectually distasteful to those trying to break out of the historically dominated phylogenetic paradigm of late nineteenth-century morphology. However, if each gene starts life anew in its present combinations, then predictions of the sort that Mendelian theory embodied were reliable and highly useful, especially to breeders. Of course, from our present, postevolutionary synthesis perspective, we recognize that genes do have histories, and though they are not "contaminated" by their particular combinations, they do undergo change of a sort that is cumulative— for example, the additive effect of mutations, crossovers, or insertions—for which there is no exact counterpart in the chemical world (though radioactive decay might be considered an analogous case of historical change within the life history of atoms of a radioactive element).

A key feature of the Mendelian scheme that gave it great currency in the promulgation of the "new biology" was that like atoms, which are discrete and whose effects, at least, are measurable, Mendelian genes could be treated quantitatively and statistically. The doyen of American cytology and development, Edmund Beecher Wilson, wrote:

> [The results of Morgan and his group] are indeed staggering—to a certain type of mind even harder to assimilate than those which physicists are now asking us to accept concerning the structure of atoms. . . . They make possible precise, qualitative prediction concerning the outcome of new experiments. In those respects they are employed in the same way as the exact concepts of the chemist or the physicist and they may, I think, lay claim to a validity of the same kind even if it be not yet quite of the same degree.[39]

Thus, two other conditions of the "new biology" were fulfilled by the Mendelian paradigm: the genetic data were quantitative and, through the use of simple statistics, could be manipulated to show regular, underlying processes and to make general predictions. The early Mendelian schemes of mono- and dihybrid crosses fit easily into this sort of simple approach. Later developments, especially those involving quantitative inheritance (i.e., traits that are not phenotypically discrete, such as a range of kernel colors in wheat), required additional hypotheses—those of modifying factors or multiple alleles acting epistatically. There were numerous debates about these auxiliary hypotheses, and some geneticists at least—such as Castle and Goldschmidt—found the constant addition of new, ancillary hypotheses a sign of weakness in the theory itself.

Growing out of the discrete and mathematical precision embodied in Mendelism was the mechanistic view of the whole organism—its phenotype—as a sum of separate parts, the genes (a mosaic, if you will) of discrete entities (genes at one level, phenotypic traits at another). A corollary of any mosaic conception is that the parts are dissociable and can be studied individually by the analytical method. Thus, as Bateson put it when addressing the New York Horticultural Congress in 1902: "The organism is a collection of traits. We can pull out yellowness and plug in greenness, pull out tallness and plug in dwarfness."[40]

What Bateson had in mind was the mosaic nature of the organism's collage of traits. As it was interpreted initially by Bateson and others, Mendelism of course fulfilled this criterion admirably. It is important to keep in mind that in contrast particularly to American Mendelians, Bateson remained skeptical about giving Mendel's factors a specific material designation, such as chromosomes or even atomistic genes; his mosaic view of the organism was limited to the characters, or traits, themselves.) The complex of characters that made up the organism could not be studied as an entirety. Indeed, that had been one of the major problems encountered in the pre-Mendelian breeding programs: breeders had tried to follow the transmission of too many traits at a time. Mendel's success had been based largely in his willingness to focus on one, two, or at the most three traits at a time. These traits were then assumed (especially in Mendel's own formulation) to act independently (as in the "law of independent assortment"). But this independent action could never be discovered without analysis of the breeding process into individual traits.

A consequence of this sort of analytical dissection was the creation of a new dialectic at the disciplinary level: the dissociation of the holistic process of reproduction into two separate processes: transmission (genetics) and development (embryology). That this split occurred at the professional as well as the conceptual level highlights the pervasive way in which the mechanistic materialist approach dominated the organization of the "new biology." Both geneticists and embryologists, from the 1920s onward, encountered considerable difficulty in applying Mendelian principles to problems of embryological differentiation or, conversely, in understanding genes as agents controlling developmental processes. As surely as the organism was analyzed into separate genes, the cell into nucleus and cytoplasm, and development into heredity and environment, so the professional division of labor into genetics and embryology split the formerly integrated view of the reproductive process that had held sway for the better part of the last half of the nineteenth century. It was a conceptual block that continued to be an issue among biologists until the 1980s and 1990s, when the application of molecular genetics to development promised to bring the two fields back together in the new disciplinary synthesis of "evo-devo."

The mechanistic and analytical method also provided an important incentive to treat Darwinian theory in terms of discrete Mendelian genes, as evidenced in the work of R. A. Fisher, R. H. Lock, J. B. S. Haldane, and other pioneers of what came to be known as the evolutionary synthesis. Fisher was the most explicit, and most overtly mechanistic, in his analysis of populations as collections of discrete Mendelian genes, and his analysis did indeed provide a powerful tool for advancing the Darwinian cause at a time when the efficacy of natural selection was being questioned on all fronts.[41] The investigation of natural selection via analysis of populations into discrete genes existing in different frequencies, Fisher boasted in 1922, "may be compared to the analytic treatment of the Theory of Gases, in which it is possible to make the most varied assumptions as to the accidental circumstances, and even the essential nature of the individual molecules, and yet to develop the general laws as to the behavior of gases, leaving but a few fundamental constants to be determined by experiment."[42]

The analytic separation of genotype and phenotype at the individual level, introduced by Johannsen in 1911, was completed by Fisher at the population level by statistical means from 1918 onward. In Fisher's population models the phenotype disappeared completely, leaving only the discrete Mendelian genes (grouped into gene "pools") as the objects of evolutionary change. A species or population became an aggregate of genes interacting randomly much as atoms or molecules in an idealized gas do. The somewhat derogatory appellation "beanbag genetics," applied to Fisher's kinetic theory by Ernst Mayr, clearly reflects the mechanistic basis on which Fisher's approach seemed to be based: whole populations (indeed, species) were abstracted into discrete, atomized components (not even individual organisms, but genes encountering one another by chance in a random universe). This was billiard-ball physics at its most extreme applied to a biological process. Although Fisher did discuss the interaction of genes in producing composite traits, such a process did not occupy center stage in his worldview. While he knew full well that many cases of epistasis and complex interactions occurred, his legacy remained that of one gene-one trait.

More field-oriented evolutionists such as Mayr and Dobzhansky came to view Fisherian population genetics as oversimplified to the point of ignoring critical factors that affected organisms in nature (population size and structure, breeding patterns, and so on). Nevertheless, historically, Fisher's approach gave evolutionary theory the prospects for a rigorous foundation for the first time. Evolutionary processes such as selection, migration, and adaptation (including the considerable British debates on mimicry and warning coloration) could now be treated quantitatively, mathematically, and, most important, through experimental analysis (for example, using population cages in the laboratory, where environmental conditions could be changed at will). Natural selection no longer remained in the pejorative realm of speculation and fantasy.

In short, then, introduction of the mechanistic materialist approach into genetics led to the expansion of many areas of biological work that previously seemed mired in purely descriptive and speculative method-ologies. With the work of Fisher, Wright, Haldane, and other mathematical population geneticists, the problem had come full circle: Darwinian theory, once considered the paragon of old-fashioned, nontestable, and nonrigorous biology, had now been placed on as solid an epistemological footing as the kinetic theory of gases. However, what remained outside the pale of this synthesis was embryonic development. In concluding this section, I would like to suggest that it was the strong emphasis on mech-anistic materialism—its atomism and reduction of the organism to a mosaic of genes in the cell nucleus and then to a mosaic of traits in the whole organism—that set it apart from development in the mind of embryologists, at least those of the Spemann school. Spemann was not opposed to genetics so much as he thought it irrelevant to the issues con-cerned with embryonic development: cell movements, differentiation, and formation of body plans. Harrison echoed the same notion, including the specific observation that geneticists had attributed all the importance to the genes in the cell nucleus, and relegated the cytoplasm to a virtually insignificant position.[43] The two fields had moved in different directions conceptually, methodologically, and professionally. It is to these latter two dimensions that we now turn.

Maintaining the Split Between Genetics and Development: Epistemic and Nonepistemic Factors

Once development and genetics diverged into separate professional domains, a new set of dialectical processes—both epistemic and nonepistemic—emerged to maintain and even intensify the separation. Among the epistemic was the powerful distinction between those who saw biological processes as wholes, not atomized into separate parts (what became known, as discussed earlier, as "holism"), and those who saw biological processes in a mechanistic light, and who thought that "holism" represented fuzzy and mystical thinking akin to vitalism. As Anne Harrington has shown so masterfully, the apparent triumph of mechanistic biology (not only in genetics but also in physiology, animal behavior, and psychology) prompted a reaction in the earlier twentieth century that tried to emphasize the "whole" organism (much as the triumph of reductionistic molecular genetics and biology in the 1980s featured the ultraholistic "Gaia hypothesis" of the Earth as an organism) which more often than not drifted into mysticism and metaphysics. Particularly centered in Germany (though with advocates elsewhere, including Henri Bergson and Pierre Teilhard de Chardin in France, and General J. C. Smuts in South Africa), its claims for the "holistic" and vital properties of living organisms seemed all too reminiscent of Haeckel's recapitulationism and back even further to the *Naturphilosophen* of the early nineteenth century.

As biology was proving itself to be a scientific enterprise worthy of standing on the same epistemological footing as physics and chemistry, claims that sounded like throwbacks to an earlier, discarded stage of immaturity augmented the professional split between genetics and development. Geneticists rode high on the wave of success in applying a highly mechanistic account of hereditary transmission that bypassed what embryologists saw as the most important process in biology: embryonic growth and differentiation. Since this divide was also generational, the incompatibility of the viewpoints became intensified and at times highly polemical.[44] While segments of the genetics community continued to profess some interest in developmental problems (as in developmental or physiological genetics), few embryologists reciprocated. Each community of workers had its own paradigms, techniques, model organisms, journals, and professional organizations. These factors, as I will discuss below, continued to widen the gap between the two fields.

Another factor that widened the genetics-embryology gap was what I will call the "tyranny of methodology." I mean by this the cluster of techniques, manipulations, equipment, and organisms that come to characterize any specific line of research, and that, as they become more specialized (and sometimes intricate), become increasingly inaccessible to outsiders. For example, the methods of mapping gene positions, including shorthand notations for map distances, units of size, the variety of special stocks developed against which new mutants could be tested, and cytological techniques, became highly specialized parcels of the geneticists' professional baggage. These methods and the intellectual reasoning behind them became increasingly inaccessible to embryologists and others outside the field. Similarly, the methods of egg constriction with fine needles and hair loops; transplantation of embryonic *Alagen* from one part of an embryo to another, or to different embryos; methods of culturing and maintaining embryos; and the fact that much of the experimental work was tied to the breeding season of amphibians, sharply differentiated the work of embryologists from that of geneticists.

Drosophila geneticists could be relatively clumsy in their handling of individual organisms (Morgan reportedly squashed flies on the microscope stage after scoring them), but experimental embryologists had to treat their embryos with great delicacy and care. Embryologists also had to deal with rampant infection of their cultures, something that *Drosophila* geneticists faced only with yeast contamination of the flies' food. The approach to the experimental organism involved two very different mind-sets and a boundary over which it became increasingly difficult for members of each group to pass. Geneticists and embryologists became more and more "adapted" to their own model organisms and methods of experimenting on them. In the sociological sense, "evolutionary divergence" became intensified. What makes this a tyranny is that each specialty increasingly became controlled by its own assemblage of methodologies. These all served, unwittingly, as barriers to crossing disciplinary lines.

The embryologist Viktor Hamburger has given personal testimony to these sorts of difficulties. In the late 1930s he tried to work on developmental genetics of the "creeper" trait in chickens (the dominant lethal gene, Cp, leads to serious limb deformities during development) under the tutelage of his friend and colleague Walter Landauer, and Landauer's colleague Dorothea Rudnick. Hamburger had previously made the switch from the amphibian to the chick embryo, but the methods (extirpation and transplantation) were the same. But Hamburger found the analysis of development in transplants of limb primordia of different genotypes unclear (heterozygous and homozygous; i.e., Cp/cp and Cp/Cp, respectively)—developmental differences did not map onto genotypic differences in any regular way—and, most of all, not illuminating from the embryologist's point of view.[45] He ultimately abandoned this line of work, returning to the study of patterns of limb innervation during embryogenesis (work that ultimately led him to collaboration with Rita Levi-Montalcini and the subsequent identification of "nerve growth factor," or NGF).[46]

The other major factor that came into play between genetics and embryology was the more extraneous, but very real, problem of professional terrain. With the rise of developmental genetics (as discussed earlier in the work of L. C. Dunn, Salome Glücksohn-Waelsch, Walter Lindauer, C. H. Waddington, and others), some significant attempts at boundary-crossing were made, and indeed, the details of T-locus in mice (Dunn, Glücksohn-Waelsch) and the creeper fowl (Lindauer) revealed some clear hopes that a more integrated theory of genetics and development was imminent. The interest, however, came more from geneticists than from embryologists. Indeed, as geneticists eyed working in areas of developmental biology, embryologists became increasingly nervous. The phenomenal success of genetics from the mid-1910s through the 1930s gave it a prestige and power that embryology, for all its many advances, lacked. For example, the budget of the Carnegie Institution of Washington's Department of Embryology (housed at Johns Hopkins University) in 1920 was $43,128, while that of its Department of Genetics (at Cold Spring Harbor), was $78,343 (not including the Eugenics Record Office, which was devoted exclusively to human heredity).[47] By 1928 the differential was even greater: $16,551 for Embryology, $211,203 for Genetics (also not including the endowment for the Eugenics Record Office, which had been combined with the Department of Genetics).[48] Much of this was clearly due to the role genetics was seen as having for agriculture; still, embryologists could not but feel to some extent that they were poor second cousins to their brilliant, if younger, disciplinary offshoot.

This disciplinary boundary concern was perhaps nowhere more clearly and explicitly stated than by Ross G. Harrison in his address as retiring vice-president of the Zoology Section of the American Association for the Advancement of Science in December 1936. The address is noteworthy for its balanced view of genetics from the embryologist's point of view. It begins with generous praise for recent advances in genetics: "The location of genes in the chromosomes, the proof of their linear order, the association of somatic characters with definite points in the chromosomes, in short, the whole development of the gene theory is one of the most spectacular and amazing achievements of biology in our times."[49]

Harrison goes on to point out that the merger of genetics and embryology has also made great strides, suggesting that the two fields are not so far apart as some investigators seem to feel: "The liaison between genetics and embryology is now established. . . ."[50] However, in an explicitly economic, even colonialist, metaphor, Harrison sounds the embryologist's warning cry about the potential danger of this liaison:

... the predicted gold rush to our own territory is upon us and times are strenuous again. . . . Now that the necessity of relating the data of genetics to embryology is generally recognized and the "Wanderlust" of geneticists is beginning to urge them in our direction, it may not be inappropriate to point out a danger in this threatened invasion. The prestige of success enjoyed by the gene theory might easily become a hindrance to the understanding of development by directing our attention solely to the genom [*sic*], whereas cell movements, differentiation and in fact all developmental processes are actually affected by the cytoplasm. Already we have theories that refer the processes of development to genic action and regard the whole performance as no more than the realization of the potencies of the genes. Such theories are altogether too one-sided.[51]

Echoing the more holistic view of embryologists in general (it is important to keep in mind that Harrison was trained in Germany and maintained close ties with his German colleagues, especially Spemann), Harrison cited a remark by E. E. Just at a symposium earlier the same day in which Just said he was "more interested in the back than in the bristles on the back, and more in eyes than in eye color."[52] Just's reference was to a widespread belief that the sorts of traits which geneticists were concerned about were minute and inconsequential variants that had little or nothing to do with fundamental characteristics such as body plans. Just felt that geneticists lost sight of the larger developmental patterns which comprise the most basic component of the developmental process. Development produces organisms, not merely eyes, backs, or wings. Coupled with the distinction between trivial and fundamental characters was the idea (to which Harrison did not subscribe) that specific and trivial traits were determined by the nucleus (the realm of the gene), while general patterns (eyes as opposed to eye color) were determined by the cytoplasm. As artificial as such claims may have been, even at the time, they were real, and reflect the degree to which genetics was still seen as mechanical and atomistic, and thus peripheral to the main problems of development. Territorial suspicions are not new to science, and they seem to have played a major role in maintaining the divergence, once it began to develop, between embryology and genetics in the first half of the twentieth century.

SUMMARY AND CONCLUSION

This chapter has traced the influence of a variety of factors on the breakdown of the first widespread "evo-devo" program—Ernst Haeckel's recapitulation theory (the "biogenetic law")—and its subsequent split into three separate fields of inquiry in the early twentieth century: genetics, embryology, and evolution. I have cast this historical development in terms of dialectics at two quite different epistemological levels: the historical/sociological and the biological. Historically, reaction against the first "evo-devo" program was really a reaction against Haeckel and his largely descriptive methods combined with unbridled speculation. But it was also a reaction against what was seen as the limited explanatory power of historical causality in biology. Younger biologists were rejecting historical causality in favor of mechanical or proximate causality (physiological for phylogenetic) explanations, and an analytic (mechanistic) for a synthetic (holistic) methodology.

The result, most immediately in the form of Roux's program for *Entwicklungsmechanik*, was to emphasize focused, analytical *experiments* on the mechanical causes of differentiation, and to test the different predictions from the mosaic versus self-regulating hypotheses. While those tests produced ambiguous results, the *methodology* of experimentation, combined with an analytic (as opposed to synthetic) frame of mind, turned biology in a wholly new direction. Applied to the older Haeckelian program, analytic methods emphasized the investigation of more restricted problems: embryonic differentiation rather than phylogeny, patterns of transmission rather than hypothetical constructs about the structure of the germ plasm that could not be tested. This changing epistemology led, by the 1920s, to the separation of the study of heredity (by then known as genetics), development, and evolution into separate fields of inquiry, each with its own set of questions, methodologies, techniques, instrumentation, model organisms, journals, and social communities.

The program originally set out in Haeckel's version of "evo-devo," which encompassed all three areas as one unified problem, was now split into separate fields of inquiry. Although workers on all sides of the divide recognized the importance of integration, most early attempts met with benign neglect (in the work of George Beadle and Boris Ephrussi, and Ivan Schmalhausen) if not open hostility (as in the work of Richard Goldschmidt and C. H. Waddington). The one synthetic development that did take place—the application of Mendelian genetics to Darwinian theory beginning in the early 1930s—was successful enough in its own right, but it did not include embryology. Ernst Mayr has argued that the evolutionists invited embryologists to join the synthesis, but the latter showed little interest. This chapter has shown that among embryologists, the genetics program (and its carryover into the evolutionary synthesis) was largely viewed as irrelevant to their own research issues. Only with the availability of the tools and conceptual framework provided by molecular genetics/biology from the 1980s onward did the problems of embryonic development begin to reenter the picture in a broad and experimentally testable framework.

At the level of biological theory itself, a dialectical relationship between focusing on the organism as a whole, integrated entity, versus viewing it as a machine composed of so many separate parts, animated much of the history of the field in the twentieth century (and certainly prior to that as well). In some sense the history of twentieth-century biology can be written as the constant juxtaposition of these views of organic life: with now one, now the other, gaining ascendance in the research community (with variations in time frame from country to country, and influenced by national differences in scientific tradition). The grand synthetic theories of the late nineteenth century, centered in Germany, developed in part as a reaction to the medical materialists of midcentury (also centered in Germany), and were in turn reacted to by a younger generation of biologists epitomized by Jacques Loeb and his *Mechanistic Conception of Life*.[53] The mechanistic view that came to be so clearly incorporated into much of classical genetics in turn provoked a dialectical return to such organicist or holistic claims as Jakob von Uexküll's *Umwelt* theory (that organisms and their environments made up an inextricable totality or "supra organism"), or Driesch's "entelechy" (an immaterial guiding force directing organic processes toward desired ends), which fought a rearguard action against the analytical approach. That such organicist ideas failed to gain much prominence (except to be held up for ridicule or dismissal) is partly a result of their often mystical formulation and invocation of metaphysical forces. It is also a result of the fact that other than proposing an alternative, but abstract, way to view organic processes, they did not offer any concrete experimental research program.

However impractical holistic claims may have been, they emphasized an important aspect of the biological world: organisms are not machines in any real sense of the word. As useful as the analogy has been in various contexts, it ultimately misses something critical that each generation of biologists has repeatedly had to rediscover: there are qualities exhibited by living organisms, from single-cell to multicellular forms, that simply have no counterparts in machines as we know them. There are *emergent properties* at a level of complexity (for example, reproduction, self-regulation and repair, evolution) not duplicated by machines. These properties, and the various levels of organization in the living system at which they emerge, are perhaps more clearly and explicitly recognized among biologists today than they ever were in the past. One of the most widely used and successful introductory biology textbooks, Neil Campbell and Jane Reece's *Biology* (6th ed., 2002), has a whole section devoted to levels of organization and the principle of emergent properties, which had few counterparts during the most recent "age of mechanistic biology" from the 1960s through the late 1980s.[54]

If I may expand a bit further, it seems to me that the historical dialectic between analysis and synthesis, reductionism and holism, should not be too surprising, since analysis is required to identify the elements of any complex system, but once those elements are identified and some aspects of their characteristics are noted, full understanding can come only when they are reintegrated and studied as a whole system. The development of limb primordia into fully differentiated limbs can be studied in isolation, through transplantation experiments. But the integration of a limb into the central nervous system with the ability to function in walking or flying can be studied only in conjunction with the developing nervous system—that is, in its overall organismic context. One characteristic that is lost in studying limbs in isolation is their innervation and control by the central nervous system. Components such as limb primordia display additional, or modified, characteristics in the context of the whole where they occur. Both synthesis and analysis are thus required for any full understanding of a biological system. We may berate Haeckel for his excesses in the synthetic line, but it should be kept in mind that Roux, Driesch, and many other early experimental embryologists were his students, and that their new line of research developed out of his—as opponents rather than loyal followers.

The integration we refer to today as "evo-devo" is one that will take time to bear fruit. Some of this is, of course, technical. But it is also conceptual. After at least three generations of pursuing separate problems and separate methodologies, it will take some time to reformulate the questions of how the evolution of complex genetic systems that undergo embryonic development could have occurred. And although I know that historical predictions are suspect, I have no doubt that such a reintegration will occur.

NOTES

1. Gould, 1977, p. 1.

2. Plesset et al., 2000, p. 45.

3. Nyhart, 2002; Gilbert, 1991a; Burian et al., 1991; Allen, 1974, 1991.

4. Allen, 1974.

5. Gilbert, 1991a, pp. 188–203.

6. Gould, 1977, pp. 167ff.

7. Allen, 1978, chaps. 1 and 2.

8. Harrington, 1996.

9. Graham, 1987; Allen, 1991.

10. Gould, 1977, p. 78.

11. Nyhart, 2002.

12. Ibid.

13. Morgan, 1891.

14. Ibid., p. 1.

15. Roux, 1895, p. 120. Emphasis in original.

16. Gould, 1977, p. 83.

17. Nyhart, 2002, p. 8.

18. Burdach, 1817, p. 19. Quoted in Nyhart, 2002, p. 8.

19. Roux, 1889, p. 27.

20. Pauly, 1987.

21. McEvoy, 1986, pp. 79–87; Taylor, 1951, pp. 426–427. A similar point has been made for the economic forces of agriculture as they influenced the development of genetics (Kimmelman, 1983; Kloppenburg, 1988; Lewontin and Berlan, 1986; Palladino, 1990, 1993, 1994) and for specific subfields such as entomology (Palladino, 1996).

22. Lillie, 1944, pp. 24–25.

23. Gilbert, 1991b; Maienschein, 1991; Mocek, 1974.

24. Machamer et al., 2000.

25. Mocek, 1998, pp. 250–255.

26. Roux, 1888; Driesch, 1891.

27. Spemann, 1901, p. 256; quoted from Hamburger, 1988, p. 14.

28. Hamburger, 1988, p. 11.

29. Hamburger, 1999, p. 231.

30. Spemann, 1943; quoted from Hamburger, 1999, p. 242.

31. Robinson, 1979.

32. Whether or not such an explicitly atomistic view was inherent in Mendel's own conception has been debated with great fervor in recent years. Vitezslav Orel, Mendel's most recent biographer, and the geneticist Daniel Hartl have argued that Mendel did think of factors as material entities (Hartl and Orel, 1992), while Robert Olby (1979) has argued that Mendel was primarily an empiricist unconcerned about the physical nature or even existence of his *Anlagen*. According to this view, Mendel's major interest was the patterns formed during the specific process of hybridization, not in constructing either an epistemology of hereditary units or in discovering a fundamental theory of heredity. For a summary of this debate and its consequences for current evo-devo thinking, see Allen, 2003.

33. Bateson, 1901; reprinted in Punnett, 1928, vol. 2, p. 1.

34. Castle, 1919, p. 127.

35. East, 1912, p. 633.

36. Jennings, 1920; Davenport, 1906.

37. De Vries, 1889: p. 9. De Vries wrote: "Wie die Physik und die Chemie auf die Moleküle und die Atome zurückgehen, so haben die biologischen Wissenschaften zu diesen Einheiten durchzudringen, um aus ihren Verbindungen die Erscheinungen der lebenden Welt zu erklären." Dunn (1965, p. 341) cites this same passage, but from the English translation, and attributes it to the "Introduction," whereas it actually comes from the introductory section of chap. 1.

38. Johannsen, 1911, p. 139.

39. Wilson, 1923, p. 15.

40. Bateson, 1902; quoted in Levins and Lewontin, 1985, p. 180.

41. Bowler, 1983; Allen, 1983, pp. 89–92; Provine, 1971, pp. 140–150.

42. Fisher, 1922, pp. 321–322; quoted in Provine, 2001 [1971], p. 149.

43. Harrison, 1937, p. 372.

44. Allen, 1978; Nyhart, 2002.

45. Hamburger, 1941.

46. Allen, 2004; Kirk and Allen, 2002.

47. Carnegie Institution of Washington, 1920, p. 18.

48. Carnegie Institution of Washington, 1928, p. 18.

49. Harrison, 1937, p. 371.

50. Ibid.

51. Ibid., p. 372.

52. Ibid., p. 372.

53. Loeb, 1912.

54. Campbell and Reece, 2002, pp. 2–4.

REFERENCES

Allen, Garland E. 1974. "Opposition to the Mendelian-Chromosome Theory: The Physiological and Developmental Genetics of Richard Goldschmidt." *Journal of the History of Biology* 7: 49–92.

———. 1978. *Life Science in the Twentieth Century.* New York: Cambridge University Press.

———. 1983. "The Several Faces of Darwin: Materialism in Nineteenth- and Twentieth-Century Evolutionary Theory." In D. S. Bendall (ed.), *Evolution from Molecules to Men*, pp. 81–201. Cambridge: Cambridge University Press.

———. 1991. "Mechanistic and Dialectical Materialism in 20th Century Evolutionary Theory: The Work of Ivan I. Schmalhausen." In Leonard Warren and Hilary Koprowski (eds.), *New Perspectives on Evolution*, pp. 15–36. New York: Wiley-Liss.

———. 2003. "Mendel and Modern Genetics." *Endeavour* 27: 63–68. For a more detailed version of this article, see the author's 2004. Mendelien Genetics and Postgenomics: The Legacy for Today." *Ludus Vitalis* 12 (No. 21) pp. 213–236.

———. 2004. "A Pact with The Embryo: Viktor Hamburger, Holistic and Mechanistic Philosophy in the Development of Neuroembryology, 1927–1955." *Journal of The History of Biology* 37: 421–475.

Amundson, Ron. 2000. "Embryology and Evolution, 1920–1960." *History and Philosophy of the Life Sciences* 22: 335–352.

Bateson, William. 1901. "Introductory Note to the Translation of *Experiments in Plant Hybridization* by Gregor Mendel." *Journal of the Royal Horticultural Society* 26: 1–3. Reprinted in Reginald C. Punnett (ed.), *Scientific Papers of William Bateson*, 2 vols. Cambridge: Cambridge University Press.

———. 1902. *Mendel's Principles of Heredity: A Defence.* Cambridge: Cambridge University Press.

Bertan, Jean-Pierre, and Richard Lewontin. 1986. "The Political Economy of Hybrid Corn." *Monthly Review* 38: 35–47.

Bowler, Peter. 1983. *The Eclipse of Darwinism.* Baltimore: Johns Hopkins University Press.

Burdach, K. F. 1817. *Über die Aufgabe der Morphologie.* Leipzig: Dyk.

Burian, Richard M., and Denis Thieffry. 2000. "Introduction: From Embryology to Developmental Biology." Special issue of *History and Philosophy of the Life Sciences* 22: 313–324.

Campbell, Neil, and Jane Reece. 2002. *Biology: Concepts & Connections.* Menlo Park, Calif.: Benjamin Cummings.

Carnegie Institution of Washington. 1920. "Report of the President (Robert S. Woodward)." *Carnegie Institution of Washington Yearbook,* no. 19. Washington, D.C.: Carnegie Institution.

———. 1928. "Report of the President (John C. Merriam)." *Carnegie Institution of Washington Yearbook,* no. 27. Washington, D.C.: Carnegie Institution.

Castle, William E. 1919. "Piebald Rats and the Theory of Genes." *Proceedings of the National Academy of Sciences USA* 5: 126–130.

Davenport, Charles B. 1906. *Inheritance in Poultry.* Carnegie Institution of Washington Publication 52. Washington, D.C.: Carnegie Institution.

De Vries, Hugo. 1889. *Intracelluläre Pangenesis.* Jena: Gustav Fischer.

Driesch, Hans. 1891. "Entwicklungsmechanische Studien I. Der Werth der beiden ersten Furchungszellen in der Echinodermenentwicklung. Experimentelle Erzeugung von Teil- und Doppelbildungen." *Zeitschrift für wissenschaftliche Zoologie* 53: 160–183. English translation by L. Metzger, Martha Hamburger, Viktor Hamburger, and Thomas S. Hall reprinted in Benjamin Willier and Jane Oppenheimer (eds.), *Foundations of Experimental Embryology,* pp. 40–50. (Englewood Cliffs, N.J.: Prentice-Hall, 1964).

———. 1894. *Analytische Theorie der organischen Entwicklung.* Leipzig: Wilhelm Engelmann.

East, Edward M. 1912. "The Mendelian Notation as a Description of Physiological Facts." *American Naturalist* 46: 633–695.

Fisher, Ronald A. 1922. "On the Dominance Ratio." *Proceedings of the Royal Society of Edinburgh* 52L: 321–341.

Gilbert, Scott. 1991a. "Induction and the Origins of Developmental Genetics." In Scott Gilbert (ed.), *A Conceptual History of Modern Embryology,* pp. 181–206. New York: Plenum Press.

———. 1991b. *A Conceptual History of Modern Embryology.* Vol. 7 of *Developmental Biology: A Comprehensive Synthesis,* Leon W. Browder, ed. New York, Plenum Press.

Gould, Stephen Jay. 1977. *Ontogeny and Phylogeny.* Cambridge, Mass.: Belknap Press of Harvard University Press.

Graham, Loren. 1987. *Science, Philosophy, and Human Behavior in the Soviet Union.* New York: Columbia University Press.

Haeckel, Ernst. 1866. *Generelle Morphologie der Organismen: Allgemeine Grundzüge der organischen Formen-Wissenschaft, mechanisch begründet durch die von Charles Darwin reformirte Descendenz-Theorie.* 2 vols. Berlin: Georg Reimer.

———. 1874a. *Anthropogenie; oder, Entwickelungsgeschichte des Menschen. Keimes- und Stammesgeschichte,* 4 vols. Leipzig, W. Engelmann.

———. 1874b. "Die Gastraea-Theorie, die phylogenetische Klassification des Tierreiches und Homologie der Keimblätter." *Jena Zeitschrift für Naturwissenschaften* 8: 1–55.

Hamburger, Viktor. 1941. "Transplantation of Limb Primordia of Homozygous and Heterozygous Chondrodystrophic ('Creeper') Chick Embryos." *Physiological Zoology* 14: 355–364.

———. 1960. *A Manual of Experimental Embryology,* Rev. ed. Chicago: University of Chicago Press.

———. 1988. *The Heritage of Experimental Embryology: Hans Spemann and the Organizer.* New York: Oxford University Press.

———. 1999. "Hans Spemann on Vitalism in Biology: Translation of a Portion of Spemann's *Autobiography.*" *Journal of the History of Biology* 32: 231–243.

Harrington, Anne. 1996. *Reenchanted Science: Holism in German Culture from Wilhelm II to Hitler.* Princeton, N.J.: Princeton University Press.

Harrison, Ross G. 1937. "Embryology and Its Relations." *Science* 85: 369–374.

Hartl, Daniel, and Vitezslav Orel. 1992. "What Did Mendel Think He Had Discovered?" *Genetics* 131: 245–253.

Holtfreter, Johannes, and Viktor Hamburger. 1955. "Amphibans." In B. H. Willier, Paul Weiss, and Viktor Hamburger (eds.), *Analysis of Development,* sec. VI, chap. 1, pp. 230–296. Philadelphia: W. B. Saunders.

Jennings, Herbert Spencer. 1920. *Life and Death: Heredity and Evolution in Unicellular Organisms.* Boston: R. G. Badger.

Johannsen, Wilhelm. 1911. "The Genotype Conception of Heredity." *American Naturalist* 45 (No. 531): 129–159.

Kimmelman, Barbara. 1983. "The American Breeders' Association: Genetics and Eugenics in an Agricultural Context." *Social Studies of Science* 13: 163–204.

Kirk, David L., and Garland E. Allen. 2001. "Viktor Hamburger: A Prepared, Persistent and Deserving Mind Favored by Many 'Fortuities.'" *Developmental Dynamics* 222: 545–551.

Kloppenburg, Jack R. 1988. *First the Seed: The Political Economy of Plant Biotechnology,* 1492–2000. Cambridge: Cambridge University Press.

Levins, Richard, and Richard Lewontin. 1985. *The Dialectical Biologist.* Cambridge, Mass.: Harvard University Press.

Lewontin, Richard, and Jean-Pierre Berlan. 1986. "Technology, Research, and the Penetration of Capital: The Case of U.S. Agriculture." *Monthly Review* 38 (July–August): 21–34.

Lillie, Frank R. 1944. *The Woods Hole Marine Biological Laboratory.* Chicago: University of Chicago Press.

Loeb, Jacques. 1912. *The Mechanistic Conception of Life: Biological Essays.* Chicago: University of Chicago Press. Reprinted with an introduction and editorial commentary by Donald Fleming (Cambridge, Mass.: Belknap Press of Harvard University Press, 1964).

Machamer, Peter, Lindley Darden, and Carl Craver. 2000. "Thinking About Mechanisms." *Philosophy of Science* 67: 1–25.

Maienschein, Jane. 1991. "The Origin of *Entwicklungsmechanik.*" In Scott F. Gilbert (ed.), *A Conceptual History of Modern Embryology,* pp. 43–61. New York: Plenum Press.

McEvoy, A. F. 1986. *The Fisherman's Problem: Ecology and Law in the California Fisheries, 1850–1980.* Cambridge: Cambridge University Press.

Mocek, Reinhard. 1974. *Wilhelm Roux—Hans Driesch: Zur Geschichte der Entwicklusngsphysiologie der Tiere.* Jena: Gustav Fischer.

———. 1998. *Die werdende Form: Eine Geschichte der kausalen Form.* Marburg: Basilisken Presse.

Morgan, Thomas Hunt. 1891. "The Embryology and Phylogeny of the Pycnogonids." Ph.D. dissertation, Johns Hopkins University.

———. 1916. *Critique of the Theory of Evolution.* Princeton, N.J.: Princeton University Press.

Nieto, M. Angelo, and Pat Simpson. 2002. "Pattern Formation and Developmental Mechanisms: Gene Families and Developmental Diversity." *Current Opinion in Genetics and Development* 12: 383–385.

Nyhart, Lynn. 1995. *Biology Takes Form: Animal Morphology and the German Universities, 1800–1900.* Chicago: University of Chicago Press.

———. 2002. "Learning from History: Morphology's Challenges in Germany ca. 1900." *Journal of Morphology* 252: 2–14.

Olby, Robert. 1979. "Mendel no Mendelian?" *History of Science* 17: 53–72.

Palladino, Paolo. 1990. "The Political Economy of Applied Research: Plant Breeding in Great Britain, 1910–1940." *Minerva* 28: 446–468.

———. 1993. "Between Craft and Science: Plant Breeding, Mendelian Genetics, and British Universities, 1900–1920." *Technology and Culture* 34: 300–323.

————. 1994. "Wizards and Devotees: On the Mendelian Theory of Inheritance and the Professionalization of Agricultural Science in Great Britain and the United States, 1880–1930." *History of Science* 32: 409–444.

————. 1996. *Entomology, Ecology and Agriculture: The Making of Scientific Careers in North America 1885–1985.* Amsterdam: Harwood Academic Publishers.

Pauly, Philip J. 1987. *Controlling Life: Jacques Loeb and the Engineering Ideal in Biology.* Berkeley: University of California Press.

Plesset, Judith, Samuel Scheiner, and Susan Singer. 2000. "Evolution and Development at the National Science Foundation." *Genesis* 28: 45–46.

Provine, William B. 1971. *The Origins of Theoretical Population Genetics.* Chicago: University of Chicago Press. With a new "Afterword" [2001].

Punnett, Reginald C. (ed.). 1928. *Scientific Papers of William Bateson*, 2 vols. Cambridge: Cambridge University Press.

Ramachandra, Nallur B., Ruth D. Gates, Peter Ladurner, David K. Jacobs, and Volker Hartenstein. 2002. "Embryonic Development in the Primitive Bilaterian *Neochilidia fusca*: Normal Morphogenesis and Isolation of POU Genes Brn-1 and Brn-3." *Development, Genes and Evolution* 212: 55–69.

Robinson, Gloria. 1979. *A Prelude to Genetics: Theories of a Material Substance of Heredity, Darwin to Weismann.* Lawrence, Kan.: Coronado Press.

Roux, Wilhelm. 1888. "Beiträge zur Entwickelungsmechanik des Embryo V. Über die kunstliche Hervorbringung halber Embryonen durch Zerstörung einer der beiden ersten Furchungzellen, sowie über die Nachentwickelung (Postgeneration) der fehlenden Körperhälfte." *Virchow's Archiv für pathologische Anatomie und Physiologie und klinische Medezin* 114: 113–153. English version translated by Hans Laufer in Benjamin Willier and Jane Oppenheimer (eds.), *Foundations of Experimental Embryology*, pp. 4–37. (Englewood Cliffs, N.J.: Prentice-Hall, 1964).

————. 1889. "Die Entwickelungsmechanik der Organismen, eine anatomische Wissenschaft der Zukunft. Festrede." Reprinted in Wilhelm Roux, *Gesammelte Abhandlungen über Entwickelungsmechanik der Organismen*, vol. 2, pp. 24–54. (Leipzig: Wilhelm Engelmann, 1897).

————. 1895. "The Problems, Methods and Scope of Developmental Mechanics." Introduction to *Archiv für Entwickelungsmechanik der Organismen* 1(1890): 1–42. Translated by William Morton Wheeler and published in *Biological Lectures Delivered at the Marine Biological Laboratory, Woods Hole, Massachusetts, in the Summer Session of 1894*, pp. 107–146 (Boston: Ginn, 1895).

Rudnick, Dorothea, and Viktor Hamburger. 1940. "On the Identification of Segregated Phenotypes in Progeny from Creeper Fowl Matings." *Genetics* 25: 215–224.

Sinnott, Edmund W. 1936. "A Developmental Analysis of Inherited Shape Differences in Cucurbit Fruits." *American Naturalist* 70: 245–254.

Spemann, Hans. 1901. "Entwicklungsphysiologische Studien am Tritonei I." *Roux's Archiv für Entwicklungsmechanik der Organismen* 12: 224–264.

———. 1903. "Entwicklungsphysiologische Studien am Tritonei III." *Roux's Archiv für Entwicklungsmechanik* 16: 551–631.

———. 1924. "Vererbung und Entwicklungsmechanik." *Naturwissenschaften* 12: 65–79.

———. 1943. *Forschung und Leben*, F. W. Spemann, ed. Stuttgart: Adolf Spemann.

Taylor, Harden F. 1951. *Survey of Marine Fisheries of North Carolina.* Chapel Hill: University of North Carolina Press.

Wilson, Edmund B. 1904. "Experimental Studies in Germinal Localization. II. Experiments on the Cleavage-Mosaic in Patella and Dentalium." *Journal of Experimental Zoology* 1, (August): 197–268.

———. 1923. *The Physical Basis of Life.* New Haven: Yale University Press.

7

The Cell as the Basis for Heredity, Development, and Evolution: Richard Goldschmidt's Program of Physiological Genetics

Marsha L. Richmond

The rediscovery of Mendel's laws of heredity in 1900 led early twentieth-century biologists to reexamine the empirical basis and conceptual understanding of heredity, development, and evolution. Biologists had long viewed heredity and development as intimately intertwined aspects of the life history of organisms. Development, for example, was simply the continuation of processes set in motion by generation.[1] The proliferation of detailed studies of cell division after 1880, however, began to complicate this picture. Once material bodies in the cell nucleus—the chromosomes—began to be identified as the bearers of heredity, the relationship between heredity and development became problematic.

Theorists such as August Weismann used these new cytological findings to arrive at a view of heredity that was ostensibly sundered from development. According to Weismann's germ-plasm theory, heredity involved the transmission of the germ plasm from one generation to the next; moreover, the germ line was sequestered from the events of individual development. "With the mental separation of the organism into a segment of transmission and one of development," Frederick Churchill noted, "heredity as a process became restricted to the function of transmission between generations of organisms."[2]

Weismann himself attempted to bridge this gap by formulating a theory that linked heredity to the events of embryogenesis. The differentiation of individual somatic cells into various types of tissues and organs, in his view, resulted from the successive parceling out of qualitatively different combinations of particles inherited from the germ plasm. Yet development in Weismann's scheme was nonetheless subservient to the transmission of hereditary particles.[3]

Mendel's explanation of the statistical outcome of hybrid crosses likewise assumed the presence in the germ cells of randomly assorting and segregating hereditary "factors." Such an assumption fitted well with the

tenets of the "new heredity" of the turn of the twentieth century, although
not completely with Weismann's views.[4] Well coordinated with the concerns of cytologists, Mendelian hybridization experiments focused on
aspects of factoral transmission, sexual reproduction, and the production of
organic variation. Left behind, however, was the nineteenth-century belief
of the intimate connection between heredity, development, and evolution.
The new Mendelian "preformationism" did not address how discrete
"factors" or "genes" could direct the epigenetic processes of development
seemingly associated with quantitative differences in activity rather than
qualitative differences in substance.[5]

Certainly, some early Mendelians, especially William Bateson and
Hugo de Vries, attempted to maintain the close linkage between heredity,
development, and evolution.[6] However, after the publication of *The
Mechanism of Mendelian Heredity* by Thomas Hunt Morgan, Alfred H.
Sturtevant, Herman J. Müller, and Calvin B. Bridges in 1915, the new field
of genetics increasingly focused on problems connected with gene transmission. The Mendelian-chromosome theory of heredity assumed that the
chromosomes were "the bearers of the Mendelian factors" and that heredity was "a problem concerning the cell, the egg, and the sperm."[7] Development, by implication, was not a concern of geneticists but a problem
for embryologists.[8] The failure of the new genetics to address development
was lamented by Theodor Boveri, who wrote shortly before his death:
"About heredity itself, however—that is, the question of how the given
constellation in the zygote leads to the hereditary effects investigated by
the geneticist—we know nothing at all."[9]

Not all geneticists, however, were willing to accept such a break from
the traditional understanding of the connection between heredity, development, and evolution. Foremost among these was the German geneticist
Richard Goldschmidt (1878–1958). Goldschmidt's interest in problems
connected with heredity and development was evident long before he
turned to genetics in 1909 (see figure 7.1). Trained in the morphological
tradition under Otto Bütschli at Heidelberg and Richard Hertwig at
Munich, Goldschmidt had early demonstrated a pronounced interest in
cell studies. He particularly focused on exploring how the hereditary material directed metabolic activities in the cytoplasm. Indeed, his frustration
at the failure of morphological methodological tools to provide definitive
answers to this inquiry prompted his migration into genetics in 1909. He
hoped that the experimental techniques offered by combining breeding
experiments with cytological analysis would be a promising means by
which to probe the nature of heredity at the cellular level.[10]

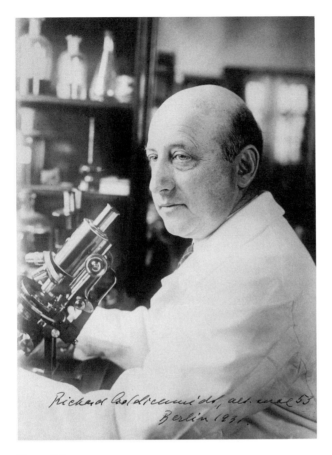

Figure 7.1
Richard Goldschmidt (Berlin, 1931). Curt Stern Photographic Collection, American
Philosophical Society, Philadelphia.

In 1909, Goldschmidt initiated a Mendelian study of sex determination in the gypsy moth, *Lymantria dispar*. His early papers show that he initially adopted a factorial view of sex determination, assuming that the inheritance of qualitatively different male and female sex factors could explain the development of alternative secondary sex characters in organisms. However, the discovery of a new phenomenon—the apparent "changing dominance" in sex determination, with female moths developing male secondary sex characters—suggested that the inheritance of sex factors per se did not determine the developmental outcome. Rather, there appeared to be an intervening mediating process conjoining factorial inheritance and character development. Goldschmidt recognized that this case offered a promising means of shedding light on the problem Weismann posed: explaining the mechanism by which the hereditary factors controlled the developmental of external characters. He pursued this problem for the remainder of his career.

To explain how genetically male or female moths could develop characters of the opposite sex, Goldschmidt devised a "labile" conception of the gene, one that stressed *quantitative* rather than *qualitative* properties of the hereditary factors. In a stream of papers beginning in 1911, Goldschmidt developed these ideas into a mature "physiological" theory of genetics published in *Physiologische Theorie der Vererbung* (1927). The crux of his conception was the assumption that genes were either enzymes or enzyme-like entities able to stimulate and direct biochemical and physiological activities within the cell cytoplasm, and hence within the organism as a whole. As he noted in a 1933 synopsis of his views, "the action of the genes in controlling development is to be understood as working through the control of reactions of definite velocities, properly in tune with each other and thus guaranteeing the same event always to occur at the same time and at the same place."[11] Goldschmidt's epigenetic conception of gene action thus retained the synthetic vision of late nineteenth-century biologists, yet he recast the presumed linkage between heredity, development, and evolution within the contemporary empirical and conceptual matrix of genetics, cytology, and biochemistry.

Goldschmidt's view of the "quantitative" gene contrasted with the qualitative conception held by mainstream geneticists, particularly those influenced by the work of the Morgan school. His early, vocal criticism of their interpretation of the phenomenon of crossing-over soon marginalized him from the advancing front of "transmission" genetics.[12] The rivalry between Goldschmidt and the Morgan group continued unabated throughout the classical period of genetics. Indeed, after he launched a two-pronged attack on the particulate notion of the gene in 1939, only a few years after immigrating to the United States, and, soon thereafter, on gradualistic evolution upheld by the neo-Darwinian founders of the evolutionary synthesis, Goldschmidt increasingly became labeled a "heretic."[13] To this day adjectives such as "maverick," "unorthodox," "obstructionist," and "controversial" often precede Goldschmidt's name in the historical literature.[14]

It appears, however, that in assessing Goldschmidt's work in genetics and evolution, historians have been overly influenced by the Nobel

Prize-winning work of the Morgan school and the successes of contributors to the evolutionary synthesis. In part the dismissal of Goldschmidt's views can be traced to the continuing influence of early histories of genetics written by geneticists closely associated with the Morgan program.[15] Certainly Goldschmidt believed in the value of introducing unifying theories into biology at a time when cautious empiricism generally held sway.[16] But to diminish Goldschmidt's contributions to the history of genetics on the basis of his criticism of transmission genetics (and *Drosophila* genetics in particular) fails to capture the vibrancy of the many lines of work carried out in genetics during the classical period.[17] Indeed, as we shall see, Goldschmidt's program of physiological genetics stimulated a number of developmentally minded geneticists and developmental biologists in the 1930s and 1940s, and it may also have moved Morgan's followers to pursue certain developmental questions.

The same is true for Goldschmidt's evolutionary views. Ernst Mayr has rejected the claim that Goldschmidt's saltationist views and typological thinking about species "affected the actual theory formation by the participants in the evolutionary synthesis."[18] Yet even Mayr concedes that Goldschmidt's developmentalist perspective was provocative at the time: "Goldschmidt brought an entirely new kind of thinking into his evolutionary arguments. No one else dealt with the functional-developmental aspects as did Goldschmidt."[19] Goldschmidt's advocacy of the importance of investigating the physiological action of genes in developmental processes, and his suggestion that the evolution of new species may be the product of a saltatory hereditary mechanism, certainly stimulated the thinking of other contemporary "outsiders" such as Julian Huxley, John Burdon, Sanderson Haldane, and Gavin de Beer.[20] More recently, these ideas have been cited by Stephen Jay Gould in support of the Eldredge-Gould theory of punctuated equilibrium.[21] Goldschmidt's developmentalist perspective has indeed been lauded by many working in "Evo-Devo"—the new synthesis of evolutionary and developmental biology—as an early example of the view that alterations in genes or their time of action (or chromosomal rearrangements) could affect early developmental processes, resulting in major evolutionary change.[22]

While it would certainly be wrong to view Goldschmidt as a precursor of contemporary research programs in Evo-Devo, nonetheless his early attempt to identify the functions of interactions between genes and the cytoplasm bears a striking resemblance to the aims and assumptions of modern-day theorists. Goldschmidt's lifelong interest in problems of gene expression during differentiation and development was intimately tied to his understanding of the evolutionary process. It is true, however, that the terms in which he conceived these interconnections are different from those of today. As Rudolf Raff notes, "The lack of knowledge about the nature of genes (they were still commonly thought to be analogous to enzymes) and the lack of any clear model of gene activation prevented this discussion from providing a fruitful basis for research at that time."[23] Moreover, Goldschmidt's own experimental approach did not bear fruit; advances in the nascent field of "biochemical genetics" took a path different from that pursued by Goldschmidt in the 1930s.[24] Nonetheless, his empirical findings and theoretical analysis should not be undervalued. His attempt to synthesize the data of genetics, development, and evolution into a unified conceptual framework had an impact on his contemporaries, and his ideas continue to seem remarkably fresh today. This chapter will examine the reception of Goldschmidt's program of physiological genetics by developmentally minded biologists as well as transmission geneticists. It aims to show that rather than being a fringe figure in genetics and evolutionary studies, Goldschmidt left a considerable legacy for both that continues to resonate today.

GOLDSCHMIDT'S VIEW OF CELL PHYSIOLOGY

At his core, Goldschmidt was a cell biologist. He was trained by two of Germany's most preeminent cytologists, Otto Bütschli at Heidelberg and Richard Hertwig at Munich. As a young man, he established himself as a clever experimentalist with pronounced tendencies toward theoretical biology.[25] In his research on cells, he applied his morphological training to understanding the nuclear control of cell differentiation. During his detailed study of nerve cell development in the nematode *Ascaris*, Goldschmidt arrived at a theory that explained the nuclear control of heredity and cell metabolic activities. The "chromidial theory," developed between 1905 and 1910 and based on an extensive study of both protozoan and metazoan cells, followed lines earlier suggested by Weismann. Goldschmidt posited two different kinds of nuclear chromatin—*idiochromatin* (sequestered in the nucleus and responsible for reproduction) and *trophochromatin* (extruded from the nucleus into the cytoplasm and directing somatic activities in the cell). Goldschmidt believed the cytoplasmic "chromidia" and "chromidial apparatus" observable under the microscope were indications of the action of the "vegetative" trophochromatin.[26] The main point to note about this early work is Goldschmidt's focus on the problem of the nuclear control of cytoplasmic differentiation and his uninhibited theorization founded on detailed empirical investigation.

Goldschmidt's chromidial theory was widely discussed by biologists in Europe and the United States well into the 1920s.[27] The geneticist Theophilus Shickel Painter, for example, noted the way in which the chromidial theory influenced his own early thinking about the cell. In a 1943 letter to Goldschmidt, Painter wrote:

> I came into zoology via the protozoa route and in the days when it was still common to speak of tropho- and idiochromatin. I have never entirely forgotten these basic concepts and my thinking in cytology has been influenced by this early training. I read your 1904 article several years ago and I recognized then, as I do now, that what you and others like you were trying to explain was the dominant role played by the nucleus in cell physiology. Somehow you had to implement the relation in morphological terms. And in reading this paper one must keep in mind the setting and the interpretations given known facts as of that date.[28]

Painter presumably was not the only young biologist to be influenced by such a physicochemical approach to understanding cell regulatory processes.

The inability of the current techniques in morphology to extend and validate the chromidial theory frustrated Goldschmidt and prompted him to search for a more promising experimental methodology with which to tackle such problems. After reading Wilhelm Johannsen's *Elemente der exakten Erblichkeitslehre* in 1909, Goldschmidt recognized that the new experimental approach offered by Mendelism might reveal how heredity controls development by means of analyzing the outcome of hybrid crosses and hypothesizing about the events occurring inside the cell nucleus and cytoplasm.[29] His belief was not unfounded. As T. H. Morgan noted in 1925, "Genetics has proved a more refined instrument in analyzing the constitution of the germinal material than direct observation of the germ cells themselves."[30] Indeed, throughout his long career in genetics, Goldschmidt maintained his early aim of conceptualizing the connections between heredity, development, and evolution by speculating about the events occurring inside the cell nucleus and cytoplasm on the basis of recent empirical findings.

THE MAKING OF GOLDSCHMIDT'S PHYSIOLOGICAL THEORY OF HEREDITY

Between 1910 and 1935, Goldschmidt focused on studying the mechanism and physiology of sex determination in the gypsy moth, *Lymantria dispar* (see figure 7.2).[31] Following up on earlier accounts of the production of abnor-

Figure 7.2
Female intersexes, in a series from true females (top left) to females almost indistin-
guishable from true males (bottom right). Richard Goldschmidt, *The Mechanism and
Physiology of Sex Determination*, translated by William J. Dakin (London: Methuen, 1923),
p. 83.

mal sexual morphs in varietal crosses between European species of *Lymantria*, Goldschmidt designed controlled crosses between European races and varieties indigenous to Japan. He discovered similar sexual abnormalities. Crosses between certain races produced moths exhibiting a mixture of male and female secondary sexual characteristics. Initially identifying these forms as gynandromorphs, Goldschmidt drew up a Mendelian factorial analysis to account for the appearance of secondary sexual characteristics of the opposite sex. He assumed that individuals carried both male and female sex factors and that normally one sex factor masked the other through an epistatic/hypostatic relationship (that is, a dominant/recessive relationship between two nonallelic factors).[32] By 1915, having examined subsequent generations, Goldschmidt realized that the abnormal moths were not true gynandromorphs but, rather, what he termed "intersexes."[33] They did not exhibit a mixture of male and female traits but appeared to have begun development as male or female, then subsequently switched to exhibit organs or parts characteristic of the opposite sex.

To explain the production of intersexes, Goldschmidt was prompted to introduce new "labile" assumptions into the accepted Mendelian analysis. Like others who had attempted to develop a Mendelian interpretation of sex determination, Goldschmidt accepted the following premises:

1. Each individual possesses both male and female sex factors.

2. The genetic basis of sex determination is set at fertilization.

3. The normal mechanism for the distribution of sex factors can be represented as XX versus XY (similar to a Mendelian backcross).

However, to explain the changing dominance of one sex factor over another during development, he added several novel assumptions stipulating how the factors influence development:

4. Genes are enzymes or other autocatalytic entities that act quantitatively in development.

5. The quantity of the sex factors present in the X chromosome is inherited, and different geographical races of a species vary in the quantitative values of their sex factors.

6. The sex that ultimately develops depends on the quantitative relationship, or *valency*, of the interacting inherited sex factors.

7. In hybrid crosses, a quantitative imbalance between sex factors can arise, resulting in the production of intersexes.

8. Intersexes arise when the quantitative difference in sex factors exceeds an "epistatic minimum" (characteristic for each species. and one factor no longer dominates the other, such that characters associated with the opposite develop.

Figure 7.3
Richard Goldschmidt in the Osborn Zoological Lab at Yale University (1915). Curt
Stern Photographic Collection, American Philosophical Society, Philadelphia.

Goldschmidt outlined these assumptions in a series of papers published after 1914. Hampered in extending his experimental program owing
to the outbreak of World War I, and unable to return to Germany following a year's research in Japan, Goldschmidt used the opportunity of
working as a visiting scientist in the United States to reflect on the implications of his breeding results (see figure 7.3).[34] He published a synthetic
account of his views on sex determination after returning to Germany. In
Mechanismus und Physiologie der Geschlechtsbestimmung, published in 1920,
Goldschmidt systematically laid out his *Lymantria* results and put forward
an account not only of the mechanism of sex determination but also a
physiological explanation based on analyzing the phenomena associated
with intersex moths.[35] Schematized, his explanation of the production of
intersexuality ran as follows:

1. In hybrids with an imbalance of the inherited sex factors, development is
initially governed by the "genetic" sex.
2. At some stage, however, a turning point is reached such that a "switch-
over reaction" occurs in which the factor that surpasses the epistatic minimum

forces "the alternative processes of differentiation to run in the direction of the other sex."[36]

3. Different degrees of intersexuality in moths can thus be explained on the basis of alterations occurring in the "timing" of developmental reactions: the earlier a "switch-over" reaction occurs, the greater the degree of observed intersexuality.[37]

Goldschmidt continued to elaborate and extend the physiological components of his theory after setting up his laboratory in the animal genetics division of the new Kaiser Wilhelm Institute for Biology in Berlin-Dahlem. By 1927, he believed he had all the elements in place to present a general physiological theory of heredity, and his *Physiologische Theorie der Vererbung* built on his explanatory scheme for the physiology of sex determination, with a few additions. He elaborated on how the quantitative gene transmitted through (nuclear) heredity became operational throughout (somatic) development. His conception, initially labeled the "mass law of reaction velocities" and later simply the "time law of development," assumed that genes control the rate of reactions of the myriad chemical processes occurring in the differentiating cell. The *Physiologische Theorie der Vererbung* thus presented the mature theory of physiological genetics that Goldschmidt had been developing for almost two decades. Its tenets can be summarized as follows:

1. Genes control independent chains of reactions that at one point produce something that leads to differentiation.

2. The quantitative imbalance between the valencies of two genes corresponds to differences in the velocities of two chains of reactions, and hence the rate of reaction.

3. The reaction that proceeds at a higher velocity controls the developmental process, including sexual differentiation.

4. The turning points in such reactions are the points at which control over the reaction product changes; expressed graphically, the turning points are the points of intersection of the two reaction curves (see figure 7.4).[38]

These tenets, Goldschmidt claimed, were empirically derived facts, not hypothetical assumptions. Only the notion that the end products of the gene-controlled reactions were particular "determining materials" (*Determinationsstoffe*), which he tentatively identified as hormones, was a "hypothesis."[39] But this hypothesis was simply a logical extension of what must be presumed about the biochemical properties of the gene: that incredibly small amounts must exert an extraordinary effect; that the amount must be conserved throughout cell division; and that the

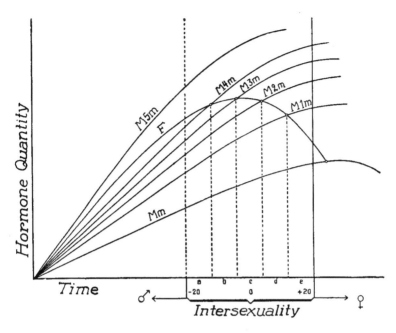

Figure 7.4
The time law of intersexuality. The F curve illustrates the production of female hormones and Mm the normal curve of male hormones. Different grades of intersexuality are produced from hybrid crosses, in which M_1m, M_2m, etc., indicate altered
quantities of male hormones. The point of intersection of the respective F and M
curves represents the "turning point" in the developmental process after which male
secondary sex characters come to predominate. From Richard Goldschmidt, *The Mechanism and Physiology of Sex Determination*, translated by William J. Dakin (London:
Methuen, 1923), p. 95.

quantity of the gene is proportional to the speed of a certain reaction.[40]
In short, Goldschmidt's basic understanding that the quantity of the gene
was linked to speeds of reaction and the ultimate production of "formative stuffs" presented an all-encompassing theory of the physiological
action of genes.[41] But because variation could be explained by assuming
fluctuations in gene quantity, Goldschmidt's theory also touched upon
certain critical aspects underlying the evolutionary process.

EVOLUTIONARY IMPLICATIONS OF GOLDSCHMIDT'S PHYSIOLOGICAL THEORY

Goldschmidt noted that *Physiologische Theorie der Vererbung* was essentially a
revision of his earlier monograph, *Die quantitativen Grundlage von Vererbung*

und Artbildung, although the discussion of the bearing of his theory on aspects of evolution had been updated.[42] In 1918, on the basis of his research into geographic variation in *Lymantria monaca*, Goldschmidt reported that the pigmentation of hybrid caterpillar larvae was controlled by multiple alleles, which he assumed were different quantities of the enzyme-gene: "If our conclusions regarding the nature of multiple allelomorphs are accepted," he explained, "it must lead to a different intellectual attitude toward the problem of variability of genes, which is so important for evolution."[43]

Applying this notion to development, slight quantitative variations in the quantity of enzyme-genes would result in alterations of the developmental processes they catalyze and, hence, phenotypic modifications of varying degrees. Among a series of multiple alleles, such quantitative variations offered the potential for evolutionary change. The selection of plus or minus gene variation, Goldschmidt noted, "can, therefore, change the quantity of the gene, and also, therefore, the somatic characters caused by quantitative differences in the gene, until the physiological limit is reached," the limit being a physiologically unviable form.[44]

Goldschmidt noted that "the first step in the differentiation of species which occurs in nature seems to be the formation of the geographic races." Indeed, it was in this light that intersexuality in *Lymantria* could best be understood, "as a step toward increasing incompatibility."[45] His notion of "incompatibility" resembles the modern view that hybrid sterility in crosses between geographic varieties is the first stage in the production of incipient species. In *Physiologische Theorie der Vererbung*, Goldschmidt further elaborated this idea, going beyond the implication of phenotypic characters depending on multiple alleles:

> The contemplation of the theory as a whole shows even more, however. Such a finely tuned system, perhaps comparable to a complicated manufacturing process utilizing a conveyor belt, must have a high degree of stability, since shifts without a complete disturbance of the whole system are conceivable only within a narrow range. The realization based on recent work in genetics that the stability of species is greater than their variability now becomes more comprehensible, along with the fact that the combination of very different systems in an individual [hybrid] is impossible. On the other hand, it also explains the view, which in itself breaks new ground, that it is not the large mutations that are species forming—they disturb the finely-tuned harmony of the chains of reactions—but rather the small changes, operating on all possible parts [of the organism], that creates an entirely new harmonic system.[46]

This statement is of particular interest, given that historians have generally assumed, based on Goldschmidt's subsequent view that only macromutations can account for speciation and the evolution of higher taxa, that he was a lifelong opponent of neo-Darwinism. Only later, in the mid-1930s, did Goldschmidt radically revise his conceptions.[47] His study of geographic variation in races of *Lymantria* indigenous to Japan also convinced him that there was a precise match between the timing of the life cycle and the sex-differentiated reactions in the developing moth. Adaptive advantage, therefore, helped to explain why closely related races possessed different valencies of the sex factors (considered as an example of a multiple allelomorphic series). Hence, environmental influences appeared to play an indirect rather than a direct role in evolution.

Such views well indicate that evolutionary problems were always in the forefront of Goldschmidt's thinking about heredity and development. Any theory of genetics not only had to explain the phenomena of gene transmission and account for developmental processes but also to contribute to an understanding of the mechanism of evolution. It is only in such a light that one can fully appreciate the confidence that Goldschmidt placed in his concept of the gene, both in the period under discussion and after 1940, when his evolving conception of the gene and the chromosome led to fundamental alterations in his developmental and evolutionary views.[48]

THE RECEPTION OF GOLDSCHMIDT'S PHYSIOLOGICAL THEORY OF GENETICS

Between 1920 and 1935, when his study of sex determination drew to a close, Goldschmidt made a heroic effort to provide a synthesis of genetics and developmental studies. In five books and close to seventy articles he outlined his views on sex determination, the phenomenon of intersexuality, and his physiological theory of heredity, pointing out the implications of his views for evolutionary problems. This outpouring of empirical data and theoretical analysis established Goldschmidt as a leading figure in genetics. However, his efforts to introduce dynamic principles into genetics were hailed as promising in the higher quest to understand the genetic control of development. As Julian Huxley noted in his 1923 review of the English translation of *Mechanismus und Physiologie der Geschlechtsbestimmung*, "If his correlation of the rate of production of the substance with the amount of some initial ferment contained in the gene, and this amount with the 'potency' of an allelomorph in a multiple series,—if this is substantiated, we acquire a new outlook into the relation between Mendelian genes and their mode of action in development."[49] Huxley was but one of a number of physiologically and developmentally minded biologists attracted to Goldschmidt's views.

Not all biologists, however, were as receptive. As Goldschmidt readily acknowledged in the opening pages of *Physiologische Theorie der Vererbung*, his views faced considerable opposition among transmission geneticists. "There are probably many reasons for this," he surmised with a tinge of resentment, then continued:

> First, my experimental work deviates in the kinds of questions it raises, as in the solutions it proposes, from the dominant methods of genetics, which are limited to a factoral analysis, so much so that orthodox Mendelians knew nothing of them from the beginning. . . . Another reason for the inattention probably lies in the coincidence that these works appeared at the same time as the striking discoveries of the *Drosophila* researchers. These have steered the interest of geneticists almost exclusively to the problem of the mechanism of heredity, and undoubtedly rightly so, since this problem has found through the remarkable accomplishments of the Morgan school an almost complete and unexpected solution.[50]

Despite these successes, however, Goldschmidt cautioned that transmission genetics should not be regarded as presenting a *theory* of heredity. A true theory needed to link the mechanism of gene transmission with the physiology of development; that is, it needed to "explain how the specific developmental processes can be initiated in their typically defined pathways from their starting point—the hereditary material [*Erbstoffe*]—and how the individual paths of development are combined, acting in unison, to produce a whole—the finished organism." It was on this basis that he dismissed what he regarded as the overly empirical and misguided eschewal of the role of theory in biology, quoting in support Ernst Rutherford's recent statement in *The Electrical Structure of Matter*: "experiment without imagination [*Phantasie*] or imagination unsupported by experiment can achieve but little; the most prominent advances arise from the happy combination of both."[51] It is not surprising, then, that Morgan and his followers were not receptive to the message presented in Goldschmidt's 1927 book.

CRITICISM OF GOLDSCHMIDT'S VIEWS BY THE MORGAN SCHOOL

Fundamental philosophical differences in interpreting genetic phenomena between Goldschmidt and the Morgan school first surfaced in 1917 in their opposing interpretations of crossing-over.[52] The Morgan school's particulate notion of the gene contrasted starkly with Goldschmidt's quantitative conception. Raphael Falk characterized these two competing approaches as reductionist and holistic, respectively, and claimed they were "incommensurable and incapable of communicating with one another."[53]

Certainly the same alternative stances can be recognized in the two groups' opposing interpretations of new data and problems. For example, although he did not mention Goldschmidt by name, Morgan made their respective differences quite clear in his widely read 1917 paper "The Theory of the Gene." In addressing the critics of the Mendelian-chromosome theory of heredity, Morgan mentioned those who charged that "the factorial interpretation is not physiological but only 'static,' whereas all really scientific explanations are 'dynamic.' " In response to this charge, he "urged the importance of keeping apart, *for the present at least*, the questions connected with the distribution of the genes in succeeding generations from questions connected with the physiological action of the genetic factors during development, because the embryological data have too often been confused in premature attempts to interpret the genetic data." He also rejected the gene-as-enzyme hypothesis, stating, "There is not the slightest reason to identify the substance produced in the testes with the substance of the gene; the chemical composition of the internal secretion may be entirely different from that of the gene, the latter producing its result in conjunction with substances resulting from other genes."[54]

Almost a decade later, in his 1926 Silliman lecture, published under the title *The Theory of the Gene*, Morgan referred to Goldschmidt's sex determination data only in passing. Yet he roundly criticized Goldschmidt's time law theory of intersexuality:

When examined in detail doubts arise, since it is bound up with assumptions concerning enzymes that are philosophical rather than chemical. Moreover, the male- and the female-producing factors are identified as the genes themselves. Such an interpretation of the process is at present purely speculative. Furthermore, his basal assumption, namely, that whichever enzyme starts first, it is overtaken later by the other competing enzyme, really begs the entire question, since this is not a recognized feature of enzyme behavior.[55]

Goldschmidt certainly recognized this apparent conceptual impasse. He wrote to Julian Huxley in 1920: "It is a shame that Morgan and his students, who are all unusually clever and experienced investigators, are stuck in such a narrow interpretation of hereditary phenomena and fight with hands and feet against a new, particularly physiological idea, that could enliven the Mendelian mechanism, which is gradually becoming boring."[56] Goldschmidt undoubtedly infuriated the Morgan group in a number of ways, most prominently by attempting to interpret their findings in light of his quantitative view of the gene.[57] This was particularly true in the case of Bridges's study of triploid intersexes and Sturtevant's work on bar-eye mutations in *Drosophila*.[58]

An insight into the group's reaction to *Physiologische Theorie der Vererbung* comes from the correspondence of Curt Stern, an assistant in Goldschmidt's lab who was a postdoctoral student with Morgan at Columbia in the mid-1920s and again at Cal Tech in 1932–1933. In May 1927, Franz Schrader wrote to Stern: "My knowledge of the reception of Goldschmidt's latest book is limited to an hours visit to the Columbia lab about a day after the book had arrived. Only the Boss and Sturt showed any signs of being affected, and apparently were disgusted especially at Goldschmidt's treatment of the Bar data. Goldschmidt's references to the fact that they have come around to his more general view on sex of course always makes them sore."[59] Sturtevant confided to Stern that he had reservations about Goldschmidt's work:

> So far as the *Lymantria* data are concerned, I must admit insufficient familiarity to make a really adequate criticism. Some points I do know, however, make me very doubtful. (a) The data are not quantitative in the first place, and a large subjective element must enter into any attempt to make a quantitative analysis. (b) Very little study of the first detailed account of the genetic results shows that the same experiment is often reported as giving two or three different results. So far as I know, these discrepancies are in no case serious in their consequences, but they greatly weaken my confidence in all the data. (c) At least in the case of larval patterns, the data seem to me to directly contradict the multiple allelomorph interpretation.[60]

Characteristically, Bridges referred to Goldschmidt's challenge in a humorous vein. In January 1927 (before the publication of Goldschmidt's book) he wrote to Stern:

> Give my best regards to Goldschmidt if you can without getting bitten. Tell him I don't claim to be very original, only to blunder along interpreting my own data and *developing* what ideas come to hand in that connection. We all do that. Everything that Goldschmidt claims to be

original is further elaborations of current ideas. He can't claim as much as he thinks but, I shall not contradict his claims.[61]

The antagonism of the Morgan group grew even greater after Goldschmidt claimed in 1929 to have produced phenotypic traits resembling mutations in *Drosophila* after subjecting the developing larvae to extreme temperatures. Franz Schrader chided Stern in 1931 for advising the group not to publish a refutation of these findings: "They have come around to so many of Goldschmidt's views after laughing at him that I think a good fall is just about coming to them."[62]

It is clear that the Morgan school and transmission geneticists in general rejected Goldschmidt's basic assumption of the quantitative gene and were critical of his physiological theory of heredity. Yet it would be wrong to assume, on this basis, that Goldschmidt's views were not influential. Several prominent geneticists and biologists were receptive to his views, even if perhaps critical of certain aspects of his explanation. Indeed, a group of influential British developmental biologists was greatly stimulated by Goldschmidt's physiological perspective and adopted elements of his approach into their own thinking about the genetic control of development. Hence, any assessment of Goldschmidt's place in the history of genetics needs to consider the wider developmental and genetics community beyond the confines of American transmission genetics based on *Drosophila* breeding.

Favorable Responses to Goldschmidt's Physiological Theory of Heredity

The antipathy of the Morgan group toward Goldschmidt's attempt to introduce physiological components into genetics was not shared by all geneticists, particularly those interested in developmental problems. In the United States, for example, members of William Castle's group at the Bussey Institution who were interested in the physiology of gene action as a result of studying the genetics of pigmentation were receptive to Goldschmidt's views.[63] The young Sewall Wright was particularly intrigued by Goldschmidt's gene-as-enzyme theory. Having interacted with Goldschmidt during his stay at the Bussey in 1915, Wright found Goldschmidt's biochemical and physiological conception of the gene appealing because it applied to his own study of coat color inheritance in guinea pigs.

In a 1927 paper written prior to the appearance of Goldschmidt's *Physiologische Theorie der Vererbung*, Wright presented, according to William

Provine, a "basic view of gene action" that was similar to Goldschmidt's perspective.[64] Wright was especially interested by Goldschmidt's suggestion that genes somehow controlled the rate of reactions in cells. In his 1934 paper "Physiological and Evolutionary Theories of Dominance," Wright cited Goldschmidt's "beautiful examples in Lepidoptera" to support the notion that "a genetic character difference is first manifested as a difference in rate of reaction, strongly suggestive of enzyme control." In criticizing Ronald A. Fisher's theory of dominance, Wright referred to Goldschmidt's "evidence that differences which on first analysis depend on regulation of time of onset or cessation of a reaction, on further analysis can be shown to depend on rates."[65] Such ideas stimulated Wright's attempt to introduce physiological considerations into the field of evolution. Indeed, he explicitly stated that "little progress [in evolutionary theory] can be made without development of a physiological theory."[66]

By far, however, it was British biologists who responded most sympathetically to Goldschmidt's physiological views, including Julian Sorrell Huxley, John Burdon Sanderson Haldane, and Conrad Hal Waddington. Bateson may have exerted some influence on encouraging a developmentalist perspective among British biologists interested in heredity. Robert Olby has noted that Bateson's study of "heredity and variation" gave a central place to development, and for this reason it does not completely map onto the field of "genetics" as it emerged after 1910.[67] But it was Huxley who was especially influential in promoting Goldschmidt's ideas in the United Kingdom.

Huxley and Goldschmidt first met in the summer of 1916 when both were in the United States. After the war, Huxley wrote to Goldschmidt: "We used to meet at Wood's Hole in 1916, when our two countries were at war; the fact that we could yet then meet as we did makes me hope that there should & can be no obstacle to our resuming relations now that that hateful period is over."[68] Goldschmidt responded that he was pleased to have received Huxley's letter, and added, "I am also of the view that the world's future will be furthered if the internationality of science can again be established."[69] Their exchange of letters continued until 1955, shortly before Goldschmidt's death.

Huxley's recent work on sex ratios in fish had focused his attention on problems pertaining to sex determination, prompting him to approach Goldschmidt. He wrote to Goldschmidt, "I find that your ideas correspond with those to which I had arrived in very many ways—& I had arrived at mine from a quite different angle." Huxley conceived of sex as resulting from a balance between "some agency" residing in the cytoplasm that exerted its effect on the X-chromosomes, the "strength" of which could vary. Similar to Goldschmidt's findings in *Lymantria*, Huxley observed a kind of sexual transformation in fish, and he queried Goldschmidt about details of his experiments and interpretation. Indeed, he was sufficiently interested to ask whether Goldschmidt would be willing to provide him with *Lymantria* stocks so that he could try out some of his ideas on an experimental system which permitted "quantitative work on the sex-problem in a way impossible with most other forms." Goldschmidt readily agreed, and promised to send Huxley egg batches in the fall. This he did, but Huxley's experience with breeding *Lymantria* came to naught after the stocks repeatedly succumbed to illness. He later told Goldschmidt that he had "wanted to see whether the degree of intersexuality in intersexual crosses could be modified by temperature, feeding, &c."[70]

Sympathetic to Goldschmidt's approach to genetical problems, Huxley sided with him against attacks from the Morgan school. He responded, for example, to Sturtevant's announcement of the discovery of intersex forms in *Drosophila*, claiming that "the intersexes are females, modified by a recessive autosomal mutant gene that causes them to show male parts," and hence were the product of an "intersex gene" not connected with the sex chromosomes. Sturtevant suggested that this finding cast doubt on Goldschmidt's and others' assumption that "the normal sex-determining mechanism itself was failing to function as usual." In response, Huxley pointed out the important differences between Sturtevant's material and the other cases of intersexuality. Largely focusing on Goldschmidt's *Lymantria* intersexes, Huxley concluded that "the burden of proof, in the present state of our knowledge, lies even more on the upholders of gene-produced intersexuality than on the upholders of the balance theory, but quite possibly both are right."[71] Huxley's response pleased Goldschmidt, since, as he told Huxley, he was unable to respond himself, since "it would not at the moment be opportune for me to do so. I completely agree with what you generally say, and would myself have written something similar."[72] Yet Huxley kept open the channels of communication with the Morgan group, through both correspondence and visits.

In August 1921, upon his return from the Oxford University expedition to Spitsbergen, Huxley visited Berlin to see Goldschmidt's laboratory and breeding setup at the Kaiser Wilhelm Institute of Biology (see figure 7.5). Two years later, while on a lecture tour in England, Goldschmidt was Huxley's guest in Oxford, and the two attended a gathering of the British Genetical Society at Tring, where Goldschmidt presented his results and interpretation of the mechanism of sex determination.

Figure 7.5
Richard Goldschmidt Lab, Kaiser Wilhelm Institute for Biology: *Lymantria* cultures
and technicians. Curt Stern Photographic Collection, American Philosophical Society,
Philadelphia.

When the English translation of Goldschmidt's *Mechanism and Physiology of Sex Determination* appeared in 1923, Huxley reviewed it favorably in *Nature*. His correspondence indicates that he discussed Goldschmidt's work with his colleagues, most notably Haldane, Edmund Briscoe Ford, Francis Albert Eley Crew, and Gavin de Beer. It is clear, then, that Huxley was profoundly stimulated by Goldschmidt's dynamic approach to genetics and that he sought to publicize and extend this program of research in England.

The extent to which Huxley was influenced by Goldschmidt's views is revealed in a letter of January 1922, in which he told Goldschmidt:

> I have been re-reading your book—& may I say how much indebted I feel to you for having brought this dynamic idea into genetics, & fused genetics with *Entwicklungsmechanik*? There are details over which I disagree—but the fundamental idea, of correlating genes, hormones, enzymes, & rates of growth & differentiation is one of the most fruitful, to my mind, that has entered biology in the last decade.[73]

As Robert Olby pointed out, Huxley was a member of the British tradition, inherited from Bateson and D'Arcy Wentworth Thompson, which
maintained "that many characteristics of organisms are determined by
purely physical considerations" and were products of "physical forces or
chemical reactions."[74] Huxley's insistence that genetics be concerned with
developmental problems can be seen in his and E. B. Ford's study of the
genetic control of eye color pigmentation in the shrimp *Gammarus*.

Huxley seized upon Goldschmidt's notion of rate genes as a means
of relating genetics to embryology. Having discovered that the eye color
of mutant forms of *Gammarus* changed with age from red to black, Huxley
and Ford began to investigate the phenomenon in detail, stimulated by
Goldschmidt's postulate of "genes controlling the rate of production of
sex-differentiating substances." They hoped to pursue this idea by providing additional examples showing "how fruitfully the factorial theory of
heredity can be extended by thinking of the genes, at least in a large
number of cases, as influencing not merely the character but also the rate
of developmental processes" (p. 132). In the case of *Gammarus*, they found
that the transformation was "dependent upon a single genetic factor controlling the *rate* of production of black pigment; this rate in its turn appears
to be correlated with the time of onset of deposition, and the *final density*
obtained" (p. 114). However, the authors also noted that Goldschmidt's
"attempts to reduce all differences between the action of factors to differences in the rates at which they affect various processes in the developing organism," while of great value, were "over-simplified" (p. 128).
Huxley thus aimed to extend this line of research to "launch a concerted
attack on the nature of gene action" through a program of physiological
genetics that "would bring together genetics, growth, and development."[75]
However, changes in his personal life brought this plan to a halt in
1932. Yet, as Frederick Churchill has convincingly argued, this work continued to exert an influence on British embryology through the publication of Huxley and de Beer's *The Elements of Experimental Embryology*
(1934).[76]

In this text, in addition to stressing the role of axial gradients and
the organizer in directing early embryogenesis, Huxley and De Beer integrated physiological genetics into their synthesis. Indeed, the text concludes by noting that "the epigenetic analysis of development is pointing
the way to a large extension of the field of heredity, in the shape of physiological genetics. It is only through a study of development that it will
be possible to understand what the term 'genetic characters' really stands
for—in other words, what are the basic processes involved in the action

of a particular Mendelian gene." In this way, the "fruitful contacts" between experimental embryology and physiology, "notably in the field of hormone action, with genetics, and with growth studies," would enrich the "search for the physico-chemical bases of the empirical biological principles."[77] Moreover, Huxley continued to stress the importance of linking the physiological approach to gene functioning as well as to evolutionary questions. This he did implicitly in *Problems of Relative Growth*, in which he explored the implications of allometry.[78] But in *Evolution: The Modern Synthesis*, Huxley explicitly stressed "that a study of the effects of genes during development is as essential for an understanding of evolution as are [*sic*] the study of mutation and selection." He also openly acknowledged his debt to both Goldschmidt and Morgan.[79]

Others were perhaps not as profoundly influenced by Goldschmidt's work as Huxley, but nonetheless assimilated his views. Haldane, for example, cited Goldschmidt's 1916 paper on sex determination in one of his earliest publications in genetics, a 1920 paper on sex ratios submitted to the *Journal of Genetics*. However, after reviewing the piece, Bateson advised him also to cite Goldschmidt's recent monographs *Die quantitative Grundlage von Vererbung und Artbildung* and *Mechanismus und Physiologie der Geschlechtsbestimmung*. Haldane replied that he was not familiar with these works but would read them.[80] A year later, Huxley served as an intermediary in sounding out Goldschmidt's opinion about "Haldane's rule," which postulated "that when hybridization leads to a marked preponderance of one sex over the other, (or to a marked abnormality of sexes) the preponderating (or normal) sex is the homozygous."[81] In his 1932 paper "The Time of Action of Genes, and Its Bearing on Some Evolutionary Problems," Haldane indeed referred to Goldschmidt's 1927 book, and its imprint on his elaboration of the biochemical nature of gene action in development and the evolutionary implications of "the time of action of genes" is inescapable.[82]

F. A. E. Crew of Edinburgh, who held one of the few professorships of genetics in Britain at the time, was a supporter of Goldschmidt's interpretation of sex determination (see figure 7.6). A friend and correspondent of Huxley's, Crew was well aware of Goldschmidt's work and used his "hypothesis" of sex determination in the early 1920s to interpret his own findings of sex reversal or intersexuality in domestic fowl.[83] In his 1925 textbook on animal genetics, Crew also applied Goldschmidt's balance view of sex to explain intersexuality in mammals, accepted his concept of the "potency" of the hereditary factors, and used a time-rate graph to provide a "speculative interpretation of the phenomena along the

Figure 7.6
Richard Goldschmidt with F. A. E. Crew (Edinburgh, 1923). Curt Stern Photographic
Collection, American Philosophical Society, Philadelphia.

lines suggested by Goldschmidt."[84] Moreover, he was not swayed by the
criticism of Goldschmidt's theory coming from the Morgan group.

In his 1933 book on sex determination, Crew expressed support for
a quantitative view of sex factors: "Though many have objected strongly
to this concept, it is a fact that his theory is one which accounts most
simply for the thoroughly orderly and consistent series of results obtained
through controlled experimentation with the geographical races of *Lymantria*. The great advantage of this theory of sex-determination is that while
it can more or less easily accommodate facts which relate to other forms
than *Lymantria*, alternative theories cannot so easily accommodate the facts
concerning *Lymantria*." Referring to Bridges's work on intersexuality in
Drosophila, Crew noted: "There can be no doubt that the phenomenon of
intersexuality is identical both in the case of *Lymantria* and in the case of
Drosophila, and the only essential difference between the two interpretations is that in Goldschmidt's there is the concept of a definite number
of gene molecules, whereas in that of Bridges there is the notion of a
definite number of completely linked genes. In the present state of our
knowledge it is permissible for each of us to choose the alternative that

he prefers."[85] Clearly, his preference was for an explanation that provided a physiological perspective.

The geologist-turned-biologist C. H. Waddington, although eventually becoming a critic of Goldschmidt's physiological explanation of development, nonetheless built on it in constructing his own mature approach to the problem of the genetic control of development. Coming to developmental studies as a young research student in the early 1930s, Waddington, drawing on the established British developmental tradition, began to explore implications of Hans Spemann and Otto Mangold's 1924 discovery of the organizer. After joining the Strangeways Laboratory at Cambridge in 1929 and developing a technique to study induction in avian and mammalian embryos, Waddington traveled to Amsterdam the following year to present his results at the International Congress of Experimental Cytology. There he met Goldschmidt, who invited him to come to Berlin to work in his laboratory. This Waddington did in 1931 with the support of a Rockefeller medical fellowship.[86]

While in Germany, Waddington also worked in Spemann's laboratory at Freiburg and then with Mangold at the Kaiser Wilhelm Institute in Berlin. Although excited by Spemann's work on the organizer, Waddington dismissed his vitalistic interpretation, convinced that the organizer could be assumed to be a chemical substance, and hence subject to an "atomistic" explanation. Between 1933 and 1937, Waddington worked on problems connected with the biochemistry of induction in collaboration with Joseph Needham and Dorothy Needham. Coming to believe that epigenetic development ultimately depended on the activity of genes, in June 1938 Waddington left for the United States to work with geneticists at Cold Spring Harbor, Columbia University, and the California Institute of Technology on the problem of the genetic control of wing development in *Drosophila*.[87]

Waddington's general textbook of genetics appeared in 1939, while he was in the United States. In the section "Genetics and Development," Waddington suggested that the "mode of action of genes during development" falls into "the general field of investigation of how an adult organism arises from the individuals of the previous generation." Hence, "some at least of the results of experimental embryology are essential for a full understanding of genetics."[88] In this section he frequently mentioned Goldschmidt's work. For example, in discussing "Gene Controlled Processes," Waddington cited Goldschmidt's study of the genetic control of pigmentation in *Lymantria* larvae alongside Ford and Huxley's work on *Gammarus*. He presented Goldschmidt's study of wing pattern inheritance and provided a full description of Goldschmidt's analysis of sex determination in *Lymantria*, as well as including seventeen of his publications in the bibliography. After describing Goldschmidt's quantitative notion of the sex factors and their connection with enzymes, Waddington noted that "this suggestion is to be regarded as a working hypothesis subsidiary to the main theory." He pointed out a few problems encountered by the gene-enzyme assumption. For example, he noted: "It is by no means clear that hypo-hypermorphs always differ only in quantity since it is easy, for instance, to imagine two enzymes which differ in chemical nature but which catalyze the same reaction with different efficiencies, so that their effects would differ quantitatively though they differed qualitatively themselves."[89] Yet overall Waddington generally expressed support for Goldschmidt's quantitative view.

In publications directed at a specialist audience, however, Waddington was less sanguine. In presenting the results of his own work on the genetics of wing pattern formation in fruit flies,[90] he harshly criticized the "sloppiness as well as explanatory inadequacy" of Goldschmidt's proposed physiological mechanisms. In 1941, in a paper published in *Nature*, Waddington launched a frontal attack on Goldschmidt in his role as the "most prominent biologist who has attempted to describe biological organization in terms which are at once developmental and not too far removed from the genetical concepts employed by students of evolution."[91] While praising Goldschmidt's introduction of the idea that genes controlled development by altering the rate of cellular reactions, he asserted that it was now outmoded. "This fruitful idea was successful in directing the attention of many geneticists to developmental problems, and has been the stimulus to much valuable work. It in no way diminishes the historical importance of Goldschmidt's theory to point out, after this lapse of time, that it is actually no more than the statement of the general notion of materialism in a four-dimensional world. So long as one considers development in material terms, there are only two things a gene could do: alter the velocity of a reaction (which includes arresting it completely), or initiate a new reaction; and the second of these can always be looked on as a secondary consequence of an alteration in the rate of some earlier reaction." In this way, Waddington brought attention to his own recent book, *Organisers and Genes*, in which he presented "some suggestions for an elaboration of Goldschmidt's hypothesis."[92]

Goldschmidt was certainly not pleased by Waddington's attack, especially after he had welcomed him into his laboratory. Goldschmidt appealed to Haldane in the matter, but Haldane declined to get involved:

You will realize that as I do not claim to be an expert on insect development, I cannot judge between you and Waddington in the controversy. Nor, of course, can I condemn him for what he did in your laboratory without hearing his version, and as your letter is a private one I cannot very well ask his opinion. However, it seems reasonable that a vertebrate embryologist should make some mistakes in his interpretation of insect material.[93]

Despite criticism directed at Goldschmidt's physiological theory of heredity, Waddington's own line of theorizing certainly built on that of Goldschmidt. As Edward Yoxen has noted, from the time he turned to the life sciences in the 1920s, Waddington expressed a strong "tendency to criticise existing concepts as imprecise."[94] Hence, his strong criticism of Goldschmidt's views may have been calculated to draw attention to his own, "more advanced" ideas. Certainly, Waddington frequently cited Goldschmidt in developing his own views with regard to a dynamic theory of genetics.[95]

Among this group of biologists, Joseph Needham was not as concerned as was Waddington and the others about the connection between genes and development. As Olby has noted, "Needham's own work was concerned with fusing biochemistry with *Entwicklungsmechanik*, but not with cytology and histochemistry."[96] Indeed, in 1934 Needham suggested that one of the major problems confronting biology was "the fusion of the two great realms of morphology and biochemistry or biophysics."[97] In his 1950 textbook on this topic, however, Needham favorably reviewed Goldschmidt's "classical" work on sex determination: "The sex factors of the nucleus, in Goldschmidt's view, produce male and female hormones, and development follows whichever of these is produced in excess. Hence, according to the speeds of the metabolic reactions producing the substances, the intersex switch-over point will occur at various times." Reflecting his biochemical orientation, Needham added, "Unfortunately, nothing whatever is known of the nature of the substances in question," citing recent unsuccessful attempts to influence insect differentiation by applying mammalian hormones.[98] That Needham would refer to this line of work at such a late date illustrates the enduring impact of Goldschmidt's work on sex determination and physiological genetics in Britain.

PHYSIOLOGICAL GENETICS AND EVOLUTION

Among the few biologists in the 1930s to attempt to apply recent developments in genetics and embryology to illuminating evolutionary questions is Gavin De Beer. Stimulated by Goldschmidt's idea of genes governing rates of reaction as a means of controlling embryonic development, De Beer applied this notion to evolution in *Embryology and Ancestors*, first published in 1930 and better known by the title *Embryos and Ancestors* used in subsequent editions. In this book De Beer focused attention on how "the genes are playing their part in company with the external factors in moulding the animal through the successive stages of ontogeny." This question, he noted, was just beginning to be asked, but "the lines on which it is to be answered were first indicated by Goldschmidt." Goldschmidt's study of the genetic control of intersexuality and of caterpillar coloration, bolstered by similar findings by Huxley and Ford and others, led De Beer to conclude "that by acting at different rates, the genes can alter the time at which certain structures appear."[99] He concluded that "heterochrony"—changes in the timing of organ formation in development—was a major source of evolutionary change. This idea was built on the assumption that "the internal factors exert their effects at certain definite rates" and that "modification of the rate of action of the internal factors in successive ontogenies will result in heterochrony."[100] De Beer's influential book turned Haeckel's famous dictum that phylogeny causes ontogeny on its head. It also focused attention on the process of heterochrony, which, as Raff has noted, "has had a profound effect on macroevolutionary thought."[101] Hence, Goldschmidt's ideas stimulated both De Beer's early work in embryology and his later thinking about evolution.

Waddington, too, cited Goldschmidt's *Physiologische Theorie der Vererbung* in his own consideration of how the notion of developmental systems impacted evolution theory. Like Wright, Waddington fully recognized "that a theory of evolution requires, as a fundamental part of it, some theory of development."[102] However, he was critical of Goldschmidt's view of the nature of systemic variations underlying phyletic change.[103] Ernst Mayr was likewise stimulated to elaborate his evolutionary views in reaction against Goldschmidt's "proposed solution of speciation through systemic mutations." Indeed, Mayr generalized this case, stating that "there are literally scores of cases in the history of science where a pioneer in posing a problem arrived at the wrong solution but where opposition to this solution led to the right solution."[104]

CONCLUSION

In the intellectual nexus of British developmental biology in the late 1920s and 1930s, then, it is clear that Goldschmidt's functionalist view of the gene excited considerable attention. His explanation of how genes could regulate chemical reactions in the cell and alter the timing of developmental events promoted an active program of exploration among a significant cohort of British biologists that included geneticists, embryologists, biochemists, developmental biologists, and theorists. It may not be entirely coincidental that mainstream geneticists increasingly turned to lines of work relating to developmental genetics after the publication of *Physiologische Theorie der Vererbung*. Goldschmidt's suggestions about how these ideas related to evolution also generated considerable attention, both pro and con.

Historical assessments of Goldschmidt's program of physiological genetics, then, should consider not simply its impact on genetics but also the stimulus it provided to other biological disciplines, including developmental biology, embryology, and evolution studies. Certainly, the empirical results of Goldschmidt's *Lymantria* work appear modest compared with the remarkable successes of the *Drosophila* research program of the Morgan school. Historians have noted significant differences in the lines of research carried on by Goldschmidt and the Morgan group. Robert Kohler has highlighted the ways in which *Drosophila* research generated a particular experimental culture and practices that led to social networks and allegiances not transferable to *Lymantria*.[105] Jonathan Harwood has identified fundamental institutional factors that constrained the disciplinary development of genetics in Germany, and others have noted similar institutional constraints operating on genetics in Britain.[106] Garland Allen has contrasted Goldschmidt's greater acceptance of the role of theory construction in biology compared with the Morgan school's high standards of empirical verification.[107] And Michael Dietrich has pointed out how Goldschmidt's later criticism of the corpuscular gene ultimately represented a political attack on the authority of the Morgan school.[108]

All of these approaches provide fundamental insights into the historical development of genetics. However, as I have attempted to show, Goldschmidt never considered himself just a geneticist. Rather, from the outset, he aimed to use the methodology of breeding analysis offered by genetics to solve the "grand problems" of biology: the interrelationships of heredity, development, and evolution. Hence, to assess his career fully, one must broaden the scope of historical analysis beyond genetics to examine the wider impact his views had on other areas of biology. From this vantage point, as we have seen, Goldschmidt's empirical results and theoretical views had a far-ranging impact. However much he may have been a "fringe figure" in genetics or the evolutionary synthesis, he was certainly not one in developmental biology.

One measure of the success of a research program, moreover, comes from gauging the influence it exerts on future generations. In Germany, it appears that Goldschmidt's work in physiological genetics attracted considerable interest and helped direct younger biologists into the field. The geneticist Ernst Caspari, for example, noted the impact of Goldschmidt's work on his distinguished career in genetics. Instrumental in opening new paths for studying developmental genetics in the 1930s through his analysis of pigment formation in insects, Caspari admitted that the impetus for this research came via Goldschmidt's writings:

> Most influential all through his life were his numerous books, textbooks as well as monographs in which he developed his theories. I can take myself as an example: my decision to study zoology was strongly influenced by Goldschmidt's popular book "Ascaris" and I owe my introduction to genetics to his "Einführung in die Vererbungswissenschaft" at a time when no courses in genetics were offered at most German Universities. I read "Physiologische Theorie der Vererbung" as a graduate student and it was instrumental in directing my interest to developmental genetics and suggesting ways to approach its problems.[109]

Goldschmidt's dynamical views greatly stimulated the thinking of contemporary biologists. Caspari recalled:

> Goldschmidt's "quantitative theory of genetics," offered for the first time a consistent model which accounted for gene structure and gene action in a simple and straightforward manner. . . . Thus, the theory offered a monumental concrete and coherent picture of the nature of genes and mutations, and of the action of genes in development. It appeared attractive to students because it united concepts from genetics, embryology and biochemistry and in this way offered a more comprehensive view of life processes than the competing theory based on interaction at the gene level only.[110]

More than his empirical findings, it was Goldschmidt's attempt to provide a "comprehensive view of life processes" that captured the imagination of so many biologists of his time, and that continues to attract interest today.

Goldschmidt's contributions to the development of twentieth-century biology, therefore, rest as much on theoretical ideas as on empirical accomplishments. His approach to genetics emerged out of a productive experimental tradition in late nineteenth-century Germany that encouraged theorists to relate their findings to "the big picture." More specifically, Goldschmidt's approach to the problems of heredity, development, and evolution grew out of the earlier work of Boveri, whose chromosome theory also focused on the developmental role of genes.[111] While this tradition had little practical affect on the development of transmission genetics, it influenced a number of leading biologists who were receptive to Goldschmidt's dynamic conception of gene action. As we have seen, they included Wright in America, Caspari and other developmental geneticists in Germany, and Huxley, De Beer, Crew, Haldane, Needham, and Waddington in Britain, all of whom attempted in various ways to synthesize a physiological understanding of heredity into current thinking about developmental, biochemical, and evolutionary problems.

By the late 1930s, even mainstream transmission geneticists began to move beyond the "static" approach of classic genetics to develop new lines of attack on the physiology of gene action. As Boris Ephrussi declared in his 1938 paper outlining his work on developmental genetics, "What geneticists want to know is what a gene is and how it works."[112] This statement characterizes Goldschmidt's life aim particularly well. From the beginning of his work in genetics, he set his sights on understanding "what a gene is and how it works," searching for ways to explore how genes control developmental processes, and then to link this understanding to possible evolutionary mechanisms. While a number of factors contributed to his particular style of genetics, it is also important to note that his physiological conception of genetics can be traced to the basic ideas of cell structure and function he developed as a young man, and specifically to the view of the nuclear control of heredity, metabolism, and development depicted in the chromidial theory. While he continually modified his views on the basis of new findings, he nonetheless strived throughout his career to fit the empirical data of genetics into the entire edifice of biological theory, founded on understanding the cell as the basis for heredity and development.[113]

Though such an approach came into conflict with what Waddington called "an aggressively anti-metaphysical period" in science, Goldschmidt's holistic vision of life fueled his own work in science and stimulated an important cohort of contemporary biologists.[114] As Stern wrote to Goldschmidt in 1958, on the occasion of his eightieth birthday, "To be Goldschmidtian has become a general trend in genetics."[115] Indeed, Goldschmidt's approach to problems of heredity, development, and evolution continues to resonate today, even if those currently engaged in attacking these problems are armed with a different set of empirical data and theoretical assumptions.

Notes

1. Darwin is a good case in point. As Rasmus Winther notes, "For Darwin, the study of heredity required the study of development" (Winther, 2000, p. 447).

2. Churchill, 1987, 361. See also Robinson, 1979.

3. Winther argues that Weismann's own views must be distinguished from the tradition of Weismannism, which regarded heredity and development as separate processes, noting that "development, heredity, and variation were intimately linked for Weismann" (Winther, 2001, p. 520).

4. Churchill, 1999. See also Gilbert, 1978.

5. Falk, 1995, pp. 225–226: "A central insight of Mendel was his ability to deal with the problem of inheritance as transmission, without being distracted by the problem of inheritance as a problem of development. He used traits merely as markers of transmission, disregarding their morphology and their developmental history or mechanics."

6. Ibid., pp. 225–235.

7. Morgan et al., 1915, pp. viii–ix.

8. For good discussions of this problem, see Allen, 1985; and Gilbert, 1988.

9. Boveri, 1918, p. 465.

10. Richmond, 1989, 1986.

11. Goldschmidt, 1933, p. 543.

12. Richmond and Dietrich, 2002.

13. Dietrich, 2003. Betty Smocovitis, however, traces this appellation to a more recent date: "It was Stephen Jay Gould and others who were to portray Goldschmidt as a 'heretic' and an antihero." Smocovitis, 1996, p. 161, n. 193.

14. See Bowler, 1984, p. 265: "The concept of evolution by discontinuous mutations moving in a particular direction was still supported in the 1930s by a few unorthodox geneticists (e.g., Goldschmidt, 1940)." See also Allen, 1974, and Piternick, 1980. Stephen Jay Gould recalled that when he was a graduate student at Columbia University in the early 1960s, Goldschmidt's name was frequently mentioned by his professors, but only in derogatory terms, indicating to graduate students "that Richard B. Goldschmidt was not to be taken seriously as an evolutionary biologist" (Gould, 2002, p. 452).

15. Dunn, 1991; Sturtevant, 1965.

16. See Goldschmidt's defense of theorizing in Goldschmidt, 1927, pp. 2–3. See also Allen, 1974.

17. See, for example, Rader, 1998.

18. Mayr, 1997, p. 33.

19. Ibid., p. 33.

20. Smocovitis, 1996; Dietrich, 1995.

21. Eldredge and Gould, 1972; Gould and Eldredge, 1993; Gould, 1980, 1982.

22. Raff and Kaufman, 1991; Raff, 1996; Gilbert et al., 1996.

23. Raff, 1996, p. 13.

24. Kay, 1993, p. 128; Kohler, 1994, chap. 7.

25. Richmond, 1986.

26. Richmond, 1989.

27. Conklin, 1924; Cowdry, 1924; Wilson, 1925, pp. 700–706.

28. T. S. Painter to Goldschmidt (December 23, 1943), Goldschmidt Papers, Bancroft Library, University of California at Berkeley.

29. Goldschmidt, 1960, p. 75.

30. Morgan, 1925, p. 693.

31. Richmond, 1986.

32. On the background of the concept of epistasis developed by the Bateson school, see Richmond, 2001.

33. Goldschmidt, 1915, 1916.

34. For the circumstances surrounding Goldschmidt's stay in the United States following the outbreak of World War I, see Goldschmidt, 1960, chap. 7.

35. Goldschmidt, 1920a.

36. Goldschmidt, 1923, p. 91.

37. Goldschmidt, 1927, chap. 3.

38. Ibid., pp. 15–16.

39. Ibid., pp. 24–25.

40. Ibid., p. 32.

41. Ibid., p. 39.

42. Ibid., p. 2; Goldschmidt, 1920b.

43. Goldschmidt, 1918, pp. 28, 40.

44. Ibid., p. 40. As we shall see, Goldschmidt further developed this idea after 1933.

45. Ibid., pp. 41, 42.

46. Goldschmidt, 1927, pp. 47–48.

47. For the development of Goldschmidt's mature evolutionary views, see Dietrich, 1995, 2000a, 2003. See also Mayr, 1997, pp. 30–33.

48. Dietrich, 2000b.

49. Huxley, 1923, p. 928.

50. Goldschmidt, 1927, pp. 1–2.

51. Ibid., p. 3.

52. Richmond and Dietrich, 2002.

53. Falk, 1986, p. 155.

54. Morgan, 1917, pp. 543, 513, 535.

55. Morgan, 1926, pp. 245–246.

56. Goldschmidt to Huxley (May 27, 1920), Julian Sorrell Huxley Papers, Woodson Research Center, Fondren Library, Rice University. Translation is mine.

57. See, for example, Goldschmidt's discussion of the support for his quantitative nature of gene mutations, and problems with the Morgan school's qualitative interpretation, in Goldschmidt, 1927, pp. 91–95.

58. Richmond, 1986, and in progress.

59. Franz Schrader to Stern (May 5, 1927), Curt Stern Papers, American Philosophical Society, Philadelphia (hereafter APS).

60. Sturtevant to Stern (October 4, 1928), Stern Papers, APS.

61. Bridges to Stern (January 13, 1927), Stern Papers, APS. Bridges is referring to Goldschmidt's priority claim for the balance theory of sex determination. See Dunn, 1991, pp. 182–183.

62. Schrader to Stern (March 14, 1931), Stern Papers, APS. Schrader prefaced this comment thus: "It has become known that you warned Schultz not to publish his anti-Goldschmidt findings in re the temperature mutations. The general opinion at Woods Hole last summer was that that was not exactly the right thing to do, in view of the loudly voiced conviction of the Pasadena people that Goldschmidt had done very bad work on it. I think myself that you should have let them go right ahead and make damn fools of themselves." For a fuller treatment of this episode, see Dietrich, 2000c, p. 1143.

63. Provine, 1986; Rader, 1998.

64. Provine, 1986, p. 119.

65. Wright, 1934, p. 35.

66. Ibid., p. 29. See also Provine, 1986, p. 301.

67. Olby, 1992. For a survey of British genetics in the interwar period, see Marie, 2004.

68. Huxley to Goldschmidt (April 8, 1920), Goldschmidt Papers, Bancroft Library, University of California at Berkeley.

69. Goldschmidt to J. S. Huxley (April 15, 1920), Julian Huxley Papers, Fondren Library, Rice University.

70. Huxley to Goldschmidt (June 1, 1920), Goldschmidt Papers, Bancroft Library, University of California at Berkeley.

71. Sturtevant, 1920; Huxley, 1920, p. 59.

72. Goldschmidt to Huxley (May 27, 1920), Julian Huxley Papers, Fondren Library, Rice University.

73. Huxley to Goldschmidt (January 6, 1922), Goldschmidt Papers, Bancroft Library, University of California at Berkeley.

74. Olby, 1992, p. 62.

75. Ford and Huxley, 1927, pp. 114, 128, 132, 1929. See also Ridley, 1985, p. 60.

76. Churchill, 1992.

77. Huxley and De Beer, 1934, p. 442.

78. Huxley, 1932. For an informative discussion of the intellectual background to this work, see Churchill 1993.

79. Huxley, 1942, p. 8. See also Hamburger, 1980, pp. 99–100; Ridley, 1985, pp. 62–63; Witkowski, 1992, pp. 93–99.

80. Haldane to Bateson (November 24, 1920), Bateson to Haldane (December 1, 1920), Haldane to Bateson (n.d. [December 1920]), Bateson Letters, John Innes Foundation Historical Collections, John Innes Centre, Norwich, U.K.

81. Huxley to Goldschmidt (June 1, 1920), Goldschmidt Papers, Bancroft Library, University of California at Berkeley. See also Haldane, 1922.

82. Haldane, 1932.

83. Crew, 1923a, 1923b.

84. Crew, 1925, pp. 232–233.

85. Crew, 1933, pp. 101–103.

86. Yoxen, 1986. Waddington visited Goldschmidt's lab in July 1931 and again in July 1933. See copy of Goldschmidt's guest book, Stern Papers, APS.

87. Robertson 1977, pp. 597–598, 579.

88. Waddington, 1939a, p. 137.

89. Ibid., p. 219.

90. Waddington, 1939b. See also Waddington,1940a.

91. Waddington, 1941, p. 108.

92. Ibid., pp. 108, 109. Waddington, 1940b.

93. Haldane to Goldschmidt (December 11, 1940), Goldschmidt Papers, Bancroft Library, University of California at Berkeley.

94. Yoxen, 1986, p. 310.

95. See, for example, the discussion of "epigenetics" and his notion of the "epigenotype" to denote the "whole complex of development" that lies between the genotype and the phenotype in Waddington, 1942.

96. Olby, 1986, p. 299.

97. Needham, 1934, pp. 275–276.

98. Needham, 1950, p. 318.

99. De Beer, 1954, pp. 17–19, 20.

100. Ibid., p. 107.

101. Raff, 1996, p. 256. See also Gould, 1977.

102. Waddington, 1941, p. 108.

103. See Goldschmidt, 1940.

104. Mayr, 1982, p. 381.

105. Kohler, 1994.

106. Harwood, 1993; Lewis, 1969.

107. Allen, 1978, 1974.

108. Dietrich, 2000c, p. 1144: "Questioning the gene, questioned the conceptual framework for *Drosophila* genetics. In doing so it questioned Morgan's authority and the authority of his successor (Sturtevant) and his associates to conceptualize the field."

109. Caspari, 1980, p. 20.

110. Ibid.

111. L. C. Dunn, in an unpublished manuscript titled "Physiological Genetics" (Dunn Papers, APS), places great emphasis on the significance of Boveri's contributions:

"When I started out to write about the general evidences of the control of development by nuclear factors I had not intended to trace the roots of all of them to 19th century origins. But that's the way in which it has turned out. Nor had I realized how frequently the ideas generated by one man would be associated with these beginnings. But Theodor Boveri has come to occupy that position, at least in my present view. To me he appears to have replaced Weismann as the originator of basic ideas in the pre-Mendelian era" (p. 16). Scott Gilbert suggests that Goldschmidt, along with Ernest Everett Just, sought "to place American genetics into the German type of developmental physiology." See Gilbert, 1988, p. 325.

112. Ephrussi, 1938, p. 5.

113. See, for example, Goldschmidt, 1918, which concludes thus: "Progress of science follows of course a slowly ascending, wavy curve, with always recurring valleys. But viewed from some distance, the waves disappear and only the upward trend remains visible. Such is also the case with our knowledge of the methods and means of evolution."

114. Waddington, 1975, p. 10. Waddington argues "that a scientist's metaphysical beliefs are not mere epiphenomena, but have a definite and ascertainable influence on the work he produces" (p. 7). This viewpoint appears particularly applicable to Goldschmidt.

115. Stern to Goldschmidt (April 5, 1958), Stern Papers, APS.

REFERENCES

Allen, Garland E. 1974. "Opposition to the Mendelian-Chromosome Theory: The Physiological and Developmental Genetics of Richard Goldschmidt." *Journal of the History of Biology* 7: 49–92.

———. 1978. *Thomas Hunt Morgan: The Man and His Science.* Princeton, N.J.: Princeton University Press.

———. 1985. "Heredity Under an Embryological Paradigm: The Case of Genetics and Embryology." *Biological Bulletin* 168 (suppl.): 107–121.

Boveri, Theodor. 1918. "Zwei Fehlerquellen bei Merogonieversuchen und die Entwicklungsfähigkeit merogonischer, partiell-merogonischer Seeigelbastarde." *Archiv für Entwicklungsmechanik der Organismen* 44: 417–471.

Bowler, Peter J. 1984. *Evolution: The History of an Idea.* Berkeley: University of California Press.

Caspari, Ernst W. 1980. "An Evaluation of Goldschmidt's Work After Twenty Years." In *Richard Goldschmidt: Controversial Geneticist and Creative Biologist*, ed. Leonie K. Piternick, pp. 19–23. Basel and Boston: Birkhäuser.

Churchill, Frederick B. 1987. "From Heredity Theory to *Vererbung*: The Transmission Problem, 1850–1915." *Isis* 78: 361.

————. 1992. "*The Elements of Experimental Embryology*: A Synthesis for Animal Development." In *Julian Huxley: Biologist and Statesman of Science*, ed. C. Kenneth Waters and Albert van Helden, pp. 107–126. Houston, Tex.: Rice University Press.

————. 1993. "On the Road to the *k* Constant: A Historical Introduction." In Julian S. Huxley, *Problems of Relative Growth*, repr. ed., pp. xix–xlv. Baltimore: Johns Hopkins University Press.

————. 1999. "August Weismann in a Mendelian World." Paper delivered at the History of Science Society annual meeting, Pittsburgh.

Conklin, Edwin G. 1924. "Cellular Differentiation." In *General Cytology: A Textbook of Cellular Structure and Function for Students of Biology and Medicine*, ed. Edmund V. Cowdry et al., pp. 539–607. Chicago: University of Chicago Press.

Cowdry, Edmund V. 1924. "Cellular Constituents—Mitochondria, Golgi Apparatus, and Chromidial Substance." In *General Cytology: A Textbook of Cellular Structure and Function for Students of Biology and Medicine*, ed. E. V. Cowdry et al., pp. 318–382. Chicago: University of Chicago Press.

Crew, F. A. E. 1923a. "Studies in Intersexuality. I. A Peculiar Type of Developmental Intersexuality in the Male of Domesticated Mammals." *Proceedings of the Royal Society of London* B 95: 90–109.

————. 1923b. "Studies in Intersexuality. II. Sex-Reversal in the Fowl." *Proceedings of the Royal Society of London* B 95: 256–278.

————. 1925. *Animal Genetics: An Introduction to the Science of Animal Breeding.* Edinburgh and London: Oliver and Boyd.

————. 1933. *Sex Determination.* London: Methuen.

De Beer, Gavin. 1954. *Embryos and Ancestors*, rev. ed. Oxford: Clarendon Press.

Dietrich, Michael R. 1995. "Richard Goldschmidt's 'Heresies' and the Evolutionary Synthesis." *Journal of the History of Biology* 7: 431–461.

————. 2000a. "From Hopeful Monsters to Homeotic Effects: Richard Goldschmidt's Integration of Development, Evolution, and Genetics." *American Zoologist* 40: 28–37.

————. 2000b. "From Gene to Genetic Hierarchy: Richard Goldschmidt and the Problem of the Gene." In *The Concept of the Gene in Development and Evolution: Historical and Epistemological Perspectives*, ed. Peter Beurton, Raphael Falk and Hans-Jörg Rheinberger, pp. 91–114. Cambridge: Cambridge University Press.

————. 2000c. "The Problem of the Gene." *Comptes rendus de l'Académie des Sciences de Paris* 323: 1139–1146.

————. 2003. "Richard Goldschmidt: Hopeful Monsters and Other 'Heresies.' " *Nature Reviews Genetics* 4: 68–74.

Dunn, L. C. 1991. *A Short History of Genetics: The Development of Some of the Main Lines of Thought, 1864–1939.* Ames: Iowa State University Press. First published 1965.

Eldredge, Niles, and Gould, Stephen Jay. 1972. "Punctuated Equilibria: An Alternative to Phyletic Gradualism." In *Models in Paleobiology*, ed. T. J. M. Schopf, pp. 82–115. San Francisco: Freeman, Cooper.

Ephrussi, Boris. 1938. "Aspects of the Physiology of Gene Action." *American Naturalist* 72: 5–23.

Falk, Raphael. 1986. "What Is a Gene?" *Studies in the History and Philosophy of Science* 17: 133–173.

————. 1995. "The Struggle of Genetics for Independence." *Journal of the History of Biology* 28: 219–246.

Ford, E. B., and Huxley, Julian S. 1927. "Mendelian Genes and Rates of Development in *Gammarus Chevreuxi*." *British Journal for Experimental Biology* 5: 112–134.

————. 1929. "Genetic Rate-Factors in *Gammarus*." *Archiv für Entwicklungsmechanik der Organismen* 117: 67–79.

Gilbert, Scott F. 1978. "The Embryological Origins of the Gene Theory." *Journal of the History of Biology* 11: 307–351.

————. 1988. "Cellular Politics: Ernest Everett Just, Richard B. Goldschmidt, and the Attempt to Reconcile Embryology and Genetics." In *The American Development of Biology*, ed. Ron Rainger, Keith Benson, and Jane Maienschein, pp. 311–346. Philadelphia: University of Pennsylvania Press.

Gilbert, Scott F., Opitz, J., and Raff, R. A. 1996. "Resynthesizing Evolutionary and Developmental Biology." *Developmental Biology* 173: 357–372.

Goldschmidt, Richard. 1915. "Vorläufige Mitteilung über weitere Versuche zur Vererbung und Bestimmung des Geschlechts." *Biologische Centralblatt* 35: 565–570.

————. 1916. "A Preliminary Report on Further Experiments in Inheritance and Determination of Sex." *Proceedings of the National Academy of Sciences USA* 2: 53–58.

————. 1918. "A Preliminary Report on Some Genetic Experiments Concerning Evolution." *American Naturalist* 52: 28–50.

————. 1920a. *Mechanismus und Physiologie der Geschlechtsbestimmung*. Berlin: Gebruder Borntraeger.

————. 1920b. *Die quantitative Grundlage von Vererbung und Artbildung*. Berlin: Julius Springer.

————. 1923. *The Mechanism and Physiology of Sex Determination*, trans. William J. Dakin. London: Methuen; New York: George H. Doran.

————. 1927. *Physiologische Theorie der Vererbung*. Berlin: Julius Springer.

————. 1933. "Some Aspects of Evolution." *Science* 78: 539–547.

————. 1940. *The Material Basis of Evolution*. New Haven, Conn.: Yale University Press.

Goldschmidt, Richard B. 1960. *In and Out of the Ivory Tower: The Autobiography of Richard B. Goldschmidt*. Seattle: University of Washington Press.

Gould, Stephen Jay. 1977. *Ontogeny and Phylogeny*. Cambridge, Mass.: Belknap Press of Harvard University Press.

———. 1980. "The Hopeful Monster Revisited." In Gould's *The Panda's Thumb*. New York: Norton.

———. 1982. "The Uses of Heresy: An Introduction to Richard Goldschmidt's *The Material Basis of Evolution*." In Richard Goldschmidt, *The Material Basis of Evolution*. New Haven, Conn.: Yale University Press.

———. 2002. *The Structure of Evolutionary Theory*. Cambridge, Mass.: Belknap Press of Harvard University Press.

Gould, Stephen Jay, and Eldredge, Niles. 1993. "Punctuated Equilibrium Comes of Age." *Nature* 366: 223–227.

Haldane, J. B. S. 1922. "Sex Ratio and Unisexual Sterility in Hybrid Animals." *Journal of Genetics* 12: 101–109.

———. 1932. "The Time of Action of Genes, and Its Bearing on Some Evolutionary Problems." *American Naturalist* 66: 5–24.

Hamburger, Viktor. 1980. "Embryology and the Modern Synthesis in Evolutionary Theory." In *The Evolutionary Synthesis: Perspectives on the Unification of Biology*, ed. Ernst Mayr and William B. Provine. Cambridge, Mass.: Harvard University Press.

Harwood, Jonathan. 1993. *Styles of Scientific Thought: The German Genetics Community, 1900–1933*. Chicago: University of Chicago Press.

Huxley, Julian S. 1920. "Intersexes in *Drosophila* and Different Types of Intersexuality." *Science* 52: 59–60.

———. 1923. "The Physiology of Sex-Determination." *Nature* 112: 927–930.

———. 1932. *Problems of Relative Growth*. London: Dial Press.

———. 1942. *Evolution: The Modern Synthesis*. London: George Allen & Unwin.

Huxley, Julian S., and De Beer, Gavin R. 1934. *The Elements of Experimental Embryology*. Cambridge: Cambridge University Press.

Kay, Lily E. 1993. *The Molecular Vision of Life: Caltech, the Rockefeller Foundation, and the Rise of the New Biology*. New York: Oxford University Press.

Kohler, Robert E. 1994. *Lords of the Fly: Drosophila Genetics and the Experimental Life*. Chicago: University of Chicago Press.

Lewis, D. 1969. "Preface: The Genetical Society—the First Fifty Years." In John Jinks (ed.), *Fifty Years of Genetics: Proceedings of a Symposium Held at the 160th Meeting of the Genetical Society on the 50th Anniversary of Its Foundation*, pp. 1–7. Edinburgh: Oliver and Boyd.

Marie, Jennifer. 2004. "The Importance of Place: A History of Genetics in 1930s Britain." Ph.D. dissertation, University College, London.

Mayr, Ernst. 1982. *The Growth of Biological Thought: Diversity, Evolution, and Inheritance.* Cambridge, Mass.: Belknap Press of Harvard University Press.

———. 1997. "Goldschmidt and the Evolutionary Synthesis: A Response." *Journal of the History of Biology* 30: 30–33.

Morgan, Thomas Hunt. 1917. "The Theory of the Gene." *American Naturalist* 51: 513–544.

———. 1924. "Mendelian Heredity in Relation to Cytology." In *General Cytology: A Textbook of Cellular Structure and Function for Students of Biology and Medicine*, ed. E. V. Cowdry et al., pp. 693–734. Chicago: University of Chicago Press.

———. 1926. *The Theory of the Gene.* New Haven, Conn.: Yale University Press.

Morgan, T. H., Sturtevant, A. H., Müller, H. G., and Bridges, C. B. 1915. *The Mechanism of Mendelian Heredity.* New York: Holt.

Needham, Joseph. 1934. "Morphology and Biochemistry." *Nature* 134: 275–276.

———. 1950. *Biochemistry and Morphogenesis.* Cambridge: Cambridge University Press. First published 1942.

Olby, Robert. 1986. "Structural and Dynamic Explanations in the World of Neglected Dimensions." In *A History of Embryology*, ed. J. J. Horder, J. A. Witkowski, and C. C. Wylie, pp. 275–308. Cambridge: Cambridge University Press.

———. 1992. "Huxley's Place in Twentieth-Century Biology." In *Julian Huxley: Biologist and Statesman of Science*, ed. C. Kenneth Waters and Albert van Helden, pp. 53–75. Houston, Tex.: Rice University Press.

Piternick, Leonie K. (Ed.). 1980. *Richard Goldschmidt: Controversial Geneticist and Creative Biologist. A Critical Review of His Contributions.* Basel and Boston: Birkhäuser.

Provine, William. 1986. *Sewall Wright and Evolutionary Biology.* Chicago: University of Chicago Press.

Rader, Karen A. 1998. " 'The Mouse People': Murine Genetics Work at the Bussey Institution, 1909–1936." *Journal of the History of Biology* 31: 327–354.

Raff, Rudolf A. 1996. *The Shape of Life: Genes, Development, and the Evolution of Animal Form.* Chicago: University of Chicago Press.

Raff, Rudolf A., and Kaufman, Thomas C. 1991. *Embryos, Genes, and Evolution: The Developmental-Genetic Basis of Evolutionary Change.* Bloomington: Indiana University Press. First published New York: Macmillan, 1983.

Richmond, Marsha. 1986. "Richard Goldschmidt and Sex Determination: The Growth of German Genetics, 1900–1935." Ph.D. dissertation, Indiana University, UMI no. 8707816.

Richmond, Marsha L. 1989. "Protozoa as Precursors of Metazoa: German Cell Theory and Its Critics at the Turn of the Century." *Journal of the History of Biology* 22: 243–276.

———. 2001. "Women in the Early History of Genetics: William Bateson and the Newnham College Mendelians, 1900–1910." *Isis* 92: 55–90.

———. In progress. "*The Making of a Heretic: Richard Goldschmidt and Physiological Genetics*," book manuscript in preparation.

Richmond, Marsha L., and Dietrich, Michael R. 2002. "Richard Goldschmidt and the Crossing-over Controversy." *Genetics* 161: 477–482.

Ridley, Mark. 1985. "Embryology and Classical Zoology in Great Britain." In *A History of Embryology*, ed. J. J. Horder, J. A. Witkowski, and C. C. Wylie, pp. 35–67. Cambridge: Cambridge University Press.

Robertson, Alan. 1977. "Conrad Hal Waddington, 8 November 1905–1926 September 1975." *Biographical Memoirs of Fellows of the Royal Society* 23: 575–622.

Robinson, Gloria. 1979. *A Prelude to Genetics: Theories of a Material Substance of Heredity, Darwin to Weismann*. Lawrence, Kan.: Coronado Press.

Smocovitis, Vassiliki Betty. 1996. *Unifying Biology: The Evolutionary Synthesis and Evolutionary Biology*. Princeton, N.J.: Princeton University Press.

Sturtevant, Alfred H. 1920. "Intersexes in *Drosophila Simulans*," *Science* 51: 325–327.

———. 1965. *A History of Genetics*. New York: Harper & Row.

Waddington, C. H. 1939a. *An Introduction to Modern Genetics*. New York: Macmillan.

———. 1939b. "Preliminary Notes on the Development of the Wings in Normal Mutant Strains of *Drosophila*." *Proceedings of the National Academy of Sciences USA* 25: 299–307.

———. 1940a. "The Genetic Control of Wing Development in *Drosophila*," *Journal of Genetics* 41: 75–139.

———. 1940b. *Organisers and Genes*. Cambridge: Cambridge University Press.

———. 1941. "Evolution of Developmental Systems." *Nature* 147: 108–110.

———. 1942. "The Epigenotype." *Endeavour* 1: 18–20.

———. 1975. "The Practical Consequences of Metaphysical Beliefs on a Biologist's Work. An Autobiographical Note." In C. H. Waddington, *The Evolution of an Evolutionist*. Ithaca, N.Y.: Cornell University Press.

Wilson, Edmund Beecher. 1925. *The Cell in Development and Heredity*, 3rd ed. New York: Macmillan.

Winther, Rasmus G. 2000. "Darwin on Variation and Heredity." *Journal of the History of Biology* 33: 447.

————. 2001. "August Weismann on Germ-Plasm Variation." *Journal of the History of Biology* 34: 517–555.

Witkowski, J. A. 1992. "Julian Huxley in the Laboratory: Embracing Inquisitiveness and Widespread Curiosity." In *Julian Huxley: Biologist and Statesman of Science*, ed. C. Kenneth Waters and Albert van Helden, pp. 79–103. Houston, Tex.: Rice University Press.

Wright, Sewall. 1934. "Physiological and Evolutionary Theories of Dominance." *American Naturalist* 68: 24–53.

Yoxen, Edward. 1986. "Form and Strategy in Biology: Reflections on the Career of C. H. Waddington." In *A History of Embryology*, ed. J. J. Horder, J. A. Witkowski, and C. C. Wylie, pp. 309–329. Cambridge: Cambridge University Press.

II

ROOTS AND PROBLEMS OF EVOLUTIONARY DEVELOPMENTAL BIOLOGY

8

THE RELATIONS BETWEEN COMPARATIVE EMBRYOLOGY,
MORPHOLOGY, AND SYSTEMATICS: AN AMERICAN
PERSPECTIVE

John P. Wourms

This chapter will provide an American perspective on the origin and history of comparative evolutionary embryology and its subsequent incorporation into evolutionary developmental biology or "Evo-Devo." In order to do this, it will be necessary to trace the interrelationships and historical development of embryology, morphology, and systematics from the early days of the Republic to the present.

Comparative evolutionary embryology had its origin in the mid-late nineteenth century when embryology, which was descriptive and to varying degrees also comparative, split into two embryologies. Comparative evolutionary embryology embraced an evolutionary theme, whereas experimental embryology, which became developmental biology, adapted a physiological and experimental theme. The former emphasized distal causality; the latter, proximal causality. During the twentieth century, comparative evolutionary embryology did not die out, as is commonly thought; rather, it survived and coexisted with developmental biology. Comparative evolutionary embryology incorporated selected elements from developmental biology and other disciplines such as molecular developmental genetics, cladistics, and molecular phylogeny into the synthesis that is evolutionary developmental biology. What is presented here is in many ways an extended outline, for it would require more than a short paper to do justice to this subject and weave together the many disparate themes.

The three fields of zoology that will be considered here were major components in Louis Agassiz's reformulation of the Meckel-Serres Law that there is a threefold parallelism which relates systematics to embryology, geological history, and morphology.[1] Subsequently, others amended this relationship to include an evolutionary perspective. It is appropriate to begin with working definitions and some commentary on embryology, morphology, and systematics.

Embryology, which dates from the time of Aristotle, was revived in the sixteenth and seventeenth centuries.[2] Classical embryology deals primarily with the sequence of events that is involved in the formation of a new individual organism from a fertilized egg. It includes the period from fertilization to hatching or birth, depending on whether the organism is oviparous or viviparous. Some attention was paid to postembryonic events, such as metamorphosis and regeneration. Embryology was formalized as a science in the period 1830–1850 by Von Baer and other members of the Dollinger school.[3] Several issues and foci of interest emerged early on: the embryological origin of new individuals in various animal species; a comparison of the overall process of embryological development as well as the embryological development of specific organs or structures among different animal species; and the possible relationship between the ontogeny of individual species and the evolution/origin of different animal species. From the time of its formulation to the present, embryology in its many forms has been involved in a continuous intellectual synthesis.[4]

Many definitions of morphology have been given. Through the years the term has lost its precision and come to be a vague catch-all. Ghiselin (1991) defines morphology as "the study of what is called 'form.'" Morphology, conceived as the science of form, encompasses the whole range of formal properties, including shapes and symmetries of the entire body as well as organs. This includes molecular morphology as well as behavioral characters such as activities and movements, and social organization. This definition corresponds to an earlier one by Conklin (1927), who stated that morphology deals with the forms and structures of living things and is studied by methods of observation, comparison, development, and experiment. Others have defined morphology as "the study of form and function." Ghiselin, however, prefers to reserve this definition for evolutionary physiological (or functional) anatomy.

The definition of systematics also poses historical, semantic, and scientific problems. Simply put, systematics attempts to establish relationships among organisms and to express these relationships in some system of classification. Ideally, the relationship reflects degrees of kindredship. Morphology, in the sense that it is used by Ghiselin (1991)—from molecules to organisms and their behavior—as well as embryology and life history characteristics (Cohen, 1984), has been used to provide criteria in systematics.[5]

EARLY DAYS OF AMERICAN NATURAL SCIENCES

To provide an American perspective on the interrelationship of embryology to morphology and systematics, it is necessary to consider briefly the history of natural sciences in North America and the United States. The study of natural history and natural sciences in the United States dates back to the colonial era and the early days of the Republic (Catesby, Abbott, Bartram, Wilson, Audubon, and DeKay, among others). Learned societies (e.g., the American Philosophical Society, 1780), local natural history societies (e.g., the Lyceum of Natural History, New York, 1817, and the Boston Society of Natural History, 1833), and museums (e.g., the Academy of Natural Sciences, Philadelphia, 1812, and the Smithsonian, 1847) came into being, and they fostered research and publication in natural history, zoology, and kindred subjects.

The natural scientists of early America faced the formidable task of comprehending the relatively unknown flora and fauna of a largely unexplored continent and its surrounding oceans. Out of necessity, the identification of both the native flora and fauna as well as new species encountered in overseas expeditions took priority over areas of inquiry such as morphology and embryology. Table 8.1 lists some of the more important expeditions and natural history surveys. Examples of this type of work include Wilson (1810) and Audubon (1831–1839) on birds; Say (1824–1828, 1830–1834) on insects and mollusks; and Holbrook (1836–1840) on North American reptiles and amphibians. The Lewis and Clark expedition of 1804 added 122 new species of animals. Storer (1839) described the fishes and reptiles, and Gould (1841) described the invertebrates, as part of the natural history survey of Massachusetts. The thirty-volume *Natural History Survey of New York* (1842–1894) included finely illustrated volumes on mollusks, fishes, reptiles, birds, and mammals by DeKay. In some instances, the discovery of a previously unknown group of organisms, such as the viviparous surfperches (family Embiotocidae), led to major morphological and embryological studies.[6]

What of embryology and morphology? Examination of the publications of nine major learned societies listed in Meisel (1924–1929) revealed that in the period 1768–1865, 102 publications out of a total of 4,607 could be categorized as embryology, *sensu latu* (see table 8.2). Most of these appeared after 1850. For all intents and purposes, virtually no embryology was done prior to 1840. There were, however, exceptions, such as Mitchell's (1803) study of the anatomy of a fetal shark. Embryological publications began to appear in the 1840s. There are several reasons for this state of affairs. First, embryology was a new and emerging science that had its roots in Europe. Prior to 1840, there were no trained practitioners of embryology in the United States. In the 1840s and 1850s, embryological research and publication began in earnest through the efforts of

Table 8.1
Expeditions and Surveys

Exploration of Western America
Lewis and Clark (1803–1806)
Long (Rocky Mountains, 1819–1820)
Frémont (California, Oregon, 1843–1844)
U.S. Survey of California and Oregon (1853–1855)
Pacific Railroad Surveys (1853–1855)

State Natural History Surveys
Natural History Survey, New York (1836)
Zoological and Botanical Survey, Massachusetts (1838)

Marine and Deep Sea
Wilkes, U.S. Exploring Expedition (1840)
Perry, U.S. Japan Expedition (1852–1854)
North Pacific Exploring Expedition (1853)
U.S. Coast Survey (Pourtales, 1853)
U.S. Fish Commission (1871)
Blake Cruises (1877–1880)
Albatross Cruises (1882–1931)

Foreign
Herndon, Valley of the Amazon (1851)
Thayer, Brazil-Amazon Expedition (1865–1866)

European-educated Americans such as Jefferies Wyman[7] and Waldo Burnett,[8] and of European émigrés, especially Louis Agassiz.[9]

In the first half of the nineteenth century, Boston vied with Philadelphia to be the intellectual center of the United States. Subsequently, it achieved that position. Boston at that time was a perfect venue for Louis Agassiz. His arrival and his subsequent activities, combined with the efforts of American scholars and an existing program in the natural sciences (the Boston Society of Natural History, the Lowell Institute, Harvard, and the Harvard Medical School), triggered an explosive flowering of zoological sciences, including morphology and embryology. American scholars such as Wyman and Burnett continued to be active. Burnett (1853) produced the first book-length, (circa 200 pages) treatment of the cell in English. He also published much on embryology and cellular histology.[10] Wyman was professor of anatomy at Harvard Medical School and a friend and colleague of Oliver Wendell Holmes. After completing his medical degree in Boston, he studied in Paris with the physiologists Flourens and Magendie and the zoologists DeBlainville, St. Hilaire, Valenciennes, Dumeril, and

Table 8.2
History of Embryological Publication in America to 1865

Institution/Journal	Date of First Paper	Ratio Embryo/ Total	First Publication
Amer. Phil. Soc. (1768); 97 years	1775	3/360	Hare and Skinner, Life Hist. Silkworm
Amer. Acad. Arts and Sciences (1780); 85 years	1850	16/270	Burnett, Comp. Spermatogenesis
Acad. Nat. Sci, Phil. (1812); 53 years	1841	15/945	Morris, Insect Development
Lyceum Nat. Hist, N.Y. (1817); 48 years	1858	2/420	Edward, Repro. in *Actinia*
Amer. J. Sci. & Arts (1818); 47 years	1849	15/1350	Desor, Albumen in Ovarian Eggs
Bost. Soc. Nat. Hist (1833); 32 years	1847	41/782	Wyman, Dogfish Shark Fetus
Smithsonian (1847); 18 years	———	0/300	———
Elliott Soc. (S.C.) (1853); 12 years	1856	5/40	McCrady, Develop. Cnidaria
Calif. Acad. Sci. (1853); 12 years	1855	3/140	Ayres, Conjoined Twin Dogs

Source: Data from Meisel (1924–1929).

Milne Edwards. Wyman published extensively in comparative morphology, including physical anthropology (e.g., early studies of the gorilla and chimpanzee), as well as in comparative embryology (e.g., development of the viviparous four-eyed fish *Anableps* and of the skate).[11] In addition, during this period many American scholars, among them Bowditch, were trained in morphology and embryology in the laboratories of Ludwig and Leuckert in Germany.[12] In the words of Alpheus Packard (1886), "The atmosphere in Boston was pervaded by high scientific aims and warmed with a zeal for scientific discovery."

FROM LOUIS AGASSIZ TO COMPARATIVE EVOLUTIONARY EMBRYOLOGY

Much has been written about Louis Agassiz.[13] Here, he will be considered in terms of his influence on the progress of zoology and embryology in America. In her assessment of Agassiz as an embryologist, Oppenheimer (1986) concluded that Agassiz made a strong mark on American zoology and was important in the development of embryology. One of her major points is that he "brought embryology with him from Europe," opening up the field and encouraging others to enter it. In examining his influence, some of the points that will be considered relevant are the following:

1. As an outstanding lecturer, he popularized both zoology and embryology to a wide audience, some of whom were motivated to enter the field.

2. He founded the Museum of Comparative Zoology at Harvard and revolutionized the study of zoology in the United States in terms of both research and training of students.

3. He provided a philosophical basis for the study of zoology by introducing the threefold parallelism set of relationships subsequently amended by others to include evolution, which was embraced by students and associates, and was used well into the twentieth century.

4. Through his own students and protégés, but mostly through the efforts of Alexander Agassiz, his colleague E. L. Mark, and a host of their students, protégés, and associates, a Cambridge school of embryology, one of whose strong points was comparative evolutionary embryology, came into existence and has persisted to the present. Boston-Cambridge became the center for embryological and morphological research in the United States. Students who were trained in this school carried with them the research agenda and philosophy of the Cambridge school as they dispersed across America, accepting academic appointments, founding or expanding natural history museums, and founding marine laboratories. In short, their research philosophy and agenda became nationwide.

What was Agassiz's threefold parallelism, and how did it lead to a research agenda in comparative evolutionary embryology in America? The possibility of a relationship between the sequence of embryonic stages in the development of individual organisms and the ordering of all living organisms from lowest to highest in a scale of perfection had been a matter of speculation since the time of Aristotle. Meckel in the 1820s, and Serres somewhat later, formalized this concept into the Meckel-Serres Law, which stated that there is a parallelism between the stages of ontogeny and the stages of the *scala natura*.[14] In the 1840s and 1850s, Agassiz expanded the Meckel-Serres Law into a three fold parallelism by interpolation of the fossil record. According to Agassiz (1857, 1962 ed., p. 114), "The phases of development of all living animals correspond to the order of succession of their extinct representatives in past geological times. As far as this goes, the oldest representatives of every class may then be considered as embryonic types of their respective orders or families among the living."

The embryological aspects of Agassiz's reformulation are better glimpsed in two of his earlier works. In the *Principles of Zoology*,[15] Agassiz states, "The importance of embryology to the study of systematic zoology cannot be questioned. For evidently, if the formation of the organs in the embryo takes place in an order corresponding to their importance, this succession must of itself furnish a criterion of their relative value in classification. Thus, those peculiarities that first appear should be considered of higher value than those that appear later. In this respect, the division of the animal kingdom into four types, the vertebrates, the articulates, the mollusks, and the radiates, corresponds perfectly with the gradations displayed by embryology." This theme was reiterated in his *Twelve Lectures on Comparative Embryology*.[16] In the fourth lecture, Agassiz discusses in detail the critical role of embryology in the study of systematics. This was the first book on comparative embryology in the United States. As the title implies, it was based on a series of public lectures at the Lowell Institute. The book was well received. In the preface, the publisher states that embryology may be considered entirely new in this country and that although few persons have entered it, it is destined to have considerable influence on the future progress of zoology through its use to improve the classification of animals.

Agassiz's introduction of geological time into the Meckel-Serres Law was potentially valuable, but his interpretation was marred by an essentialist, theistic, nonevolutionary point of view. Others were intrigued by the Meckel-Serres Law and the work of Von Baer. After a prolonged gestation, Darwin's *Origin of Species* was published in 1859. Darwin believed that evidence from embryology, together with morphology, provided the strongest support for his theory of evolution. He listed five sets of facts in embryology that are puzzling unless explained in an evolutionary context. He cited Agassiz's extension of the Meckel-Serres Law and was aware of Von Baer's work. In a letter to Asa Gray, Darwin stated that he "considers embryology to be the strongest single class of facts in favor of change of forms."[17] It is worth noting that Darwin had some firsthand experience with these issues because of his barnacle studies. It had been only recently that Thompson had discovered that larval barnacles were almost identical to the larvae of crabs and other crustaceans. Thus, barnacles were

Table 8.3
Pioneer Works in Comparative Embryology

American

L. Agassiz (1849) *Lectures on Comparative Embryology*

Burnett (1853) *"The Cell: Its Physiology, Pathology, and Philosophy, as Deduced from Original Investigations"*

Clark (1865) *The Origin of Life, and the Mode of Development of Animals*

Packard (1876) *Life Histories of Animals, Including Man; or, Outlines of Comparative Embryology*

European

Von Baer (1828) *Ueber Entwicklungsgeschichte der Tiere*

Bischoff (1842) *Entwicklungsgeschichte der Saugerthiere und des Menschen*

Darwin (1859) *Origin of Species*

Müller (1864) *Für Darwin*

Haeckel (1866) *Generelle Morphologie der Organismen*

Balfour (1880–1881) *Treatise on Comparative Embryology*

---

now recognized as being crustaceans rather than mollusks (as Cuvier had thought).

Fritz Müller (1864) and Ernst Haeckel (1866) transformed the Meckel-Serres statement into an evolutionary statement that came to be known as the biogenetic law. In one version,[18] Haeckel states, "Ontogeny is a concise and compressed recapitulation of phylogeny, conditioned by laws of heredity and adaptation." Research discoveries that favored this concept soon emerged. In 1871, Kowalevsky discovered that tunicate larvae have a notochord and form their neural tube much as *Amphioxus*, a primitive chordate, does. As a result, tunicates were now recognized as chordates, even though the postmetamorphic adult lacks a notochord. What was to become comparative evolutionary embryology slowly emerged as a discipline in the period 1850–1870 to become a dominant force in the zoological sciences until the early twentieth century. (See table 8.3 for some of the pioneer works.) The history of comparative evolutionary embryology has been considered elsewhere;[19] it will not be repeated. Here, the emphasis will be on (1) research and training in comparative evolutionary embryology at Harvard's Museum of Comparative Zoology (MCZ) and in Boston; (2) the diaspora of students and protégés trained in comparative evolutionary embryology across the United States and its consequences; and (3) the survival of comparative evolutionary embryology in the twentieth century and its transformation into evolutionary developmental biology.

THE RISE OF COMPARATIVE EVOLUTIONARY EMBRYOLOGY AT THE MCZ

From the time of his appointment as a professor in Harvard's Lawrence Scientific School, Louis Agassiz nourished the idea of building a museum of comparative zoology, embryology, and paleontology. This plan was realized in 1859 with the founding of Harvard's Museum of Comparative Zoology.[20] The MCZ, first under the leadership of Louis Agassiz, and subsequently of Alexander Agassiz and E. L. Mark, became the major U.S. center for training research scholars in zoology, especially comparative embryology and, later, comparative evolutionary embryology. This work was carried on in conjunction with scholars at other institutions, such as the Harvard Medical School and the Boston Society of Natural History. It was the emphasis on embryology that distinguished zoological research in Boston from that carried out in New York, Philadelphia, and Washington, D.C. The results were extraordinary, in terms of research accomplishments and of channeling the future course of zoology in the United States.[21] Some interesting insights can be gained through the historical accounts of Packard (1876) and Coe (1918).

In 1876, Alpheus Packard, a student of the elder Agassiz, assessed the progress of American zoology during the past century. He recognized three major epochs of activity: systematic zoology, morphological and embryological zoology, and evolution. The epoch of embryology began with the arrival of Louis Agassiz in 1846 and his lectures on comparative embryology. It continued and increased in scope with founding of the MCZ and the embryological research activities of Louis and Alexander Agassiz, their students, and their associates. Packard stated that the scope and quality of embryological work done by American students from about 1850 to 1876 exceeded what had been done in England and France and was "only second to the embryological studies in Germany, the mother of developmental zoology." To support his assessment, Packard (1876) listed some of the publications of the Boston-Cambridge school during this period:

1. L. Agassiz and H. J. Clark, embryology of cnidarians and turtles
2. W. Burnett, first book-length treatment of the cell, and embryology of insects
3. J. McCrady, cnidarians
4. E. Desor, nemertean development, Desor's larva
5. C. Girard, viviparous surfperch and invertebrates

6. A. Agassiz, embryology of echinoderms, alternation of generations of a worm, early stages of annelids, first account of metamorphosis of the tornaria larva of *Balanoglossus*, ctenophore embryology

7. J. Wyman, planarian regeneration, fish viviparity, development of the skate and midwife toad

8. A. Hyatt, embryology of ammonites

9. E. S. Morse, early stages of the brachiopod *Terebratulina* (which established that they were not mollusks)

10. A. Packard, development of *Limulus* with a trilobite-like larva

11. Scudder, Packard, Hagen, and others on insect development

12. W. K. Brooks, development of *Salpa*, a tunicate.

Wesley Coe (1918), a distinguished invertebrate zoologist, concurred with Packard and extended the analysis beyond 1876 in a more detailed treatment. He discussed in detail the effect of Louis Agassiz on the course of zoology and how this influence was passed down through his students to "the younger generations of zoologists." In addition to the periods of systematic zoology and morphology/embryology, Coe distinguished two others. During the period of evolution (1870–1890), the morphological and embryological studies of 1847–1870 continued, but now with reference to their bearing on evolutionary problems. This approach culminated between 1885 and 1890 "under the guidance of Whitman, Mark, Minot, Brooks, Kingsley, and E. B. Wilson." The period of experimental biology began in 1890 "through the application of experimental methods in the various branches of the modern science of biology."

Research and the training of future research scholars at Harvard and the MCZ can be divided into two phases. The first phase, in which Louis Agassiz was the prime mover, extended from 1850 until Agassiz's death in 1873. As noted previously, there was a progressive shift from comparative embryology to comparative evolutionary embryology in spite of Agassiz's opposition to Darwinian evolution. The second phase involved Alexander Agassiz and E. L. Mark and their students and associates, and extended from 1873 to Alexander Agassiz's death in 1910. The MCZ has continued as a leading center of zoology under the leadership of Henshaw, Barbour, Romer, Mayr, and others.[22]

This second phase is of particular interest. During this period, Alexander Agassiz, because of his own research activities and the research of his assistants and associates, his administrative abilities, and the use of his own wealth to fund the MCZ and scientific research, helped solidify the MCZ's preeminence. He did not do it alone. E. L. Mark was another key player at the MCZ. The early students of Louis Agassiz, such as Hyatt and Packard, remained active. About this time, there was also an influx into the Boston-Cambridge area of American scholars who either had received their doctorates in Germany, especially in the laboratory of Leuckart (e.g., E. L. Mark, C. O. Whitman, W. Patten) or had been visiting students in German laboratories (e.g., J. W. Fewkes and C. S. Minot). Some of them (e.g., Mark, Whitman, and Fewkes) established themselves at the MCZ, and others (e.g., C. S. Minot) took another course. Minot received his doctorate from Harvard and trained in a laboratory in Germany. He became professor of histology and embryology at the Harvard Medical School, where he established a program in morphology with a strong embryological emphasis. He founded the *American Journal of Anatomy*, wrote the first American textbook on human embryology, and produced a definitive bibliography of vertebrate embryology.[23] He also was a trustee of the Marine Biological Laboratory at Woods Hole (MBL).[24]

Alexander Agassiz's research activities fall into three categories: (1) comparative evolutionary embryology, (2) biological oceanography, and (3) biology and formation of coral reefs.[25] His embryological research, which was done early in his career, beginning in the late 1850s and extending to 1890, is of primary interest. His oceanographic work, especially the cruises of the *Albatross*, which earned him the accolade "Prince of Oceanographers," also is of interest because it involved students and associates from the MCZ.

Alexander Agassiz studied under his father, Louis Agassiz. He had degrees in engineering and natural history from Harvard. His first major scientific paper on the viviparous surfperches of California appeared in 1861 and was followed by over 250 more.[26] Much of his early work (1850–1870) on marine invertebrates and their development was done at the first marine lab in the United States, a shedlike building at the Agassiz cottage in Nahant, Massachusetts. His research and field observations led to the publication in 1865 of *Seaside Studies in Natural History*, which he wrote with his stepmother, Elizabeth Cary Agassiz. It is the first American guide to coastal marine life and contains a wealth of information on the development of intertidal and inshore fauna of the area. Among the highlights of this period were "Embryology of the Starfish" (1877), a handsomely illustrated monograph that provided a definitive account of development from fertilization through postmetamorphic juveniles, and "North American Acalephae" (1865), a study of the medusoid jellyfishes of the Atlantic and Pacific coasts, illustrated with 360 figures that Agassiz drew from life.

Alexander Agassiz was considered one of the leading American embryologists in the period 1860–1890. In 1866 he was elected to the National Academy of Sciences, and in 1879 he was awarded the Prix Serres for his research on the development of cnidarians, ctenophores, annelids, echinoderms, enteropneusts, and fishes. He was the first non-Frenchman so honored. He espoused an integrative, organismal approach to development within the conceptual framework of what is now known as the life history model. He helped establish a tradition of painstakingly detailed observation of living embryos, larvae, and postmetamorphic juveniles combined with state-of-the-art microscopy. This approach later became associated with Brooks, Whitman, and their students, and culminated in the Johns Hopkins-MBL school. He fostered an embryological research agenda at the MCZ, aided by E. L. Mark, who joined the Harvard faculty in 1877.

Mark, the first American student to obtain a doctorate under Rudolf Leuckart, brought a cytological and histological approach. With Minot and Whitman, he was responsible for the introduction of advanced European microscopic techniques. Whitman (1885), while working as Agassiz's research associate on teleost development, wrote his *Methods of Research in Microscopical Anatomy and Embryology*, the first American work on the subject. To further embryological research, Agassiz (1) ensured access to research material via the Newport lab; (2) acquired microscopes and apparatus needed for advanced microscopy; (3) assembled a research library; (4) introduced photography and microphotography; and (5) fostered publication of an in-house series of well-illustrated embryological monographs. To provide easy access to the scarce European literature, Agassiz, Mark, Faxon, and Fewkes published a series of selected illustrations from embryological monographs (i.e., "normal stages of development") with accompanying bibliographies. In addition, E. L. Mark and others translated several German reference works, notably Korschelt and Heider's (1895–1900) four-volume *Text-book of the Embryology of Vertebrates*.

Alexander Agassiz explored a number of embryological topics, among which are the following:

1. Cnidarians: alternation of generations, embryonic and larval development, tentacle differentiation, morphological variation, and growth rate of corals
2. Ctenophores: elegant comparative studies that accurately depicted unilateral (heart-shaped) cleavage and the segregation of the comb-forming cortical cytoplasm in micromeres by unequal cell division. He advocated splitting the ctenophores off from the cnidarians
3. Annelids: alternation of generations, embryonic development, and metamorphosis

4. Echinoderms: an enormous body of work on echinoderm development that involved artificial fertilization and experimental hybridization; description of starfish development from fertilization through postmetamorphic juveniles with elegant illustrations from living material; viviparity and brooding; homology of pedicellariae and simple spines; postmetamorphic studies of growth and development carried out on a comparative basis

5. Enteropneusts: description of the metamorphosis of the tornaria larva and its identification as a stage in the development of *Balanoglossus*

6. Fishes: cellular analysis of early teleost development (with C. O. Whitman); comparative studies of the development and metamorphosis of pelagic and demersal fish larvae; metamorphosis of flounder larvae and different modes of eye migration to the dorsal surface; first observations on the development of the *Lepisosteus*, a primitive bony fish.

7. Theoretical studies: Agassiz criticized Haeckel's theories in terms of both factual evidence and its interpretation. Unlike his father, Alexander Agassiz accepted the evolutionary paradigm. In 1880, he presented a remarkable address to the annual AAAS meeting, titled "Paleontological and Embryological Development," in which he presaged the role of developmental constraints in evolution, questioned the constancy of evolutionary rates, and took the first steps toward a cladistic approach to taxonomy.

The establishment of his Newport Marine Zoological Laboratory was another important contribution that Alexander Agassiz made to further zoological and embryological research. This was the first permanent marine research laboratory in the United States and served as a model for a number of marine laboratories that were founded by students and associates who had worked at Newport (figure 8.1). As noted above, the Agassiz cottage at Nahant was the first marine lab. In the summer of 1873, Louis Agassiz conducted the Penikese School at a seasonal marine station on an island near Cape Cod. Alexander Agassiz taught embryology there. The Penikese School was a glorious experiment in instruction carried out under quite primitive conditions. The elder Agassiz died late in 1873, followed shortly thereafter by Alexander's wife, Anna. A second and last session of the school was held in 1874. Attempts to move the Penikese operation to Woods Hole failed.

Because his memories of Nahant were too distressing, Alexander built a home at Newport, Rhode Island. He had a room fitted out as a laboratory in which W. K. Brooks carried out his study on the tunicate *Salpa* in 1875. In 1877, Alexander constructed a separate building that became the Newport Marine Laboratory. Unlike Penikese with its emphasis on instruction, this was a research facility for students of marine

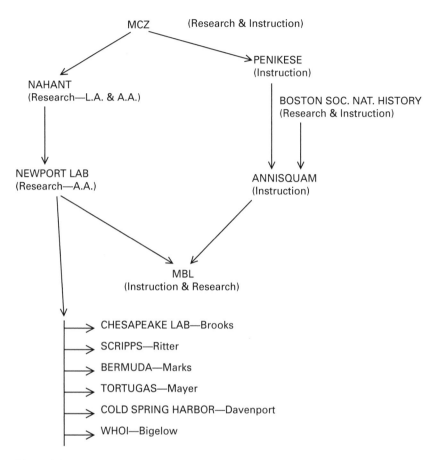

Figure 8.1
Newport as a model for U.S. Marine Labs.

zoology. The original emphasis was on the development of fishes and marine invertebrates. According to Agassiz, usually from six to eight students were in attendance. After its enlargement in 1891 and as Alexander Agassiz's interests shifted to deep-sea and coral reef biology, access to the laboratory was limited to his research assistants. In this context, Agassiz stated (in an unpublished manuscript), "I found it impossible to devote to them [students] the attention they required."

The laboratory was superbly designed and appointed. Illustrated accounts appear in the *Annual Report of the Museum of Comparative Zoology* for 1877–1878 and for 1891–1892, and elsewhere.[27] In architecture and design, the Newport lab was distinctly American and differed from European labs such as the Stazione Zoologica at Naples. It served as a physical and intellectual model for other American marine labs, incorporating many features that are now common, such as the following:

1. A well-designed flowing seawater system with aeration
2. access to a rich local fauna
3. facilities and boats for both inshore and offshore collecting
4. custom glassware and aquariums
5. a reference library
6. experienced assistants
7. facilities for photography
8. good research microscopes
9. knowledge of the local fauna and practical experience in its study using techniques such as artificial fertilization and experimental hybridization.

The Newport lab was the premier American research lab from the time of its inception until the U.S. Fish Commission Laboratory and the MBL attained maturity in the 1890s. Among those who did research at Newport were Alexander Agassiz, W. K. Brooks, W. E. Castle, C. B. Davenport, W. A. Faxon, J. W. Fewkes, J. H. Gerould, H. S. Jennings, C. H. Kofoid, E. L. Mark, A. G. Mayer, H. V. Neal, E. A. Nunn (Mrs. C. O. Whitman), G. H. Parker, W. E. Ritter, C. O. Whitman, and E. B. Wilson.[28] There is ample evidence that all of these individuals were involved in some form of embryological or developmental research. Several of them were instrumental in founding other marine laboratories: Brooks (Chesapeake lab), Davenport (Cold Spring Harbor), Mayer (Tortugas), Mark (Bermuda), Ritter (Scripps), and Whitman (MBL). The history and nature of other marine labs in New England—Annisquam, the Fish Commission lab, and the MBL—have been dealt with elsewhere.[29]

The Fish Commission lab did not acquire a permanent location at Woods Hole until 1884–1885. Its initial function was to remedy the decline in fisheries by exploring a program in fish egg hatching and stocking.[30] The study of fish development was a major aspect of its mission. From 1880 to 1886, John Ryder, one of the best fish embryologists in the late nineteenth century, was on the staff.[31] Its facilities, which were superior to those of the nascent MBL, attracted some of the Harvard students and research scholars after Alexander Agassiz shifted the research direction of the Newport lab. He paid the Harvard subscription for "a table" at the Fish Commission lab.[32]

At this distance in time, it is difficult to ascertain the intellectual pedigrees and the scope and nature of the interactions among those who were trained at the MCZ and who comprised the Cambridge school. This

Table 8.4
Students of Louis Agassiz

A. Agassiz	J. LeConte
J. Allen	E. Morse
A. Bickmore	A. Packard
H. Clark	F. Putnam
W. Faxon	S. Scudder
W. Fewkes	N. Shaler
S. Garman	A. Verrill
A. Hyatt	B. Wilder
D. S. Jordan	

difficulty is due at least in part to the ambiguous position of Alexander
Agassiz. Although he became head of the MCZ, he was not a member of
the Harvard faculty. Officially, he did not have any students. In reality, he
served as mentor to many of the students of Louis Agassiz and E. L. Mark,
and as a postdoctoral mentor and collaborator of many students who had
taken their degrees at Harvard.[33] To clarify the interactions within the Cambridge school, it will be necessary to identify the students of Louis Agassiz
and E. L. Mark and ascertain which of them were "students," protégés, or
associates of Alexander Agassiz. A list of seventeen of the more renowned
students of Louis Agassiz, based on information in Winsor (1991), is given
in table 8.4. Parker's *Mark Anniversary Volume* (1903) provides a list of thirty
of E. L. Mark's students, though not all of them were professional biologists (e.g., Theodore Roosevelt; see table 8.5). Based upon participation in
the Newport lab, collaboration on publications, positions as research assistants, and specific acknowledgments, it has been possible to draw up a list
of seventeen zoologists who could be considered students, protégés, or associates of Alexander Agassiz (table 8.6). Six of these individuals—Bigelow,
Brooks, Davenport, Mayer, Ritter, and Whitman—played leading roles in
establishing centers of academic excellence and marine laboratories.[34]

THE DIASPORA

After completing their training at the MCZ, members of the Boston-
Cambridge school of zoology sought and obtained positions at universities and scientific institutions. The diaspora of the MCZ students had a
profound affect on the progress of American zoology because they were

Table 8.5
Students of E. L. Mark (to 1903)

G. M. Allan	C. Juday
F. W. Bancroft	C. A. Kofoid
T. Barbour	F. T. Lewis
H. B. Bigelow	W. A. Locy
R. P. Bigelow	A. G. Mayer
E. A. Burt	W. J. Moenkhaus
W. E. Castle	G. H. Parker
C. B. Davenport	W. Patten
C. R. Eastman	J. Reighard
C. H. Eigenmann	W. E. Ritter
J. H. Gerould	T. Roosevelt
S. Goto	R. M. Strong
I. H. Hyde	T. W. Vaughn
R. T. Jackson	R. M. Yerkes
H. S. Jennings	H. P. Johnson

Source: Parker (1903).

Table 8.6
Students, Protégés, and Associates of Alexander Agassiz

H. B. Bigelow	C. A. Kofoid
W. K. Brooks	A. G. Mayer
W. E. Castle	H. V. Neal
H. L. Clark	G. H. Parker
C. B. Davenport	W. Ritter
W. Faxon	H. B. Ward
J. W. Fewkes	C. O. Whitman
J. H. Gerould	W. M. Woodworth
H. S. Jennings	

Table 8.7

Functional Association of Universities, Natural History Museums, and Marine Laboratories: Programs in Morphology, Embryology, and Systematics

University	Museum	Marine Laboratory
Harvard	Museum of Comparative Zoology	Newport, MBL
Yale	Peabody Museum	MBL
Columbia	Amer. Mus. Nat. Hist.	MBL, N.Y. Aquarium N.Y. Zool. Soc.
Pennsylvania	Acad. Nat. Sciences	MBL
Johns Hopkins	Smithsonian	Chesapeake Zool. Lab, MBL
Chicago	Field Museum	MBL, Shedd Aquarium
Stanford	Nat. Hist. Museum . Cal. Acad. Sci	Hopkins Marine Station
U. California (Berkeley)	Museum Vert. Zool. Cal. Acad. Sci.	Scripps

research oriented and brought with them a research agenda firmly rooted in comparative zoology and comparative evolutionary embryology. The diaspora coincided with, and may have influenced, the transition from the traditional American college system to the emergent research universities with doctoral programs.[35] Through the efforts of the young scholars, a distinctive threefold pattern tended to emerge. As research-oriented zoology departments were established at universities, young scholars would either establish an university-sponsored museum of natural history or forge links with a newly established or extant museum of natural history within the same community. In addition, the zoology departments and their founders either established their own marine labs or became sponsoring members of a multi-institutional marine lab. Eight examples of this pattern are given in table 8.7.

In New York, Bickmore founded the American Museum of Natural History. Subsequently, H. Osborne became its president and established a research oriented zoology program at Columbia. At Johns Hopkins, Brooks and Martin established a preeminent department and Brooks established the Chesapeake Zoological Laboratory. Whitman did the same at Chicago and was instrumental in establishing the MBL. David Starr Jordan, the first president of Stanford University, set up a strong zoology department with its own natural history museum and a marine lab, Hopkins Marine Station. Jordan had only a brief connection to Louis Agassiz at Penikese, but that was sufficient for him to switch from botany to ichthyology. At Stanford, he gathered together students and colleagues to establish what has been called the Jordan school of systematic ichthyology. It became the dominant force in American ichthyology and has persisted to the present.[36] Ritter and Kofoid were instrumental in establishing the zoology program at the University of California at Berkeley and its Museum of Vertebrate Zoology. Ritter established a marine laboratory at La Jolla that became Scripps Institute of Oceanography.[37]

The close relationships that were established among these three types of institutions permitted a type of institutional division of labor. Morphological and systematic studies, especially with a field orientation, tended to be carried out in the museums and marine labs, whereas the university departments tended to become oriented toward lab-based, physiological research. This arrangement had both advantages and disadvantages. Where the museums and marine labs were independent entities, the university-based zoology departments were relieved of the massive expense of establishing and maintaining collections of specimens, expeditions, and separate marine labs. The division of intellectual labor and its conduct in separate venues facilitated the separation of experimental embryology from comparative evolutionary embryology. On the other hand, the separate venue of museums and marine labs provided a safe haven for comparative evolutionary embryology.

Mention has already been made of the achievements of some of the former students of the Boston-Cambridge school. It is not feasible to discuss all of them. Rather, two well-known figures, W. K. Brooks and C. O. Whitman, will be considered in terms of their effect on the field of zoology and their roles in training the next generation of zoologists and embryologists. Brooks was called to the newly organized Johns Hopkins University, where with Newell Martin he was responsible for establishing the preeminence of the graduate program in zoology.[38] As previously noted, he established the Chesapeake Zoological Lab, which in its early years led a peripatetic existence.[39] At Hopkins, Brooks established a research program in morphology in which he and his students engaged in descriptive studies of morphology and embryology. According to Benson,[40] Brooks emphasized the need for careful comparative studies of the embryos of related taxa, using the latest microscopic techniques, prior to formulating any ancestral relationships between the embryos. His intellectual approach reflected his training at the MCZ, where descriptive embryological and morphological methods were used to examine systematic relationships in an evolutionary context. The emphasis on microscopic technique also was a product of training at MCZ, where Mark and Whitman had introduced the most advanced microtechniques from Germany. The results were readily apparent in Alexander Agassiz's exquisite studies of echinoderm and ctenophore development and in Whitman's work on the early stages of teleost development. From 1877 to 1908, forty-three students of Brooks were awarded their doctorates, among them E. B. Wilson, T. H. Morgan, E. G. Conklin, R. Harrison, E. W. Gudger, and H. V. Wilson. E. B. Wilson, whose early descriptive study of the sea pansy *Renilla* was very well received, and Morgan joined the faculty at Columbia. There, Wilson codified the study of cytology in his magnum opus *The Cell*, and retained research interests in the cellular and evolutionary aspects of development. Morgan, whose early work had been on the embryology of sea spiders (pycnogonids), a curious arthropod group, became a pioneer and leader of American genetics with his studies of *Drosophila*. Gudger joined the American Museum of Natural History, where he continued his studies of fish development from a comparative and evolutionary viewpoint. H. V. Wilson settled at the University of North Carolina and became director of the Chesapeake lab, whose permanent location was at Beaufort, North Carolina. His major early work had been on the embryology of the sea bass.[41] He is better known for his study of sponges, in which he used both descriptive and experimental approaches (i.e., cell dissociation).[42]

Ross Harrison, who spent most of his career at Yale, began his research on invertebrate development. Somewhat later, between 1901 and 1907, he pioneered the technique of tissue culture in order to resolve questions related to neuronal growth in teleost embryos.[43] Subsequently, he focused his efforts on amphibian development. One of his students was V. C. Twitty of Stanford, who was one of the mentors of the present writer. (This aside is introduced to make the point that depending on academic pedigree, the current senior generation of zoologists is only three to four academic generations removed from Alexander Agassiz.) Although most of the research of Harrison and Twitty was experimental, they also did morphological work, and in their later years they emphasized organismal studies. E. G. Conklin received his degree in 1891, and after stints at several colleges, he joined the Princeton faculty in 1908.[44] He is perhaps best known for his cell lineage studies of the snail *Crepidula* and the ascidian *Styela*, as well as the embryology of *Amphioxus*. He used both descriptive and experimental methods. Conklin maintained an active interest in evolutionary questions that mirrored his training with Brooks in the MCZ tradition. In the introduction to his book *A Synopsis of the General Morphology of Animals* (1927), Conklin states, "The subject of animal morphology is here dealt with from the genetical (embryological and evolutionary) point of view because it is easier to understand complicated structures when they are seen in the process of becoming and also because by this method, fundamental resemblances, or homologies, are more readily appreciated." Somewhat farther on, he points out that the comparison of these morphological resemblances or differences among organisms is the basis for establishing systematic relationships.

C. O. Whitman first came in contact with the Agassiz tradition as a student at the two summer sessions of the Penikese school. He was influenced by the two Agassizes, E. S. Morse, and A. S. Packard to pursue a career in zoology with a special interest in embryology. After Penikese, he spent some time at Anton Dohrn's laboratory in Naples and then proceeded to Germany. At Leipzig, he studied with Rudolph Leuckart and was awarded his doctorate in 1878. After a short time at Johns Hopkins, he replaced E. S. Morse as professor of zoology at the University of Tokyo. Upon his return to America, he was a research associate of Alexander Agassiz at the MCZ from 1882 to 1886. While at the MCZ, he published *Methods of Research in Microscopical Anatomy and Embryology* (1885), which was the first book of its kind in English and introduced the new, advanced microtechniques of Europe to the United States. Subsequently, he served as director of the Allis Lake Laboratory and as a professor at Clark University. In 1892, he became the first professor and head of the department of zoology, as well as curator of the zoological museum, at the University of Chicago. He remained at Chicago for the rest of his life. At Chicago, Whitman was able to attract an outstanding faculty. He was chosen as the first director of the newly established Marine Biological Lab at Woods Hole in 1888 and served in this capacity until 1900. He founded the *Journal of Morphology* and the forerunner of the *Biological Bulletin*. With colleagues, he founded the American Morphological Society, which became the American Society of Zoologists (and is now the Society for Integrative and Comparative Biology).[45]

Davenport (1917) states that Whitman's research activities fell into three categories. The first is the invertebrate period (ca. 1879–1899), during which he carried out his pioneering research on the organization of the egg and cell-lineage studies in the embryo of the leech *Clepsine*. The second period was devoted to vertebrates. This included his work with Alexander Agassiz on fish eggs and his own work on amphibians and the primitive fish *Amia*. His collaboration with Agassiz resolved a long-standing question in teleost embryology: the origin of the nuclei in the periblast (yolk syncytial layer). They found that the nuclei are derived entirely from the marginal cells of the blastodisc.[46] The third and last period was a mixture of animal behavior, genetics, and evolution. In this phase, he began his studies on heredity and evolution of pigeons.[47] On the whole, Whitman appears to have been concerned with morphology, embryology, and evolution. According to Davenport (1917, p. 25), "He was not very cordial to developmental mechanics, and was critical of the enthusiastic rush to the mutation theory and Mendelism."

Lillie (1911) provides a list of Whitman's doctoral students. There were a total of forty-four, three at Clark and forty-one at Chicago. Six of his students were women. His students embraced a broad spectrum of research activities, including ecology (V. E. Shelford); oceanography (H. C. Bumpus); entomology, especially ants (W. M. Wheeler); physiology (B. Lillie); and embryology/morphology (F. R. Lillie, C. Clapp, A. L. Treadwell, W. J. Moenkhaus, H. H. Newman, and C. Zeleny). Moenkhaus and Newman did some of the early work on the embryonic development of hybrid fishes. Zeleny and Treadwell worked on aspects of the cytoplasm (i.e., localization and cell lineage). Cornelia Clapp was active at the MBL and did important work on the toadfish. Frank Lillie worked on aspects of fertilization, cytoplasmic localization in eggs, and cell lineage.[48] Lillie joined the faculty at Chicago and became an integral part of the zoology program. He was also active at the MBL, serving as director from 1908 to 1925, and as president of the corporation from 1926 to 1942.[49] Lillie trained many students, one of whom was the embryologist B. H. Willier. Willier, who went on to Johns Hopkins. Among his students were C. Markert, J. Saunders, and J. P. Trinkaus of Yale. One of Trinkaus's students was N. K. Wessells of Stanford.

There is purpose to this brief exercise in academic genealogy. Depending on the lineage, many of the current senior generation of zoologists are only three to six academic generations removed from the MCZ of Alexander Agassiz. In the case of the present writer, one mentor was N. K. Wessells (Wessells, Trinkaus, Willier, Lillie, Whitman, Alexander Agassiz), and another mentor was V. C. Twitty (Twitty, Harrison, Brooks, Alexander Agassiz). In the late 1960s, the present writer had the pleasure of conversing with Henry Bigelow, then an emeritus professor at Harvard (Bigelow, Alexander Agassiz). Although many zoologists share many common academic lineages, it does not necessarily follow that their research philosophies, attitudes, and agendas are held in common. Nonetheless, reasonable assessment of evidence indicates that research agendas and philosophies in comparative evolutionary embryology and morphology have been transmitted through successive generations to the present time.

REFUGES OF COMPARATIVE EVOLUTIONARY EMBRYOLOGY

As the divide between comparative evolutionary embryology and experimental embryology/developmental biology widened, the venues in which the two branches of embryology were studied tended to become physically separated. Experimental embryology/developmental biology was centered in the biological science or zoology departments of research universities. Comparative evolutionary embryology tended to be centered at marine labs of universities, natural history museums (whether independent or affiliated with a university), and fisheries research stations. Some of the reasons that these institutions became refuges for comparative evolutionary embryology are to be found in their mission, the nature of their research, and the advantages and disadvantages imposed by the organisms that were being studied. Moreover, these institutions offered a congenial collegial atmosphere because they housed scholars with research interests in systematic zoology, morphology, ecology, and, in some instances, paleontology.

Natural history museums, by definition, have a specific charge to carry out comparative and evolutionary studies of the world's flora and fauna, both extant and extinct. Their business is to study the systematic and phylogenetic relationships of plants and animals, to establish and maintain reference collections for testing the validity of established classifications, and for identifying and classifying new species. They also serve as a base of operations for expeditions designed to study live organisms under field conditions. A number of modern museums have facilities for maintaining live animals, or else have close working arrangements with zoos and public aquariums.

One of the best examples of a museum providing a safe haven for comparative evolutionary embryology is the American Museum of Natural History (AMNH) in New York city.[50] The AMNH was conceived by Albert S. Bickmore, a student of Louis Agassiz, while still a student at the MCZ. The MCZ was its model. The AMNH came into existence in 1869–1870 and opened in 1877. Bickmore served as its director. During its early years, it suffered financially. Morris K. Jesup became museum president in 1881 and began to institute major changes. One of the most significant of these changes was the appointment of Henry F. Osborn as curator of vertebrate paleontology in 1890. More or less simultaneously, Osborn became Da Costa professor at Columbia University. His mission was to develop a first-rate department of biology at Columbia.

Osborn taught vertebrate morphology and evolution at Columbia while developing a program in vertebrate paleontology at the museum.

At the time of his appointment, Osborn was a professor of comparative anatomy at Princeton, where he had received his D.Sc. He had done postgraduate work in Europe, where he had studied with T. H. Huxley and F. M. Balfour. His course work with Huxley entailed the study of comparative morphology with the aim of establishing evolutionary relationships. In similar fashion, his embryological studies with Balfour emphasized evolutionary concerns. Osborn's early work was on the embryology of marsupials and placental mammals, especially their fetal membranes. He and his students also worked on the embryology of amphibians and reptiles. Subsequently, his interests shifted to vertebrate paleontology.[51]

At Columbia, Osborn was responsible for building a first-class department of zoology. In addition to Bashford Dean and James McGregor, he brought in other outstanding biologists such as E. B. Wilson and T. H. Morgan, and raised money for graduate fellowships, a library, and a new building. He established a closely knit departmental program of teaching and research in areas of vertebrate/invertebrate zoology, morphology, and embryology. Recognizing the importance of marine biology, Osborn became active in the administration of the MBL and also established a Columbia table at the Naples Marine Station. He became president of the newly formed New York Zoological Society in 1893 and was able to appoint W. T. Hornaday as its first director. Simultaneously, he was building a program in vertebrate paleontology at the AMNH. Osborn's activities were part of his vision to establish an interacting nexus of universities and research institutes in New York. He succeeded, and it is still very active. Between 1895 and 1899, he gave up undergraduate teaching and resigned his administrative positions at Columbia in order to concentrate his efforts at the AMNH. Nonetheless, from 1892 to 1909, Osborn taught and directed graduate students at Columbia, among whom were W. D. Matthew, W. K. Gregory, R. S. Lull, and F. B. Sumner. In 1908, Osborn was appointed president of the AMNH, and served until 1930.

Under Osborn's direction, the AMNH, with its connections to Columbia and the New York Zoological Society (zoo and aquarium), flourished and became an epicenter for research in systematic zoology, paleontology, evolution, and comparative embryology. Its reputation for scholarly achievement has continued to grow. Some comments about its organization are in order. In addition to the standard museum departments, the AMNH at various times had departments of comparative anatomy, experimental biology, animal behavior, and fish genetics. It maintained its

Table 8.8
Organization of AMNH

Standard Departments

Vertebrate paleontology
Osborn, Matthews, Barnum, Andrews, Simpson

Ichthyology
Dean,★ Gudger,★ Breder,★ Nicholas, Nelson, Atz,★ W. Beebe,★ D. Rosen,★ P. Rasquin★

Comparative anatomy
W. K. Gregory★

Experimental biology
G. K. Noble★

Ornithology
J. Allen, E. Mayr, F. Chapman

Animal behavior
L. Aronson, T. C. Schneirla, P. Cahn,★ E. Shaw, E. Clark

Fish genetics (with New York Zoological Society)
M. Gordon, D. Rosen,★ K. Kallman

Invertebrates
L. H. Hyman, N. Eldredge, H. Stunkard

Lerner Marine Lab
P. Gilbert★

★Involved in aspects of embryological research.

Lerner Marine Lab at Bimini and had field stations in Florida and Arizona (see table 8.8). The AMNH has been home to many renowned scholars. Osborn, Matthews, R. C. Andrews (dinosaur eggs and embryos), and G. S. Simpson (evolutionary theory) were in vertebrate paleontology. Libby Hyman wrote her classic six-volume text on invertebrates at the AMNH. Ernst Mayr was curator of ornithology prior to moving to the MCZ. Gregory produced his classic works on evolution and fish skulls. He also trained Romer and Breder. Embryological work, often with graduate students from Columbia and New York University, was carried out in several different departments (e.g., Noble in experimental biology; Gregory in anatomy; Dean, Gudger, and others in ichthyology; and Gordon in genetics).

Evolution, always a dominant theme, cut across departmental boundaries (e.g., Osborn, Dean, Gregory, Mayr, Simpson, Eldredge). Beginning with Bashford Dean, a long tradition of comparative evolutionary embryological studies of fishes was established at the AMNH. Dean, who had been encouraged in his study of zoology by Edward Morse, took his degree in geology at Columbia with Newberry. He joined the Columbia faculty in 1890 and remained there until 1912. He became the first curator of fishes at the AMNH and remained there from 1903 to 1928.[52] Dean studied primitive fishes. His first book, *Fishes, Living and Fossil* (1895), was concerned with the morphology and systematic/evolutionary relationships of living and fossil fishes, especially in light of what was then known of their embryology. Dean produced pioneering monographs on the embryonic development of the hagfish (an agnathan) and chimaeras (sister group of the sharks and rays), as well as studies of the sturgeon, gar, and bowfin.[53] Major studies on the anatomy, reproduction, and embryonic development of the primitive frilled shark *Chlamydoselachus*, and the horned shark *Heterodontus*, were completed posthumously by E. W. Gudger, who was a student of W. K. Brooks, and B. Smith, and published in the seven-volume *Bashford Dean Memorial Volume of Archaic Fishes*.[54] Dean obtained hagfish and chimera embryonic material at Hopkins Marine Station, and the frilled and horned shark material at the Misaki Marine Station in Japan. He tried unsuccessfully to obtain embryos of *Polypterus*, a forerunner of modern teleost fishes.[55]

Studies of fish embryology were continued by Gudger, C. M. Breder, and others. Breder and Rosen published *Modes of Reproduction in Fishes* (1966), an invaluable bibliographic study. William Beebe published the definitive study on the stalk-eyed larvae of the deep-sea fish *Idiacanthus* (1934). Perry Gilbert used the facilities at the Lerner Marine Lab in the early phase of his studies of shark reproduction and development. Stocks of platys in Myron Gordon's fish genetics lab, and of tilapia and blind cave fish in Lester Aronson's behavior lab, provided material for embryological studies as well as systematic and evolutionary studies from the 1940s through the 1960s.[56]

Previously unknown species of fishes and other organisms continue to be discovered. In some instances, they are of significant evolutionary interest. The Australian lungfish, *Neoceratodus forsteri*, was described by Gunther in 1870. Semon (1901) first described its development, which resembles that of amphibians.[57] Coelacanths were long thought extinct until the discovery of the living coelacanth, *Latimeria chalumnae*, in 1938. Its mode of development remained unknown until 1975, when a team at the AMNH discovered that *Latimeria* is viviparous.[58] Subsequently, Wourms et al. (1991) described the placental maternal-embryonic relationship, using embryos and maternal tissues from the AMNH specimen. Even now, nothing is known of its early embryonic development. Both lungfishes and the coelacanth are sarcopterygian fishes, and thus are part of the group that evolved into tetrapods (i.e., amphibians, reptiles, etc.).

MARINE LABORATORIES AND FISHERIES STATIONS

Beginning in the mid-nineteenth century, marine biology became popular.[59] In the mid-late part of that century, due to the *Challenger* expedition and Alexander Agassiz's activities on the *Blake* and the *Albatross*, biological oceanography and the study of deep-sea organisms became of interest.[60] Coincident with the development of marine biology and biological oceanography, and prompted by a decline in commercial fisheries, fisheries biology and fisheries research stations came into being, both in the United States and in Europe.[61] For the sake of simplicity, only marine labs and marine fisheries will be considered. This is perhaps unfortunate because it will exclude about 50 percent of all living fish species that live in freshwater, especially in the tropics. Some are of considerable interest because they are primitive or have evolved interesting features (e.g., viviparity, loss of eyes in blind cave fishes). These fishes have been the object of embryological studies at freshwater research stations (e.g., Whitman at the Allis Lake Laboratory), freshwater fisheries stations, public aquariums, or aquarium facilities within natural history museums (e.g., AMNH).

The mission and research objective of marine labs are similar to those of natural history museums, except that research is confined to marine organisms and the emphasis is on the study of living organisms in terms of their life history, embryology, behavior, physiology, and so on. Since evolution, diversity, and adaptation are prevailing themes, systematics and morphology are also emphasized. Fisheries research stations, by definition, have a more applied mission. Their research is concerned with population biology, ecology, reproductive biology, life history studies, and embryonic and larval development. Obviously, systematics provides an underpinning for the enterprise.

What sets marine labs and fisheries stations apart is the organisms with which they deal and the constraints that those organisms impose. One is usually dealing with shallow-water or deep-water marine organisms that need to be collected, often with difficulty, and maintained alive for research. The logistical problems are far more complicated than those involved in the study of "model organisms." Marine invertebrates and fishes are the prime objects of zoological interest at marine labs, while at fisheries stations, fishes and invertebrates of commercial interest are the objects of study. There are an estimated 25,000–36,000 species of fishes. By and large, relatively little was known about their life history, embryonic development, and larval stages prior to 1850–1875. This state of affairs is being very slowly remedied.[62] Much the same can be said about marine invertebrates. Their incredible phylogenetic diversity, diversity of habitat and life styles, and innumerable taxa render their study much more complex. With an increased emphasis on deep-sea biology and the integration of more efficient methods of sampling the deep sea and its fauna, not only have new organisms been discovered, but entire new ecosystems as well (e.g., thermal vents and hydrocarbon seeps and their associated fauna). What was challenging has become formidable. The complex diversity of marine organisms mandates a comparative and evolutionary approach whether it be in morphology, physiology, or embryology.

Fisheries Research Stations
The U.S. Commission of Fish and Fisheries was created by Congress in 1871 and charged with studying and recommending solutions to an apparent decline in the New England fishery. Spencer Fullerton Baird, assistant secretary at the Smithsonian Institution, was its first commissioner. He was one of the preeminent zoologists of the nineteenth century and had played an important role in fostering natural history at the Smithsonian.[63] Research activities in coastal New England began in 1871 and were conducted at various locales until 1880. A number of well-known biologists, some with MCZ associations, carried out research for the Fish Commission during this time (e.g., A. E. Verrill, A. Hyatt, A. S. Packard, T. Gill, W. G. Farlow).[64] Eventually, Woods Hole was chosen as a permanent site, and the newly constructed Woods Hole Laboratory was completed and opened for research in 1885.

At that time, not much was known about the reproduction, life histories, embryonic development, and larval stages of marine fishes and of many marine invertebrates. Specimens of pelagic and benthic organisms collected by oceanographic expeditions under the auspices of the Fish Commission provided a surfeit of riches. The work of the Fish Commission laboratory evolved into two areas: (1) systematics and ecology, and (2) life history studies, especially embryonic development and larval stages. Fishes and selected invertebrates were the objects of study. Embryology attracted considerable attention. To pursue embryological studies of fishes, Baird was able to recruit John Ryder from the Academy of Natural Sciences in Philadelphia. Ryder was associated with the Fish Commission from 1880 to 1886. In 1886, he accepted a chair of comparative histology and embryology at the University of Pennsylvania. During his tenure with the Fish Commission, he produced fifty papers on the development of fishes and twenty-nine on the oyster and its culture. In his short life (1852–1895), he published 278 papers, most of which dealt with aspects of development.[65] He was very interested in the relationship between embryology and evolution. In addition to exquisitely detailed studies of fish embryology, Ryder explored a variety of specialized developmental topics:

1. Comparative studies of viviparity in fishes
2. Origin of the amnion and placenta
3. Mechanical genesis of form (developmental biomechanics)
4. Evolutionary/developmental origins of fin patterns and tail heterocercy in fishes (this may have influenced Ross Harrison's early studies of this subject)
5. Early stages of elasmobranch development
6. First studies of *Fundulus* development
7. Inheritance of developmental modifications in the early development of goldfish varieties (perhaps a prelude to developmental genetics).[66]

The Fish Commission laboratory at Woods Hole was used by a number of people associated with the MCZ (Fewkes, Parker, Boyer, Davenport, Castle, and others) because its facilities were superior to those of the nascent MBL.[67] In the context of fisheries research, it is interesting to note that W. K. Brooks carried out a series of studies on the morphology, reproduction, and embryology of the oyster in order to provide a scientific basis for successful oyster farming. He summarized his work in his book on the oyster, first published in 1891.[68] Several of his students did research for the Fish Commission, and one, F. H. Herrick, published the definitive study of the American lobster, including its development (Herrick, 1895). Fisheries research and oceanographic studies with an embryological emphasis also were pursued in Europe: McIntosh and Masterman (1897), Ehrenbaum (1905 & 1909), Sanzo (1910), Lo Bianco (1931–1933), and Russell (1976), who summarized his own work from 1924 to 1976. Among the interesting research that emerged from these efforts were Schmidt's studies of the leptocephalus larva of the eel.[69] In addition to fisheries research, there were modern oceanographic studies, such as the Dana expeditions and Thorson's (1946) monumental study of the larvae of bottom-dwelling invertebrates.

Broad-based, scientifically oriented fisheries research with an emphasis on life history studies and comparative embryological research extended well into the twentieth century. The two world wars and the Great

Depression had an adverse effect on fisheries research and embryological research in general, especially in Europe. Resources were much more limited and research was restricted to the important food fishes. In the early 1950s, research efforts were revived and greatly expanded as fisheries biologists once again recognized the importance of life history and embryological studies. The expansion took place on a global basis. In the United States, activities were centered on the Pacific coast.[70] Following the collapse of the Pacific sardine populations in the late 1940s, the California Cooperative Oceanic Fisheries Investigation (CALCOFI) program was set up. In the context of this program, Elbert H. Ahlstrom built up a research team at the Fisheries Center in La Jolla, close to the Scripps lab, that played a leading role in fisheries research based on ontogeny.[71]

Ahlstrom and his team, as well as other investigators with a similar outlook, were interested in and concentrated their research efforts on the study of the embryological development, early life history, and larval stages of marine teleosts. Larval fish are often quite different from the adult stage and often differ from the larvae of other species within the same genus and family. Each species has a morphologically distinct larva. Larval stages are highly specialized, and many have curious evolutionary adaptations (e.g., stalked eyes or a trailing gut appendage. Larvae are well adapted for life in a planktonic realm. The larva and its adult perform in "two quite separate evolutionary theatres."[72] The Ahlstrom Memorial Symposium, titled *Ontogeny and Systematics of Fishes*, emphasized larval evolution in terms of ontogeny, systematics, and phylogeny.[73] In this volume, Kendall et al. (1984) list forty contributions in the period 1930–1984 in which ontogenetic characters have been used to formulate systematic and evolutionary relationships in twenty-five taxa (families through orders of marine teleosts). Similar research, spearheaded by Mansueti, was being carried out on the east coast of the United States, especially in the Mid-Atlantic Bight.[74] Research on the development of fishes has continued (e.g., a 1,505-page, multiauthor volume on the early stages of fishes in the California Current region.[75] This work is both comprehensive and comparative, treating almost 150 families of fishes. Surprisingly, this research, with its strong evolutionary implications, has attracted little attention outside of the circle of fishery biologists.

Marine Laboratories

The history of marine labs in the United States has been dealt with extensively, so it would be redundant to repeat it here.[76] In discussing Harvard's MCZ and the diaspora of its students and protégés, several early marine labs were considered (figure 8.1). In most instances, their founders were involved in embryological research: Alexander Agassiz at Newport, Brooks at the Chesapeake Zoological Lab, Whitman at the MBL, Ritter at Scripps, and Mayer at Tortugas. In the case of Hopkins Marine Station and Woods Hole Oceanographic Institution, both Gilbert and Bigelow were ichthyologists. A number of other marine labs have come into existence since those early days. Dexter (1988) lists fifty of the best-known American labs, most of which are discussed in Vernberg (1963).

The zoological mission of marine labs, as originally conceived, is to study marine invertebrates and fishes in the context of the marine ecosystem. Preference is given to the study of living organisms. The research agenda can be seen to have an advantage in Ritter's report on the progress of zoological research at Scripps.[77] He divided the research into two major categories. First, field studies and systematics were needed to acquire a "speaking acquaintance" with the fauna. Second, in-depth biological studies were designed to acquire "a deeper knowledge" of the organisms. These studies were specified as ecology, morphology and physiology, reproduction and development, adaptation, evolution, and behavior. This viewpoint was common in the early years. Maienschein (1988, p. 22), in assessing the early days of the MBL, states: "Whitman and the MBL researchers regarded the place as a center for both morphology and physiological study, with a strong emphasis on study of individual organisms to address questions of heredity, development, and evolution." She also documents a move away from traditional embryological concerns toward biochemistry and chemical physiology and, in more recent times, molecular biology. In similar fashion, the graduate-level summer course in the comparative embryology of marine invertebrates that was a premiere attraction at the MBL for many decades[78] and provided training for several generations of embryologists, is no longer offered there, but has been retained elsewhere.

It is important to realize that while each marine lab has its own distinctive persona, they also fall into one or the other of two broad categories. Marine labs either are independent entities, such as the MBL and the Mount Desert Island Biological Laboratory, or are part of a university, such Hopkins Marine Station (Stanford University), Scripps (University of California), and the Friday Harbor Laboratory (University of Washington). With the passage of time, especially in the latter half of the twentieth century, the mission and research agendas of these two classes of institutions tended to diverge. University-affiliated marine labs have adhered to a more traditional mission and research agenda. Allowing for technological advances, their mission and research agendas would be similar to those set forth by Ritter at Scripps. On the other hand, independent marine labs such as the MBL, which is unique, have gone in the direction of "big science," following a reductionist trend and emphasizing research of interest to the biomedical community.

Some comments on the university-affiliated marine labs are in order. As noted, the founder of Scripps was Ritter, an embryologist, who was succeeded by T. W. Vaughn, another MCZ scholar, who was interested in coral reefs. Although there was a shift in emphasis to oceanography, a strong cadre of biologists was on the faculty. Carl Hubbs carried out a major program in ichthyology involving systematics, morphology, and embryology. Recently, Holland and Holland have done outstanding evo-devo research with *Amphioxus*. At Stanford, C. H. Gilbert, an ichthyologist who was the first director of Hopkins Marine Station, was succeeded by W. K. Fisher, an invertebrate zoologist who served from 1918 to 1943. Important studies in systematics, life history, and development of fishes were carried out by Rolf Bolin and his students. Hopkins became a mecca for the study of invertebrate zoology for a number of decades, first through Ricketts and Calvin (1939) and then through the efforts of D. P. Abbott in invertebrate zoology that culminated in the magisterial volume *Intertidal Invertebrates of California* (Morris et al., 1980).

In similar fashion, A. C. Giese and his students carried out an internationally recognized research program in ecological physiology and reproductive biology of invertebrates that culminated in the nine-volume treatise *Reproduction of Marine Invertebrates* (Giese et al., 1974–1991). The Friday Harbor Laboratory of the University of Washington was founded in 1904 by Trevor Kincaid and T. C. Frye under the name Puget Sound Biological Station. Originally it was a summer instructional operation. In 1924, it moved to its present location, had a laboratory building constructed, and acquired its present name.[79] In 1963, at least fifteen faculty from the university were on the staff and engaged in research. Robert Fernald was associated with the lab from the 1940s until his death in 1983. He served as acting director (1958–1960) and director (1960–1972). Fernald and his students had an active research program in comparative marine invertebrate embryology, and they taught a graduate course of the same name that is still offered (Strathmann, 1987). Relatively recently, Strathmann, with the aid of former students and colleagues of Fernald, produced a major study on the reproduction and development of marine invertebrates. It is reasonable to conclude that marine labs, especially those affiliated with a university, provided a safe haven for scholars conducting research in comparative evolutionary embryology. This was vital during the period of apparent stasis of interest in the field.

THE TWO EMBRYOLOGIES

The history of embryology/developmental biology has been chronicled in a number of accounts.[80] The conventional view is that embryology, which began as descriptive embryology, acquired comparative and evolutionary inflections and was transformed into experimental embryology, which eventually became developmental biology. There are ample grounds to reexamine this thesis. There seems to be an emerging consensus that in the latter part of the nineteenth century, embryology, which at the time was comparative and evolutionary, split into two embryologies: (1) comparative evolutionary embryology and (2) experimental embryology/ *Entwicklungsmechanik*. Both embryologies have survived to the present. Experimental embryology became developmental biology. Comparative evolutionary embryology incorporated aspects of other disciplines, such as molecular developmental genetics, to become evolutionary developmental biology (figure 8.2). Here, selected aspects of this history will be reiterated. The objective is to show that the two embryologies coexisted throughout the twentieth century. Developmental biology became dominant, but comparative evolutionary embryology, albeit marginalized, survived. Recently, with the revival of interest in evolutionary problems and the extension of molecular developmental genetics into areas of profound evolutionary significance, comparative evolutionary embryology has been transformed into evolutionary developmental biology.[81]

Earlier in this account, the rise of comparative evolutionary embryology was traced, especially in an American context. In brief, modern embryology began with Von Baer and the Dollinger school in 1830s. Soon it became both descriptive and comparative. Von Baer and others, including Louis Agassiz, considered patterns of embryonic development in what could be deemed "quasi-evolutionary" terms. Darwin felt that embryology provided evidence for his theory of evolution. Haeckel, in the late 1860s, promulgated his biogenetic law (i.e., ontogeny recapitulates phylogeny). This became something of a paradigm for those working in comparative evolutionary embryology. Others, such as F. M. Balfour and Alexander Agassiz, pursued embryological research with an evolutionary theme, although along lines somewhat different from those of Haeckel. There were a number of problems with Haeckel's biogenetic law that soon became apparent. One major problem was that he failed to realize that

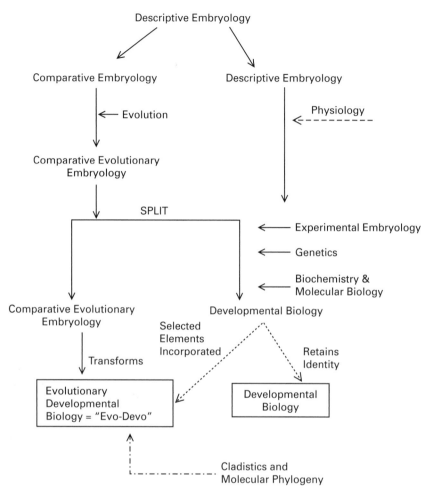

Figure 8.2
The two embryologies and the origin of Evo-Devo.

evolutionary changes in the developmental process produced phylogenies, not the other way around. Another problem was a logistic one. In an era of emerging science, there seemed to be little or no way of testing evolutionary hypotheses and constructs. There was also an increased emphasis on experimental methods in science and on proximal causality in scientific explanation. In addition, there seemed to have been a cultural reaction to Haeckel's rhetorical excesses. Some embryologists began to move in a different conceptual and methodological direction.

Various terms, such as experimental embryology, *Entwicklungsmechanik*, and analysis of development, have been used to categorize a set of conceptual and methodological approaches (physiology, biomechanics, biochemistry, cell biology) that are used to study the proximal causes of embryonic development. These approaches have a strong experimental bias. It is difficult to say when experimental approaches to the study of development were introduced.[82] The conventional viewpoint traces the origin of experimental/analytical embryology to His's dissatisfaction with the causal explanations provided by Haeckel's biogenetic law and the need for a mechanical explanation of morphological change, as well as Roux's surgical experiments on amphibian eggs in the latter part of the century. Experimental studies of development, however, date back much further.[83] For example, in the eighteenth century, Spallanzani worked on insemination, Trembley on *Hydra* regeneration, and Broussonet on regeneration in fishes.

In the second quarter of the nineteenth century, experimental physiology, under the aegis of Claude Bernard and Johannes Müller, became a formal science. Müller is a pivotal figure because he was both a physiologist and an embryologist. He "rediscovered" and investigated the yolk sac placenta of sharks.[84] Most likely, Müller and his many students were responsible for transferring an analytical-experimental approach from physiology to embryology in the mid-nineteenth century. Among Müller's students were E. Haeckel, W. His, R. Virchow, and E. DuBois-Reymond. In the early stages of his career, even Haeckel engaged in experimental work. He divided the larva of the siphonophore *Crystallodes*, in two and obtained two smaller but perfect larvae.[85] In the 1860s, Ransom carried out a remarkable series of studies on contractility and peristaltic waves in the yolk cytoplasmic layer of teleost eggs. He performed elaborate experiments to ascertain the effects of metabolic poisons, temperature, electricity, carbon dioxide, and oxygen on contractility and cytokinesis. He also isolated a myosin-like protein from fish, amphibian, and avian eggs.[86] The first attempt to analyze the mechanism of embryo formation in teleost eggs was undertaken by Lereboullet, who used different temperature regimes, chemicals, and pressure to perturb development.[87] The contributions of His and Roux are well known. Because these two embryologists wrote several extensive works setting forth their theories and concepts, their names have been linked to the origins of analytical embryology. Roux adapted a narrowly defined conceptual approach in his *Entwick-lungsmechanik*, which was reinforced by the highly mechanistic views of Loeb. At this point, the study of embryology was splintered, and the two embryologies, comparative evolutionary embryology and experimental embryology/developmental biology, diverged.

Both embryologies survived. Developmental biology became dominant for a variety of reasons, one of which was the success of its analytical/experimental research methodology, which permitted it to incorporate research approaches from other branches of science, such as biochemistry. Comparative evolutionary embryology survived and, in a modest way, flourished in a variety of refuges in the period 1900–2000.

The state of affairs with respect to the two embryologies in the United States from about 1950 to 1975 can best be appreciated by examining the twenty-fifth symposium of the Society for Developmental Biology, titled *Major Problems in Developmental Biology*.[88] This is the major organization that represents the field of embryology, *sensu latu*, in the United States. The symposium contains a number of papers on subjects considered to be the major problems facing the research community. In addition, in her introductory paper on the growth and development of developmental biology, Oppenheimer (1967) discusses the history of this field from about 1938 to 1966. In assessing the lines that define the "new developmental biology," she states that the twentieth-century concept of the unity of biochemistry supplants the eighteenth-century concept of type and the nineteenth-century concept of unity of descent as a synthesizing scheme. She points out that developmental biology has come into being as a result of the merger of experimental/analytical embryology with other fields, such as genetics and biochemistry, and, through this merger, the study of macromolecules, proteins, enzymes, cells and organelles, metabolic pathways, immune reactions, microbes, protozoans, and fungi. In her overview, there is no mention of comparative evolutionary embryology or the role of embryology in the study of evolution.

None of the papers in the symposium volume bear on evolution. In point of fact, the term "evolution" does not even appear in the index. This symposium accurately reflects the attitudes that prevailed in developmental biology at this time. To link the study of development with evolution, as in comparative evolutionary embryology, was considered irrelevant at best. More often, the concept was greeted with outright antagonism. Given this attitude, it is not surprising that embryology/development did not contribute to the Modern Synthesis; mainstream developmental biology was not interested.[89] The dominant culture of developmental biology, with its control of academic positions, training of graduate students, and sources of research funding, increasingly marginalized comparative evolutionary embryology and its practitioners.

Although comparative evolutionary embryology was marginalized, it existed in parallel with developmental biology. It survived in the museums, marine labs, and fisheries research stations previously discussed because its goals coincided well with missions of these institutions and the organisms that comprised their research agenda. Moreover, Europe seems to have been more hospitable to embryological and morphological studies that addressed evolutionary problems, such as those of Schmalhausen (1949) and other members of the Russian school of evolutionary morphology. By way of demonstrating this point, table 8.9 lists important books and monographs on comparative evolutionary embryology and evolutionary developmental biology published between 1880 and 2001. With the exception of the war years and early postwar years (1940–1950), these works appeared with a consistent regularity. They served to keep alive the spirit of inquiry and provided intellectual background for its revival.

The publication of Gould's *Ontogeny and Phylogeny* (Gould, 1977) signaled a revival of interest in comparative evolutionary embryology. A number of factors were involved. Punctuated equilibria[90] provided an alternative view for macroevolutionary change. In their paper "The Spandrels of San Marco," Gould and Lewontin (1979) reasserted the concept of developmental homology (which subsequently would be extended to include gene function). They also proposed that heterochrony (changes in the relative timing of developmental events) and allometry (differential growth) provide developmental mechanisms that could provide rapid macroevolutionary change and novelty.[91]

The rediscovery of heterochrony and allometry raises some interesting questions of intellectual history. In his book on the principles of rational comparative embryology, Eugen Schultz (1910) went into considerable detail and placed great emphasis on the role of the various types of heterochrony as a general developmental mechanism for effecting evolutionary change. His work appears to have been overlooked in western

Table 8.9
Comparative Embryology and Evo-Devo: 1880–2001

Balfour (1880–1881) *Treatise on Comparative Embryology*

Hamann (1892) *Entwicklungslehre und Darwinismus*

Korschelt & Heider (1893; 1902; 1936) *Lehrbuch der vergleichenden Entwicklungsgeschichte*

Willey (1894) *Amphioxus and The Ancestry of Vertebrates*

Keibel (1897–1938) *Normentafeln zur Entwicklungsgeschichte der Organismen*

Hertwig (1901–1906) *Handbuch der vergleichenden und experimentellen Entwicklungslehre der Wirbeltiere*

Ziegler (1902) *lb. Vergl. Entwicklung*

Schultz (1910) *Prinzipien der rationellen vergleichenden Embryologie*

MacBride (1914) *Textbook of Embryology*. Vol. I. *Invertebrata*

Russell (1916) *Form and Function*

Graham Kerr (1919) *Textbook of Embryology*. Vol. II. *Vertebrata*

Garstang (1922; 1928) Restatement of Biogenetic Law; Origin & Evolution of Larval Forms. Tunicates & Chordate Phylogen

Dawydoff (1928) *Traité d'embryologie comparée des invertébrés*

Goodrich (1930) *Structure and Development Vertebrates*

Richards (1930) *Comparative Embryology*

De Beer (1930; 1940) *Embryology and Evolution; Embryos and Ancestors*

Needham (1931) *Chemical Embryology*

Huxley (1932) *Problems of Relative Growth*

Schindewolf (1936) *Paläontologie, Entwicklungslehre und Genetik*

Haldane (1932) *Causes of Evolution*

Goldschmidt (1940) *Material Basis of Evolution*

Schmalhausen (1942) *Organism as a Whole in Development and Evolution*

Waddington (1940) *Organisers and Genes*

Pasteels (1937; 1940) *Comparative Gastrulation in Vertebrates*

Waddington (1956) *Principles of Embryology*

Waddington (1957) *Strategy of the Genes*

Nelsen (1953) *Comparative Embryology of the Vertebrates*

Berrill (1955) *Origin of the Vertebrates*

Bonner (1958) *Evolution of Development*

Kumé & Dan (1957) *Invertebrate Embryology*

Sachwatkin (1956) *Vergleichende Embryologie der niederen Wirbellosen*

Willmer (1960) *Cytology and Evolution*

Bullough (1967) *Evolution of Differentiation*

Table 8.9 (continued)

Siewing (1969) *Lehrb..d. Vergleich. Entwicklung*

Anderson (1973) *Embryology and Phylogeny in Annelids and Arthropods*

Jagersten (1972) *Evolution of the Metazoan Life Cycle*

Wolsky & Wolsky (1976) *Mechanism of Evolution*

Gould (1977) *Ontogeny and Phylogeny*

Bonner et al. (1981) *Evolution and Development*

Goodwin, Holder, Wylie (1983) *Development and Evolution*

Mlikovsky & Novak (1985) *Evolution and Morphogenesis*

Arthur (1984) *Mechanisms of Morphological Evolution: A Combined Genetic, Developmental and Ecological Approach*

Raff & Raff (1987) *Development as an Evolutionary Process*

Raff & Kaufman (1983) *Embryos, Genes and Evolution*

Thomson (1988) *Morphogenesis and Evolution*

Nitecki (1990) *Evolutionary Innovations*

Hall (1992) *Evolutionary Developmental Biology*

Raff (1996) *The Shape of Life*

Arthur (1997) *The Origin of Animal Body Plans*

Gerhart & Kirschner (1997) *Cells, Embryos, and Evolution*

Hall & Wake (1999) *The Origins and Evolution of Larval Forms*

Carroll et al. (2001) *From DNA to Diversity*

Davidson (2001) *Genomic Regulatory Systems: Development and Evolution*

Europe, although it may have influenced the Russian school of evolutionary morphologists. Credit for introducing heterochrony into evolutionary developmental studies is usually given to De Beer (1930).

In similar fashion, allometry as a developmental mechanism for explaining macroevolutionary change had been known for some time.[92] The major breakthrough took place in developmental genetics during the 1980s.[93] It was discovered that the same developmental regulatory genes, especially those of the Hox gene family, were shared by animals with different body plans (e.g., insects, fish, mice). The conserved roles in development of these shared regulatory genes are being used to investigate and hypothesize body part homologies between distantly related animals, often with different body plans (Holland and Holland, 1999). The revolutionary advances since the mid-1980s have revitalized developmental studies of evolution and transformed comparative evolutionary embryology into evolutionary developmental biology (see figure 8.2).

Acknowledgments

My research in evolutionary developmental biology has been supported
by the National Science Foundation.

Notes

1. L. Agassiz, 1857; Mayr, 1982.

2. Needham, 1959.

3. Oppenheimer, 1967; Churchill, 1991.

4. Needham, 1959; Horder et al., 1985; Hall, 1992.

5. Mayr, 1982.

6. L. Agassiz, 1853; Girard 1858.

7. Packard, 1886.

8. Wyman, 1854; Wourms, 1997.

9. Oppenheimer, 1986.

10. Wyman, 1854.

11. Packard, 1886.

12. Rosen, 1936; Rothschuh, 1973.

13. Lurie, 1960, 1974; Winsor, 1991.

14. Mayr, 1982.

15. L. Agassiz and Gould, 1848, p. 154.

16. L. Agassiz, 1849.

17. Gould, 1977, p. 70.

18. Mayr, 1982, p. 474.

19. Gould, 1977; Mayr, 1982; Horder et al. 1985.

20. Winsor, 1991.

21. Ibid.

22. Ibid.

23. Donaldson, 1915.

24. Lillie, 1944.

25. Winsor, 1991; Murray, 1911.

26. Goodale, 1912.

27. Ingersoll, 1883; G. R. Agassiz, 1913; Winsor, 1991.

28. Winsor, 1991; *Annual Reports* of the MCZ.

29. Benson, 1988a; Benson 1988b; Maienschein, 1985.

30. Allard, 1990.

31. Wourms, 1997.

32. Winsor, 1991.

33. Ibid.; Mayer, 1911; Murray, 1911.

34. Mayer, 1911; Murray, 1911; G. R. Agassiz, 1913; Goodale, 1912; Rainger, 1988; Winsor, 1991.

35. Rainger et al., 1988.

36. Hubbs, 1964.

37. Raitt and Moulton, 1967.

38. Benson, 1985.

39. Conklin, 1910; McCullough, 1969.

40. Benson, 1985, 1988a.

41. H. V. Wilson, 1891.

42. Anonymous, 1939.

43. Wourms, 1997.

44. Harvey, 1958.

45. Lillie, 1911; Morse, 1911; Davenport, 1917; Dexter, 1979.

46. Lillie, 1911; Wourms, 1997.

47. Lillie, 1911.

48. Lillie, 1944; Costello et al., 1957.

49. Watterson, 1979.

50. Saunders, 1952; Hellman, 1969; Sloan, 1980; Rainger, 1991.

51. Rainger, 1991.

52. Gudger, 1930.

53. Wourms, 1997.

54. Gudger, 1930–1942.

55. Hall, 2001.

56. Breder and Rosen, 1966.

57. Kemp, 1982.

58. Smith et al., 1975.

59. Ward, 1974.

60. Schlee, 1973.

61. Russell, 1976; Hobart, 1995.

62. Wourms, 1997.

63. Allard, 1990; Hobart, 1995.

64. Allard, 1990.

65. Allen, 1896.

66. See Moore's bibliography in ibid.

67. Allard, 1990; Winsor, 1991.

68. Brooks, 1905.

69. Russell, 1976.

70. Blaxter, 1984.

71. Ibid.; Moser, 1984.

72. Moser and Ahlstrom, 1974.

73. Moser, 1984; Cohen, 1984.

74. Jones et al., 1978.

75. Moser, 1996.

76. Dexter, 1988; Benson, 1988a, 1988b, Maienschein, 1986, 1988; Lillie, 1944; Raitt and Moulton, 1967.

77. Ritter, 1912.

78. Lillie, 1944.

79. Vernberg, 1963.

80. Needham, 1959; Oppenheimer, 1967; Horder et al., 1985; Churchill, 1991; Hall, 1992.

81. Gilbert et al., 1996.

82. Oppenheimer, 1955.

83. Gasking, 1970.

84. Wourms, 1997.

85. Haeckel, 1869.

86. Wourms, 1997.

87. Ibid.

88. Locke, 1966.

89. Hamburger, 1980.

90. Eldredge and Gould, 1972.

91. Gilbert et al., 1996.

92. Thompson, 1917; Huxley, 1932.

93. Raff, 1996.

REFERENCES

Anonymous. 1939. Henry Van Peters Wilson. *Journal of the Elisha Mitchell Scientific Society*, 55: 1–6.

Agassiz, Alexander. 1864. On the embryology of echinoderms. *Memoirs of the American Academy of Arts and Sciences*, 9: 1–30.

———. 1865. North American Acalephae. *Memoirs of the Museum of Comparative Zoology*, 1(2): 1–234.

———. 1877. Embryology of the starfish. *Memoirs of the Museum of Comparative Zoology*, 5(1):1–83.

———. 1877–1878. *Annual Report of the Museum of Comparative Zoology.* Cambridge, Mass.: MCZ.

———. 1880. Paleontological and embryological development. *Proceedings of the American Association for the Advancement of Science*, 29: 389–414.

———. 1891–1892. *Annual Report of the Museum of Comparative Zoology.* Cambridge, Mass.: MCZ.

Agassiz, Elizabeth C., and Alexander Agassiz. 1865. *Seaside Studies in Natural History: Marine Animals of Massachusetts Bay. Radiates.* Boston: Houghton Mifflin.

Agassiz, George R. 1913. *Letters and Recollections of Alexander Agassiz.* Boston: Houghton Mifflin.

Agassiz, Louis. 1849. *Twelve Lectures on Comparative Embryology.* Boston: Redding.

———. 1853. Extraordinary fishes from California constituting a new family. *American Journal of Science & Arts*, ser. 2, 16: 380–390.

———. 1857. Essay on classification. In *Contributions to the Natural History of the United States*, vol. 1, pp. 1–232. Boston: Little, Brown.

Agassiz, Louis, and Augustus A. Gould. 1848. *Principles of Zoology.* Boston: Gould, Lincoln.

Allard, Dean C. 1990. The Fish Commission Laboratory and its influence on the founding of the Marine Biological Laboratory. *Journal of the History of Biology*, 23: 251–270.

Allen, Harrison. 1896. A biographical sketch of John Adam Ryder. *Proceedings of the Academy of Natural Sciences*, 48: 222–256.

Audubon, John J. L. 1831–1839. *Ornithological Biography*, 5 vols. Edinburgh: Adam Black.

Balfour, Francis Maitland. 1880–1881. *A Treatise on Comparative Embryology*, 2 vols. London: Macmillan.

Beebe, William. 1934. Deep sea fishes of the Bermuda Oceanographic Expedition. Family Idiacanthidae. *Zoologica* (New York), 16(4): 149–241.

Benson, Keith R. 1985. American morphology in the late nineteenth century: The biology department at Johns Hopkins University. *Journal of the History of Biology*, 18: 163–205.

———. 1988a. Why American marine stations? The teaching argument. *American Zoologist*, 28: 7–14.

———. 1988b. Laboratories on the New England shore: The "somewhat different direction" of American marine biology. *New England Quarterly*, 56: 53–78.

———. 1988c. From museum research to laboratory research: The transformation of natural history into academic biology. In Ronald Rainger, Keith R. Benson, and Jane Maienschein (eds.), *The American Development of Biology*, pp. 49–83. Philadelphia: University of Pennsylvania Press.

Blaxter, J. H. S. 1984. Ontogeny, systematics and fisheries. In H. Geoffrey Moser (ed.), *Ontogeny and Systematics of Fishes*, pp. 1–6. Special publication no. 1, American Society of Ichthyologists & Herpetologists. Lawrence, Kan.: Allen Press.

Breder, Charles M., and Donn E. Rosen. 1966. *Modes of Reproduction in Fishes.* Garden City, N.Y.: Natural History Press.

Brooks, William K. 1905. *The Oyster: A Popular Summary of a Scientific Study,* 2nd ed. Baltimore: Johns Hopkins University Press.

Burnett, Waldo I. 1853. The cell: Its physiology, pathology, and philosophy, as deduced from original investigations to which are added its history and criticism. *Transactions of the American Medical Association*, 6: 645–832.

Churchill, Frederick B. 1980. The modern evolutionary synthesis and the biogenetic law. In Ernst Mayr and William B. Provine (eds.), *The Evolutionary Synthesis*, pp. 112–122. Cambridge, Mass.: Harvard University Press.

——. 1991. The rise of classical descriptive embryology. In Scott F. Gilbert (ed.), *A Conceptual History of Modern Embryology*, pp. 1–29. Baltimore: Johns Hopkins University Press.

Coe, Wesley R. 1918. A century of zoology in America. *American Journal of Science*, new ser., 46: 355–398.

Cohen, Daniel M. 1984. Ontogeny, systematics and phylogeny. In H. Geoffrey Moser (ed.), *Ontogeny and Systematics of Fishes*, pp. 7–11. Special publication no. 1, American Society of Ichthyologists & Herpetologists. Lawrence, Kan.: Allen Press.

Conklin, Edward G. 1910. Biographical memoir of William Keith Brooks. 1848–1908. *Biographical Memoirs, National Academy of Sciences*, 7: 23–88.

Conklin, Edwin G. 1927. *A Synopsis of the General Morphology of Animals*. Princeton, N. J.: Princeton University Press.

Costello, Donald P., M. E. Davidson, A. Eggers, M. H. Fox, and C. Henley. 1957. *Methods for Obtaining and Handling Marine Eggs and Embryos*. Woods Hole, Mass.: Marine Biological Laboratory.

Davenport, Charles B. 1917. The personality, heredity, and work of Charles Otis Whitman, 1843–1910. *American Naturalist*, 51: 5–30.

Dean, Bashford. 1895. *Fishes, Living and Fossil*. New York: Macmillan.

De Beer, Gavin R. 1930. *Embryology and Evolution*. Oxford: Clarendon Press.

——. 1940. *Embryos and Ancestors*. Oxford: Clarendon Press.

DeKay, James E. 1842–1844. *Zoology of New York*, 5 vols. Albany, N.Y.: Carroll and Cook.

Dexter, R. W. 1979. C. O. Whitman (1842–1910) and the American Society of Zoologists. *American Zoologist*, 19: 1251–1253.

Dexter, Ralph W. 1988. History of American marine biology and marine biology institutions. Introduction: Origins of American marine biology. *American Zoologist*, 28: 3–6.

Donaldson, Henry H. 1915. Charles Sedgwick Minot. *Proceedings of the Boston Society of Natural History*, 35: 79–93.

Ehrenbaum, Ernst. 1905–1909. Eier und Larven von Fischen. In *Nordisches Plankton zoologischer Teil*, vol. 1, pp. 1–216, 217–414. Kiel and Leipzig: Lipsius and Tischer.

Eldredge, Niles, and Stephen J. Gould. 1972. Punctuated equilibria: An alternative to phyletic gradualism. In Thomas J. M. Schopf (ed.), *Models in Paleobiology*, pp. 82–115. San Francisco: Freeman, Cooper.

Galtsoff, Paul S. 1962. *The Story of the Bureau of Commercial Fisheries Biological Laboratory, Woods Hole, Massachusetts*. Washington, DC.: U.S. Department of the Interior.

Gasking, Elizabeth. 1967. *Investigations into Generation, 1451–1828*. Baltimore: Johns Hopkins University Press.

————. 1970. *The Rise of Experimental Biology*. New York: Random House.

Ghiselin, Michael. 1991. Classical and molecular phylogenetics. *Bolettino di zoologia*, 58: 289–294.

Giese, Arthur C., John S. Pearse, and Vicki B. Pearse. 1974–1991. *Reproduction of Marine Invertebrates*, 9 vols. New York: Academic Press; Palo Alto, Calif: Blackwell Scientific.

Gilbert, Scott F. (Ed.). 1991. *A Conceptual History of Modern Embryology*. Baltimore: Johns Hopkins University Press.

Gilbert, Scott F., John M. Opitz, and Rudolph A. Raff. 1996. Resynthesizing evolutionary and developmental biology. *Developmental Biology*, 173: 357–372.

Girard, Charles. 1858. Fishes. In *Explorations and Surveys for a Railroad Route from the Mississippi River to the Pacific Ocean*, part 4. Washington, DC.: U.S. War Department.

Goodale, George L. 1912. Biographical memoir of Alexander Agassiz. *Biographical Memoirs, National Academy of Sciences*, 7: 289–305.

Gould, Augustus A. 1841. Report on the *Invertebrata of Massachusetts*. Cambridge, Mass.: Folsom, Wells, and Thurston.

Gould, Stephen J. 1977. *Ontogeny and Phylogeny*. Cambridge, Mass.: Belknap Press of Harvard University Press.

Gould, Stephen J., and Richard C. Lewontin. 1979. The spandrels of San Marco and the Panglossian paradigm. A critique of the adaptationist program. *Proceedings of the Royal Society of London*, B205: 581–598.

Gregory, William K. 1933. Fish skulls. *Transactions of the American Philosophical Society*, 23(2): 75–481.

————. 1951. *Evolution Emerging*, 2 vols. New York: Macmillan.

Gudger, Eugene W. 1930. Memorial of Bashford Dean. In *Bashford Dean Memorial Volume of Archaic Fishes*, art. I, pp. 1–42. New York: American Museum of Natural History.

————. (ed.). 1930–1942. *Bashford Dean Memorial Volume of Archaic Fishes*. New York: American Museum of Natural History.

Haeckel, Ernst. 1866. *Generelle Morphologie der Organismen*, 2 vols. Berlin: Georg Reimer.

————. 1869. *Zur Entwicklungsgeschichte der Siphonophoren*. Utrecht: C. van der Post.

Hall, Brian. 1992. *Evolutionary Developmental Biology*. London: Chapman & Hall.

————. 2001. John Samuel Budgett (1872–1904): In pursuit of *Polypterus. Bioscience*, 51: 399–403 .

Hall, Brian and Marvalee Wake (eds.). 1999. *The Origin and Evolution of Larval Forms*. San Diego: Academic Press.

Hamburger, Viktor. 1980. Embryology and the Modern Synthesis. In Ernst Mayr and William B. Provine (eds.), *The Evolutionary Synthesis*, pp. 97–112. Cambridge, Mass.: Harvard University Press.

Harvey, E. Newton. 1958. Edwin Grant Conklin. *Biographical Memoirs, National Academy of Sciences*, 31: 54–91.

Hellman, Geoffrey. 1969. *Bankers, Bones, and Beetles: The First Century of the American Museum of Natural History*. Garden City, N.Y.: Natural History Press.

Herrick, Francis H. 1895. The American lobster: A study of its habits and development. *Bulletin of the United States Fish Commission*, 1895: 1–252.

His, Wilhelm. 1874. *Unsere Körperform und das physiologische Problem ihrer Enstehung*. Leipzig: Vogel.

His, Wilhelm. 1894. Ueber mechanische Grundvorgange thierischer Formenbildung. *Archiv Für Anatomie und Physiologie*, 7: 1–80.

Hobart, W. L. 1995. *Baird's Legacy: The History and Accomplishments of NOAA's National Marine Fisheries Service, 1871–1996*. NOAA Technical Memorandum NMFS-F/SPO-19. Washington, D.C.: U.S. Department of Commerce.

Holbrook, John E. 1836–1838. *North American Herpetology*, 4 vols. Philadelphia: J. Dobson.

Holland, Nicholas D., and Linda Z. Holland. 1999. *Amphioxus* and the utility of molecular genetic data for hypothesizing body part homologies between distantly related animals. *American Zoologist*, 39: 630–640.

Horder, T. J., J. A. Witkowski, and C. C. Wylie (eds.). 1985. *A History of Embryology*. Cambridge: Cambridge University Press.

Hubbs, Carl L. 1964. History of ichthyology in the United States after 1850. *Copeia*, 1964: 42–60.

Huxley, Julian S. 1932. *Problems of Relative Growth*. London: Methuen.

Ingersoll, Ernst. 1883. Professor Agassiz's laboratory. *Century Magazine*, 26(5): 728–734.

Jones, Philip W., F. Douglas Martin, and Jerry D. Hardy, et al. 1978. *Development of Fishes of the Mid-Atlantic Bight*, 6 vols. Fort Collins, Colo.: U.S. Fish & Wildlife Service.

Kemp, Anne. 1982. The embryological development of the Queensland lungfish *Neoceratodus forsteri* (Krefft). *Memoirs of the Queensland Museum*, 20: 553–597.

Kendall, Arthur W., Elbert H. Ahlstrom, and H. Geoffrey Moser. 1984. Early life history stages of fishes and their characters. In H. Geoffrey Moser (ed.), *Ontogeny and System-*

atics of Fishes, pp. 11–22. Special publication no. 1, American Society of Ichthyologists & Herpetologists. Lawrence, Kan.: Allen Press.

Kohlstedt, Sally G. 1988. Museums on campus: A tradition of inquiry and teaching. In Ronald Rainger, Keith R. Benson, and Jane Maienschein (eds.), *The American Development of Biology*, pp. 15–47. Philadelphia: University of Pennsylvania Press.

Korschelt, Eugen, and Karl Heider. 1895–1900. *Text-book of the Embryology of Invertebrates*, trans. Edward L. Mark, W. McM. Woodworth, Matilda Bernard, and M. F. Woodward, 4 vols. London: Swan-Sonnenschein.

Kowalevsky, Alexander. 1871. Weitere Studien über die Entwicklung der einfachen Ascidien. *Archiv Für mikroskopische Anatomie*, 7: 101–130.

Lillie, Frank R. 1911. Charles Otis Whitman. *Journal of Morphology*, 22: xv–lxxvii.

———. 1944. *The Woods Hole Marine Biological Laboratory*. Chicago: University of Chicago Press.

Linton, Edwin. 1915. Reminiscences of the Woods Hole Laboratory of the Bureau of Fisheries, 1882–1889. *Science*, 41: 737–753.

Lo Bianco, Salvatore. 1931–1933. *Eggs, Larvae and Juvenile Stages of Teleostei*. Monograph 38. Fauna and Flora of the Bay of Naples. Dr. G. Bardi, Rome and R. Fridlander and Sohn, Berlin. English translation, Washington, D.C.: Smithsonian Institution Press, 1969.

Locke, Michael (ed.). 1966. *Major Problems in Developmental Biology*. New York: Academic Press.

Lurie, Edward. 1960. *Louis Agassiz: A Life in Science*. Chicago: University of Chicago Press.

———. 1974. *Nature and the American Mind: Louis Agassiz and the Culture of Science*. New York: Science History.

Maienschein, Jane. 1985. Agassiz, Hyatt, Whitman, and the birth of the Marine Biological Laboratory. *Biological Bulletin*, 168(Suppl.): 26–34.

———. 1988. History of marine laboratories: Why do research at the seashore? *American Zoologist*, 28: 15–25.

———. 1991a. *Transforming Traditions in American Biology, 1880–1915*. Baltimore: Johns Hopkins University Press.

———. 1991b. The origins of *Entwicklungsmechanik*. In Scott F. Gilbert (ed.), *A Conceptual History of Modern Embryology*, pp. 43–61. Baltimore: Johns Hopkins University Press.

———. (ed.). 1986. *Defining Biology: Lectures from the 1890s*. Cambridge, Mass.: Harvard University Press.

Marshall, N. B. 1954. *Aspects of Deep Sea Biology*. New York: Philosophical Library.

Mayer, Alfred G. 1911. Alexander Agassiz, 1835–1910. *Annual Report of the Smithsonian Institution,* 1910: 447–472.

Mayr, Ernst. 1982. *The Growth of Biological Thought: Diversity, Evolution, and Inheritance.* Cambridge, Mass.: Belknap Press of Harvard University Press.

McCullough, Dennis M. 1969. W. K. Brooks's role in the history of American biology. *Journal of the History of Biology,* 2: 411–438.

McIntosh, William C., and Arthur T. Masterman. 1897. *The Life Histories of the British Marine Food-Fishes.* London: C. J. Clay and Sons.

Meisel, Max. 1924–1929. *A Bibliography of American Natural History,* 3 vols. Brooklyn, N.Y.: Premier.

Mitchell, Samuel L. 1803. Peculiarities in the anatomy and physiology of the shark, particularly as respects the production of its young. *Medical Repository,* 2: 78–81.

Moore, John A. 1986. Science as a way of knowing—Developmental biology. *American Zoologist,* 27: 1–159.

Morgan, Thomas H. 1891. A contribution to the embryology and phylogeny of the pycnogonids. *Studies of the Biological Laboratory of Johns Hopkins University,* 5: 1–76.

Morris, Robert H., Donald P. Abbott, and Eugene C. Haderlie. 1980. *Intertidal Invertebrates of California.* Stanford, Calif.: Stanford University Press.

Morse, Edward S. 1911. Biographical memoir of Charles Otis Whitman. *Biographical Memoirs, National Academy of Sciences,* 7: 269–288.

Moser, H. Geoffrey (Ed.). 1984. *Ontogeny and Systematics of Fishes.* Special publication no. 1, American Society of Ichthyologists & Herpetologists. Lawrence, Kan.: Allen Press.

———. 1996. *The Early Stages of Fishes in the California Current Region.* California Cooperative Oceanic Fisheries Investigations, Atlas no. 33. Lawrence, Kan.: Allen Press.

Moser, H. Geoffrey, and Elbert H. Ahlstrom. 1974. Role of larval stages in systematic investigations of marine teleosts: The Myctophidae, a case study. *Bulletin of the United States Fish Commission,* 72: 391–413.

Müller, Fritz. 1864. *Für Darwin.* Leipzig: Wilhelm Engelmann.

Murray, John. 1911. Alexander Agassiz: His life and work. *Bulletin of the Museum of Comparative Zoology,* 54(3): 139–158.

Needham, Joseph. 1959. *A History of Embryology,* 2nd ed. Cambridge: Cambridge University Press.

Oppenheimer, Jane M. 1955. Problems, concepts and their history: Methods and techniques. In Benjamin H. Willier, Paul A. Weiss, and Viktor Hamburger (eds.), *Analysis of Development,* pp. 1–24, 25–38. Philadelphia: W. B. Saunders.

————. 1967. *Essays in the History of Embryology and Biology*. Cambridge, Mass.: MIT Press.

————. 1986. Louis Agassiz as an early embryologist in America. In Randolph S. Klein (ed.), *Science and Society in Early America: Essays in Honor of Whitfield J. Bell, Jr.*, pp. 393–414. Philadelphia: American Philosophical Society.

Packard, Alpheus S. 1876. *Life Histories of Animals, Including Man; or, Outlines of Comparative Embryology*. New York: Henry Holt.

————. 1886. Memoir of Jeffries Wyman, 1814–1874. *Biographical Memoirs, National Academy of Sciences*, 2: 75–126.

Parker, George H. (Ed.). 1903. *Mark Anniversary Volume, to Edward Laurens Mark*. New York: Henry Holt.

Raff, R. A. 1996. *The Shape of Life*. University of Chicago Press, Chicago.

Rainger, Ronald. 1988.Vertebrate paleontology as biology: Henry Fairfield Osborn and the American Museum of Natural History. In Ronald Rainger, Keith R. Benson, and Jane Maienschein (eds.), *The American Development of Biology*, pp. 219–256. Philadelphia: University of Pennsylvania Press.

————. 1991. *An Agenda for Antiquity*. Tuscaloosa: University of Alabama Press.

Rainger, Ronald, Keith R. Benson, and Jane Maienschein (Eds.). 1988. *The American Development of Biology*. Philadelphia: University of Pennsylvania Press.

Raitt, Helen, and Beatrice Moulton. 1967. *Scripps Institute of Oceanography: First Fifty Years*. San Diego: Ward Ritchie Press.

Ricketts, Edward F., and Jack Calvin. 1939. *Between Pacific Tides*. Stanford, Calif.: Stanford University Press.

Ritter, William E. 1912. The marine biological station of San Diego: Its history, present conditions, achievements, and aims. *University of California Publications in Zoology*, 9: 137–248.

Rosen, George. 1936. Carl Ludwig and his American students. *Bulletin of the History of Medicine*, 4: 609–650.

Rothschuh, Karl E. 1973. *History of Physiology*, trans. Guenter Risse. Huntington, N.Y.: Robert E. Krieger.

Russell, Frederick S. 1976. *The Eggs and Planktonic Stages of British Marine Fishes*. London: Academic Press.

Ryder, John A. 1884a. A contribution to the embryography of osseous fishes, with special reference to the development of the cod (*Gadus morrhus*). *Report of the United States Fish Commission*, 10: 455–605.

————. 1884b. On the preservation of embryonic materials and small organisms, together with hints upon embedding and mounting sections serially. *Report of the United States Fish Commission*, 10: 607–629.

————. 1886a. On the development of viviparous osseous fishes. *Proceedings of the United States National Museum*, 8: 128–155.

————. 1886b. The development of *Fundulus heteroclitus*. *American Naturalist*, 20: 824.

————. 1886c. On the origin of heterocercy and the evolution of the fins and fin rays of fishes. *Report of the United States Fish Commission*, 12: 981–1107.

————. 1887. On the development of osseous fishes including marine and freshwater forms. *Report of the United States Fish Commission*, 13: 489–605.

Sanzo, Luigi. 1910. Uova e larve di scomberoidi. *Rivista mens. Pesca* (Pavia), 5th ser., 12: 201–205.

Saunders, John R. 1952. *The World of Natural History, as Revealed in the American Museum of Natural History*. New York: Sheridan House.

Say, Thomas. 1824–1828. *American Entomology*, 3 vols. Philadelphia: Philadelphia Museum/S.A. Mitchell.

————. 1830–1834. *American Conchology*. New Harmony, Ind.: School Press.

Schlee, Susan. 1973. *The Edge of an Unfamiliar World: A History of Oceanography*. New York: Dutton.

Schmalhausen, Ivan I. 1949. *Factors of Evolution*, trans. Isadore Dordick, ed. Theodosius Dobzhansky, Philadelphia: Blakiston.

Schultz, Eugen. 1910. *Prinzipien der rationeller vergleichenden Embryologie*. Leipzig: Wilhelm Engelmann.

Semon, Richard. 1901. Normentafel zur Entwicklungsgeschichte des *Ceratodus forsteri*. In Franz Keibel (ed.), *Normentafeln zur Entwicklungsgeschichte der Wirbeltiere*, vol. 3, pp. 1–38, Jena: Gustav Fisher Verlag.

Sloan, Douglas. 1980. Science in New York City, 1867–1907. *Isis*, 17: 35–76.

Smith, C. Lavett, Charles S. Rand, Bob Schaeffer, and James W. Atz. 1975. *Latimeria*, the living coelacanth, is ovoviviparous. *Science*, 190: 1105–1106.

Storer, David H. 1839. *Reports on the Fishes, Reptiles, and Birds of Massachusetts*. Boston: Dutton and Wentworth.

Strathmann, Megumi. 1987. *Reproduction and Development of Marine Invertebrates of the Northern Pacific Coast*. Seattle: University of Washington Press.

Thompson, D'Arcy. 1917. *On Growth and Form*. Cambridge: Cambridge University Press.

Thorson, Gunnar. 1946. *Reproduction and Larval Development of Danish Marine Bottom Invertebrates*. Copenhagen: C. A. Reitzel.

Vernberg, F. John (Ed.). 1963. Field stations of the United States. *American Zoologist*, 3: 245–386.

Wallace, Joseph. 2000. *A Gathering of Wonders: Behind the Scenes at the American Museum of Natural History.* New York: St. Martin's Press.

Ward, Ritchie. 1974. *Into the Ocean World.* New York: Knopf.

Watterson, Ray L. 1973. Benjamin Harrison Willier: 1890–1972. His life as an outstanding biologist, embryologist, and developmental biologist. *Developmental Biology,* 34: f1-f19.

———. 1979. The striking influence of the leadership, research, and teaching of Frank R. Lillie (1870–1947) in zoology, embryology, and other biological sciences. *American Zoologist,* 19: 1275–1287.

Whitman, Charles O. 1885. *Methods of Research in Microscopical Anatomy and Embryology.* Boston: S. E. Cassino.

Wilson, Alexander. 1808–1814. *American Ornithology,* 9 vols. Philadelphia: Bradford & Inskeep.

Wilson, Henry V. 1891. The embryology of the sea bass (*Serranus atarius*). *Bulletin of the United States Fish Commission,* 9: 209–278.

Winsor, Mary P. 1991. *Reading the Shape of Nature.* Chicago: University of Chicago Press.

Wourms, John P. 1997. The rise of fish embryology in the nineteenth century. *American Zoologist,* 37: 269–310.

Wourms, John P., James W. Atz, and Dean Stribling. 1991. Viviparity and the maternal-embryonic relationship in the coelacanth *Latimeria chalumnae. Environmental Biology of Fishes,* 32: 225–248.

Wyman, Jeffries. 1854. Notice of the life and writings of the late Dr. Waldo Irving Burnett. *American Journal of Science,* 2nd ser, 18: 255–264.

Young, Craig M., and Kevin J. Eckelbarger (Eds.). 1994. *Reproduction, Larval Biology, and Recruitment of the Deep-Sea Benthos.* New York: Columbia University Press.

9

MORPHOLOGICAL AND PALEONTOLOGICAL PERSPECTIVES FOR A HISTORY OF EVO-DEVO

Alan C. Love

Exploring the history of evolutionary developmental biology (hereafter, Evo-devo) is an exciting prospect, given its current status as a cutting-edge field of research. The first and obvious question concerns where to begin searching for the materials to write a history of Evo-devo. Since this new discipline adopts a moniker that intentionally juxtaposes "evolution" and "development," individuals, disciplines, and institutional contexts relevant to the history of evolutionary studies and investigations of ontogeny put themselves forward. Each of these topics has received attention from historians, and thus there is both primary and secondary material from which to draw. For example, many historians have documented the historical trajectories of genetics and embryology, their split, and various relations (or lack thereof), especially in the first three decades of the twentieth century.[1]

Alternatively, a cue can be taken from how the contemporary "end point" of the historical study is being conceptualized. In his definitive textbook on Evo-devo, Brian Hall articulates a multifaceted picture of the composition of this new discipline:

> For evolutionary developmental biology (EDB or "evo-devo") is not merely a fusion of the fields of developmental and evolutionary biology, the grafting of a developmental perspective onto evolutionary biology, or the incorporation of an evolutionary perspective into developmental biology. EDB strives to forge a unification of genomic, developmental, organismal, population, and natural selection approaches to evolutionary change. It draws from development, evolution, palaeontology, molecular and systematic biology, but has its own set of questions, approaches and methods.[2]

This portrayal suggests that historical material from molecular biology and systematics, as well as paleontology, may be relevant for understanding how we arrived at the present situation. A related strategy that takes its point of departure from the emphases of current researchers is the exploration of one or more key concepts. Two cognate notions have been persistently foregrounded by proponents of Evo-devo: *innovation* and *novelty*. Biologists involved in research that welds evolutionary and developmental themes see their nascent synthesis as a prime venue for exploring old but unanswered questions about the origin of qualitatively new features in the history of life (e.g., body plans, jaws, feathers, and limbs): "[F]inding answers to what constitutes an evolutionary innovation . . . and how developmental mechanisms have changed in order to produce these innovations are major issues in contemporary evolutionary developmental biology."[3] At the first meeting of the Evo-devo subdivision within the Society for Integrative and Comparative Biology, this emphasis was particularly perspicuous:

> [Evo-devo] may lead to a mechanistic explanation of the origin of *evolutionary innovations* and the *origin of body plans*. . . . Evolutionary innovations and the evolution of body plans are hard to understand in population genetic terms since they involve radical changes in the genetic/developmental architecture of the phenotype. . . . evolutionary innovations are outside the scope of any current research program. Through its contribution to the solution of that question, [Evo-devo] genuinely expands the explanatory range of evolutionary theory. We think that this is the one area where [Evo-devo] will have its most lasting impact on evolutionary theory and biology in general. . . . we see in the problem of innovation and the evolution of body plans a unique opportunity for [Evo-devo] to develop its own independent identity as a research program.[4]

Recent textbooks on evolution and development devote entire chapters (or more) to novelty.[5] It should not go without notice that this conceptual emphasis is connected with an explicit challenge to the sufficiency of neo-Darwinian evolutionary theory: "quantitative change is only part of the story of evolution, for it does not address the question of the origin of discrete (qualitatively different) novelties."[6]

If we take the concepts of innovation and novelty as clues to understanding aspects of the history of Evo-devo, one preliminary observation is of immediate interest. The traditional "home" for discussions of innovation and novelty consists of comparative evolutionary embryology, morphology, and paleontology.[7] This is intimately connected with attentiveness to the relationship between evolution and development. One can quickly find paleontologists in the period from 1875–1910 utilizing the biogenetic law and biased variation due to ontogenetic trajectories in order to buttress theories of orthogenesis.[8] And though the biogenetic law now resides primarily in the dustbin of biological ideas,[9] the idea of relative timing changes occurring in development (heterochrony), thereby resulting in significant evolutionary consequences, can be observed routinely in the work of contemporary paleontologists,[10] and was always a mainstay of comparative embryological work.[11]

Utilizing concepts currently of importance to Evo-devo has the potential to offer a perspective different from that of historical investigations based on disciplines currently predominant in Evo-devo. Some recent disciplinary characterizations of Evo-devo are cast primarily in terms of developmental genetics as a bridge between genetic accounts of evolution and a molecularized embryology.[12] Assuming this vantage point leads to an exploration very different from utilizing the concepts of innovation and novelty. Besides being an alternative perspective, using a concept-based historical approach may shed light on the more recent history of Evo-devo just prior to the profound demonstration of Hox gene conservation,[13] a time when it also appears that a number of morphological and paleontological researchers were reviving an interest in the intersection between evolution and development.[14]

The aim of the present chapter is to explore morphological and paleontological perspectives for a history of Evo-devo, using the current emphasis on innovation and novelty as a guide. To that end I begin with a discussion of what was excluded from the Modern Synthesis from the perspective of morphologists rather than embryologists.[15] The next two sections are devoted to historical vignettes, from the late 1940s until the mid-1960s, of two researchers, one morphological (D. Dwight Davis) and one paleontological (W. K. Gregory), using the concepts of innovation and novelty to uncover interestingly relevant history for Evo-devo. Although the foci of my discussion, Davis's morphological research on carnivorous mammals and Gregory's encyclopedic *Evolution Emerging* (1951), provide only a partial perspective on the disciplinary contexts and conceptual proclivities of morphology and paleontology (as well as on the variegated thinking of the individual researchers themselves), I suggest that they offer a fruitful viewpoint for those wrestling with the history of Evo-devo, including its contemporary relevance.

MORPHOLOGY AND THE MODERN SYNTHESIS

Although much attention has been devoted to the exclusion of embryology from the Modern Synthesis, the possible exclusion of other domains has been less noticed. Tracking this issue with respect to morphology and paleontology provides an appropriate backdrop to a historical investigation utilizing the contemporary emphasis on innovation and novelty precisely because this emphasis is routinely juxtaposed with a challenge to the putative "reigning orthodoxy" of neo-Darwinism, which was forged in the context of the Modern Synthesis. Additionally, some current representations of Evo-devo appear to marginalize the role of morphology and paleontology. For example, Wallace Arthur's "circle of evolutionary theory" consigns to comparative anatomy (morphology) the duty of providing the "Classical Evidence of Evolutionary Pattern."[16] (figure 9.1). Significantly, Arthur does not represent morphology and most of paleontology as part of neo-Darwinism.

There are at least three discernible accounts of the relationship of morphology to the Modern Synthesis. The first is that morphology

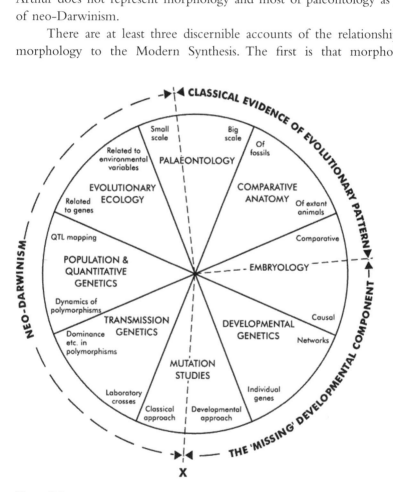

Figure 9.1
"Disciplinary 'circle' of evolutionary theory" (Arthur, 1997, p. 287). Reprinted with the permission of Cambridge University Press.

covertly contributed to the Modern Synthesis through the likes of Gavin de Beer, E. S. Goodrich, Julian Huxley, Bernard Rensch, and I. I. Schmalhausen.[17] Although it is plausible due to the proximity of such individuals with respect to the theoretical context of the Modern Synthesis, the evidence is largely circumstantial.[18] This concern coincides with a second account maintaining that although many key researchers were poised to contribute morphological perspectives to the Modern Synthesis, in matter of fact they did not. The fallout of this exclusion, intended or not, was a perception that morphological (and paleontological) research perspectives were alternative conceptions of evolutionary theory and not strictly part of the Modern Synthesis, which made these conceptual frameworks implicitly specious. At one Darwin centennial celebration, held in Chicago in 1959, the morphologist and paleontologist Everett Olson voiced a worry over being sidelined from the predominant ideas of the synthetic theory of evolution, then in its heyday:

> There exists . . . a generally silent group of students engaged in biological pursuits who tend to disagree with much of the current thought [i.e., the Modern Synthesis] but say and write little because they are not particularly interested, do not see that controversy over evolution is of any particular importance, or are so strongly in disagreement that it seems futile to undertake the monumental task of controverting the immense body of information and theory that exists in the formulation of modern thinking. . . . Wrong or right as such opinion may be, its existence is important and cannot be ignored or eliminated as a force in the study of evolution.[19]

Olson was keenly aware of the lack of "elasticity" in the synthetic theory of evolution in conjunction with its seeming ability to explain everything. The burden of proof was shifted from showing that an alternative explanation for a phenomenon was possible to demonstrating that the phenomenon could not be sufficiently handled by the synthetic theory: "Morphologists and paleontologists feel this, perhaps, more strongly than other students of biology. . . . The extent of assumption, interactions of assumptions, and the degrees of extrapolation give a sense of uneasiness when the animals and their structures are foremost in mind."[20]

The second historical account concerning the role of morphology in the Modern Synthesis, in conjunction with Olson's comments, implies that morphology (and its cobelligerent, paleontology) did have theoretical resources to offer in the formulation of the synthetic theory of evolution. A qualitatively stronger thesis, representing the third historical account, is that because morphology is descriptive, it *could not* have made a contribution to the synthetic theory: "To many it has seemed enigmatic that morphology contributed virtually nothing to the synthetic theory of evolution. . . . Morphology has contributed so little primarily because it has had so little to contribute. It is a descriptive science of form, and only when conjoined with other disciplines does it tell us anything about causes. . . . morphology tends to be the sort of discipline that will follow, rather than lead, in the development of evolutionary theory."[21] The Modern Synthesis was focused on the mechanisms of evolutionary *processes*, and morphology simply could not tackle the causal question. Ghiselin attributes the difficulty of morphology not just to its descriptive orientation but also to metaphysical trappings such as essentialism and the search for laws following the model of physical science. Coleman's companion analysis largely concurs, and he points out that morphologists failed to connect their work to genetic mechanisms.[22] Similarities in general characteristics were detailed in order to reconstruct the history of life while ignoring population diversity. Variation was an unimportant phenomenon, signaled by the metaphysical inclinations of many morphologists toward an idealistic, archetype-based philosophical framework.[23]

One way to begin adjudicating between the second and third historical accounts regarding the role of morphology in the formation and codification of the Modern Synthesis is to dissect any attempts in the American context of the late 1940s to assess a contribution to the synthetic theory of evolution from the ranks of morphology. One possible candidate is D. Dwight Davis. In his report from the 1947 Princeton meeting of the Committee on Common Problems of Genetics, Paleontology, and Systematics, Davis opened with a familiar refrain: "Recent syntheses of current evolutionary thought have, almost without exception, ignored comparative anatomy completely or considered it only very obliquely."[24] He did not explicitly challenge the synthetic theory of evolution in this respect, and acknowledged that morphologists had not thought in terms of "populations."[25] Davis perceived a remarkable congruence of ideas between population geneticists and evolutionary morphologists, and his article offered a picture of morphological research as *consistent* with the aims of other domains within the Modern Synthesis. Despite his claim of a reciprocal relationship between the findings of genetics and morphology, Davis never provided an account of why morphology is important for genetics.

A constructive account of the importance of morphology absent in Davis's 1949 article can be partially recovered in a later discussion.[26] While delineating the proper goal of comparative anatomy, Davis consciously steered clear of idealistic notions found in the phylogenetically oriented morphology of the Gegenbaur school and its conceptual progeny,[27] locating morphology's absence of contribution to the Modern Synthesis in transnational obstacles of misunderstanding. Davis believed this occurred in part because the "subtle" meanings of morphological concepts were hard to grasp outside of the German-speaking context and German anatomists failed to grasp "key" aspects of Darwinism. He was in a good position to make these comments because he and his colleague Karl Schmidt had translated the work of German-speaking morphologists into English. For example, they had a draft translation of Adolf Portmann's *Introduction to the Comparative Anatomy of Vertebrates* (1948) ready in 1951. Upon Schmidt's death (September 1957) the University of Michigan Press expressed interest in the manuscript, although it ultimately decided against publication. In response to the initial interest in the manuscript, Davis's reply touched on the problems of understanding between American and European morphologists: "I do not believe this book could be published as a straight translation, however thoroughly corrected and polished. As I recall, there are certain viewpoints that would be unacceptable in the U.S. Indeed, the whole book reflects the differences between European and American attitudes toward zoology in general."[28]

Davis rejected the claim that comparative anatomy could not contribute to questions of evolutionary causality largely because morphology addressed a higher level of organization than genetic approaches, and he advocated shifting the focus of morphologists from structural similarity to structural difference: "If comparative anatomy is to qualify as a science it must, like any other science, offer a rational explanation for the phenomena with which it deals. The phenomena of comparative anatomy are not the observed structure of vertebrates, but the observed differences between the structure of one vertebrate and another. . . . its proper goal is . . . to explain the observed variations in [the common structural] plan."[29] "Major" differences in the common structural plan attended to by comparative anatomists are based on the same mechanisms as those minor phenotypic variations described by population geneticists. The problem becomes demonstrating how the transitions occurred in a particular case. Morphology contributes to a causal understanding of evolution at a level inaccessible to laboratory genetic studies, illuminating the phenomenon of adaptation at higher levels of structural organization: "We are dealing with adaptation, with functional mechanisms, and differences in structure are meaningless unless they can be correlated with differences in function."[30]

After identifying structural differences and correlating these differences with function in relation to organismal habits and behavior, the last step is to figure out whether or not the morphological traits in question are under direct or indirect genetic control. For Davis, this was the way to "salvation" for comparative anatomy.[31]

Davis's argument is still far from complete, and thus does not fully resolve whether or not morphology had distinct theoretical resources available to contribute to the Modern Synthesis. But there is also good reason to suppose from Davis's unpublished papers that these are only glimpses of material that was meant for a book-length treatment. In the same letter to Edwin Watkins about the translation of Portmann, he wrote: "My personal opinion is that every existing textbook of comparative anatomy is based on concepts of the Nineteenth Century, and therefore cannot be expected to inspire either the student or the teacher of today. A complete re-orientation of approach, consonant with modern biological concepts in other fields, is badly needed. . . . I started to write what I conceive to be such a text, but I don't know whether I will ever finish the job."[32]

Samples of this material can be found in a collection of outlines, text, and bibliographies titled "The Comparative Morphology of Vertebrates."[33] The first section is devoted to a definition of functional morphology (in outline only) in which Davis emphasizes the difference between a static, descriptive anatomy and a dynamic, comparative anatomy that is a "shift of the major interest fr[om] the product to the process." The second part, "The problem of adaptation" (outline only), reveals that his aim is to understand the "origin of functional adaptations," especially those which mark higher taxonomic categories ("Adaptation at population [and] subspecies level vs. adaptation at supergeneric level"). The third part, "Locomotion," exists both in outline and as partially completed text. Here we find an assemblage of concerns about biomechanical constraints alongside an extended discussion of "the question of the origin (as opposed to the evolution) of paired appendages," one of the classic instances of evolutionary novelty. Another unpublished outline, "Animal locomotion," reinforces these emphases.[34] Of four ways described to study animal locomotion (phylogeny, mechanics, physiology, and origin of adaptive mechanisms), it is clear that Davis is in favor of the last, which, he explicitly notes, integrates data from developmental mechanics. Under a heading "Evolutionary mechanisms," a subpoint reads "Pop. Genetics = mech. of evol. at lowest level (Not much help in locomotion problems)[.]" This is followed by a section on the limitations of genetic experimentation for understanding the origin of suprageneric adaptations.

The emerging theme through all of this work is that the theoretical contribution of morphology to evolutionary theory is *analytical tools and conclusions at higher levels of structural organization.* Although his papers favor the second historical account—that morphology could have contributed to the Modern Synthesis, but did not—these comments and Davis's keen interest in evolutionary novelties (or the origin of adaptations at the suprageneric level), such as paired limbs and the panda's thumb, lend credence to considering him as part of a morphological perspective for the history of Evo-Devo.

The Functional Morphology of D. Dwight Davis

I have argued elsewhere that Davis's agenda for a functional approach to morphology reveals that it was not just comparative anatomy per se that was excluded from the Modern Synthesis, but rather the research perspective of typology often associated with morphology.[35] Claims by Rainer Zangerl (a vertebrate anatomist who worked on fossil turtles at the University of Chicago) that the "morphotype," which intentionally abstracts away from existing variation, is where "morphology must and can make an important contribution to the future development of ultimate theoretical thinking in biology"[36] simply fell outside the pale of Modern Synthesis orthodoxy.

Here I want to turn away from issues about the Modern Synthesis in order to draw attention to the specifics of Davis's work and how it can be considered part of a history of Evo-Devo. This is appealing for three reasons: (1) Davis was very concerned with innovations and novelties; (2) recent historical studies of comparative anatomy and morphology have focused primarily on the period prior to 1940;[37] and (3) Davis worked at the University of Chicago, where a more flexible, but less prominent, theoretical framework for morphology and paleontology quietly continued with individuals such as Davis and Olson laying the foundations for a highly interdisciplinary research context that still exists.[38]

Davis (1908–1965) was a vertebrate anatomist who became the curator of the Division of Anatomy at Chicago's Field Museum of Natural History in 1941. Although his earlier work from the 1930s focused on reptiles and amphibians, his interest in mammals (carnivores in particular) predominated during his tenure as curator. One feature that caught his attention in bears is the distinct scapula with its unusually large postscapular fossa in comparison with other carnivores.[39] This unique feature led him to consider its significance, if any, especially because he was interested in understanding the morphological similarities between bears and the giant panda. To do so required an investigation of associated muscles and the biomechanics of the shoulder—in short, the functional morphology of the shoulder architecture of bears. After considering distinct facets such as the rectangular shape of the scapula (in contrast to the fan shape in other carnivores) to anchor the needed flexor muscles of the broad neck found in bears, and making explicit comparisons with the shoulder flexors of other carnivores, as well as the relevant biomechanics, Davis identified six distinctive features of the bear shoulder. But none of these features was qualitatively different, and thus the unique shoulder architecture of bears contained no structural novelties; "None of these differences is absolute; each is merely a quantitative difference from the normal carnivore condition. There is nothing qualitatively new in the shoulder of the bear."[40] In fact, it is clearly an exaggeration of that found in its nearest known relatives, procyonids (e.g., raccoons). Functional study of the relevant structures reveals that the form of the shoulder architecture in bears is designed for climbing. This conclusion is reinforced by the convergent aspects of shoulder morphology in anteaters and armadillos adapted for digging.

Davis's study of shoulder architecture in bears was ultimately a prelude to his lifelong focus of research, the giant panda. Although many are familiar with the fact that Davis produced a massive treatise on the panda through the popular writings of Stephen Jay Gould,[41] the emphasis on imperfect adaptation and historical contingency in these articles overshadows a key aspect of Davis's investigation and why Gould would have been keenly aware of it apart from the moral he sought to impart concerning evolutionary tinkering. Gould exhibited an interest in the evolutionary significance of allometric growth early in his career.[42] This was later observable in his discussions of the history and evolutionary importance of heterochrony,[43] which served to spur many into Evo-devo types of studies. Thus it is little surprise that what Gould finds attractive about Davis's discussion of the panda and its unique opposable thumb is the postulation by Davis of the enlarged radial sesamoid resulting from a simple mutation "affecting the timing and rate of growth."[44] This was "D'Arcy Thompson's solution of reduction to a simple system of generating factors" such that a complex morphological feature can "arise as a set of automatic consequences following a simple enlargement of the radial sesamoid bone."[45] Revisiting Gould's comments directs us to look more carefully at the thematic structure of Davis's investigation of the giant panda.[46]

Davis's study was originally undertaken in order to determine the taxonomic position of the giant panda (*Ailuropoda melanoleuca*) but, as seen in the title, it expanded into a broad morphological study of evolutionary mechanisms. He was working in the functional morphological framework that emphasized the adaptive differences of vertebrate form:

> [C]omparative anatomists have scarcely begun to seek . . . adequate explanations for the differences in vertebrate structure. . . . it is of crucial importance to ask whether comparative anatomy can undertake to explain, in causal-analytical terms, the structural differences that characterize taxa among vertebrates. . . . I believe [comparative anatomy] must shift its major emphasis from the conservative features of evolution to its radical features, from the features that organisms under comparison have in common to those they do not have in common. It must seek rational explanations for these differences, drawing on data from other fields where this is necessary and possible.[47]

Davis was also not a genetic reductionist who attributed any significant morphological change to natural selection operating on the gradual accumulation of small mutations. He held that there was increasing evidence that "in vertebrates, a quite simple change in epigenetic mechanisms may have a profound and extensively different end result."[48] This prefatory remark in favor of epigenetic mechanisms is expanded into a formal goal of his investigation in the Introduction: "(4) Determination of the morphogenetic mechanisms that were involved in effecting these changes [in structure]. . . . By a judicious combination of the comparative method with the known data of mammalian epigenetics I believe it is possible to infer, with varying degrees of confidence, the true mechanisms behind many of the major structural differences that distinguish *Ailuropoda* from the true bears."[49] The Introduction also highlights Davis's acceptance of the importance of relative growth rates (allometry), geometric transformations, and rate genes as he invokes Huxley and D'Arcy Thompson, as well as Richard Goldschmidt's *Physiologische Theorie der Vererbung* (1927).[50]

Turning to specific explanations, Davis's analysis of limb proportions in giant pandas and bears demonstrates that they are not immediately explicable via attention to functional requirements. Commensurate with this finding is the discovery of allometry in these proportions (length of tibia versus length of femur) as well as in the pelvic region (breadth versus length). Davis compared these intraspecific relative growth plots for giant pandas and different bears interspecifically to draw out principles of evolutionary change in morphology: "Body proportions in the pandas and bears are not the result of selection for mechanical efficiency. Rather they reflect pleiotropic correlations with other features that have been altered

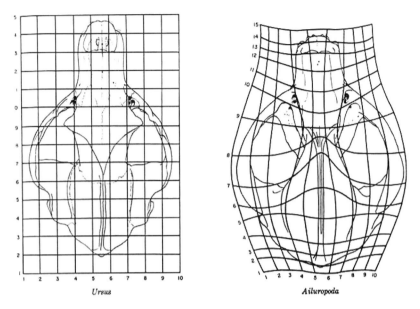

Ursus   Ailuropoda

Figure 9.2
"Skull proportion differences between *Ursus horribilis* and *Ailuropoda melanoleuca* displayed via deformed coordinates" (Davis, 1964, p. 67). Reprinted by kind permission of The Field Museum (Chicago).

through natural selection."[51] The nature of the panda skull, in contrast to that of other bears as well as of other carnivores, is elucidated via transformation grids with deformed Cartesian coordinates (figures 9.2 and 9.3). Davis concludes that *Ailuropoda* and *Ursus* show no critical differences from a generalized carnivore in longitudinal proportions of the skull, although there are differences in depth and breadth.

Despite the fact that this conclusion relies only on qualitative geometrical considerations,[52] when it is corrected for the effect of absolute size and supplemented by principles from the biomechanics of mastication (*Ailuropoda* has greater efficiency in this respect than a generalized carnivore), the differences in skull proportion can be attributed to increased herbivory among carnivores, exhibited to an extreme in the giant panda.[53] The panda skull appears to be the result of modifying a carnivore skull structure to the very different demands of a plant fiber diet. Davis had also studied the masticatory apparatus of the spectacled bear (*Tremarctos ornatus*), which has one of the most herbivorous diets among bears.[54] The skull of *Tremarctos* is also compared with *Ursus* in a transformed coordinate grid like the one for the panda (figure 9.2), and dis-

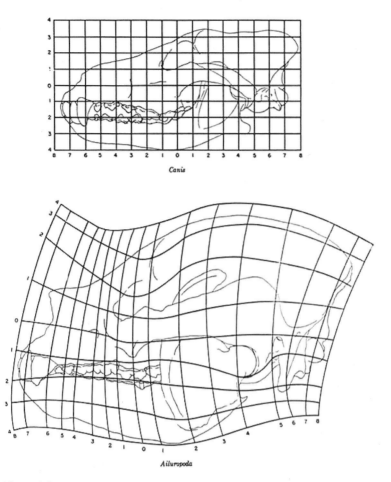

Figure 9.3
"Skull proportion differences between *Canis lupus* and *Ailuropoda melanoleuca* displayed via deformed coordinates" (Davis, 1964, p. 68). Reprinted by kind permission of The Field Museum (Chicago).

plays deformation trends similar to, though less marked than, those of *Ailuropoda*.[55] Interestingly, the digestive system of *Ailuropoda* shows only minor modifications, primarily in reduced intestinal length.[56]

In the ensuing discussion on cranial morphology, the morphogenetic mechanisms of the mammalian skull are attributed to a "mosaic of independent morphogenetic units" that become functionally integrated by selection operating on the timing of growth and differentiation, as well as external mechanical demands during ontogeny.[57] Support for this explanation is drawn from studies of cranial development in bulldogs and the details known about the ontogeny of the vertebrate limb. Davis is at pains to stress that the external mechanical demands during ontogeny are responsible for most of the specific skeletal morphology in the panda. For example, in his general discussion of osteological characters he says:

> Many of the differences between panda and bear skeletons are adaptive, but their cause is extrinsic to the bone itself; that is, they merely reflect the response of the bone tissue to external pressures, stresses, and strains, and other purely mechanical factors. In the absence of the appropriate stimulus such characters fail to appear. Among such features are the surface modeling of bones, torsions, form and extent of articular areas, and size and position of foramina. These are characteristic features of the skeleton of *Ailuropoda*, and they may be clearly adaptive in the sense of promoting the efficiency of the organism, but they are epigenetic to the bone and therefore are not the result of natural selection *on the skeleton*.[58]

In other words, natural selection did not continuously sculpt genetic differences underlying the skeleton, but preserved the epigenetic results of differing mechanosensitive interactions during ontogeny.[59]

One cherished conceptual theme of embryological approaches to evolution, both past and present, is found repeatedly in Davis's explanations—morphological gradients. They appear in his discussion of the vertebral column,[60] where "lumbosacral peculiarities" of the panda are hypothesized to be the result of an "accidental" (i.e., not due to selection) heterochrony, the expansion of the proximal ends of the ribs,[61] and in the explanation of dentition,[62] where the "field control concept" and "differential growth" allow him to reconcile the greater similarities between giant panda and raccoon premolars with the closer phylogenetic relationship between giant pandas and bears. Davis consistently states that many changes from *Ursus* to *Ailuropoda* are not the result of numerous small mutations sculpted by natural selection, but rather of a few genetic changes (possibly only two mutations for skeletal differences!) that were pleiotrop-

ically magnified through ontogenetic trajectories containing distinct morphogenetic fields.[63]

This claim did not go unnoticed. Reviewers remarked that "[t]he
numerous morphological differences between *Ailuropoda* and bears are
frankly discussed, but the evolutionary mechanisms by which Davis proposes
to explain them will probably not be accepted by those who regard natural
selection as the dominant force behind organic evolution. Davis insists that
many anatomical features in *Ailuropoda* could not be adaptive, and hence
must be due to general disturbances of morphological homeostasis. He suggests that the most profound differences between giant pandas and bears
resulted from the action of relatively few genes which had large, pleiotropic
or even catastrophic effects."[64] Davis's claim was rejected more explicitly
when his work was cited favorably in later macroevolutionary discussions,
such as that of the paleontologist Steven Stanley.[65] Russell Lande responded
specifically to Stanley's use of Davis's work:

> Similar confusion exists concerning variation in allometric growth and
> developmental fields. That morphological differences between related
> species could be explained as simple changes in a few growth gradients
> or developmental fields (as attempted by Davis 1964, for the giant
> panda) does not imply that only a few genes were involved. On the
> contrary, evidence exists that natural variation in parameters of allo-
> metric growth and developmental fields is usually influenced by multi-
> ple genetic factors acting relatively late in development [. . .]."[66]

The use of Davis's research as contrary to neo-Darwinian tenets was
also iconic in that a giant panda appeared on the cover of Stanley's
book.[67]

The pattern of explanation using morphogenesis and allometry is
not unique to Davis's study of the giant panda. For example, in his description of the mammals of the lowland rain forest of North Borneo, Davis
does not hesitate to claim that the enlarged nose morphology of the
proboscis monkey is not functionally significant and results, rather, from
a differential growth rate in the morphogenetic mechanism of the ancestral snub nose.[68] Selection need not be invoked for the unique nose of
the proboscis monkey because the morphogenesis of this feature is posi-
tively correlated with body size, which demonstrates an increasing trend
in this lineage. The invocation of morphogenetic gradients and fields; persistent use of allometric transformations and heterochronies to explain
morphological differences (including the distinct, nonosseous penis morphology of giant and lesser pandas in terms of neoteny[69]); a focus on
putative evolutionary novelties such as the panda's thumb, the ursid

scapula, and the origin of paired limbs; consistent attention to the developmental origin of the characters under consideration; an emphasis on epigenetics and mechanical "constraints" in ontogeny; and a desire to carefully state where natural selection is (and is not) causally responsible all situate Davis naturally within a history of Evo-Devo from a morphological perspective.

EVOLUTION EMERGING: PALEONTOLOGICAL APPROACHES TO EVOLUTIONARY INNOVATION

William K. Gregory (1876–1970), protégé of Henry Fairfield Osborn, was a vertebrate paleontologist at Columbia University and the American Museum of Natural History for most of the first half of the twentieth century.[70] At the end of his career he published a two-volume, synthetic account of evolutionary patterns and processes throughout the history of life.[71] *Evolution Emerging* (*EE*) is an impressive work. The first volume, which contains the textual component, runs over 700 pages (including bibliography and index). The second volume, which consists solely of detailed illustrations corresponding to the text, exceeds 1,000 pages and includes numerous foldouts. It is very unlikely that a book of the same scope and presentation could be published today (for financial reasons), let alone attempted (because of professional specialization trends).[72] The introductory chapter begins with an overarching metaphor ("The Cosmic Cinema"), invectives against anthropomorphism, an articulation of Gregory's idiosyncratic distinction between polyisomerism and anisomerism,[73] and ends with twenty-nine quatrains, of which the following two are examples.

IV

The dolphin's equal teeth have been derived
From very unequal teeth of carnivores
Unequals changing into equal parts
Are "secondary polyisomeres."

V

But equal parts to parts unequal changed
Are, on the contrary, called "anisomeres."
As when the rows of teeth almost alike
Gave rise in crocodilians to festoons.

Section headings and subdivisions are creatively titled (e.g., "The Sponge-state and Its Citizens," "The Bivalves—Brainless but Successful"), and the pace is breathtaking; the discussion begins with the emergence of life on Earth and moves through to human beings before reiterating and integrating larger philosophical themes (especially polyisomerism and anisomerism) in the last ten pages of the main text.[74] Gregory closes the text portion with an appropriate paean: "And so the cosmic kaleidoscope keeps turning round and round, slowly but endlessly dissolving old combinations while creating new patterns, new values, new opportunities."[75] Many reviewers flagged Gregory's philosophical inclinations and extensive use of colorful metaphor.[76]

Who was the audience for this comprehensive discussion of the history of life, couched in unabashed personal and philosophical reflection? One reviewer claimed that Gregory's book on the process of evolution would be of interest to geneticists, paleontologists, ornithologists, taxonomists, mycologists, ecologists, morphologists, and comparative psychologists and physiologists.[77] It was identified as research in systematics in an editorial in *Science*,[78] and Eric Jarvik cited it in regard to theories of tetrapod origins.[79] Maybe even more important, when two biologists reinvigorated the discussion of vertebrate origins,[80] a classic issue concerning innovations and novelties because of the numerous new features—both morphological (jaw and head) and embryological (neural crest cells and placodes)—that emerge at this phylogenetic juncture, Gregory's discussion in *EE* and separate articulation of his theory of vertebrate origins[81] are prominently cited.

A complete discussion of *EE* is impossible here, but my present intent is to investigate how these volumes might justifiably be understood as part of a history of Evo-devo, especially as a paleontological perspective concerned with innovation and novelty. The article on vertebrate origins just cited (Northcutt and Gans, 1983) is one obvious entry point. Another is observed in the comments of a reviewer about what is required to understand evolution:

> The synthesis, the development of understanding of the total complex of processes, will perhaps become the function of some new sort of naturalist, who can combine the gleanings of the many special sciences: of the taxonomists, morphologists and comparative physiologists who have described the present diversity of living things; of the paleontologists who have accumulated the historical documentation; of the geneticists who have dissected the mechanisms of inheritance and variation; and of the ecologists who have formulated the principles of population dynamics and analyzed the operation of environmental forces. But before this new naturalist can begin to operate, each of us must formulate, synthe-

size and clarify the content of his special science, keeping in mind the needs and perspectives of this new, emerging field. This Dr. Gregory has done, providing us with a work that each of us can use in his special field, and that all of us can use in building toward this broader under-standing of the living universe.[82]

That *EE* could serve as a stimulus toward a synthesis of biological disci-plines surrounding explaining evolutionary history exhibits a theme that resonates with Evo-devo as a multidisciplinary synthesis, similar to Brian Hall's description above. In what follows, I restrict my attention to por-tions of Gregory's text that juxtapose evolutionary and developmental themes, with particular emphasis on innovations and novelties.

Gregory's welding of evolution and development displays one con-tinuity with earlier research in that Haeckelian themes linger throughout, which can be seen in some of his specific explanations: "In the coelen-terates (hollow intestine), including the highly varied hydroids, jellyfishes (medusae), corals, seafans, etc., the body is essentially cup-like, derived from the outer and inner layers of the gastrula stage of the embryo. This cup-like condition of the gastrula stage is due to the faster growth of the smaller ectoderm cells, which grow around and enclose the larger, nutri-ment-bearing endoderm cells. Thus the adult coelenterates may be regarded as forms which have never gone far beyond the gastrula stage."[83] Other examples abound, such as the comment that horseshoe crabs liter-ally go "through" a "trilobite larval stage"[84] and the observation that ontogeny mirrors phylogeny in brachiopod valve development.[85] These recapitulatory themes bear on his understanding of the origin of innova-tions and novelties because "advanced larval forms often become the starting-point for new lines of evolution," as well as the origin of new taxa, such as spiders from scorpions via "paedogenesis,"[86] or triggerfishes through "metamorphosis or transformation."[87] By "paedogenesis" and "metamorphosis or transformation," Gregory has in mind a recapitulatory notion of heterochrony understood primarily in terms of "acceleration and retardation of growth."[88] ("According to this principle, certain ancient embryonic features which are ordinarily [*sic*] passed through before the adult stage are sometimes retained in the adult by the lagging of later growth stages wither [*sic*] in vigor or in timing."[89]) Another example is Gregory's explanation of the origin of scorpion fishes:

> However, it is not necessary to push the scorpaneid stem back to the berycoid stock, because the stages in the development of the rosefish

. . . indicate that all we need to assume is that there was a great increase in the size of the eye and an accelerated individual development of it, so that at a relatively early stage it became the dominant organ, around which the preopercular and suborbital Anlagen formed a continuous tract. . . . The subsequent downward growth of the hyomandibular and preopercular, together with the forward growth of the jaws, plus the necessary growth force of the suborbital itself, all coöperated to produce the observed result."[90]

These types of explanation are common and strewn throughout Gregory's discussions of different taxa.

These recapitulatory themes are pictorially represented in Gregory's illustration of the origin of the vertebrates, "From Animated Seed-Capsule to Motile Adult by Neoteny" (figure 9.4). His discussion of the Devonian palaeoniscoid *Cheirolepis* is a good narrative exemplar of how developmental data are enlisted to understand the origin of evolutionary novelties through an extension of a developmental trajectory: "[T]he bilateral symmetry of his body was doubtless due to the symmetrical budding of metameric segments on either side of the blastopore, with resulting pushing of the older buds toward the rear. . . . This also permitted the

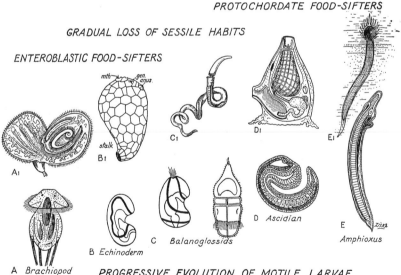

Figure 9.4
"Gregory's Haeckelian perspective on the origin of vertebrates" (Gregory, 1946, p. 362 and EE, II: p. 84). Courtesy of American Museum of Natural History.

differentiation of the head-end from the tail. The longer the period of growth and multiplication of the proto-vertebrae, the greater was the number of vertebral segments and the approach toward an eel-like larval stage. Thus a simple prolongation of [a] certain stage of growth may in the course of time induce striking changes in body form."[91] Paedogenetic events can also explain problematic characters, such as "reptilian" features in monotremes,[92] bringing resolution to phylogenetic confusion. All of these examples are a reminder that many of Gregory's concerns are derived from the vigorous debates about phylogeny, especially the origin of the vertebrates, among comparative anatomists from 1880 to 1930.[93] Lacking fossil record evidence, data from embryology and comparative anatomy were the only route to phylogenetic hypotheses.[94]

Yet Gregory's concerns were not wholly phylogenetic. In an article published in the journal *Evolution* a few years after *EE*,[95] his 1946 work (repeated in *EE*) is explicitly noted for discussing the relation between development and evolution, especially his theory that defining, novel characteristics of chordates and vertebrates (e.g., notochord and somites) could have arisen as larval specializations that were locked in via neoteny. (Gregory's hypotheses are cited alongside those of Gavin de Beer in this same article.[96]) Developmentally oriented explanations of innovations and novelties are plentiful in *EE*. In his discussion of coelenterates, Gregory poses the question of how the efficient, poisonous "dart-cell mechanism" arose. He takes it as a requirement for any explanation that it consider the developmental fact that the cnidoblast ("stinging cell"), which houses the components responsible for the mechanism ("nematocysts"), does not originate in the tentacle, but migrates from elsewhere to its eventual epidermal location. His tentative explanation includes a discussion of nematocyst cell lineage.[97] Spirality in gastropods is articulated with respect not just to shell shape or color patterns but also to the twist of the visceral nerve loop that occurs during ontogeny[98] (figure 9.5).

The origin of insect wings,[99] the notochord,[100] bone,[101] the oralobranchial cavity in cephalaspid ostracoderms,[102] the shift to a cartilaginous head and body in lampreys,[103] the origin of electric organs,[104] the Weberian apparatus in the fish order Ostariophysi (e.g., carp, catfish),[105] color patterns in teleost fishes,[106] the origin of the oddly placed "sucking disc" from the dorsal fin in sucking fishes,[107] the fanlike dental plates in lungfishes,[108] distal skeletal elements of the tetrapod limb,[109] the origin of frogs by modification of the vertebrae and ribs during larval stages,[110] the emergence of feathers,[111] the origin of the diaphragm,[112] mammalian hair, skin, and ears,[113] the transition from platypus bill to echidna snout,[114] the tran-

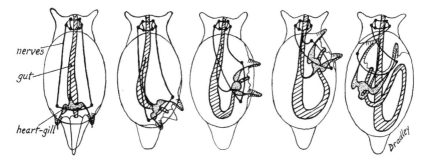

Figure 9.5
"Twisting of the nerve loop during snail development" (*EE*, II: p. 13). Courtesy of American Museum of Natural History.

sition from cynodont reptile skull to opossum skull,[115] the presence of extra zygapophysial processes in the vertebrae of Xenarthan edentates (e.g., armadillos),[116] and the cranial shape of orangutans[117] are all explored through developmental considerations.[118] There are also several discussions of mechanical forces shaping morphology during ontogeny, such as in the crests and ridges of sauropod vertebrae.[119]

In addition to gastropod ontogeny (figure 9.5),[120] it is worthwhile to note the frequent developmental representations utilized by Gregory. Despite the fact that the primary illustrations throughout are either phylogenetic, morphological, or paleontological, there are also pictorial representations for the ontogeny of brachiopods,[121] echinoderms and balanoglossids,[122] *Nereis* and *Amphioxus*,[123] ascidians,[124] silver sharks,[125] the rosefish,[126] turtles (focused on the carapace),[127] the vertebrate inner ear,[128] and marsupials and the platypus.[129] This reveals a significant theoretical slant on how evolutionary relationships are to be discerned and character transitions are to be explained.[130] This does not even touch upon the in-text descriptions of development that repeatedly occur in volume 1.[131]

In addition to specific examples, Gregory draws out general (though arguably cryptic) principles relevant to thinking about the origin of innovations and novelties. One "general principle of evolution" is described thus:

> When a new hereditary structural feature becomes widely distributed in a large population, its further progress or decline may be due to generally uniform or recurrent preëxistent forces or conditions; in other words, it may follow certain precedents, general patterns, rules or laws

observable in many similar cases. But the same new characteristic or organ, in so far as it is new, arises not from any one of the various forces or conditions that were prerequisite for its existence but from the new and repeated intersections of all these prerequisite series of forces and conditions at particular moments of time and locations in space. It is the concurrence of such unpredictable intersections of various forces and conditions at given moments that give rise to what is called chance or luck, which has operated in the world of animals with chitinous articulated skeletons much as it has in the world of articulately speaking men.[132]

Gregory seems to be driving at a general notion of "evolutionary chance" that will give unique, historical explanations for the origin of new structures and functions while incorporating preexisting constraints and general rules or patterns. In his review comments on "from ganoids to recent teleosts," he rules out Lamarckism as a possibility and utilizes embryology to explain dorso-ventral differentiation and streamlining.[133] This discussion also contains a statement that the "polyisomeres" that are "anisomerized" through adaptive specialization arise or originate from "centers of organization" or "fields of force" during ontogeny.[134] The emergence of new polyisomeres is attributed to processes of "budding" during ontogeny and "is the rule in many cases of increase in the number of vertebral segments, of teeth, fin-rods, etc."[135] The "lock and key" fit between accessory reproductive organs in opposite sexes, the interlocking relations among carpal bones in mammals, and the occlusal relations between upper and lower teeth is largely due to the ontogenetic *and* phylogenetic "buckling up, folding, invagination, evagination, etc., of an originally extended tissue layer or layers."[136]

One would be remiss not to take into account aspects of Gregory"s work besides *EE*. An interesting convergence between Davis and Gregory is found in their predilection for *functional* morphology.[137] In this respect these two individuals are not easily categorized into idealist-leaning morphologists and paleontologists trapped in typological thinking. It also becomes clear that Gregory's deployment of recapitulation was not naïve: "Since Patten's time the dogma of recapitulation has been pretty well deflated by T. H. Morgan and others. But with reference especially to the lower invertebrates (e.g., Coelenterata, Annulata, Echinodermata, Brachiopda [*sic*] and others) it is still probable that larval or young stages may often but not always retain some earlier phylogenetic features which are lost in the adults."[138] Gregory deploys the principle of recapitulation "tentatively," always being careful to highlight the methodology before proceeding.[139]

Gregory also recognizes different uses of the principle: "The embryonic development of a given *organ* or *group of organs* may or may not give reliable indications of the remote history of the animal; yet it may also give fairly indirect evidence as to the relationships of the animal to other systematic groups having similar embryonic stages or processes."[140] In some cases the embryological evidence is patently misleading, as in the inner skull of vertebrates,[141] and is therefore utilized only in conjunction with a broad array of other skeletal features to infer phylogeny. Though some of Gregory's penchant for recapitulatory explanations (either of evolutionary transitions or of phylogenetic relationships) is attributable to the milieu of his own training in paleontology at the outset of the twentieth century, his reflective deployment of these ideas over fifty years later through multiple examples directs us to see more. The link between paleontology, development, and the concepts of innovation and novelty is persistent and inescapable, compelling us to recognize figures such as Gregory, and paleontological research more generally, in any history of Evo-devo.[142]

CONCLUSIONS

In conclusion, it needs to be observed that my historical presentation is biased in its textual/exegetical orientation rather than adopting a narrative of various contextual elements that foreground social factors pertinent to these individuals and their institutional homes. For example, one would likely draw attention to the differences between the University of Chicago/Field Museum of Natural History and Columbia University/ American Museum of Natural History. Addressing the work of multiple morphological and paleontological researchers in this time period is also necessary to strengthen the conclusions I have drawn here. Further neglected components in my analysis include the culture and practice of working morphologists and paleontologists,[143] and the role of sociological dynamics among the participants.[144]

There are also connections between Davis's and Gregory's particular explanations. Davis's effort to demonstrate that the giant panda was a modified bear was in part directed at Gregory's view that the giant panda was more closely related to lesser pandas and raccoons (procyonids).[145] A key character in the dispute was dentition, and we have already observed that Davis appealed to morphogenetic fields in dental development to bolster his case. The taxonomic question is not idle for understanding the role of Davis's invocation of development in the context of alternative explanations of evolutionary change. Two reviewers captured this explicitly: "If *Ailuropoda* were to be regarded as a kind of lesser panda which had relatively recently made an adaptive shift into a new niche (involving selection for larger size and more powerful jaws and teeth), then many of the apparently non-adaptive features particularly in the locomotor complex, might be explicable on a selectionist basis."[146] More recent phylogenetic studies utilizing molecular characters appear to vindicate Davis's conclusion,[147] although whether this constitutes a "solution" to an outstanding controversy or reaffirmation of Davis's careful anatomical work is debated.[148] The lesser (red) panda remains a taxonomic anomaly within the order Carnivora.[149]

Ultimately it is necessary to broaden the scope of historical source material for Evo-devo to include examples such as the morphological work of Davis and the paleontological contributions of Gregory.[150] Functional morphology and paleontology were central disciplinary homes for discussions of innovation and novelty, whether in avian jaw articulation,[151] the vertebrate middle ear,[152] the tetrapod limb,[153] insect wings,[154] or the artiodactyl tarsus.[155] Closely connected with these types of investigations were attempts to elucidate the origin of particular groups of organisms, including eumetazoa,[156] coelenterates,[157] fishes,[158] amphibians,[159] birds,[160] and mammals,[161] all of which display a conjunction of developmental, morphological, and paleontological methodologies.[162] Comparative anatomical or embryological approaches to phylogenetic relationships also were present long after their supposed heyday.[163] Focusing on the time period that is often seen as the "golden age" of the Modern Synthesis (including its "exclusion" of embryology) is particularly revealing because the obvious activity centered on innovation and novelty helps in part to puncture the problematic notion of "quiescent periods" in the history of Evo-devo,[164] especially when articles written in this period point toward a strikingly contemporary Evo-devo research agenda.[165]

What benefit is derived from identifying Davis's giant panda study or *Evolution Emerging* as part of a history of Evo-devo via the concepts of innovation and novelty? There are at least three possibilities. First, the history of developmental genetics has received tremendous attention because it seems to be playing such a critical role in the emergence of contemporary Evo-devo, and it dovetails nicely with the genetics and embryology exclusion histories.[166] Although some advocates sensitive to the paleontological aspects of Evo-devo have cited *Evolution Emerging*,[167] Davis is not discussed, and many of the articles just mentioned are ignored, even when they seem obviously relevant. Painting a more complete picture of various historical trajectories relevant to contemporary Evo-devo is clearly a beneficial outcome of attending to these morphological and pale-ontological perspectives.[168]

Second, it is significant that neither Davis nor Gregory was hostile to the Modern Synthesis. Davis was a participant[169] and Gregory made numerous favorable remarks about the results of genetic research on evolutionary mechanisms (especially natural selection) in the work of individuals such as Sewall Wright and Theodosius Dobzhansky.[170] This point can be emphasized through the irony that one reviewer held that "if there is any weak point [in *EE*], it is in the high degree of efficacy attributed to natural selection." The point was made with respect to the origin of a particular innovation, the electric organ of fishes: "The transformation of muscle to electric tissue requires radical histologic and functional changes in the muscle. To assume sufficiently numerous successive mutations in the same direction is hardly justifiable."[171] At the same time, Davis was accused of being anti-selectionist in his explanations, as discussed in reviews of his giant panda monograph and in connection with its co-option by Steven Stanley for his book *Macroevolution*. But Davis's ire was actually directed at the seemingly hegemonic discipline of genetics, which had grown significantly through the 1950s, rather than at the Modern Synthesis itself, as seen in the Goldschmidt quote he placed on the reverse of the title page to the giant panda monograph: "The field of macrotaxonomy . . . is not directly accessible to the geneticist. . . . Here the paleontologist, the comparative anatomist, and the embryologist are supreme." The real concern lies in the recognition of different levels of organization in biological entities.

Finally, the issue of recognizing the significance of different levels of biological organization redirects our attention to evaluating what morphology and paleontology can contribute to evolutionary theory today. Arguably, higher levels of structural organization were neglected in the Modern Synthesis, and paying attention to this feature may be critical for the future unfolding of Evo-devo as a research discipline. Most of the innovations and novelties that exercise researchers tend to occur at these high levels of organization. This helps make connections between all of these approaches within Evo-devo. Morphologists and paleontologists provide the analysis of evolutionary novelties at these higher levels of organization, while developmental biologists account for the generative principles of their construction from lower levels of organization during ontogeny. Attending to this point allows a rational reconstruction of the exclusion of embryology from the Modern Synthesis. If the Modern Synthesis was inordinately focused on lower levels of organization in terms of population genetics, then researchers who worked on elucidating processes in a developing organism responsible for constructing higher levels of organization were sidelined. Put differently, the absence of morphology from the Modern Synthesis explains part of the exclusion of embryology. Historical exploration based on the concepts of innovation and novelty leads to a partial reinterpretion of one key juncture from the past that is cited for keeping evolution and development separate during the twentieth century. Now that proponents of Evo-devo are renegotiating the interrelations between evolutionary and developmental biology, morphological and paleontological historical perspectives may provide useful material for navigating ongoing research on evolutionary innovations and novelties.

ACKNOWLEDGMENTS

I would like to thank Bernadette Callery and the staff of the Carnegie Natural History Museum Library, and Armand Esai of the Field Museum of Natural History Library, for research assistance and permissions; Rudolf Raff for loaning me his original copy of *Evolution Emerging*; and the organizers of the Dibner workshop for the invitation to participate. Sabine Brauckmann, Fred Churchill, Paul Griffiths, Nick Hopwood, Manfred Laubichler, Jim Lennox, Jane Maienschein, Bob Olby, and Jeff Schwartz, as well as the participants in the Dibner workshop in Cambridge (October 18–19, 2002) and the Evolutionary Morphology Group at the University of Chicago (May 8, 2003), kindly provided helpful comments and criticism on various versions of this material. I am also grateful to John Flynn for bringing me up to date on carnivore phylogeny.

NOTES

1. Allen, 1985a, 1985b; Gilbert, 1978; Maienschein, 1987; Sapp, 1987.

2. Hall, 1999, p. xv.

3. Olsson and Hall, 1999, p. 612.

4. Wagner, Chiu, and Laubichler, 2000, pp. 820, 822.

5. Carroll, Grenier, and Weatherbee, 2001, chap. 6; Gerhart and Kirschner, 1997, chap. 5; Hall, 1999, chap. 13; West-Eberhard, 2003, part II.

6. West-Eberhard, 2003, p. 6. Some researchers have been very explicit about labeling the origination of characters (innovation and novelty) as distinct from the theory of character maintenance and diversification due to natural selection found in contemporary evolutionary theory (Müller and Newman, 2003). This is often encapsulated by

the qualitative versus quantitative distinction: "Novelty always represents a qualitative departure from the ancestral condition, not merely a quantitative one" (Müller, 2002, p. 828; see also Müller, 1990; Müller and Wagner, 2003).

7. Love, 2003; Love and Raff, 2003. An important consideration is that what we refer to as evolutionary innovations and novelties may have been captured with different terminology in the past. In fact, they have often gone under the title "the origin of higher categories/taxa" or "major transformations"—for instance, Bock, 1965; Devillers, 1965; Orton, 1955; Schaeffer, 1965; Simpson 1961. But a translation can be effected because for earlier biologists the origin of new higher categories or major transformations in evolutionary history was a problem of how particular, defining characters originated (such as the head, jaw, neural crest cells, and somites of vertebrates), which clearly falls within the purview of contemporary discussions of innovation and novelty.

8. Bowler, 1988, pp. 99–103; Gould, 1977, pp. 85–96, and 2002, pp. 365–383.

9. See Churchill in this volume.

10. McNamara, 1990, 1995, 1997.

11. E.g., Berrill, 1935, 1955; Garstang, 1928.

12. Arthur, 2002; Wilkins, 2002.

13. McGinnis and Krumlauf, 1992.

14. Alberch et al., 1979; Bonner, 1982; Gould, 1977; Riedl, 1977, 1978.

15. See Gilbert, 1994; Hamburger, 1980.

16. Arthur, 1997, p. 287.

17. See Waisbren, 1988.

18. Love, 2003.

19. Olson, 1960, p. 523.

20. Ibid., p. 530.

21. Ghiselin, 1980, p. 181.

22. Coleman, 1980.

23. See Coleman, 1976.

24. Davis, 1949a, p. 64.

25. Ibid., p. 76.

26. Davis, 1960.

27. Nyhart, 1995.

28. Letter to Edwin Watkins (March 6, 1958), Field Museum of Natural History Archives, box 1, Original Zoology, Davis, D. Dwight, MSS, original and translated works.

29. Davis, 1960, p. 46.

30. Ibid., p. 49.

31. Ibid., p. 50.

32. Letter to Edwin Watkins (March 6, 1958), Field Museum of Natural History Archives, box 1, Original Zoology, Davis, D. Dwight, MSS, original and translated works.

33. Not dated, Field Museum of Natural History Archives, box 1, Original Zoology, Davis, D. Dwight, MSS, original and translated works.

34. Not dated, Field Museum of Natural History Archives, box 1, Original Zoology, Davis, D. Dwight, MSS, original and translated works.

35. Love, 2003.

36. Zangerl, 1948, p. 372.

37. See Bowler, 1996; Maienschein, 1991; Nyhart, 1995, 2002.

38. Rainger, 1993.

39. Davis, 1949b.

40. Ibid., p. 302.

41. Gould, 1980a, 1980b, 1980c.

42. Gould, 1966.

43. Gould, 1977.

44. Gould, 1980b, p. 23.

45. Gould, 1980a, p. 43.

46. Published as vol. 3 of *Fieldiana: Zoology Memoirs* by the Chicago Natural History Museum, the 32 centimeter pages of Davis's 339-page monograph with lavishly detailed drawings (some in color) is clearly a lifework—his magnum opus. To do it full justice would require far more space than is available in the present study, so I focus primarily on those aspects which can be seen as historical source material for Evo-Devo.

47. Davis, 1964, pp. 5, 11.

48. Ibid., p. 5.

49. Ibid., p. 12.

50. Huxley, 1993 [1932]; Thompson, 1992 [1942]. There are multiple meanings of "allometry" (or different kinds of allometries) (Gayon, 2000; Gould, 1966; Strauss, 1993). Here I am using the term to broadly capture the idea of coordinated alterations in morphology due to altered rates of growth during ontogeny with evolutionary implications. The two forms most relevant in the present discussion are the growth of one part *relative* to another or to the whole organism (Huxley's *heterogony* or allometry), and *absolute* shape transformations between two different organisms (Thompson's deformed Cartesian coordinate maps). These distinctions, and the many interconnections between the work of Huxley, Thompson, and Goldschmidt, are extensively discussed elsewhere (Churchill, 1993).

51. Davis, 1964, p. 40.

52. See Huxley, 1993 [1932], chap. 4.

53. Davis, 1964, pp. 67–69.

54. Davis, 1955.

55. Davis, 1964, p. 29.

56. Ibid., pp. 216–218.

57. Ibid., p. 72. See also the discussion of myological evolution (pp. 196–198) and arteries (pp. 274–280).

58. Ibid., p. 122.

59. For an updated discussion, see Müller, 2003.

60. Davis, 1964, pp. 84–85.

61. Ibid., p. 88.

62. Ibid., pp. 127–130.

63. Ibid., pp. 122–124.

64. MacIntyre and Koopman, 1967, p. 73.

65. Stanley, 1979, pp. 55–56, 138, 157–158.

66. Lande, 1980, pp. 234–235.

67. Stanley, 1979.

68. Davis, 1962, p. 67.

69. "*From the ontogenetic standpoint this approaches the fetal condition, and represents a state of arrested development, of fetalization*' " (Davis, 1964, p. 228).

70. Rainger, 1991.

71. Gregory, 1951. (Cited hereafter as *EE* I and II.)

72. Robert Carroll's *Vertebrate Paleontology and Evolution* is an approximation, though it is much more limited in scope (Carroll, 1988).

73. *Polyisomerism* = "the state in which many homologous parts, or polyisomeres, are arranged along any primary or secondary axis, whether straight or curved." *Anisomerism* = "the state in which one or more parts are emphasized at the expense of the rest, while the original number of separate parts is usually reduced, either by fusion or by elimination" (Gregory, 1934, p. 1). Gregory's terminology was meant to apply to purely physical as well as biological entities, which was noticed and indirectly criticized by a philosopher shortly thereafter (Plochmann, 1959).

74. *EE*, I, chap. 25.

75. Ibid., p. 559.

76. Jepsen, 1951; Rand, 1951; Straus, 1954.

77. Bates, 1951, p. 393.

78. Blackwelder, 1951.

79. Jarvik, 1955.

80. Northcutt and Gans, 1983.

81. Gregory, 1946.

82. Bates, 1951, p. 394.

83. *EE*, I, p. 26.

84. Ibid., p. 61.

85. Ibid., p. 77.

86. Ibid., p. 63.

87. Ibid., pp. 196–197.

88. Ibid., p. 38.

89. Ibid., p. 363.

90. Ibid., p. 209; see "The Role of Paedogenesis in the Ancestry of Modern Urodeles," Ibid., p. 253.

91. Ibid., p. 229.

92. Ibid., pp. 363–366.

93. Bowler, 1996; Gregory, 1946.

94. *EE*, I, p. 51.

95. Orton, 1955.

96. De Beer, 1951; see Hall, 2000. Orton's article is a good example of translating past terminology concerning the origin of higher categories into contemporary terminology about innovation and novelty. The following passage illustrates the persistent ambiguity between origin of new groups/categories and the origin of novel characters. "De Beer (1951) discussed in detail the potential importance of ontogenetic changes and cited examples of various kinds of developmental modifications that might have significant evolutionary results. He treated at considerable length the idea of origin of new major groups by neoteny from highly evolved larvae. It has been speculated that even the chordates may have originated by neoteny; the notochord and somites may have been larval specializations in the ancestral stock (see Gregory, 1946, 1951; De Beer, 1951). Neoteny is, however, not the only means by which developmental modifications may have major evolutionary consequences. The entire ontogeny as a unit may shift relatively rapidly (in geological time) to a new structural plan. It has been suggested that a change of this nature, the extreme shortening of the vertebral column in both larva and adult, precipitated the origin of the frogs" (Orton, 1955, p. 81).

97. *EE*, I, pp. 26–27.

98. Ibid., p. 41.

99. Ibid., p. 66.

100. Ibid., p. 88.

101. Ibid., p. 106.

102. Ibid., p. 107.

103. Ibid., p. 108.

104. Ibid., pp. 137–138.

105. Ibid., pp. 160–161.

106. Ibid., pp. 184–191.

107. Ibid., pp. 206–207.

108. Ibid., pp. 237–238.

109. Ibid., p. 246.

110. Ibid., pp. 252–253.

111. Ibid., pp. 314–315.

112. Ibid., pp. 334–335.

113. Ibid., pp. 344–347.

114. Ibid., p. 365.

115. Ibid., pp. 366–367.

116. Ibid., p. 390.

117. Ibid., p. 480.

118. Notably, many of these putative explanations were contentious, and formed a portion of the complaints offered by reviewers (Rand, 1951, pp. 437–438).

119. *EE*, I, p. 304; see also pp. 301–302.

120. *EE*, II, p. 13.

121. Ibid., p. 60.

122. Ibid., p. 65.

123. Ibid., p. 69.

124. Ibid., p. 70.

125. Ibid., p. 123.

126. Ibid., p. 297.

127. Ibid., p. 397.

128. Ibid., pp. 597, 610, 613.

129. Ibid., p. 654.

130. A cluster of these illustrations (predictably) falls within the discussion of the origin of vertebrates. Importantly, they are not found *only* in this discussion.

131. See *EE*, I, pp. 79ff.

132. Ibid., p. 61.

133. Ibid., pp. 226–227.

134. Ibid., pp. 228, 553.

135. Ibid., p. 246.

136. Ibid., p. 510.

137. Rainger, 1991, chap. 9. Rainger highlights the fact that Gregory's research program was just as much morphological as it was paleontological (pp. 220–224).

138. Gregory, 1950, p. 170.

139. A good example is his discussion of early brachiopods without shells (*EE*, I, pp. 77–78).

140. Ibid., p. 85 [my emphasis]; see Gould, 1977, pp. 170–175.

141. *EE*, I, pp. 260–261.

142. See Hall, 2002.

143. See Pickering, 1992.

144. Gerson, in this volume.

145. Gregory, 1936.

146. MacIntyre and Koopman, 1967, p. 73.

147. Goldman, Giri, and O'Brien, 1989; O'Brien 1987; O'Brien et al., 1985; Sarich, 1973; Zhang and Ryder, 1993.

148. See the exchange of correspondence between John Flynn, André Wyss, and Stephen J. O'Brien in *Scientific American* (June 1988).

149. Flynn et al., 2000; Flynn and Nedbal, 1998; Flynn, Neff, and Tedford, 1988.

150. There are many morphological and paleontological perspectives from the European context that are critical roots for contemporary Evo-devo. (See, e.g., Riedl, 1977, 1978).

151. Bock, 1959.

152. Tumarkin, 1955; Watson, 1953.

153. Westoll, 1943.

154. Forbes, 1943.

155. Schaeffer, 1948. Importantly, these are the sources for Ernst Mayr's seminal revisiting of the topic during this period (Mayr, 1960).

156. Hanson, 1958.

157. Hand, 1959.

158. Romer, 1946.

159. Szarski, 1962.

160. Bock, 1965.

161. Olson, 1959.

162. See also, Schaeffer, 1965.

163. Dillon, 1965; Marcus, 1957.

164. Arthur, 1997, p. 86.

165. E.g., Devillers, 1965. Devillers marshals developmental evidence for morphogenetic mechanisms that may have been involved in three different evolutionary transitions: the origin of the cerebral inductor complex, the origin of paired fins and skull bone variation in fishes, and digit/fibula reduction in tetrapods. Commensurate with the investigation of Davis, the morphogenetic field concept plays a prominent role.

166. Love, 2003; see Burian, 2000; Burian, Gayon, and Zallen, 1991; Gilbert, 1991, 1994, 2000.

167. Hall, 1999; Raff, 1996. Neither author attempts to explore Gregory's work in detail. Brian Hall makes the connection between paleontology and Evo-devo very explicit in a recent review article, although Gregory's contribution is not discussed (Hall, 2002).

168. Importantly, comparative anatomy is distinguished from paleontology in some conceptualizations of Evo-devo (Hall, 1999; Wagner, Chiu, and Laubichler, 2000). The attention given to analytical techniques such as morphometrics within Evo-devo (Roth and Mercer, 2000) also drives us back to morphological research and a long-standing connection between morphometrics and ontogeny (Olson and Miller, 1999 [1958], chap. 7).

169. Davis, 1949a.

170. *EE*, I, pp. 280–281, 454, 510, 513, 536, 557, 559.

171. Rand, 1951, p. 438.

REFERENCES

Alberch, P., S. J. Gould, G. F. Oster, and D. B. Wake. 1979. "Size and Shape in Ontogeny and Phylogeny." *Paleobiology* 5: 296–317.

Allen, G. E. 1985a. "Heredity Under an Embryological Paradigm: The Case of Genetics and Embryology." *Biological Bulletin* (supp.): 107–121.

———. 1985b. "T. H. Morgan and the Split Between Embryology and Genetics, 1910–35." In *A History of Embryology: The Eighth Symposium of the British Society for Developmental Biology*, ed. T. J. Horder, J. A. Witkowski, and C. C. Wylie, pp. 113–146. Cambridge: Cambridge University Press.

Arthur, W. 1997. *The Origin of Animal Body Plans: A Study in Evolutionary Developmental Biology*. New York: Cambridge University Press.

———. 2002. "The Emerging Conceptual Framework of Evolutionary Developmental Biology." *Nature* 415: 757–764.

Bates, M. 1951. "Paleontology and Evolution." *American Naturalist* 85: 393–394.

Berrill, N. J. 1935. "Studies in Tunicate Development. Part III.—Differential Retardation and Acceleration." *Philosophical Transactions of the Royal Society of London* B225: 255–326.

———. 1955. *The Origin of Vertebrates*. Oxford: Clarendon Press.

Blackwelder, R. E. 1951. "Systematics in Zoology." *Science* 113: 3.

Bock, W. J. 1959. "Preadaptation and Multiple Evolutionary Pathways." *Evolution* 13: 194–211.

————. 1965. "The Role of Adaptive Mechanisms in the Origin of Higher Levels of Organization." *Systematic Zoology* 14(4): 272–287.

Bonner, J. T., ed. 1982. *Evolution and Development*. Berlin: Springer Verlag.

Bowler, P. J. 1988. *The Non-Darwinian Revolution: Reinterpreting a Historical Myth*. Baltimore: Johns Hopkins University Press.

————. 1996. *Life's Splendid Drama: Evolutionary Biology and the Reconstruction of Life's Ancestry, 1860–1940*. Chicago: University of Chicago Press.

Burian, R. M. 2000. "General Introduction to the Symposium on Evolutionary Developmental Biology: Paradigms, Problems, and Prospects." *American Zoologist* 40: 711–717.

Burian, R. M., J. Gayon, and D. T. Zallen. 1991. "Boris Ephrussi and the Synthesis of Genetics and Embryology." In *Developmental Biology: A Comprehensive Synthesis, vol. 7, A Conceptual History of Modern Embryology*, ed. S. F. Gilbert, pp. 207–227. New York: Plenum Press.

Carroll, R. L. 1988. *Vertebrate Paleontology and Evolution*. New York: W. H. Freeman.

Carroll, S. B., J. K. Grenier, and S. D. Weatherbee. 2001. *From DNA to Diversity: Molecular Genetics and the Evolution of Animal Design*. Malden, Mass.: Blackwell Science.

Churchill, F. B. 1993. "On the Road to the *k* Constant: A Historical Introduction." In *Julian Huxley, Problems of Relative Growth*, pp. xix–xlv. Baltimore: Johns Hopkins University Press.

Coleman, W. 1976. "Morphology Between Type Concept and Descent Theory." *Journal of the History of Medicine and Allied Sciences* 31: 149–175.

————. 1980. "Morphology in the Evolutionary Synthesis." In *The Evolutionary Synthesis: Perspectives on the Unification of Biology*, ed. E. Mayr and W. B. Provine, pp. 174–180. Cambridge, Mass.: Harvard University Press.

Davis, D. D. 1949a. "Comparative Anatomy and the Evolution of Vertebrates." In *Genetics, Paleontology, and Evolution*, ed. G. L. Jepsen, E. Mayr and G. G. Simpson, pp. 64–89. Princeton, N. J.: Princeton University Press.

————. 1949b. "The Shoulder Architecture of Bears and Other Carnivores." *Fieldiana: Zoology* 31: 285–305.

————. 1955. "Masticatory Apparatus in the Spectacled Bear *Tremarctos ornatus*." *Fieldiana: Zoology* 37: 25–46.

————. 1960. "The Proper Goal of Comparative Anatomy." In *Proceedings of the Centenary and Bicentenary Congress of Biology Singapore, December 2–9, 1958*, ed. R. D. Purchon, pp. 44–50. Singapore: University of Malaya Press.

————. 1962. *Mammals of the Lowland Rain-Forest of North Borneo*, Bulletin of the Singapore National Museum, no. 31. Singapore: Lee Kim Heng.

————. 1964. *The Giant Panda: A Morphological Study of Evolutionary Mechanisms. Fieldiana: Zoology Memoirs*, vol. 3. Chicago: Chicago Natural History Museum.

De Beer, G. R. 1951. *Embryos and Ancestors*, rev. ed. Oxford: Oxford University Press.

Devillers, C. 1965. "The Role of Morphogenesis in the Origin of Higher Levels of Organization." *Systematic Zoology* 14(4): 259–271.

Dillon, L. S. 1965. "The Hydrocoel and the Ancestry of the Chordates." *Evolution* 19(3): 436–446.

Flynn, J. J., and M. A. Nedbal. 1998. "Phylogeny of the Carnivora (Mammalia): Congruence vs. Incompatibility Among Multiple Data Sets." *Molecular Phylogenetics and Evolution* 9(3): 414–426.

Flynn, J. J., M. A. Nedbal, J. W. Dragoo, and R. L. Honeycutt. 2000. "Whence the Red Panda?" *Molecular Phylogenetics and Evolution* 17(2): 190–199.

Flynn, J. J., N. A. Neff, and R. H. Tedford. 1988. "Phylogeny of the Carnivora." In *The Phylogeny and Classification of the Tetrapods, vol. 2, Mammals*, ed. M. J. Benton, pp. 73–116. Oxford: Clarendon Press.

Forbes, W. T. M. 1943. "The Origin of Wings and Venational Types in Insects." *American Midland Naturalist* 29(2): 381–405.

Garstang, W. 1928. "The Morphology of the Tunicata, and Its Bearings on the Phylogeny of the Chordata." *Quarterly Journal of Microscopical Science* 72: 51–187.

Gayon, J. 2000. "History of the Concept of Allometry." *American Zoologist* 40: 748–758.

Gerhart, J., and M. Kirschner. 1997. *Cells, Embryos, and Evolution: Toward a Cellular and Developmental Understanding of Phenotypic Variation and Evolutionary Adaptability*. Malden, Mass.: Blackwell Science.

Ghiselin, M. T. 1980. "The Failure of Morphology to Assimilate Darwinism." In *The Evolutionary Synthesis: Perspectives on the Unification of Biology*, ed. E. Mayr and W. B. Provine, pp. 180–193. Cambridge, Mass.: Harvard University Press.

Gilbert, S. F. 1978. "The Embryological Origins of the Gene Theory." *Journal of the History of Biology* 11: 307–351.

————. 1991. "Induction and the Origin of Developmental Genetics." In *Developmental Biology: A Comprehensive Synthesis, vol. 7, A Conceptual History of Modern Embryology*, ed. S. F. Gilbert, pp. 181–206. New York: Plenum Press.

————. 1994. "Dobzhansky, Waddington, and Schmalhausen: Embryology and the Modern Synthesis." In *The Evolution of Theodosius Dobzhansky: Essays on His Life and Thought in Russia and America*, ed. M. B. Adams, pp. 143–154. Princeton, N. J.: Princeton University Press.

————. 2000. "Diachronic Biology Meets Evo-Devo: C. H. Waddington's Approach to Evolutionary Developmental Biology." *American Zoologist* 40: 729–737.

Goldman, D., P. R. Giri, and S. J. O'Brien. 1989. "Molecular Genetic-Distance Estimates Among the Ursidae as Indicated by One- and Two-Dimensional Protein Electrophoresis." *Evolution* 43(2): 282–295.

Gould, S. J. 1966. "Allometry and Size in Ontogeny and Phylogeny." *Biological Reviews of the Cambridge Philosophical Society* 41: 587–640.

———. 1977. *Ontogeny and Phylogeny.* Cambridge, Mass.: Belknap Press of Harvard University Press.

———. 1980a. "Double Trouble." In Stephen J. Gould, *The Panda's Thumb: More Reflections in Natural History,* pp. 35–44. New York: Norton.

———. 1980b. "The Panda's Thumb." In Stephen J. Gould, *The Panda's Thumb: More Reflections in Natural History,* pp. 19–26. New York: Norton.

———. 1980c. "Senseless Signs of History." In Stephen J. Gould, *The Panda's Thumb: More Reflections in Natural History,* pp. 27–34. New York: Norton.

———. 2002. *The Structure of Evolutionary Theory.* Cambridge, Mass.: Belknap Press of Harvard University Press.

Gregory, W. K. 1934. "Polyisomerism and Anisomerism in Cranial and Dental Evolution Among Vertebrates." *Proceedings of the National Academy of Sciences USA* 20: 1–9.

———. 1936. "On the Phylogenetic Relationships of the Giant Panda (*Ailuropoda*) to Other Arctoid Carnivora." *American Museum Novitates* 878: 1–29.

———. 1946. "The Roles of Motile Larvae and Fixed Adults in the Origin of the Vertebrates." *Quarterly Review of Biology* 21(4): 348–364.

———. 1950. "Parallel and Diverging Evolution in Vertebrates and Arthropods." *Evolution* 4: 164–171.

———. 1951. *Evolution Emerging: A Survey of Changing Patterns from Primeval Life to Man.* New York: Macmillan.

Hall, B. K. 1999. *Evolutionary Developmental Biology,* 2nd ed. Dordrecht: Kluwer Academic.

———. 2000. "Balfour, Garstang, and de Beer: The first Century of Evolutionary Embryology." *American Zoologist* 40: 718–728.

———. 2002. "Palaeontology and Evolutionary Developmental Biology: A Science of the Nineteenth and Twenty-first Centuries." *Palaeontology* 45(4): 647–669.

Hamburger, V. 1980. "Embryology and the Modern Synthesis in Evolutionary Theory." In *The Evolutionary Synthesis: Perspectives on the Unification of Biology,* ed. E. Mayr and W. B. Provine, pp. 97–112. Cambridge, Mass.: Harvard University Press.

Hand, C. 1959. "On the Origin and Phylogeny of Coelenterates." *Systematic Zoology* 8(4): 191–202.

Hanson, E. D. 1958. "On the Origin of Eumetazoa." *Systematic Zoology* 7(1): 16–47.

Huxley, J. S. 1993 [1932]. *Problems of Relative Growth*. Baltimore: Johns Hopkins University Press.

Jarvik, E. 1955. "The Oldest Tetrapods and Their Forerunners." *Scientific Monthly* 80(3): 141–154.

Jepsen, G. L. 1951. "Book Review, Evolution: *Evolution Emerging. A Survey of Changing Patterns from Primeval Life to Man*." *Quarterly Review of Biology* 26(4): 382–384.

Lande, R. 1980. "Microevolution in Relation to Macroevolution." *Paleobiology* 6(2): 233–238.

Love, A. C. 2003. "Evolutionary Morphology, Innovation, and the Synthesis of Evolutionary and Developmental Biology." *Biology and Philosophy* 18(2): 309–345.

Love, A. C., and R. A. Raff. 2003. "Knowing Your Ancestors: Themes in the History of Evo-Devo." *Evolution & Development* 5(4): 327–330.

MacIntyre, G., and K. Koopman. 1967. "The Giant Panda: A Morphological Study of Evolutionary Mechanisms." *Quarterly Review of Biology* 42(1): 72–73.

Maienschein, J. 1987. "Heredity/Development in the United States, Circa 1900." *History and Philosophy of the Life Sciences* 9: 79–93.

———. 1991. *Transforming Traditions in American Biology, 1880–1915*. Baltimore: Johns Hopkins University Press.

Marcus, E. 1957. "On the Evolution of Animal Phyla." *Quarterly Review of Biology* 33(1): 24–58.

Mayr, E. 1960. "The Emergence of Evolutionary Novelties." In *Evolution After Darwin, vol. 1, The Evolution of Life: Its Origin, History and Future*, ed. S. Tax, pp. 349–380. Chicago: University of Chicago Press.

McGinnis, W., and R. Krumlauf. 1992. "Homeobox Genes and Axial Patterning." *Cell* 68(2): 283–302.

McNamara, K. J. 1990. "The Role of Heterochrony in Evolutionary Trends." In *Evolutionary Trends*, ed. K. J. McNamara, pp. 59–74. Tucson: University of Arizona Press.

———. 1997. *Shapes of Time: The Evolution of Growth and Development*. Baltimore: Johns Hopkins University Press.

———. ed. 1995. *Evolutionary Change and Heterochrony*. New York: Wiley.

Müller, G. B. 1990. "Developmental Mechanisms at the Origin of Morphological Novelty: A Side-effect Hypothesis." In *Evolutionary Innovations*, ed. M. H. Nitecki, pp. 99–130. Chicago: University of Chicago Press.

———. 2002. "Novelty and Key Innovation." In *Encyclopedia of Evolution*, ed. M. D. Pagel, vol. 2, pp. 827–830. Oxford: Oxford University Press.

————. 2003. "Embryonic Motility: Environmental Influences and Evolutionary Innovation." *Evolution & Development* 5(1): 56–60.

Müller, G. B., and S. A. Newman. 2003. "Origination of Organismal Form: The Forgotten Cause in Evolutionary Theory." In *Origination of Organismal Form: Beyond the Gene in Developmental and Evolutionary Biology*, ed. G. B. Müller and S. A. Newman, pp. 3–10. Cambridge, Mass.: MIT Press.

Müller, G. B., and G. P. Wagner. 2003. "Innovation." In *Key Concepts and Approaches in Evolutionary Developmental Biology*, ed. B. K. Hall and W. M. Olsson, pp. 218–227. Cambridge, Mass.: Harvard University Press.

Northcutt, R. G., and C. Gans. 1983. "The Genesis of Neural Crest and Epidermal Placodes: A Reinterpretation of Vertebrate Origins." *Quarterly Review of Biology* 58(1): 1–28.

Nyhart, L. K. 1995. *Biology Takes Form: Animal Morphology and the German Universities, 1800–1900*. Chicago: University of Chicago Press.

————. 2002. "Learning from History: Morphology's Challenges in Germany ca. 1900." *Journal of Morphology* 252: 2–14.

O'Brien, S. J. 1987. "The Ancestry of the Giant Panda." *Scientific American* (November): 102–107.

O'Brien, S. J., W. G. Nash, D. E. Wildt, M. E. Bush, and R. E. Benveniste. 1985. "A Molecular Solution to the Riddle of the Giant Panda's Phylogeny." *Nature* 317: 140–144.

Olson, E. C. 1959. "The Evolution of Mammalian Characters." *Evolution* 13: 344–353.

————. 1960. "Morphology, Paleontology, and Evolution." In *Evolution After Darwin, vol. 1, The Evolution of Life: Its Origin, History and Future*, ed. S. Tax, pp. 523–545. Chicago: University of Chicago Press.

Olson, E. C., and R. L. Miller. 1999 [1958]. *Morphological Integration*. Chicago: University of Chicago Press.

Olsson, L., and B. K. Hall. 1999. "Introduction to the Symposium: Developmental and Evolutionary Perspectives on Major Transformations in Body Organization." *American Zoologist* 39: 612–666.

Orton, G. L. 1955. "The Role of Ontogeny in Systematics and Evolution." *Evolution* 9(1): 75–83.

Pickering, A. 1992. "From Science as Knowledge to Science as Practice." In *Science as Practice and Culture*, ed. A. Pickering, pp. 1–26. Chicago: University of Chicago Press.

Plochmann, G. K. 1959. "Darwin or Spencer?" *Science* 130: 1452–1456.

Raff, R. A. 1996. *The Shape of Life: Genes, Development and the Evolution of Animal Form*. Chicago: University of Chicago Press.

Rainger, R. 1991. *An Agenda for Antiquity: Henry Fairfield Osborn & Vertebrate Paleontology at the American Museum of Natural History, 1890–1935.* Tuscaloosa: University of Alabama Press.

————. 1993. "Biology, Geology, or Neither, or Both: Vertebrate Paleontology at the University of Chicago, 1892–1950." *Perspectives on Science* 1(3): 478–519.

Rand, H. W. 1951. "The Cosmic Cinema." *Science* 113: 434–438.

Riedl, R. 1977. "A Systems-Analytical Approach to Macro-Evolutionary Phenomena." *Quarterly Review of Biology* 52(4): 351–370.

————. 1978. *Order in Living Organisms: A Systems Analysis of Evolution,* trans. R. P. S. Jeffries, New York: Wiley.

Romer, A. S. 1946. "The Early Evolution of Fishes." *Quarterly Review of Biology* 21(1): 33–69.

Roth, V. L., and J. M. Mercer. 2000. "Morphometrics in Development and Evolution." *American Zoologist* 40: 801–810.

Sapp, J. 1987. *Beyond the Gene: Cytoplasmic Inheritance and the Struggle for Authority in Genetics.* New York: Oxford University Press.

Sarich, V. M. 1973. "The Giant Panda Is a Bear." *Nature* 245: 218–220.

Schaeffer, B. 1948. "The Origin of a Mammalian Ordinal Character." *Evolution* 2: 164–175.

————. 1965. "The Role of Experimentation in the Origin of Higher Levels of Organization." *Systematic Zoology* 14(4): 318–336.

Simpson, G. G. 1961. "Some Problems of Vertebrate Paleontology." *Science* 133: 1679–1689.

Stanley, S. M. 1979. *Macroevolution: Pattern and Process.* San Francisco: W. H. Freeman.

Straus, W. L. 1954. "Book Reviews, Physical Anthropology: *Evolution Emerging: A Survey of Changing Patterns from Primeval Life to Man.*" *American Anthropologist* 56(1): 147–148.

Strauss, R. E. 1993. "The Study of Allometry Since Huxley." In *Problems of Relative Growth,* pp. xlvii–lxxv. Baltimore: Johns Hopkins University Press.

Szarski, H. 1962. "The Origin of Amphibia." *Quarterly Review of Biology* 37(3): 189–241.

Thompson, D'A. W. 1992 [1942]. *On Growth and Form,* rev. ed. New York: Dover.

Tumarkin, A. 1955. "On the Evolution of the Auditory Conducting Apparatus: A New Theory Based on Functional Considerations." *Evolution* 9: 221–243.

Wagner, G. P., C.-H. Chiu, and M. Laubichler. 2000. "Developmental Evolution as a Mechanistic Science: The Inference from Developmental Mechanisms to Evolutionary Processes." *American Zoologist* 40: 819–831.

Waisbren, S. J. 1988. "The Importance of Morphology in the Evolutionary Synthesis as Demonstrated by the Contributions of the Oxford Group: Goodrich, Huxley, and de Beer." *Journal of the History of Biology* 21: 291–330.

Watson, D. M. S. 1953. "The Evolution of the Mammalian Ear." *Evolution* 7: 159–177.

West-Eberhard, M. J. 2003. *Developmental Plasticity and Evolution.* New York: Oxford University Press.

Westoll, T. S. 1943. "The Origin of the Primitive Tetrapod Limb." *Proceedings of the Royal Society of London* B131: 373–393.

Wilkins, A. S. 2002. *The Evolution of Developmental Pathways.* Sunderland, Mass.: Sinauer.

Zangerl, R. 1948. "The Methods of Comparative Anatomy and Its Contribution to the Study of Evolution." *Evolution* 2: 351–374.

Zhang, Y.-P., and O. A. Ryder. 1993. "Mitochondrial DNA Sequence Evolution in the Arctoidea." *Proceedings of the National Academy of Sciences USA*, 90(20): 9557–9561.

10

Echoes of Haeckel? Reentrenching Development in
Evolution

William C. Wimsatt

The publication of Gould's *Ontogeny and Phylogeny* in 1977 marked and helped to initiate a new visible effort by some evolutionists—mostly paleontologists, systematists, and developmental biologists—to restore development to its proper place in an evolutionary synthesis. In the United States, reactions among population biologists (the then dominant alliance of population genetics and community ecology) to claims that one needed to take development into account were often hostile—even when these claims were quite benign. This was most directly a reflection of strong gene-centered reductionist currents according to which anything in development, any organic adaptation, or anything at higher levels was just to be cashed out in terms of selection coefficients acting on the genic level, or dismissed as muddled thinking.

But not all sought a seamless integration of development into the "Modern Synthesis." Skeptics toward claims of the gene-centered theory of population genetics had often rallied around arguments for the recognition of "internal factors" of evolution, antedating Gould's work by nearly two decades. These factors arose in different contexts and referred to diverse traditions: morphology, developmental biology, and general systems theory. Here I sketch elements of the conceptual context for these objections as exemplified by the reception of Lancelot Law Whyte's advocacy of "internal factors in evolution." A decade later Rupert Riedl's notion of "burden" crystallized many of these intuitive objections into a coherent theory—an alternative vision which came much closer to the observed phenomena and structures that developmentalists and morphologists saw. It was in many ways more satisfying than the principled but abstract claims of contemporary population genetics to provide an adequate and complete account of microevolution that also covered macroevolution.[1] I discuss Riedl's systematic and sophisticated account at more length, and suggest several reasons why Riedl was not more influential. But the directions he indicated have come to seem increasingly important since. I briefly indicate the subsequent and ongoing development of similar ideas by Arthur, and by Wimsatt and Schank, who both moved closer to a synthesis with evolutionary genetics.

DEVELOPMENT AND MACRO-EVOLUTION AS CHATTELS OF POPULATION GENETICS

The mid-1960s saw a purifying wave of criticism directed against evolutionary ideas that were not directly rooted in population genetics. George Williams's influential 1966 book became the first to argue the omnipotence of the "gene's eye view of evolution" that was pointed to by the groundbreaking work of W. D. Hamilton (1964). Many evolutionary discussions made reference to "the survival of the species" or the action of selection on groups of organisms. Williams argued that all of it was unjustified and sought to expunge it. His book rapidly became required reading for graduate students in evolutionary biology across the country, and engendered a wave of reductionist thinking that has lasted until the present.

But Williams saw other enemies (Wimsatt 1970). He viewed Waddington's (1957) selection experiments on "genetic assimilation" of the bithorax phenotype as a potential challenge to neo-Darwinism, which it was not. Most others now treat it as a genetically acceptable interpretation of the important "Baldwin effect." Williams took great pains to dismiss it as lacking in evolutionary potential. And he simply ignored everything else in *The Strategy of the Genes*, including Waddington's increasingly influential notions of canalization.[2] As late as 1986, the population geneticist Bruce Wallace invoked Weismann to argue that development had nothing to contribute to population genetics, continuing misinterpretations then endemic among population geneticists (Griesemer and Wimsatt 1989). These were representative views among population geneticists. Similar views were argued—or more commonly asserted—to me in conversations through the 1980s. The works of Hamilton, Maynard Smith, and others appeared to cut directly against all higher-level units of selection. Not only groups but even organisms were read off as just different "environments" of the gene.

On this view, knowledge of development had one, and only one, point of relevance to evolutionary theory. It could help to determine the value of parameters—specifiable net selection coefficients for alleles and their mutants in populations of organisms, and thus provide inputs for population genetic theory. Developmental features were like any other adaptations, and not worthy of special note. The macro-evolutionary concerns of comparative embryology and structural biology were similarly dismissed: what would do for the short run, it seemed to population geneticists, would do for the long run, which was, after all, just a bunch of short runs set end to end. So the answers population genetics provided for microevolution would do for macroevolution as well.

But not all was well for the project of doing the population genetics of whole organisms—even more obviously so once one considered development. A tightly integrated system acting over time would involve the action and interaction of multiple factors (and therefore multiple genes) at multiple developing levels of organization. The temporal and qualitative diversity of their action (and thus of their interactions) would require population genetic models immensely more complex than those pursued or even considered up to that point. These models would have to deal with massive epistatic (interactive nonadditive, nonlinear, and asymmetric) networks of causal relations between genes.

Population geneticists could not handle selection models for even a few loci without drastic simplifications, much less the 10^5 to 10^6 loci assumed (in 1975) for whole genotypes. Lewontin's influential discussion of dimensionality showed how rapidly the complexity of the dynamical equations for change grew with the number of loci. It grew even more rapidly if the epistatic interactions characteristic of development were not simplified out of existence (1974, table 56, p. 283, Wimsatt 1980). Lewontin also made pioneering use of computer simulations for up to thirty-two loci, but these push back the limits of complexity only a small part of the way: Learning how to make the right simplifications in theory, simulation models, or experimental designs is still a hard-won skill. And there was no sense of how to put development systematically into this picture. One could not derive, compute, or simulate one's way out of the problem.

From the late 1950s on, there were more attempts to argue the role of development in evolution and to redraw qualitative connections with evolutionary theory.[3] Waddington's important work (1957) was particularly influential. But other classical embryologists were moved by the new progress in genetics. The second edition of Balinsky's classic embryology text (1965) added a page (537) closing a section on gene expression to argue the inevitability of differential magnitudes of effect of gene expression at different stages in development. Earlier changes ("those which are not lethal") have later effects, whereas later changes could not have earlier effects. He uses this assymmetry to argue for a relative conservatism of earler features and to explain von Baer's law.[4] Here he comes halfway to anticipating "burden" or "generative entrenchment," which generate similar but larger assymmetries.

I will consider the vortex of shifting opinion on the role of development primarily through two figures, Lancelot Law Whyte and Rupert Riedl, who urged the importance of "internal factors" in evolution, and then track the growing developments of the approach initiated by Riedl. Whyte got "lots of press," perhaps more than he merited. He was a well-connected "public intellectual," but not a scientist. His work was thin and "suggestive" at best. That he got a hearing from many respected scientists is a measure of the need many felt for a fuller account of biological organization and the mysterious complexities of development than were promised by the major theories of the time. What has survived (see discussion of Arthur below) was his continuing insistence that selection must act to ensure consistency with internal features as well as the challenges of the environment.

Rupert Riedl, a respected European morphologist, presented arguments, analyses and theory that deserved much closer attention. He offered a well-developed account of biological organization, apparently well-anchored in information theory, and used rich insights from morphology driven by his theory of "burden" to link gene-action to development, biological organization, and evolution. His well-founded intuitions relate closely to methods and analyses now pivotally important to theories and practice of evolutionary developmental biology. Yet his work drew little attention, even as concerns with development were beginning to move to center stage for evolutionary studies. There are things to be learned from both of these paradoxical receptions.

Wallace Arthur and I independently discovered and developed the mechanisms motivating Riedl's robust intuitions. Each of us was well along before others pointed out similarities between our respective accounts, and of both with those of Riedl. After outlining Riedl's approach, assessing its limitations and the (somewhat different) reasons why it did not get more serious attention, I will sketch our respective developments. Collectively they generate a powerful approach to various evolutionary phenomena that complements genetic methods when they are available, but can proceed without them.

INTERNAL FACTORS IN EVOLUTION

Lancelot Law Whyte (1965) occupied a social locus in science more reminiscent of the 18th century. Quoted widely among a group of older biologists and conceptual fellow travelers, he was most often called an "independent philosopher-scientist" as he appeared in conference proceedings with the likes of W. H. Thorpe, Paul Weiss, and Ludwig von Bertalanffy. His fans touted hierarchial organization, and were suspicious of reductionism. He attended and hosted conferences (at least four of them), often urging deep methodological or foundational changes in the natural and biological sciences. Reputable people attended them, and often were invited to reach beyond their domains. Whyte's views are less important for themselves than as indicators of the times. Some reputable biologists "around the edges" remained dissatisfied with what population genetics and the then current state of developmental biology could provide as explanations for the evolution of complex biological organization. Many hoped that new abstract and general theory could be constructed to fill the gap.

The animal ethologist W. H. Thorpe took Whyte seriously:

> I personally found Waddington a little "cagey" concerning the significance of biological organization at a relatively low molecular level. J. B. S. Haldane was himself forced to the conclusion that it was almost certain that facts will be discovered to show that the theory of natural selection is not fully adequate to account for evolution. Waddington does not seem to have much to say for the idea of "internal" factors in evolution as set forth by L. L. Whyte but I still think that [Whyte's] book contains some important points which geneticists as a whole have failed to stress sufficiently—and I was glad to see that Paul Weiss also thinks that Whyte makes a significant contribution. The essence of Whyte's book is the conclusion (shared among others, by Haldane, Spurway, and von Bertalanffy) that the conditions of biological organization restrict . . . the possible avenues of evolutionary change from a given starting point. (Koestler and Smythies, 1970, pp. 431–432).

A decade earlier, Whyte had marshaled various complaints about overlooked factors. In a one-page (!) research report to *Science* (1960), he urged that "internal factors" imposed selection before traditionally recognized external factors could act, and were important to evolution. He claimed that they were not captured by existing theory and were ignored by establishment theorists. Responding in a long letter to *Science*, Richard Lewontin (already a rising star in population genetics) and Ernst Caspari (a senior statesman of developmental genetics and Lewontin's colleague at Rochester) replied that Whyte's note brought nothing that was both new and correct. Dobzhansky, Schmalhausen, and others had dealt with development and with constraints, and they suggested that it was a mistake to draw a boundary line between internal and external anyway. (True, but arguably not addressing Whyte's point.[5]) The whole interchange (Lewontin, Caspari, and Whyte, 1960) would not have convinced anyone to change sides.[6] In 1965 Whyte put together a book with a sketchy history of the problem (quotes from notables who "recognized" it), how he came to it, and responses to his critics.

But his attempt to sketch a theory of these internal factors was less successful than his attempt to convince some that there was a problem: To think about the "internal factors" he emphasized, he sought general algebraic "Coordinative Conditions, (C.C.) . . . expressing the biological spatio-temporal coordination rules of ordering which must be satisfied (to within a threshold) by the internal parts and processes of any cellular organism capable of developing and surviving in some environment." (Whyte, 1965, p. 35)

Whyte sought something like a master equation that defined an "invariant characteristic configuration," but also an equilibrium tendency, and covered not only homeostasis, but also developmental trajectories (pp. 35–36).

Coordinative conditions are the "general conditions which are satisfied by all organisms, but they can be met in countless contrasted particular manners corresponding to the hierarchy of phyla, classes, orders, families, genera, species, and variations (p. 37).

These were fundamental expressions of molecular 3-dimensional and kinematic constraints, so were "bottom up" (presumably why he thought of them as universal).

Whyte pictured a master equation which could be specialized in different ways, to understand "the precise animate and inanimate conditions of the competition for survival, say of one of the earliest birds of 130 million years ago" (p. 38). He expressed a kind of faith in the reach of a geometric, infinitely detailed, all-embracing mathematical theory that was (and is) foreign to biology: a LaPlacean demon as organismic biologist. This attempt at general theory was quintessential "physics envy." It was immensely reductionistic, formal, and structural, but because Whyte spoke of invariant relations rather than atomistic components, it was not seen as such and so had appeal to others suspicious of mechanism. It was deterministic rather than stochastic unlike modern evolutionary theories—a kind of biological structuralism reflecting deterministic fractal orders on all scales.[7]

Whyte was politely tolerated, but appears not to have been taken seriously by most professional biologists in the fields his comments addressed. (He got perhaps more of a hearing in England than in the more "empiricist" United States). Reading his arguments one marvels that anyone took him seriously, even when he was "on the right side." Imprecise and general complaints and a hope of mathematical equations do not a theory make.

But it was characteristic of the time that some would listen to him and perhaps take him seriously. It has persisted to this day. Even though his positive account was stillborn, he is widely (and incorrectly, as Rasmussen (1991) and Churchill (this volume) demonstrate) credited with having first pointed out the importance of "internal factors" in evolution.

This was a period of ferment with hopes that new post-war high-tech developments—information theory (Quastler 1953, Gatlin 1967), hierarchy theory (Koestler and Smythies 1970, Weiss 1971, Pattee 1973), general systems theory (Von Bertalanffy 1968, and others),[8] and game theory (Lewontin 1961, Maynard Smith 1982) might provide new tools to think about biological organization and its evolution, as well as complex organizations everywhere. Waddington's Villa Serbelloni Conferences, published in four volumes as *Towards a Theoretical Biology* between 1968 and 1972 were the highest-profile systematic explorations of these themes, but there were many such conferences from the late 1950s to the mid-1970s, each with a volume, and each looking for a new angle. Development fit happily here as another deep problem which people hoped to attack with new systemic theory, and many of those with tastes for general systemic theories started with development.[9]

As luck would have it, I met Whyte. Stanley Shostak, a developmental biologist at the University of Pittsburgh, hosted him in the fall of 1967, while I was a student in Shostak's course. I had read widely in the "complex systems" literature in the preceding 4 years,[10] and remembered Whyte's lecture as mildly interesting but unimpressive, and almost totally free of data. Shostak had a reception for Whyte in his lab and I stayed and talked with him. His interests and approaches were uncommonly far from empirical issues, (almost "philosophical"). I discovered that he was a writer not a developmental biologist. I bought his book (from him!), read it, and was still underimpressed,[11] and astounded that a biology department would have such a speaker.

I recently asked Shostak how Whyte came to give the seminar. He thought that perhaps Whyte had volunteered to talk while traveling

through, and was taken up on it. Shostak (who has a taste for scientific variety) volunteered to act as his host. I asked him how well he thought Whyte was received more generally. He said: "In 1973 I took a sabbatical with Michael Chance, Whyte's nephew. Michael intimated that Whyte was [angry] that I didn't keep up a correspondence with him, since, I gather, I gave him the only friendly reception he received in the US." (E-mail, October 11, 2002, telephone conversation, October 12, 2002).
   I return to this incident below.

EXPLANATORY INTERLUDE

The rest of this chapter is concerned with a particular kind of dynamical feature of development. This is sometimes read as an internal constraint on selection or development, though I argue (Wimsatt 1986, 2002) that it is a mistake to see it as solely or even primarily internal. But it is a deep and general architectural feature of developmental life cycles. It is a selective constraint, because violating it in increasing degrees produces, probabilistically, increasingly intense stabilizing selection. One could easily see it (as apparently both Riedl and Arthur did; I do not) as the kind of factor that Whyte sought.[12] The deepest significance of "burden," "generative entrenchment," and "morphogenetic tree theory" is that they address features of causal relationships in nature that inextricably yoke development and evolution together. They thus denote an absolutely central feature of evolutionary developmental biology, and properly, of evolutionary processes in general.
   The key intuition motivating Riedl's notion of "burden" (or my "generative entrenchment," or the development of Arthur's "morphogenetic tree" theory) is the following. Suppose you have a way of determining the sequential dependencies of the development and expression of elements (entities, processes, structures, behaviors—all traits) throughout the life cycle of an organism. Even without knowing the detailed organization of these dependencies, there should be a fairly robust trend: things with relatively little depending upon them should more commonly be modifiable without having far-reaching or large effects than those with many other things depending on them. With more downstream effects, there is a greater chance that something will go seriously wrong and the net effect will be seriously maladaptive. These assymmetries in size and probability of maladaptive effects makes more deeply burdened or generatively entrenched things generally more evolutionarily conservative. Degree of entrenchment should correlate inversely with rate of evolu-

tionary change. Finally, as argued by Riedl, Arthur, and myself, traits earlier in development are more commonly more burdened than traits appearing later. This usually produces greater evolutionary conservatism at earlier stages and predicts and explains the phenomena of von Baer's laws, giving greater similarity and taxonomic generality for (most) features appearing earlier in development. This we have in common, though we came to these intuitions along different trajectories, and have applied them to somewhat more divergent ranges and classes of problems.

BURDENING NEO-DARWINISM WITH MORPHOLOGY AND GENERAL SYSTEMS THEORY: RUPERT RIEDL

Shostak's reading of Whyte's reception in America sets the stage for Rupert Riedl's far more significant contributions through the concept of "burden." The bare idea of burden or generative entrenchment is less significant than the framework for evaluating and applying it, how it interacts with other factors in development and evolution, and the problems he convincingly addressed with it. His theoretical apparatus gave him purchase on many crucial problems of evolution, development, and the structure of phylogeny and biological organization. On these grounds, Whyte got nowhere, and it is surprising that he garnered so much attention. On these grounds it is also surprising that Riedl didn't get much more attention. But the puzzles of their receptions are related.

Rupert Riedl (1975) was a comparative morphologist who studied with Konrad Lorenz. Most of his scientific career has been in Vienna, but he spent several years in the U.S. in the late 1960s and early 1970s at the University of North Carolina. His large edited volume on the *Fauna and Flora of the Adriatic* (Riedl, 1970) appeared while he was there, and most of the 1975 book was apparently also written there. This gave him a local following for his developing theory there and at Duke.[13] He was well respected in Europe,[14] and got his book translated quickly (1978), though perhaps not entirely literally.[15]

Whereas Whyte had many complaints but no theory and no examples, Riedl had a rich theory with multiple intersecting and mutually supporting conceptual and theoretical tools illustrated with diverse and deep empirical examples. Inspired by the progress in numerical taxonomy, he also made serious attempts to operationalize his theory. Although his primary examples were from morphology, he had the tools to work on dependency in processes generally, and was explicitly aware of their broader applicability to gene action, physiology, behavior, or even to cognition, technology, and culture. His book is a major attempt to use "burden" to explain everything from why phylogenies and biological organization are hierarchical (in different but related arguments), to why we think hierarchically, to the forms of our culture and our artifacts.

But despite these sound scientific virtues, there were problems. Riedl's theory began with information theory and preached broad scope and high theoretical tone, sometimes verging on philosophy. It promised a new "general morphology" to a shifting audience more appreciative of the explanatory power of population genetics.[16] Population genetics had gained a strong hegemony over evolutionary studies in the fifteen years since Whyte had first sounded his complaints: new theory seen as competing was ignored or received a hostile reception. (Development was in the former category, higher-level units of selection in the latter.) General systems theorists had provided another decade of vacuous pronouncements and increasingly repetitive "suggestive discussions." By the mid-1970s most biologists were suspicious of anyone using this vocabulary,[17] not to mention that of nineteenth century structural biology. Riedl adopted both.

THE ARCHITECTURE OF RIEDL'S THEORY

Riedl's theory is developed in a quasi-foundational way, using information theory to characterize "decision" strategies of the genes. He pays close attention to how minimal units are combined, and how to quantify burdens put upon them by the higher level or dependent parts and interactions, ultimately to aggregate these upwards, embracing biological order at all levels. He begins with a discussion of concepts of order, information, and probability. In chapter 2, he develops a notion of a decision hierarchy, and the amount of information it contains, and (in chapter 3) the molecular basis of such order (the kind of gene control switches one can find and how they can be organized). These chapters cover ninety-five dense pages, too long for anyone not already convinced of the value of this approach. The flavor is formal and reductionist, providing many definitions and a bewildering array of symbols (three pages of them!), though relatively few equations.

The next four chapters develop and articulate four major patterns (or dimensions) of order. These provide a matrix for the dependency relations quantified as the theoretical construct of "burden." This framework gives tools for rich descriptions of biological organization from a developmental perspective. It allows qualitative and sometimes quantitative predictions of the relative evolutionary stability of different traits, and to articulates the claims Riedl makes about the hierarchical orders of nature.

The first pattern of order is the notion of a standardized part: one repeated many times, becoming a module that is both fixed in basic architecture as other things come to depend on it, and modified elaboratively as it is co-opted for different functions. Macromolecules and cells are standardized parts, as are vertebrae, kidneys, and cerebral hemispheres. The burden on a standardized part is a product of its hierarchical position (how many things above it and depending upon it), the number of parts of that type (the relative fraction of the organism devoted to it),[18] and the number of functions these standardized parts serve in different higher-level structures. This discussion bears upon the concept of serial homology and several kinds of modularity emerging as important in current discussions of evolution and development (Schlosser and Wagner 2004).

Here and in subsequent chapters he opens with a series of thought experiments. He first considers a case that meets the conditions for being a "standardized part" to a high degree, and then relaxes the conditions to exhibit a range of patterns arrayed along that dimension. Thus for the standard part he considers (in figure 10.1) there is a series of abstract cases arrayed along a dimension from complete standardization of structure and position, to gradual weakening of standardized position or arrangement, to form of the parts, and finally to size.

He then turns to more biological realism (figures 23–27 in the book), showing variation in standard parts within and across species, using teeth, hair, flagellae, vertebrae, and bones in the hands of different quadrupeds. Throughout the book, there are easily five rich detailed comparative figures of morphological structures for each one as abstract as this, but this abstract figure suggests the generality, abstraction, and foundational conceptual character that Riedl sought. In places, his presentation is more like philosophy than science. But his discussion illuminates the problem, so at least here, philosophical analysis is yoked effectively with scientific analysis. His development in some ways anticipates the lovely conceptual and scientific work of Dan McShea (2000, McShea and Venit 2001) on the notion of a "part" for use in his comparative analyses on the evolution of complexity.

The next pattern of order is hierarchy; in taxonomy, in biological organization, and in our thought processes. This is the longest chapter, over sixty pages, and it contains the richest development and application of "burden." Riedl begins with a taxonomy without hierarchy, adding in features to give one, and comparing the phylogenetic trees one would get. The illustrative taxonomic hierarchies are done with Sokal's imaginary

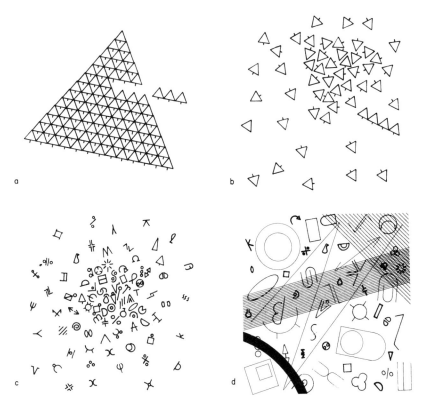

Figure 10.1
A graphical attempt to dissolve standard-part order. (a) Standardization of structure is complete and standardization of position is almost complete. (b) Standardization of structure is complete, but standardization of position has almost disappeared. (c) Standardization of structure is disappearing, remaining only as the standard characters of "symbol," "size," and "line thickness." (d) Even standardization of size and line thickness has disappeared. After Riedl 1978, figure 21 a–d, p. 96.

"animalcules" (see figure 10.2). This demonstration of a taxonomy for imaginary organisms without burden illustrates how strongly our phylogenetic data show the strong influence of burden in the production of new types in evolution. Riedl considers rank hierarchies at length, as well as control hierarchies, but his discussion is richest in recognizing the complexity of the problem of biological organization.

It is possible to describe an organism as an inclusive hierarchy of cells of different types (as one might do if focusing on the "standard part hierarchy"), but as Riedl notes, this would be seriously misleading:

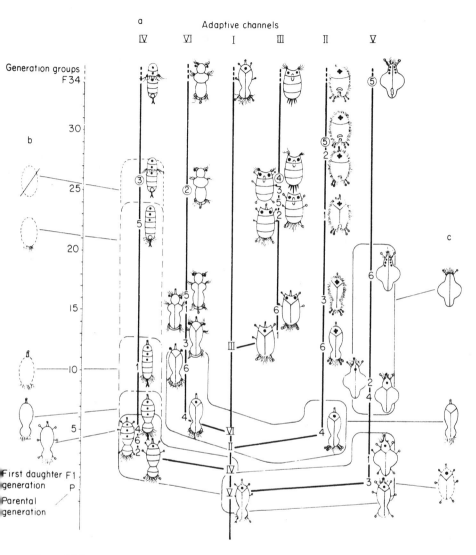

Figure 10.2
Disappearance of phyletic similarity, given equal prospects of mutational change in all features. (a) The occupation and adaptation to adaptive channels II–VI, as decided by the dice, is reached by accident in 34 generations. Decisions causing alterations are marked according to generation and channel. (b) As a result, at each step a sixth of the resemblance to the original form is lost. (c) Four agreeing character states occur in a chain of three, or at most four, of the closest related mutants. After 34 steps no relationships can be recognized. After Riedl 1978, figure 32 a–c, p. 120.

The arrangement of systems . . . is complicated. It is not a mosaic, but a matter of organization. . . . It may seem inconsistent that the chains of hierarchy of homologous systems are included within each other like boxes within a box, while the limits of neighboring systems can overlap or even coincide. This overlap is not accidental, however, but merely a consequence of functional stratification. Within themselves, the systems are hierarchically arranged, box within box, when we are thinking of partial functions, (supportive system, vertebral column, cervical vertebral column, atlas, ontoid process, ventral articular facet) (figure 10.3). But they overlap when the functions cut across each other (nervous system, musculature, limb, skin, vascular system). (Riedl 1978, p. 125)[19]

On the preceding page, Riedl advanced (in another context) a partial solution to understanding the source of this complexity, and something of its character: *"The step-wise fixation of features is therefore crucial for every hierarchy, and indeed the definitive fixation of newly added features."* (1978, p. 124, his italics).[20]

He is advocating evolution through "a patchwork of layered kluges" (Wimsatt, 2007, ch.7), evolution as a series of layered contingencies added to development, with ones near the surface sometimes being removed or substantially reworked, but things sufficiently layered over increasingly and irreversibly frozen in by developmentally downstream adaptive modifiers, and possibly modestly transformed, especially if co-opted also for different functional systems.

The picture of evolution as a succession of kluges reaches back to Darwin's discussion of the adaptations of orchids, and has had multiple prior partial elaborations in the literature (e.g., Jacob, 1977). Against the background of Gould and Lewontin (1979), Gould and Vrba (1982) elaborated it enough to come to own it in the public eye. This process confounds efforts to find a neat, well-behaved tree-like organization for functional systems, and for it to be seen simply as an optimizing process (Wimsatt, 2002c). This continually functioning and functional midden of organic design is the home of burden and generative entrenchment, and why it is so difficult to analyze.

For his third pattern of order, interdependence, Riedl's thought experiment reminds us that Hieronymus Bosch "combined known structures to produce unknown, absurd, impossible forms" (1978, p. 180). Figure 52 in the book depicts Bosch's menagerie of ill-fitted organisms, parallels Empedocles' first selectionist evolutionary or developmental theory. And Aristotle found Empedocles' solutions as much "against nature" (and contrary to development) as we find those of Bosch. The parts of organic

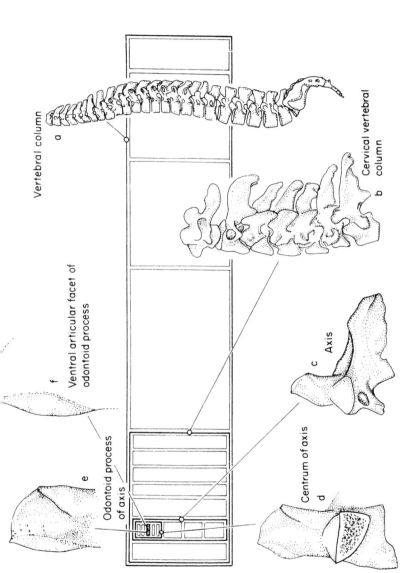

Figure 10.3
The hierarchial organization of homologues. The hierarchical arrangement of homologues illustrated by a hierarchical series in the human skeletal system. The series consists of five cadre homologues (vertebral column to odontoid process, shown as thickly drawn rectangles in the diagram) and a minimal homologue (ventral articular facet of the odontoid process, stippled in diagram). After Riedl 1978, figure 11 a–f, p. 40.

systems obviously are adaptively interrelated, and it is the functional futility of the way that familiar parts are misassembled that is so striking in Bosch. Adaptive organization (and the burden induced by disorganized aggregations) are again manifest. But interdependence has other consequences. The causal mappings between gene and character, if unconstrained, would make evolution orders of magnitude slower or impossible.[21] So Riedl discusses pleiotrophy, one gene affecting multiple characters, and polygeny, multiple genes affecting single characters. Why do they not produce a thicket of overlapping causal and adaptive connections together frustrating any attempts at change? (This is a central topics of current evolutionary developmental biology (See Schlosser and Wagner, 2004), but was long ago raised as a vexing and unsolved problem within population genetics by Franklin and Lewontin 1970). Riedl's answer (p. 185) is that the network is organized so as to show purposive interdependence. This would satisfy no one today. From this it does not follow without further argument that "All successful change is organized," as he titles a section on page 189. He invokes D'Arcy Thompson's "transformation grids," where multiple characters change together in a coordinated way to produce a new morph adapted to a different ecological niche.[22]

The topic of coordinated change leads him to homeotic mutants, including *bithorax*. While the mutant is not functional, it is a partial throwback to the ancestral four-winged state. There is a coordinated change of many characters with change in segment identity, and many of the changes that do occur make coordinated functional sense. Analysis of the homeobox mutants, beginning in the late 1970s, gave developmental genetics its largest push toward theoretical centrality, and drew a host of new researchers to it. Comparing the homeobox systems across various phyla has given striking examples of processes of segment specification and differentiation, of co-opted function in new contexts. At the same time they have shown relative conservation of developmentally generative systems, as well as the impact, however achieved, of evolutionarily macroscopic architectural changes accompanying duplication and differentiation of the *Hox* clusters throughout the animal kingdom.

Riedl's fourth and final pattern of order is "tradative" order, which "adds the time axis to the three principles already discussed" (p. 203). For a situation totally lacking traditive order he considers the Greek myth of Daphne metamorphosing into a tree. (1978, figure 61, p. 203). Daphne (or her descendants) could no more evolve into a tree than she could metamorphose into one. (Even with an excessive number of friendly macromutations, the distance is just too great!) Riedl considers the inheritance and evolutionary stability of developmental control structures as so central in function that their burden makes them stabler than all else. This too is partially confirmed in the latest work (Davidson 2001). Indeed, where elsewhere he has recognized that it is von Baer rather than Haeckel who is correct, here he backslides: "At the beginning of this chapter, I proposed to take Haeckel literally. Accordingly, I now postulate that the biogenetic law holds not only for the pattern of events, but for the decisions behind them. The epigenetic system is a brief recapitulation of its own origin." (p. 212, his italics)

Why play Haeckel, even as a working hypothesis? Riedl has already noted that if things are quickly frozen in, one would get a picture with near-absolute recapitulation. It is now generally recognized that significant evolutionary changes even within lower taxa appear to act more upon control than on structural genes, so it might seem that fixation at the higher taxon levels must involve fixation of control functions at higher levels. And it is plausible to do so, Wallace Arthur (1984, 1988, 1997) builds his theory primarily around genes involved in the control of development.

A DEVELOPED EXAMPLE

What does this all come to? Riedl presents many figures and tabular summaries keyed to his taxonomic and morphological work. Figure 10.4 shows variations in the number of homonomous bones in the hands (forward limbs) of different tetrapods. He argues that successsive evolution in the mammalian lineage has fixed features in order, generally from more centrally located to more peripheral. In this delightful example the dependence of peripheral on more central is literal and visually obvious, Moreover, since Darwin and Haeckel the tree has been the root metaphor both for dependencies in development and for phylogenetic descent (Griesemer and Wimsatt 1989, Richards 1996).

This evolution and complex dependency of parts results in a trajectory of change from first innovation (if preserved) to increasingly deep fixation,

> . . . the tricodonts continue the mammalian line as the first primitive mammals. The number of fingers is fixated (apart from reduction) and there is extensive fixation of the carpals. Even in aquatic mammals there is no increase in the number of fingers (figure 10.4, whale). Even the

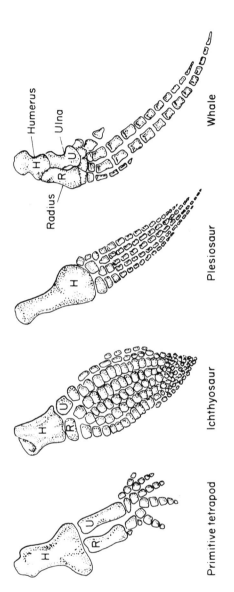

Figure 10.4
Change in the number of homonomous bones in the tetrapod hand. Ichthyosaurs, plesiosaurs, and whales all show an increase in the number of finger bones (polyphalangy). Ichthyosaurs show an increase in the number of fingers (polydactyly), while whales show a decrease. After Riedl 1978, figure 27, p. 109.

number of spinal nerves that take part in the brachial plexus is limited, which again is connected with the high fixation of the number of vertebrae. Only the distal end of autopodium remains free, as expressed by the fact that the number of phalanges may increase.

In summary, therefore, the constancy of the features increases step by step in the course of evolution. According to the burden carried the constancy spreads outwards peripherally from a central position. Deeper lying features leave the zone of free adaptability as they are burdened by new features, in a state of free experiment, which builds on them. (Riedl, 1978, p. 158)

Riedl presents his summary analysis of and diagrams this particular example, the tetrapod limbs in the various dimensions of table E (figure 10.5). Here Riedl plots evolutionary trajectories of four different "fixation layers" (deepest: paired appendages through shallowest: girdles and nerves) as they change from variable to fixed (in five stages). He tracks their complexity (number of parts), integration, functional position (i.e., depth), prospect of substitution, taxonomic representation (level and constancy), and mode of alteration. He intends that we understand the pattern of this particular case (layering of fixation layers, and their multi-dimensional trajectories through time) as generic.

Of course there are exceptions (also noted): the reduction in number of fingers in marine mammals and in the evolution of the horse. And yet there are comprehensible reasons in deeper grounds for the possibility of reduction and the impossibility of increasing number of fingers: the impossibility of increased number of fingers, presumably linked to the conservatism in number of spinal nerves and number of vertebrae. This coordinated picture and table allow one to understand his otherwise almost impossibly dense table.

PROBLEMS WITH RIEDL'S ACCOUNT AND OTHER FACTORS AFFECTING ITS RECEPTION

Riedl did remarkably well. There are some genuine failures, other things that are easy to criticize (probably unfairly), and not a few ways in which his main flaw was to be out of step with the times, or unwilling to stay within his own discipline. His most visible sins (to American empiricist scientists) was to mix philosophy with science, and to make the common mistake of scientists: thinking that philosophy differs from science only in being broader, looser, less testable, and less disciplined. But let us consider substantive criticism first.

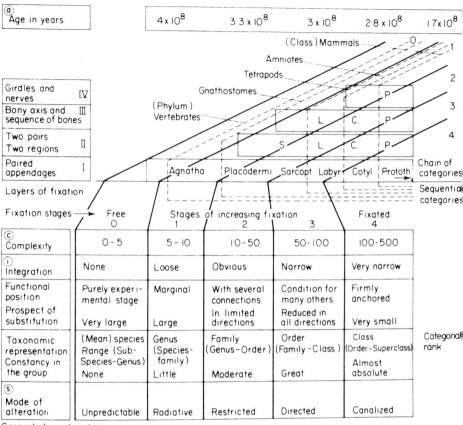

Figure 10.5
The fixation path of a group of features as illustrated by the mammalian limb. The figure illustrates the building up, step by step, of the fixation layers I to IV in passing through fixation stages 0 to 4 of certain limb features, within the systematic groups closest to the ancestors of present-day mammals. In the sequence of these groups, the class or subclass names are better known and have been used instead of the family names. Thus the agnatha correspond to the Astraspidae; the placoderms, to the Arctolepidae; the sarcopterygians, to the Osteolepidae; the labyrinthodonts, to the Elpistostegidae; the cotylosaurs, to the Romeridae; and the prototheres to the Triconodotidae. After Riedl 1978, table E, p. 157.

Riedl's most important problem is that his system predicts too much fixity through accumulation of burden over time. This may indicate too strong a focus on vertebrate evolution (see Raff, 1996; Arthur, 1997) and leaves him open to criticism based on cases where early developmental stages appear to have been quite changeable (Sander, 1983; Elinson, 1987; Doboule, 1994; and, most elaborately, Raff, 1996). Such counterexamples may be limited to specific phyla in ways suggesting specific mechanisms. Analysis of such cases could lead to refinements in the theory of burden, which in different ways, is part of what both Arthur and I have tried to do. Riedl also fails to consider other important kinds of organizational features which produce more complex relationships with burden. (They appear in discussion, but with no indication that they can systematically reduce burden or entrenchment.) This is particularly clear at four points that have deeper theoretical reach.

First, although he discusses feedback and homeostasis in various contexts, Riedl seems not to recognize (1978, pp. 80, 82, 197–199) that Waddington's canalization is a mechanism for producing fixity or stabilization distinct from that of burden, not just the fixation resulting from burden. (Riedl's own definition of canalization is quirky and out of step here.) Indeed, canalization in Waddington's sense can oppose or reduce burden. It covers a broader category of homeostatic processes in which deviation from a canalized developmental trajectory is counteracted by internal adjustment mechanisms in ontogeny. Burden causing biological selection stabilizes only intergenerationally. Perturbation of something highly burdened through mutation, teratogenesis, or other causes, produces seriously maladaptive downstream consequences, and "deviant" organisms are removed by stabilizing selection. But here the (maladaptive) expression is *not* canalized in Waddington's sense.

Moreover, a state canalized at one stage of development can *unburden* or free upstream features to vary if the regulatory mechanisms can produce the canalized state across variations in the supposedly burdened feature. In "genetic canalization" (Lerner 1954), the phenotype in question is produced by a wide variety of possible genotypes, whose variable features are thereby unburdened (or disentrenched) relative to that trait. Gilbert, Opitz, and Raff 1996 suggest a large role for this, securing invariance or organizational modes at higher levels (e.g., that of the developmental field) across a variety of lower level genetic variations. Sexual reproduction should select for this kind of "canalization": the need to maintain reasonable levels of heritability in the face of recombination must be constantly selecting out genotypes which deviate too far

(Wimsatt, 1981, 1987:26, 2001, Stearns, 1994.) Wagner's recent review of robustness and evolvability (2005) discusses cases like this at all levels of organization.

With canalization and burden distinguished, we can recognize interesting evolutionary relations between them (Wimsatt, 1999b, 2002a). Canalized features should be more likely to become generatively entrenched than variable ones because their stability makes it both more likely and adaptive to accumulate modifiers that depend on their presence to function properly (Wimsatt and Schank, 1988; Wimsatt, 2001).

What about the reverse? If something is burdened, is it adaptive to canalize it? This ought to depend on the costs or difficulty of fixing different kinds of failures, and the benefits (and probabilities) of doing so: perturbations of deeply entrenched things might favor quitting early and trying again—apoptosis and spontaneous abortion. Whether or when one can go from entrenchment to canalization seems worth investigating, but not to be taken for granted.

Finally, since stability or invariance may in different contexts suggest either canalization *or* entrenchment, they are easily confused. Evidence of stabilization, without further qualification, counts as disjunctive evidence for either (or both!) against other kinds of hypotheses, but differential evidence for neither against the other. Thus, the main extant analysis for "innateness" unpacks it in terms of developmental canalization using as evidence the relative stability of the trait across different individuals and environments. A stronger case can be made for analyzing these phenomena in terms of generative entrenchment (Wimsatt, 1986, 2002a).

Second, Riedl suggests that presence of a greater number of standardized units indicates a larger burden, but the relationship is more complex. One must consider the impact of designed levels of redundancy in duplicated units. A greater number of standardized units may indicate greater required capacity for cells or organs of that type (marking a higher total burden), or serve primarily to increase redundancy or reserve capacity, thus reducing burden on individual units, and acting primarily to reduce the failure probability of that organ system.[23] In organic design multiple units may overlap in serving various functions (functional multiplexing), so design for reliability and redundancy can make burden very difficult to evaluate. One must know both how individual units can fail and how they are organized. (Wimsatt and Schank [1988, pp. 251–254] consider several kinds of [idealized] redundancy). If cellular reproduction is blocked in the only stem cell for that type, the burden should be larger the more cells should have been produced in that lineage. But what if extra cells represent redundancy and extra capacity in the liver, or alternative biochemical pathways allow metabolism under different circumstances? Redundancy is an important and common feature of organic design, and a crucial way of allowing change in otherwise fixed components—and thus of escaping burden. This is why gene duplication and subsequent diversification of function of the duplicated elements has played such an important—indeed, pivotal—role in evolution.

Third, Riedl considers both standard parts and interdependence, but fails to notice that the evolution of modularity may simultaneously produce greater standardization and reduced interdependence. Increased evolutionary near-decomposeability (Simon, 1962) or "quasi-independence" (Lewontin, 1978) may increase the modifiability of some parts of the phenotype in different directions, and decrease burden. Varieties of modularity and conditions under which they can evolve has become a hot topic in evolutionary developmental biology. Schank and Wimsatt 2000 discuss reduction of generative entrenchment through modularity, and Schlosser and Wagner ed. 2004 reviews the rapidly diversifying recent work on modularity in evolution and development.

Fourth, Riedl's conception of the fixity produced by burden is too internalized. (Lewontin and Caspari made this complaint against Whyte in 1960, and internalism is now much less taken for granted.) Evolution acts to increase fitness and its heritability. Heritability and fitness are both relational properties. There is no reason why increased heritability of increased fitness must be secured by the fixation of genes, or if it is, that it is anything other than a means for securing heritability of fitness (Wimsatt 1986). Burden or entrenchment should fall jointly on environmental and organismal features. Reception of the right environmental inputs can be as generatively entrenched as having all of the particularly pivotal genes, as the phenomenon of "phenocopy" demonstrates. The roles of genes are taken for granted, so one must continually emphasize the environment: the more we learn about developmental genetics the more the environment is seen as playing an architectural role in the developmental program. This is why Scott Gilbert (2001) now urges Eco-evo-devo as the right parsing for many problems internalized by Evo-devo. Along with this goes Griesemer's emphasis on the importance of reproduction rather than replication, and the current popularity of developmental systems theory (Oyama, Griffiths, and Gray 2001), and niche construction (Odling-Smee,

Laland, and Feldman 2003). The views defended by Riedl, Arthur, Wimsatt, and Schank would make relative generative entrenchment a better predictor of long-range evolutionary stability than either genetics or environment alone.

Finally, Riedl's attempts to operationalize his theory enter contentious territory. Doing so in these complex domains inevitably requires compromises, simplifications, and will fail for hard cases. These invite easy criticism, as the furor of competition between alternative approaches to taxonomy has amply illustrated (Hull 1988). The complex dependence of burden on different modes of causal interaction and organization in development could make any attempt to quantify burden appear hopelessly oversimplified. This is a dilemma, because one must operationalize to apply the theory.

But we must also consider how the quantified measures are used. It would be problematic if Riedl's conclusions depended too strongly on detailed calculations of burden. But he used numerical indices primarily illustratively, and I have not seen any cases where he argued a point with finicky detailed numerical calculations. He most commonly made comparative judgments. These are usually more robust, especially when applied to organisms from the same lineage, avoiding heterogeneous variations that would endanger cross-phylogenetic comparisons. Even ordinal comparisons can be compromised by systematic errors (some made possible by the theoretical problems discussed above), but Riedl does remarkably well here.

So why did Riedl not get more serious consideration? This is not the nineteenth century: few evolutionary or developmental biologists could (or would have been motivated to) follow Riedl in German. Their first exposure to his book would have been Riedl's paper in *Quarterly Review of Biology* in 1977, presaging the English translation. Riedl could have taken burden, illustrated it with a few key examples, and suggested how it articulated and fleshed out the four basic patterns of organization, (standard-parts, hierarchy, interdependence, and tradition), through which he develops the rest of the book. But he did not. Instead, he attempts a fifteen-fold condensation of the book, skipping few theoretical or philosophical points, but eliminating over 300 rich and illuminating diagrams, illustrations, and tables, and detailed coordinated examples which make his intricate theoretical arguments far more comprehensible. (The paper, unlike the book, contained no illustrations!) His examples are much more accessible than the theory, and certainly more accessible than the almost totally non-specific and general presentation of his article. They could have convinced many readers there was something systematic there worth attended to, even if they didn't follow or chose to reject many of Riedl's claims.

What *didn't* Riedl leave out of the article announcing the book? Attempts to show that his hierarchical view of nature applied directly to thought and culture, and ultimately to the cosmos—all as extensions of the nature of evolutionary systems. He comments on the nature of biology, the natural sciences, epistemology, and the place of man in nature—not in one metaphorical sentence at the end but in separate long paragraphs, where he hints at having solved each of the major philosophical problems in these areas. And it doesn't help that he cites with approval the usual suspects we met in discussing Whyte: Weiss (three books) Koestler and Smythies (1970), and Whyte (1965). A few more reputable authors intermingle with Hume and Kant, (reputable, but not in a science paper!), on the way to Bergson and Teilhard de Chardin. In the strongly empiricist ideology of that time, and even today, such references (especially the last two) would discredit a work as serious science. "Real scientists" would barely admit to knowing who these people were—or perhaps just well enough not to read anything in which so many of them are cited.

WALLACE ARTHUR AND MORPHOGENETIC TREE THEORY

As an undergraduate in biology, Wallace Arthur had an interest in morphology which led him to study left- and right-coiling varieties of the snail *Cepaea* and wonder how the morphological switch from one form to the other could work. He did ecology and population genetics in graduate school, and embarked on a normal career trajectory for a new population biologist—having nothing to do with development. Then an accident brought him back: when searching by page number for an article by Kimura in *Scientific American* in 1979, he got the wrong month and ended up looking at *hox* genes and compartments in *Drosophila* development in an article by Garcia-Bellido, Lawrence, and Morata. The rich systematic patterns of segment specification re-interested him in morphology, led to an interest in the origin of *Bauplans*, and ultimately to his groundbreaking 1984 book: *Mechanisms Of Morphological Evolution: A Combined Genetic, Developmental and Ecological Approach*. He went on to write two other books developing his account of morphological evolution as affected by "morphogenetic tree theory." His account of developmental dependency is now the most fully developed such account for biological evolution, and its interaction with other organizational factors is richly

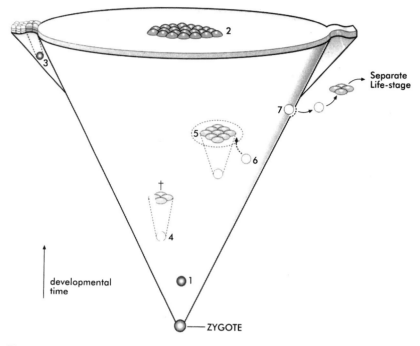

Figure 10.6
A more complex view of the "developing organism as inverted cone" model. (1) Cell in which gene A is first switched on. The descendants of this cell (2) form the center of the disk. (3) Cell in which gene A is reused to mark the proximal end of the appendage. (4) Cell (in which gene B is active) that proliferates slowly and whose descendants die, resulting in a negligible effect of gene B on the adult disk, despite its earliness of activity. (5) Clone of cells in which gene C (marking a mobile morphogen) is active. This gene's region of effect is greater than its region of expression. (6) Cell that migrates and comes under the influence of the product of gene C (7) Cell leading to another life stage, when the disk is considered to be a larval stage of a complex life cycle. After Arthur 1997, figure 8-3, p. 196.

developed. He is now justifiably recognized as its major spokesman. (See, e.g., Arthur, 2002).

His picture of developmental dependency is elaborated in an "inverted cone diagram" (1997, fig. 8–3, p. 196; figure 10.6), reflecting the spreading dependencies in development in ways suggested by but not limited to cellular descent diagrams.[24] This figure also reflects the qualifications required to deal with the various outcomes possible for cells and their products—many of them elaborated and more fully documented in the intervening 13 years.

Unlike those who came to evolutionary developmental biology through morphology or developmental genetics, Arthur brought a population geneticist's perspective on evolutionary change. One must ask how large "organized" mutations (such as the *Bithorax* series) happen, but one had also to wonder how such major changes could be sufficiently adaptive when they first occurred to get established in a population. His basic proposal has remained the same through all three books. He focused on developmentally significant control genes (or "D-genes"), and proposed a special mode (or set of circumstances) through which they might become established (his "n-selection"). Almost all D-gene morphological mutations would be lethal. Even if one were viable and if it had consequences that were immediately salutary, a morphological mutant would not be adapted to the existing functional order, so commonly it should show a net decrease in fitness. If it survived to become established, a period of integrative "fine tuning," through added "downstream modifiers" would then increasingly entrench the new D-gene.

But this is not yet an adequate account: if it were not to be outcompeted before it had a chance to improve its "tuning", the new mutation had to occur in a small isolated population, and in other ways be in a period of relaxed competition. There are several ways to do this. It might initiate exploitation of a new niche, or be in a colonizing situation, or recent ecological disaster might have removed competition and cleared new opportunities (figure 10.7). Our simulations (Wimsatt and Schank, 2004) point to relaxed competition and small population size as critical factors in making changes of larger selective effect, confirming at least parts of Arthur's proposed mechanism, and doing so in a manner consistent with the "shifting balance" picture of Sewall Wright (1931).

Arthur retained the population geneticists' concerns in his middle (1988) and latest (1997) book on this topic, as well as his concentration on morphological evolution, but the population genetics has grown successively more muted as the role and knowledge of developmental genetics have grown. Arthur's 1984 book is in some ways like Morgan's 1934 book: two separated discussions of fields that one knew must be combined but without knowing quite how. Though much more had been learned in the intervening 50 years, the gap was still large. In (1997), Arthur is able to review and selectively to illustrate a great deal of developmental genetics that was totally unknown in (1984), showing the interactive complexity of development, while still confirming the broad outlines of the picture one would expect: greater freedom to change later in development than earlier, but with various mechanisms of escape from any simple pattern. His discussions

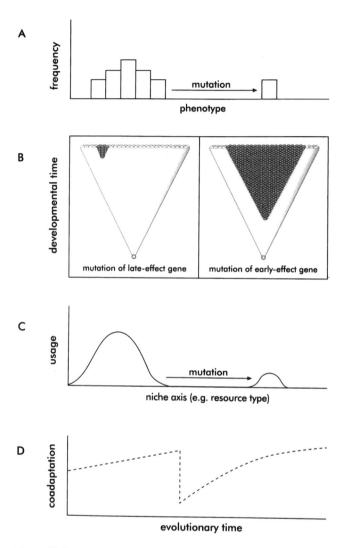

Figure 10.7
A model of the invasion of a new niche in early animal evolution. (A) Production of
a cluster of mutant progeny by mutation; (B) effect of this mutation on development
(right), contrasted with the effect of a less major mutation (inverted cone model,
section, affected cells shaded); (C) niche shift that is brought about fortuitously by the
mutation; (D) pattern of change in internal coadaptation over time. After Arthur 1997,
figure 9-4, p. 230.

look much less like the schematic developmental cascades (or cell lineages) of his first book. Instead his discussions are full of detailed developmental mechanisms or sketches of them, the richness of interactive regulatory mechanisms strengthening both the claims for massive developmental dependencies, and the revealed difficulty of teasing out the exact details through multiple levels of organization of interacting systems. Even without a complete account (and he shows an interesting new healthy skepticism for whether a complete account is required), he has gone a long way towards delivering on the kind of "bottom up" sketches hinted at in the early chapters of Riedl's 1978 book. Intriguingly, as the detail grows, one sees more regulatory loops, switches, modulations harking back (with new components and tools) to the heyday of systems talk and cybernetics. Riedl's systems theory approach is thus also confirmed. Other perspectives even to Davidson's resolute reductionism (2001) confirm this picture: Davidson's molecular garden is populated by genetic control circuits.

Arthur has now contributed more to such a theory of entrenchment, yoking developmental processes to the dimensions of evolutionary stasis and change in biology, than any other writer, both by his theoretical work and through his professional activity. As his acquaintance with the developmental genetics has grown, so has the visibility and perceived importance of the fixed pivots and directional biases generated by developmental dependencies. He has also become more central to the discipline, becoming one of the editors of its new journal, *Evolution and Development*, writing a broad and deep review of the theoretical structure of the discipline in *Nature* in 2002, and even a popular account in 2004. In 1984, his book had something of the character of a quirky outsider-strange and speculative despite its obvious competence, in part because these domains were incommensurable and he was trying to tie them together. By 1997, his most recent book had become another centrally important addition, and the only one combining developmental and ecological arguments from the perspective of a population biologist. Part of what had changed was the books, but part of what had changed was the emergence of a new discipline that made the earlier book relevant.

Rudy Raff (1996, p. 21) has a revealing list characterizing differences in perspective between population biologists and developmental biologists, but Raff's development of the material still largely follows the latter. Developmental geneticists now listened because Arthur had become a native speaker, even if he had not initially joined their experimental explorations of developmental hardware. The theoretical perspective of burden, morphogenetic tree theory, and generative entrenchment has come of age.

Arthur has continued to introduce new models and metaphors to better capture the perceived changes in architecture. Not giving up on developmental dependency, his newer work expands to recognize its interactions with other modes of organization, a necessary elaboration and enrichment of the theory. He now (2002) talks about "genetic reprogramming" to better capture the changing dynamics as older components are reused in new contexts (a point emphasized by both Jacob (1977) and by Gould (1989) and now common currency in talk about evolution.) Arthur now (2004, personal communication 2003) talks about "biased variation" rather than constraint, as a more dynamical picture of the way that development affects evolutionary direction. This "reprogramming" is perhaps the most striking new pattern to come out of the enormous databases produced as spin-offs of the human genome project. Predictably, the ability to do phylogenies of particular genes has been enormously enhanced; tracking the retuning of function down descendant lineages. But the ability take a sequence and look for all close matches across organisms and across functional systems has revealed a wealth of related genes in functional systems quite unlike the presumed home of the sequence with which the search was started. And this has reinforced the perceived importance of gene duplication and subsequent divergence of function of the new redundant copies. Their discovered sequence similarity comes as a genuine surprise to investigators not expecting the degree of functional elaboration they have found. An awareness of redundancy qualifies the entrenchment, because an additional copy allows change of one or the other without pain of failure, but a discovery of common descent of such diverse genes reinforces the picture of evolution as a series of layered and incremental kluges, with rich dependent linkages throughout.

In his 1997 book, Arthur also generated major clarifications of the confusions surrounding Von Baer's laws—now regarded as patterns, rather than laws. He suggests (p. 258ff) that we distinguish ontogenetic and phylogenetic patterns in morphology, lineage, and genetics, yielding six hierarchies to be related, not two. They come the closest to fitting von Baer patterns, he finds, for the genetic hierarchies, and he also argues (in a point paralleling Darcy Thompson's comment on his own transformation grids) that von Baer relations will often fit better locally, within a sufficiently small phylum, or for one particular life stage, or for life-stages separately in a complex life-cycle. Von Baer's laws get a closer analysis at his hands than they have anywhere else. "Von Baerian divergence is a taxonomic statement [and he argues a fairly restricted one] and not a generally applicable law." (1997, p. 275)

Redeveloping Evolution—Wimsatt, Schank, and Generative Entrenchment

I hit on the idea of the developmental lock (DL) in the winter of 1972/3, when I was teaching Herbert Simon's 1962 classic, "The Architecture of Complexity", to a class on philosophy of psychology. Simon argues for the importance of hierarchy in adaptive organization, and that more complex systems are made out of simpler ones. Simon even refers to the biogenetic law, but this formulation didn't follow from his models. He had justified an argument for systems that grow through aggregation by putting together stable subsystems, not for ones that developed by growing and differentiating (Wimsatt 1974). The developmental lock (figure 10.8) is a hybrid of two different model "locks" considered by Simon. Unlike them, it could apply to developmental systems. I quickly realized that it predicted a pattern of greater conservatism in evolution for features expressed earlier in development.[25] I first presented it (with work on reliability in hierarchial series-parallel networks) at a conference of the American Academy of Arts and Sciences in April 1974. The conference papers were not published, and the idealizations in the model appeared so extreme that I thought it wise to develop it further and look for convincing applications before publishing it.

I tried it in classes and conferences and collected data and ideas for a decade. I didn't think of morphology or *Bauplans* initially: my first target was elsewhere. I saw generative entrenchment (or GE) as a way to capture developmentally the phenomena for which the innate-acquired distinction is normally invoked (in animal behavior, psychology, linguistics and philosophy),[26] and similar relations in scientific and technological evolution, in which some assumptions, principles, or technologies become increasingly fixed, standardized, foundational and unquestioned as other things are built upon them. GE is first discussed in print (four pages of a paper on units of selection) in 1981. A joint paper with Robert Glassman on early developmental plasticity in 1984[27] considered other ramifications. An extended treatment of GE and my first ("developmental lock") model appeared in 1986 in a paper suggesting broader application, but focused to give alternative tools to dissect the innate-acquired distinction.[28]

This paper led quickly to three others by or with students: Nicholas Rasmussen (1987) used series-parallel networks of simple "developmental locks" (suggested as a possibility in the 1984 paper and by the data) to analyze relative times of action and dependency relations of twenty-two genes in *Drosophila* development. This demonstrated that generative

Number of combinations = 10^{10} =

$$10 \times 10 \times 10 \times 10 \times 10 \times 10 \times 10 \times 10 \times 10 \times 10$$

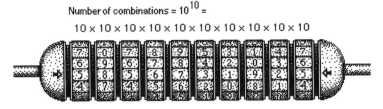

Figure 1a: Simon's (1962) "complex lock":
10 wheels with 10 positions per wheel. In the "complex lock" the correct combination is only discoverable as a complete solution. (No clues are given for
partial solutions.) Expected number of trials = $10^{10}/2 = 5 \times 10^9$.

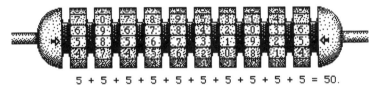

$$5 + 5 + 5 + 5 + 5 + 5 + 5 + 5 + 5 + 5 = 50.$$

Figure 1b: Simon's "simple lock":
Just as above, but a faint "click" is heard when each wheel is turned to its
correct position, thus allowing independent partial solutions. Expected number
of trials = 50.

(The advantage of near decomposeability in problem-solutions is the ratio of
expected number of trials for the two locks = $5 \times 10^9 / 50 = 10^8$!)

lower contexts (components)   higher contexts (fitness)

1   2   3 ········· m ··························10

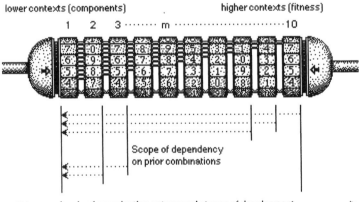

Scope of dependency
on prior combinations

≫≫ ··········· levels of organization or temporal stages of development ···················→

Figure 1c: Wimsatt's Developmental Lock:
(Simple if worked from left to right, but complex if worked from right to left).
Suppose a "click" is emitted by each wheel when it is in its "right" position,
but what position is "right" is a function of the actual positions (whether
right or not) of any wheels to the left of it, so that a change in position of
any wheel randomly resets the combinations of all wheels to the right of it.
(from Wimsatt, 1986)

Figure 10.8
The developmental lock. After Wimsatt 1986, p. 193.

entrenchment could be used as a theoretical tool to infer the structure of developmental circuits. Rasmussen's paper added new clarity and operational criteria for how one could apply the model, and for the first time yielded new data and highly specific predictions from it (see figure 10.9).[29]

Jeffrey Schank had wanted to test some of Stuart Kauffman's (1985) ideas on the evolution of gene control networks while modeling the effects of differential generative entrenchment, thus testing my ideas at the same time. This initiated a major focus of our research. Kauffman's networks exhibited differential GE (also a generic property—manifestly so in Kauffman's own figures—but one not utilized or included in his arguments). These network models also allowed dropping some of the constraining assumptions of the "developmental lock" models. They also for the first time made it possible to construct population genetic style evolutionary models incorporating the effects of generative entrenchment, thus connecting developmental genetics and population biology.

We pushed the theoretical analysis further with simulations (Schank and Wimsatt, 1988), modeling populations of twenty-gene, twenty-connection "gene-control networks" (the size Kauffman used), and then with more realistic 100-gene, 100-connection models (Wimsatt and Schank, 1988) and later with up to 260 connections with different parameters and architectures (extensive simulations done in 1987–1991, partly described in Wimsatt and Schank, 2004). With software designed and written by Schank, we confirmed, qualified, and contested different of Kauffman's conclusions. Kauffman had claimed from his simulations that selection could not maintain large multilocus systems in the face of mutation, so that self-organization was required to explain biological organization. (His arguments were general and theoretical and derived from those in the 1960s and 1970s over "genetic load." They were not targeted against specific complex adaptations). We specifically considered diverse "genetic load" considerations, and made new proposals that led to strikingly different outcomes.

Our work showed that with more realistic models,[30] many many more loci could be maintained by selection than Kauffman had supposed. His arguments against the selectionist paradigm were based on fragile or incorrect assumptions. In Nature, solutions most likely involved both selection and self-organization working in concert, with GE structuring selection. More interestingly, the number of loci maintainable by selection appeared to be augmented by *differential* GE, a result not anticipated in any of the discussions of load. Our simulations demonstrated first that GE'd gene activities were differentially preserved in evolution, with the losses

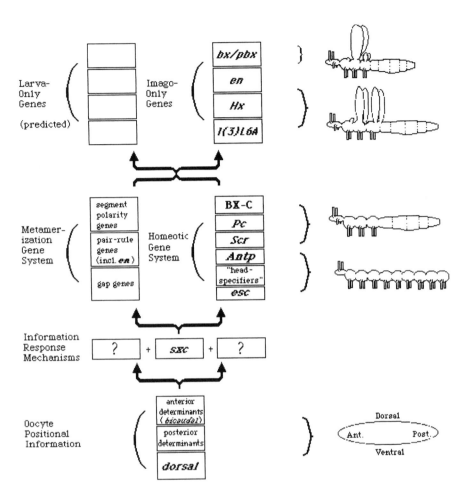

Figure 10.9
A summary of the generative entrenchment relations between those genes and gene functions of the *Drosophila* developmental system which are discussed in the text, with greater entrenchment toward the bottom of the figure. To the left of each group of genes is the name of the major subsystem to which they belong. Arrows indicate direction of information flow in "program." To the right are representations of the forms generated by the indicated gene functions. These approximate the sequence of fruit fly evolution. After Rasmussen 1987, figure 3, p. 293.

occurring completely among less entrenched connections. They also suggested that GE was crucial to the evolution of increasing complexity. It appeared (Wimsatt and Schank, 1988) that four factors acting over evolutionary time (two driven by GE) could inflate the number of genes maintainable by selection (and thus the size of evolving genomes) by two to five orders of magnitude.[31]

We argued conceptually as well as with simulations. Differential generative entrenchment is a generic property in Kauffman's sense: it is statistically inevitable, spontaneously generated, and does not require selection to maintain it (Schank and Wimsatt, 1988). But standard processes of selection and mutation in finite populations made it inevitable for a second reason (Wimsatt and Schank, 1988): even if one *could* start with a structure of gene consequences which were totally egalitarian or symmetric, randomly targeted mutation or selection would "break symmetry," creating differential GE in the structure, and conditions for further differential preservation, elaboration, and entrenchment of some of the circuit elements—amplifying the original differences. This put all subsequent arguments about differential generative entrenchment on a new footing: it was not a property which might be there or not, but an absolutely fundamental feature of the evolutionary process. In later papers, (Wimsatt, 1999, 2001) I added two minimal and generic developmental assumptions to "Darwin's principles" (Lewontin, 1970). These generated additional intuitive features of living things such as the existence of life cycles, and provided a foundation for generative entrenchment models as general as population genetics—indeed more so, since they didn't require genes.

After nearly a decade of working on other things, a return to GE has generated eight new papers in the last seven years, with further ramifications sketched below. Two dealt more broadly with theoretical arguments and simulations involving the evolution of modularity and generative entrenchment (Schank and Wimsatt, 2000; Wimsatt and Schank, 2004). The latter also incidentally gave partial confirmation for Arthur's proposed mechanism for the incorporation of new "deep" mutations. Two other papers returned to the innate-acquired distinction (Wimsatt, 1999b, 2002a). Four more dealt with the necessity and advantages of incorporating development (and generative entrenchment) into the analysis of cultural change, examinations of cultural inheritance, and explorations of what allowed the rapid rate of cultural evolution and the larger proportion of deeper and more revolutionary cultural changes than apparently found in biology (Wimsatt, 1999a, 2001, 2002b; Wimsatt and Griesemer, 2006).[32]

This put new rigor into attempts (including Riedl's) to apply burden to cultural evolution. Generative entrenchment is absolutely required to understand both the ontogenetic acquisition and the evolution of complex cumulative cultural skills, and the maintenance of complex cultural institutions, organizations, and material, normative, and ideational structures. Other studies (so far, with mainly peripheral application of GE) have dealt with the evolution of particular cultural entities (Weismann diagrams in Griesemer and Wimsatt 1989, and Punnett squares in Wimsatt 2006), or outlined the scheme for how to see GE in cultural evolution (Callebaut, 1993, pp. 379–382, 425–429; Wimsatt, in process).

Culture is not biology: the patterns of biological inheritance turned out historically to be relatively easy to unravel—much easier that is than the mysteries of development. Cultural patterns of inheritance are far more labile and variable, with changing modes of parentage and proportions and numbers of contributing parents. There are multiple hereditary channels, with some acting as control structures for others. All of this is rendered possible because our cultural genomes are assembled over our lifetimes, not given in one bolus at the beginning (Wimsatt, 1999a). As a result, our partially assembled cultural genome acts as a selective agent in the assembly and interpretation of the rest. In biology, heredity, development, and selection are at least relatively separable in most contexts, a fact that enormously eases their dissection, analysis, and modeling. But the sequential and interactive assembly of our cultural genomes makes development, heredity, and selection almost impossibly confounded for culture. However there is a compensation: if heredity is hard to tease out and more context-dependent for culture, development must be comparatively easier to analyze than it is for biology. After all, cultural traits must be learnable and transmissible, so the means for developing complex cultural traits must be there and accessible or we couldn't do it. Generative entrenchment (which after all yields stable invariances in pattern across generations) may be able to act as the organizing processes for cultural evolution like hereditary processes have for biology (Wimsatt, and Griesemer 2006; Griesemer, 2000). Let us hope so.

NOTES

1. Riedl's work on "burden" published in book form in 1975 in German, but not in English until 1978, the year after Gould's. It never became nearly as well known in the United States. Sadly, Riedl died on September 18, 2005, while this volume was in preparation. All figures in this chapter reprinted with permissions.

2. One might suppose that he ignored development entirely, but Williams did suggest that selection should design organisms to get rapidly through dangerous stages, lingering, if at all, in stages with lesser threats. But this inference cashed developmental features immediately out in terms of consequences for selection, and his comments would have followed naturally from "age-structure" models which were familiar tools to population biologists with demographic interests.

3. Even Morgan, father of "classical genetics" advocated related views in the early 1930's, and he was not alone. Rasmussen (1991) discusses Morgan at length, and Churchill (this volume) make it clear that there were many urging views which modern evolutionary developmental biology would find familiar. But the three decades of the postwar solidification of the Modern Synthesis (Gould famously called it a "hardening"), coinciding roughly with the founding of the Society for the Study of Evolution and its journal *Evolution* in 1948 had little commerce with developmental biology or with macroevolution. This sets the tone (reflected and amplified by Williams in 1966) greeting Waddington in 1957, Whyte in 1965, and Riedl in 1978.

4. Balinsky treats together the laws of von Baer and Müller-Haeckel, which Gould (1977) takes some pains to distinguish. Raff (1996) speaks of both Haeckel and von Baer as "recapitulationist" in spite of their differences, presumably because the "hourglass" models of variation in development he favors (Sander 1983, Elinson 1987, Raff 1996) are troublesome for both.

5. This perspective made Lewontin an early hero of "developmental systems theory" when it took off in the mid-1980s. See Oyama, Griffiths and Gray, eds. 2001.

6. Though Lewontin actively sought ways to incorporate development into evolutionary theory. It figured centrally in the conference he organized at Syracuse in 1967 to promote the new population biology (Lewontin ed., 1968). He had also written an elegant paper in 1965 on the evolution of life-history strategies, and figured centrally at the Waddington Serbelloni conferences, which focused more on developmental themes than any other topic. There were ways to bring development into population biology (norms of reaction, developmental switches, and life history theory), but from today's perspectives these all seem to keep development at a distance. On the other hand, with Riedl's work, the inspired *Size and Cycle* of John Tyler Bonner (1965) and Gould's fusion of development, evolution and history of science, it is hard to see how developmental biologists could have done any better before the explosion of information from developmental genetics following from the exploitation of HOX genes beginning in the late 1970s.

7. This conception of theory parallels Sommerhoff's (1950) work of a decade before, which sought to construct a generalized analytic biology in terms of special "coenetic variables" which express the goal-directive and homeostatic conditions of biological organization. Both might have resonated with D'Arcy Thompson's (1917) earlier and broadly influential work, or Brian Goodwin's work (1994), perhaps the most extensively developed modern structuralisms, though these two works are much richer in examples, analytical tools, and detailed analyses than either Whyte or Sommerhoff. Somerhoff and Whyte never got past the abstractions, exemplifying a faith in a general theory anchored in or inspired by cybernetics and general systems theory.

8. See also the *Yearbooks of the Society for General Systems Theory* (starting in 1959), which juxtaposed papers and translations by some of the leading social scientists and theoretical biologists of the day. Its contents included a paper by Kenneth Arrow, Richard Levins' first paper on fitness set analysis, and a translation, commissioned by Donald Campbell, of Konrad Lorenz's "Kant's Categories in Light of Contemporary Biology" which helped to form notions of innateness circulating among MIT linguists and philosophers in the mid-1960s. (Jerrold Katz recommended it to me, saying that "everyone has read it," when I visited there in 1968. I had already read it.) The *Journal for Theoretical Biology*, heterogeneous in content, and only barely respectable to many empiricist biologists (in spite of a distinguished list of authors) began publication in 1961.

9. A history of these influences in biology is still to be written. A good title might be "A Bridge Too Far," for while promising and attractive to many, the imagined solutions of that period look little like the furthest extensions today.

10. My introduction to this was Frank Rosenblatt's course in "Brain Models and Mechanisms," at Cornell in the fall of 1964. His course ranged much further than the title suggests, spanning perhaps "evolution and development of complex adaptive systems." His book on Perceptrons (1962) made him grandfather of connectionism, but he encouraged us to read widely in related areas, from self-organization to artificial intelligence. Development was crucially important for Rosenblatt, who liked to criticize *ex nihilo* models which took adult perceptual activity as given but never asked how these abilities could be generated.

11. The book was "padded": 127 pages total, including a 25-page introduction, a long index, a broadly relevant list of 39 references, and immense margins bounding about 20,000 words.

12. Arthur (1997: 218–226) credits Whyte sufficiently to devote an eight-page section to "internal selection" using it to organize much of his own discussion. In Arthur's defense, Whyte was certainly for Arthur's generation and that of his teachers the most visible exponent of its importance. "Burden" was surely seen by Riedl as an internal constraint, but people coming out of the "general systems" tradition (such as Whyte) tend to be looking for failures of neo-Darwinism, and burden could have looked just too neo-Darwinian. From early on, I saw generative entrenchment as not necessarily internal (1986), in part as a tool to argue against what I saw as illegitimate uses of "innateness." Being anti-neo-Darwinian was a common theme. When Kauffman was at Chicago (1969–1973), I argued with him that he should not see self-organization as opposing, or even usually as alternative to selection, but as a co-actor with it (Wimsatt 1986). He resisted this for many years, but came partially around to it in his book (1993). In 1984 he was still criticizing generative entrenchment as "too conservative" and "neo-Darwinian." That was fine with me. I was a "broad neo-Darwinian," in Arthur's (1988) later vocabulary. Whyte's attempt to find and analyze general "coordinating conditions" was in any case a non-starter.

13. I first heard about Riedl's work from Mark Rauscher at Duke after I gave a seminar on generative entrenchment in about 1986. Then I was put off by Riedl's 1977 article and a superficial look at his book by some of the things I attribute here

to others below. I didn't pursue it again until 1996, just before I visited the Konrad Lorenz Institute and met Riedl. I was then deeply impressed by his book.

14. While Riedl was respected, he apparently had alienated many and developed more than a few enemies, which may have hindered broader acceptance of his views on the continent.

15. My German is inadequate to judge the quality of the translation. Two different people, both of who should know, have said respectively that it was "terrible," and that while not true to the original it "probably increased its readability and consistency."

16. Though not all was quiet acceptance. Lewontin's book (1974) raised serious problems for the hope of assembling a workable mathematical theory spanning genic, organismal, and population levels with any completeness. Technical problems prevented evaluation of individual genic fitness contributions. And combinatorial explosion in the complexity and number of equations arose for multilocus models when organismal fitness was epistatic. Population structure in space and time was also emerging as an important complication for real cases. Later, Lewontin's joint paper with Stephen Gould (Gould and Lewontin, 1979) contributed the detailed reasons (a list of specific mechanisms already recognized by population genetics) why pan-adaptationism and optimality assumptions should be suspect.

17. Stanley Salthe (1985) argued for the importance of hierarchy in characterizing biological and taxonomic organization throughout this period and often appealed to this broader vocabulary and more abstract theory. He confirms these intuitions, and sees them as major reasons why evolutionary biologists ignored his own work (E-mail, 4-15-03). Using the "wrong" vocabulary and sources can make gaining acceptance much more difficult. But general systems theory made far too many promises and delivered very little, and the announced opposition to neo-Darwinism of many of its devotees didn't make friends among population biologists.

18. Riedl misses something important here: the presence of more parts of a given type indicates greater importance to the organism (and thus relates to burden), but if additional parts of that type (e.g., number of functioning liver cells) give redundancy and additional capacity, this will actually reduce burden or entrenchment per unit. This relationship is treated formally as a "k-out-of-m" organ in Wimsatt and Schank 1988. I return to this below.

19. I considered this problem in my dissertation, and came up with a very similar description. I sought to characterize how the organization of functional systems and subsystems mapped onto morphological parts (1971, 1972, 1997). I then considered how the existence of multiple alternative modes of decomposition of a system into parts, all required to analyze a problem, increased the complexity both of the system and of its analysis (1974). I did not then treat entrenchment as the central feature in this context. Only more recently when revisiting this (1971) material did I address issues of generative entrenchment in functional organization systematically (2002).

20. But do not misunderstand this quote: the elaboration of hierarchy likely does turn on the layering of kluges, but the latter process does not necessary yield hierarchies. Burden facilitates the construction of hierarchies, but does not necessitate it.

21. Possible reductions of interdependence by selection to increase evolvability appears to be missed by Riedl—probably because he has such a strong tendency to emphasize functional integration. The classic statement of the need for "quasi-independence" is Lewontin 1978. Wimsatt 1981 proposed that there may be selection for increased quasi-independence. This would decrease entrenchment (Schank and Wimsatt, 2000) though conditions under which selection can actually achieve this are currently up for debate (Wagner and Altenberg, 1996). The topic of modularity is excellently and broadly reviewed in Schlosser and Wagner, 2004.

22. Surprizingly, Riedl does not consider allometric growth here as an example of coordinated change in a number of characters.

23. One could argue that with optimal design, if a system is sufficiently important to justify the cost of the added redundancy, the system would show greater burden at its loss. This would push burden or entrenchment closer to a relation of straight translatability with intensity of selection. I think that both would be better served by regarding burden or generative entrenchment as a phenotypic organizational cause of selection, than as a synonym for it.

24. Arthur makes rich use of cellular descent-tree-like diagrams for various purposes. Compare, e.g., 1984, figures 9.3 and 9.5 pp. 144–146, and figure 10.12, p. 166; and 1988, figure 2.1, p. 23, and its direct ancestor in 1984, figure 5.2, p. 74.

25. I had assumed the lock would work for evolution, but that was not at issue until the 1974 conference. Stephen Gould's paper (a precursor to one line of argument in his 1977 book) distinguished between Haeckel's idea of ontogenetic recapitulation of phylogeny, and von Baer's ideas of taxonomic divergence in development. I had assumed that the developmental lock predicted a roughly Haeckelian course for evolution. This had bothered me and Steve's paper (sent before the conference) made clear that it was incorrect. But I then realized that the DL actually predicts a von Baer-ian rather than a Haeckelian pattern. This was great news for the model, and I incorporated Gould's insight into my presentation at the conference!

26. It is a mystery to me why Riedl never applied "burden" to innateness, since it was so central to his teacher, Konrad Lorenz, and Riedl skated so widely and so close to it. This application and the developmental lock convinced Riedl that I had indeed thought of generative entrenchment on my own.

27. Glassman heard me present the idea at a conference in 1982, wanted to use it, and suggested that we co-author.

28. This paper was first presented at two conferences in 1985, one organized by William Bechtel, where Kauffman and I directly exchanged views stimulating our later directions. In July I presented it at the ICSEB Congress in Sussex, first met Wallace Arthur who had presented his theory at the same time in another session and bought his (1984). We talked again in Chicago the next year while he was visiting Gould on sabbatical. Schank first heard Kauffman at a third conference in Spring of 1985 at the Field Museum in Chicago (Wimsatt, 1987), and ran his first simulations that fall in LISP, and later in Pascal, first on a MacPlus, and later on a prototype Mac II. Growing out of the Bechtel meetings, Kauffman, Richard Burian, and I organized (largely

through Kauffman's interventions) a meeting at the Santa Fe Institute in the fall of 1989 which brought together philosophers, historians and (mostly) developmental biologists to look at the growing intersections between Evolution and Development.

29. Almost all of these predictions have survived as correct, though they have been buried in the immense volume of data emerging from the HOX research.

30. Unrealistic assumptions in Kauffman's simulations are discussed and partially eliminated in Schank and Wimsatt 1988, and Wimsatt and Schank 1988. Aside from ignoring differential GE, two are particularly crucial: (1) Most seriously, the unrealistic assumption that selection had to maintain a unique "wiring diagram," that there were no multiple alternative solutions. This assumption, shown to be false by substantial phenotypic robustness at all levels of organization (Wagner, 2005), makes selection's task much harder, and biased the outcome strongly in Kauffman's favor. Genericity, like GE, is actually a degree property (in effect, an entropy measure), not a categorical property as Kauffman's discussion seemed to suggest. Genericity (and selection) are probably most effective when intermediate degrees of genericity and selection interact symbiotically to produce organization (Wimsatt, 1986). (2) Kauffman's model was unrealistic in not being a truncation-selection model, a feature corrected in all our later simulations. This and differential GE interact strongly to lock in and maintain many more selected features (sketched in Wimsatt and Schank, 1988, fully developed in our 2004).

31. This tremendous range in estimates arises from uncertainties in data and mechanisms that act multiplicatively. Whatever the numbers, it was clear that Kauffman's estimates were far too low, and differential entrenchment was a major cause of the difference.

32. The idea that cultural evolution must be faster then biological evolution may be misleading. Our tendency to think of vertebrate evolution obscures the frustrating and often dangerous rapidity of bacterial and viral evolution, not to mention the comparatively glacial but still threateningly rapid acquisition of pesticide resistance by insects.

REFERENCES

Arthur, W. (1984). *Mechanisms of Morphological Evolution: A Combined Genetic, Developmental and Ecological Approach.* New York: Wiley.

———. (1988). *A Theory of the Evolution of Development.* New York, Wiley.

———. (1997). *The Origin of Animal Body Plans: A Study in Evolutionary Developmental Biology.* New York: Cambridge University Press.

———. (2002). The emerging conceptual framework of evolutionary developmental biology. *Nature*, 415 (February): 757–764.

———. (2004). *Biased Embryos and Evolution.* Cambridge: Cambridge University Press.

Balinsky, B. I. (1965). *An Introduction to Embryology.* 2nd ed. Philadelphia: W. B. Saunders Co.

Bonner, J. T. (1965). *Size and Cycle: An Essay on the Structure of Biology*. Princeton: Princeton University Press.

Callebaut, W. (1993). *Taking the Naturalistic Turn or: How Real Philosophy of Science Is Done*. Chicago: University of Chicago Press.

Churchill, F. (2004). Believing in the Biogenetic law: the internal critique. (this volume).

Davidson, E. H. (2001). *Genomic Regulatory Systems: Development and Evolution*. San Diego, Academic Press.

Doboule, D. (1994). Temporal colinearity and the phylotypic progression: a basis for the stability of a vertebrate Bauplan and the evolution of morphologies through heterochrony. *Development Supplement*: 135–142.

Elinson, R. P. (1987). Change in developmental patterns: Embryos of amphibians with large eggs. In *Development as an Evolutionary Process*, R. A. Raff and E. C. Raff, eds. New York: Alan R. Liss. 8: 1–21.

Franklin, I., and R. C. Lewontin (1970). Is the gene the unit of selection? *Genetics*, 65: 707–734.

Garcia-Bellido, A., P. A. Lawrence, and G. Morata (1979). Compartments in animal development. *Scientific American*, 241(1): 90–98.

Gatlin, L. (1972). *Information Theory in the Living System*. New York: Columbia University Press.

Gilbert, S. F. (2001). Ecological developmental biology: Developmental biology meets the real world. *Developmental Biology*, 233: 1–12.

Gilbert, S., J. M. Opitz, and R. Raff (1996). Resynthesizing evolutionary and developmental biology. *Developmental Biology*, 173: 357–372.

Glassmann, R. B., and W. C. Wimsatt (1984). Evolutionary Advantages and Limitations of Early Plasticity, in R. Almli and S. Finger, eds., *Early Brain Damage, volume I*, Academic Press, 35–58.

Goodwin, B. C. (1994). *How the Leopard Changed its Spots: The Evolution of Complexity*. London: Weidenfield and Nicolson.

Gould, S. J. (1977). *Ontogeny and Phylogeny*. Cambridge: Belknap Press of Harvard University Press.

Gould, S. J., and R. C. Lewontin (1979). The Spandrels of San Marco and the Panglossian Paradigm: A Critique of the Adaptationist Programme. *Proceedings of the Royal Society of London. Series B, Biological Sciences*, 205(1161): 581–598.

Gould, S. J., and E. S. Vrba (1982). Exaptation—a missing term in the science of form. *Paleobiology*, 8(1): 4–15.

Griesemer, J. R. (2000). Reproduction and the Reduction of Genetics. In P. Beurton, R. Falk, and H-J. Rheinberger (eds.), *The Concept of the Gene in Development and Evolution, Historical and Epistemological Perspectives*. Cambridge: Cambridge University Press, 240–285.

Griesemer, J. R., and W. C. Wimsatt (1989). Picturing Weismannism: A case study in conceptual evolution. In M. Ruse, ed., *What Philosophy of Biology Is* (essays for David Hull). Dordrecht: Martinus-Nijhoff, pp. 75–137.

Hamilton, W. D. (1964). The genetical evolution of social behavior, I, *Journal for Theoretical Biology*, 7: 1–16.

Hull, D. L. (1988). *Science as a Process: An Evolutionary Account of the Social and Conceptual Development of Science*. Chicago: University of Chicago Press.

Jacob, F. (1977), Evolution and tinkering. *Science*, 196: 1161–1166.

Kauffman, S. A. (1969). Metabolic stability and epigenesis in randomly constructed genetic networks. *Journal for Theoretical Biology*, 22: 437–467.

————. (1985), Self-organization, selective adaptation and its limits: A new pattern of inference in evolution and development. In D. Depew and B. Weber, eds., *Evolution at a Crossroads: The New Biology and the New Philosophy of Science*. Cambridge: MIT Press, 169–207.

————. (1993). *The Origins of Order: Self-organization and Selection in Evolution*. London: Oxford University Press.

Koestler, A., and J. R. Smythies (1970). *Beyond Reductionism: New Perspectives in the Life Sciences*. New York: Macmillan.

Lerner, I. M. (1954). *Genetic Homeostasis*. New York: Wiley.

Lewontin, R. C. (1961). Evolution and the theory of games. *Journal for Theoretical Biology*, I: 382–403.

————. (1965). The evolution of life history strategies. In W. Baker, ed., *The Genetics of Colonizing Species*. New York: Wiley, pp. 77–94.

Lewontin, R. C., ed. (1968). *Population Biology and Evolution*. Syracuse: Syracuse University Press.

Lewontin, R. C. (1970). The units of selection. *Annual Review of Ecology and Systematics*, 1: 1–18.

————. (1974). *The Genetic Basis of Evolutionary Change*. New York: Columbia University Press.

————. (1978). Adaptation. *Scientific American*, 239(3): 212–218, 220, 222 passim.

Lewontin, R. C., E. W. Caspari, et al. (1960). Developmental selection of mutations (letters). *Science*, 132(3440): 1688, 1692–1694.

Maynard Smith, J. (1982). *Evolution and the Theory of Games.* Cambridge: Cambridge University Press.

McShea, D. W. (2000). Functional complexity in organisms: Parts as proxies. *Biology and Philosophy,* 15: 641–668.

McShea, D. W., and E. P. Venit (2001). What is a part? In G. P. Wagner, ed., *The Character Concept in Evolutionary Biology.* San Diego: Academic Press: 259–284.

Morgan, T. H. (1934). *Embryology and Genetics.* New York: Columbia University Press.

Odling-Smee, F. J., K. N. Laland, and M. Feldman (2003). *Niche Construction: The Neglected Process in Evolution.* Princeton, N.J.: Princeton University Press.

Oyama, S., P. Griffiths, and R. Gray, eds. (2001). *Cycles of Contingency: Developmental Systems and Evolution.* Cambridge: MIT Press.

Pattee, H. H. (1973). *Hierarchy Theory: The Challenge of Complex Systems.* New York: George Braziller.

Quastler, H. H., L. Angenstein, et al. (1953). *Essays on the Use of Information Theory in Biology.* Urbana: University of Illinois Press.

Raff, R. A. (1996). *The Shape of Life: Genes, Development, and the Evolution of Animal Form.* Chicago: University of Chicago Press.

Rasmussen, N. (1987). A new model of developmental constraints as applied to the *Drosophila* system. *Journal of Theoretical Biology,* 127: 271–299.

———— (1991). The decline of recapitulationism in early 20th century biology: Disciplinary conflict and consensus in the battleground of theory, *Journal for the History of Biology,* 24: 51–89.

Richards, R. J. (1987). *Darwin and the Emergence of Evolutionary Theories of Mind and Behavior.* Chicago: University of Chicago Press.

———— (1992). *The Meaning of Evolution.* Chicago: University of Chicago Press.

Riedl, R., ed. (1970). *Fauna und Flora der Adria.* Hamburg: P. Parey.

Riedl, R. (1975). *Die Ordnung des Lebendigen: Systembedingungen d. Evolution.* Hamburg, Berlin: Parey.

———— (1977). A systems-analytic approach to macro-evolutionary phenomena. *Quarterly Review of Biology,* 52: 351–370.

———— (1978). *Order in Living Organisms: A Systems Analysis of Evolution.* New York: Wiley.

Rosenblatt, F. (1962). *Perceptrons: Principles of Neurodynamics and the Theory of Brain Mechanisms.* Washington: Spartan Books.

Salthe, S. (1985). *Evolving Hierarchial Systems: Their Structure and Representation.* New York: Columbia University Press.

Sander, K. (1983). The evolution of patterning mechanisms: Gleanings from insect embryogenesis and spermatogenesis. In B. C. Goodwin, N. Holder, and C. C. Wylie, eds., *Development and Evolution*. Cambridge: Cambridge University Press, 137–159.

Schank, J. C., and W. C. Wimsatt (1988). Generative entrenchment and evolution. In A. Fine and P. K. Machamer, eds., *PSA-1986*, vol. 2. East Lansing: The Philosophy of Science Association, 33–60.

——— (2000). Evolvability: Modularity and generative entrenchment. In R. Singh, C. Krimbas, D. Paul, and J. Beatty, eds., *Thinking About Evolution: Historical, Philosophical and Political Perspectives* (Festschrift for Richard Lewontin, vol. 2). Cambridge: Cambridge University Press, 322–335.

Schlosser, G., and G. P. Wagner, Eds. (2004). *Modularity in Development and Evolution*. Chicago: University of Chicago Press.

Simon, H. A. (1962). The architecture of complexity. *Proceedings of the American Philosophical Society*, 106: 467–482.

Sommerhoff, G. (1950). *Analytical Biology*. Oxford: Oxford University Press.

Stearns, S. C. (1994). The evolutionary links between fixed and variable traits. *Acta Paleontologica Polonica*, 38: 215–232.

Thompson, D'Arcy W. (1917). *On Growth and Form*. Cambridge: Cambridge University Press.

Von Bertalanffy, L. (1968). *General Systems Theory: Foundations, Development, Applications*. New York: Braziller.

Waddington, C. H. (1957). *The Strategy of the Genes; A Discussion of some Aspects of Theoretical Biology*. London: Allen & Unwin.

Waddington, C. H., ed. (1968–1972). *Towards a Theoretical Biology*, Volumes 1–4. An IUBS symposium. Chicago: Aldine.

Wagner, G. P., and Altenberg, L. (1996). Complex adaptations and the evolution of evolvability. *Evolution*, 50: 967–976.

Wallace, B. (1985). Reflections on the still-hopeful monster. *Quarterly Review of Biology*, 60(1): 31–42.

Wallace, B. (1986). Can embryologists contribute to an understanding of evolutionary mechanisms. In W. Bechtel, ed., *Integrating Scientific Disciplines*. Dordrecht: Kluver, 149–163.

Weiss, P., ed. (1971). *Hierarchially Organized Systems in Theory and Practice*. New York: Hafner.

Whyte, L. L. (1960). Developmental selection of mutations (Reports). *Science*, 132(3432): 954.

——— (1965). *Internal Factors in Evolution*. New York: Braziller.

————, A. Wilson, and D. Wilson, (1970). *Hierarchial Structures.* New York: Elsevier.

Williams, G. C. (1966). *Adaptation and Natural Selection: A Critique of some Current Evolutionary Thought.* Princeton: Princeton University Press.

Wimsatt, W. C. (1970). Review of G. C. Williams' adaptation and natural selection. *Philosophy of Science,* 37(4): 620–623.

————. (1971). Modern Science and the New Teleology-I: The Logical Structure of Function Statements. Ph.D., dissertation, The University of Pittsburgh.

————. (1972). Teleology and the logical structure of function statements. *Studies in History and Philosophy of Science,* 3: 1–80.

————. (1974). Complexity and organization. In K. F. Schaffner and R. S. Cohen, eds., *PSA-1972* (Boston Studies in the Philosophy of Science). Dordrecht: Reidel, 20: 67–86.

————. (1980). Reductionistic research strategies and their biases in the units of selection controversy. *Scientific Discovery-vol. II: Case Studies.* T. Nickles. Dordrecht: Reidel, 213–259.

————. (1981). Units of selection and the structure of the multi-level genome. In P. D. Asquith and R. N. Giere, eds., *PSA-1980,* vol. 2. Lansing, Michigan: The Philosophy of Science Association, 122–183.

————. (1986). Developmental constraints, generative entrenchment, and the innate-acquired distinction. In William Bechtel, ed. *Integrating Scientific Disciplines.* Dordrecht: Martinus-Nijhoff, 185–208.

————. (1997). Functional organization, functional analogy, and functional inference. *Evolution and Cognition,* 3(2): 2–32.

————. (1999a). Genes, Memes, and Cultural Inheritance. *Biology and Philosophy,* 14 (special issue on the influence of R. C. Lewontin): 279–310.

————. (1999b), Generativity, entrenchment, evolution, and innateness. In V. Hardcastle, ed., *Biology Meets Psychology: Philosophical Essays.* Cambridge: MIT Press, 139–179.

————. (2001). Generative entrenchment and the developmental systems approach to evolutionary processes. In S. Oyama, R. Gray, and P. Griffiths, eds., *Cycles of Contingency: Developmental Systems and Evolution.* Cambridge: MIT Press, 219–237.

————. (2002a). Evolution, entrenchment, and innateness. In T. Brown and L. Smith, eds., *Reductionism and the Developement of Knowledge.* Erlbaum: 53–81.

————. (2002b). False models as means to truer theories: The case of blending inheritance. *Philosophy of Science,* 69(3): S12–S24.

————. (2002c). Functional organization, functional inference, and functional analogy. In Robert Cummins, Andre Ariew, and Mark Perlman, eds., New York: Oxford University Press, 174–221.

————. (2006). La geometría analítica de la genética: La evolución de los cuadros de Punnett. In Edna Suarez Diaz, ed., *Variedad Sin Limites: Las Representaciones en la Ciencia*. Mexico City: Universidad Nacional Autonoma de Mexico (UNAM) y Editorial Limusa, S. A. de C. V.

————. (2007). *Re-Engineering Philosophy for Limited Beings: Piecewise Approximations to Reality*. Cambridge: Harvard University Press, in press.

————. (in process). Generative Entrenchment, Scientific Change, and the Analytic-Synthetic Distinction. [invited address, Central Division APA meetings, April 1987].

Wimsatt, W. C., and J. R. Griesemer (2006). Reproducing entrenchments to scaffold culture: The central role of development in cultural evolution. In R. Sansome and R. Brandon, eds., *Integrating Evolution and Development*. Cambridge: MIT Press, in press.

Wimsatt, W. C., and J. C. Schank (1988). Two constraints on the evolution of complex adaptations and the means for their avoidance. In M. Nitecki, ed., *Evolutionary Progress*. Chicago: The University of Chicago Press, 231–273.

————. (2004). Generative entrenchment, modularity and evolvability: When genic selection meets the whole organism." In G. Schlosser and G. Wagner, eds., *Modularity in Evolution and Development*, Chicago: University of Chicago Press, 359–394.

Wright, S. (1931). Evoluntion in Mendelian populations. *Genetics*, 16: 97–159.

11

FATE MAPS, GENE EXPRESSION MAPS, AND THE
EVIDENTIARY STRUCTURE OF EVOLUTIONARY
DEVELOPMENTAL BIOLOGY
Scott F. Gilbert

THE MAPPING HERITAGE IN EMBRYOLOGY

The mapping concept is one of the most important devices threading its way through classical embryology, developmental biology, developmental genetics, and evolutionary developmental biology. Mapping is, of course, a metaphor referring to the practice of cartography, the construction of a two-dimensional representation of a three-dimensional surface. In their public use, maps attempt to accurately scale the political or geographic dimensions of a territory so that the user of the map can obtain an overview of the political or physical landscape. Maps emphasize boundaries and distinctions—where the land terminates and the sea begins, at what point the laws of Germany end and the laws of France take effect. Mapping is thus an attempt to depict on paper the boundaries that exist in the social or natural world. Needless to say, mapmaking has always been a political act, and no representation is ever completely without its abstractions and exaggerations.[1] Unlike "normal tables," which they superficially resemble, "maps" are not merely depictions of the embryo. Rather, they are the superposition of prospective fate or gene expression upon the depiction of the (usually early) embryo. Thus, the two major types of maps used in developmental biology have been the "fate map" and the "gene expression map."

Fate maps are the attempts of embryologists to identify the cells that produce certain adult or larval structures, and to label them as such on an early-stage embryo. In other words, embryologists make a map of *what is to be* and superimpose this data *onto a map of what is yet to develop*. Examples of fate maps are shown in figure 11.1.

Cell fate-mapping started with tracing cell lineages, the identification of cell fate by observation alone.[2] This was done, for example, by E. G. Conklin at the Woods Hole Marine Biological Laboratory. Conklin's analysis of the tunicate embryo in 1905 was a landmark in the field. Often, these data were depicted as a dichotomously branching tree (figure 11.2A),

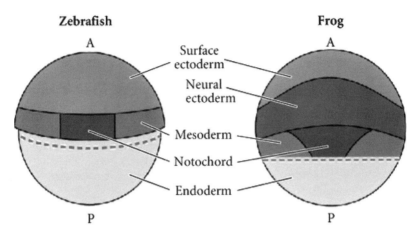

Figure 11.1
Surface fate maps of frog (*Xenopus*) and zebra fish (*Danio*) embryos at the early gastrula stage on the dorsal surface. These and the other illustrations are taken from Gilbert (2000).

thereby placing the data into the standard format of taxonomic field keys. The data of these "trees" could later be superimposed onto the early embryonic cells, and made into a fate map (figure 11.2B). In 1933, Conklin used the data collected in 1905 to make one of these summary fate maps. Cell lineage fate maps were also used by Hörstadius in the 1930s to show the fate map of the sea urchin embryo.

Walther Vogt constructed the first systematic fate maps when he applied vital dyes to the early amphibian embryo.[3] He made agar chips impregnated with a vital dye and then placed the egg in a wax depression so that he could place chips onto the egg and impart dye to a specific region of the egg or embryo (figure 11.3). In this way, he was able to mark cells early in development and see what they became later. He found that cells in similar areas gave rise to similar nearby structures. However, there were also boundary lines, such as those where ectoderm ended and mesoderm began. It is easy to see why these were called fate maps. Vogt's fate maps looked very much like globes. They had two poles— the animal pole and the vegetal pole—and an equator down the center which served as a landmark.[4] Even today, this geographic view of the amphibian egg is the most common representation. It is difficult to avoid seeing the egg as a globe, and why bother to avoid it? In amphibian embryos, the divisions are orthogonal, dividing the egg by lines of latitude and longitude. The prime meridian—literally the first cleavage plane—

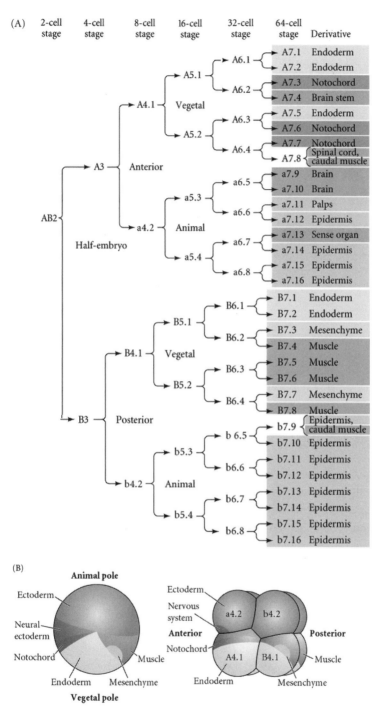

Figure 11.2
Fate map of the tunicate embryo. (A) Dichotomous branching lineage diagram of the tunicate *Styela*. (B) Fate map constructed by superimposing the fates onto the one-cell and eight-cell embryos.

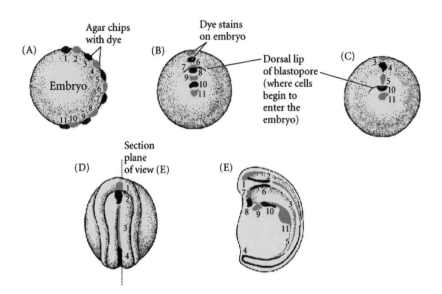

Figure 11.3
Vital staining of a newt embryo to produce a fate map. (A) Vogt's method for marking specific regions of the embryo with dye-impregnated agar chips (B–D) Dorsal surface views of successively later embryos. (E) Newt embryo dissected to show stained cells in the interior.

bisects the gray crescent and separates the embryo into its future left and right sides.

Modern investigators still make fate maps, and the preparation of a fate map is usually one of the first steps in any research program. The dyes have become better (less diffusible, longer lasting), and the resolution has attained the single-cell level.[5] We can now label individual cells of the early embryo and find out what their descendants become. Moreover, this is not of just historical interest. There are still heated debates over fate maps. The location of the blood-forming cells of *Xenopus* is contested,[6] and the endo-derm-ectoderm boundary of sea urchin embryos has only recently been determined.[7] Some researchers have criticized the attempt to place complex three-dimensional fates onto the global surface. Bauer and col-leagues[8] point out that it is not correct merely to project a later-stage fate map back to cleavage stages, and they document inconsistencies in the lit-erature that might be due to researchers' having uncritically projected later cell positions onto earlier embryonic stages.

Moreover, the geomorphic amphibian fate map of Vogt turned out not to be universally applicable to other amphibians. For years, the fate map of *Xenopus* was expected to be patterned totally according to the surface cells. After all, the rocks a mile below the surface of the United States are not the province of some other country. We even claim the air space above our nation as our own. However, this was found to be false in *Xenopus*. Here, the fate map of the *internal* cells differed from the fate map of the *external* cells.[9] The geographical assumption turned out to be wrong. Fate maps are being refined, revised, and constructed for new organisms. It has been an ongoing research project from the 1920s to the present.

Good fate maps are absolutely critical in embryology and evolutionary developmental biology. As Viktor Hamburger[10] pointed out, the first time Spemann interpreted his results of the dorsal blastopore lip transplantation, he got it wrong. Spemann had used a very primitive fate map, and he had thought that the cells above the dorsal blastopore lip were fated to become neural ectoderm, not dorsal mesoderm. It had to be pointed out to him by a colleague, Hans Peterson of Heidelburg, that a revised fate map—one that Spemann himself had devised in 1921—suggested these cells were fated to become mesoderm. That single fact completely changed the interpretation of his results, and the confirmation of that later interpretation—Spemann and Mangold's paper[11]—became one of the most important experiments in modern embryology. One of the earliest experiments in evolutionary developmental biology (and, for that matter, in the area of development genetic) was the famous Spemann and Schotte experiment[12] wherein they transplanted prospective jaw tissue between salamander and frog embryos. The frogs had newt jaws and vice versa. For the experiment to work, the position of the prospective jaw had to be carefully defined.

When fate maps are compared, evolutionary change can be inferred. This was done by Tung and colleagues,[13] who compared the vital dye-derived fate map of *Amphioxus* with the lineage-derived fate map of the tunicate *Ciona*. The results showed similarities that were unexpected on the basis of the mode of cleavage, but were expected on the basis of evolutionary phylogenies. More recently, Greg Wray and Rudy Raff[14] compared fate maps to show the changes in blastomere fate during early development of direct-developing sea urchins. They showed that the fate map of the direct-developing urchins had been changed so that the cells that would have formed larval ectoderm were forming the vestibular structures that become the adult. Moreover, the positions of neuronal precursors had been altered. The fact that these species diverged only around 10 million years ago showed that significant developmental change can occur through altering early development, something that had formerly been thought to be impossible.

GENE EXPRESSION MAPS

A related, albeit chronologically later, program is the mapping of gene expression patterns. There are many ways to map gene expression, and they usually give a direct projection upon an embryonic surface. The most common way is through in situ hybridization. Here, one can actually stain for the accumulation of a particular type of mRNA. This is an incredible ability, and it has revolutionized developmental biology. One obtains a probe that is complementary to the mRNA and labels it with a reporter molecule. The probe and its reporter accumulate only at the sites where the target mRNA has accumulated. A second way is to use a reporter gene sequence. These are genes (such as those for bacterial β-galactosidase and jellyfish green fluorescent protein) that can be readily identified, are not usually made in the experimental animal, and are fused to the regulatory region of a particular gene. Wherever the gene is usually expressed, the reporter gene will be transcribed. These techniques have brought the notion of structure back into developmental biology. In the early 1980s, the evidence of developmental genetics consisted of bands on gels. Since the 1990s, evidence has consisted of radioactive grains on microscope slides or colored products on whole-mount embryos. One now has to know the structure of the embryo.

Gene expression maps have had very important consequences. One of them was to show that units of embryonic construction need not be the same units as the adult units of function. Thus, compartments, rhombomeres, internal ribs, parasegments, and enamel knots are anatomical regions of gene expression and function that have no obvious anatomical correlate in the adult. These regions of gene-expressing cells are critical in constructing the body. Without maps of gene expression patterns, we probably would not be aware of them.

Gene expression maps can serve several functions. The most obvious is that they can be used as fate maps. For instance, if one knows that the Brachyury and *Twist* genes are expressed in *Xenopus* and *Drosophila* mesodermal cells, one can look at which cells transcribe these genes and know that they will become mesoderm. Similarly, since only the notochordal cells of *Xenopus* transcribe the *Chordin* gene, a probe to *Xenopus chordin* can be used to locate cells that produce the notochord.

These gene expression data can provide extremely useful information about cell fate. Indeed, the fine structure of the organizer did not become known until gene expression data were merged with the more detailed fate maps of *Xenopus laevis*. It is now thought that in *Xenopus*, the cells of the organizer ultimately contribute to four cell types: pharyngeal endoderm, head mesoderm (prechordal plate), dorsal mesoderm (primarily the notochord), and the chordaneural hinge.[15] The pharyngeal endoderm and prechordal plate lead the migration of the organizer tissue and appear to induce the forebrain and midbrain. The dorsal mesoderm induces the hindbrain and trunk, and the chordaneural hinge induces the tip of the tail. The expression patterns of transcription factors provided the key to how this was done.[16] Vegetal plate rotation was required to put the *Cerberus*-expressing pharyngeal endoderm cells dorsal and anterior to those cells expressing *chordin* or *Goosecoid*. This revised fate map,[17] made primarily by gene expression data, was published by Winklebauer and Schürfeld (figure 11.4), who found that the vegetal endoderm actively pushes a

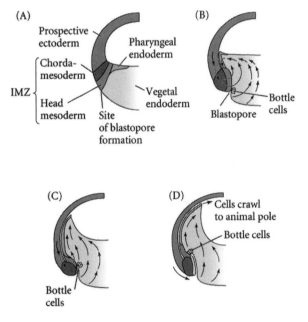

Figure 11.4
Early movements of *Xenopus* gastrulation, as shown by gene expression. The cells fated to form the pharyngeal endoderm express *Cerberus*. These cells are at the anteriormost point of the migrating epithelium, and are moved into this position by the rotation of the deeper vegetal cells. The cells fated to become head mesoderm express the *Goosecoid* gene, and they follow the pharyngeal endoderm. The cells fated to become chordamesoderm are expressing *Xbra* (*Xenopus, Brachyury*), and they follow the other two regions.

region of endoderm up ahead of the mesoderm, causing it to contact the surface cells and migrate anteriorly to the mesoderm.

Gene expression maps need not be confined to fate maps, however. They also can be used to show signaling. Indeed, one of the most important uses of gene expression mapping has been to show which cells are signaling centers. In this analysis, the gene being expressed encodes a paracrine factor that is capable of influencing the development of other cells. The zone of polarizing activity in the chick limb provides a good case. Since the 1960s, it has been known that a small block of mesodermal cells at the posterior junction of the limb bud and the body wall has a remarkable ability: when it is taken from one limb bud and placed in the anterior margin of a second limb bud, the host limb bud develops two mirror-image sets of digits. This region was called the zone of polarizing activity (ZPA). Riddle and his colleagues[18] in Cliff Tabin's laboratory showed that ZPA was defined by the expression of the Sonic Hedgehog gene. First, they showed that the Sonic Hedgehog protein is necessary and sufficient to account for the ZPA's activities. When they caused this protein to be synthesized in the anterior margin, they obtained mirror-image duplications. More interestingly, they were able to correlate the time, place, and amount of *Sonic Hedgehog* gene expression with the classically defined potency of this region to induce the mirror-image duplications.

The relationship between fate map and gene expression map has been of critical concern to evolutionary developmental biology. The temporal priority of the gene expression map over the fate map is seen when genes are functionally deleted from embryos. In these cases, the fate map changes as well[19] (figure 11.5). The cell lineage map, the fate map, and the gene expression map have been united in this type of experiment, with the gene expression map having priority and explaining the others.

This is critical in evolutionary developmental biology. One of the tenets of evo-devo has been that evolution is predicated upon hereditable changes in development. These changes are therefore usually changes in gene expression. Thus gene expression changes and continuities are the stuff from which much of evo-devo has been made. The case for the inclusion of evolutionary developmental biology into evolutionary biology has been made largely upon the changes and the continuities of gene expression maps. Thus, comparative gene expression maps have played key roles as evidence for the importance of evolutionary developmental biology.

One of the first of these comparative gene expression maps showed the homologous expression of the Hox genes and head transcription factors[20] between the protostomal arthropods (represented by *Drosophila*)

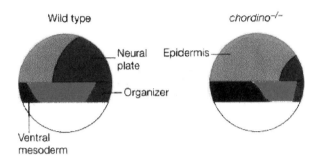

Figure 11.5
Change in fate map of the zebra fish when the Chordino gene is deleted. The Chordino product is important in forming neural (dorsal) ectoderm and dorsal mesoderm. In the absence of this gene, the epidermis expands at the expense of the neural plate, and the ventral and lateral mesoderm expand at the expense of the dorsal (organizer) mesoderm.

and the deuterostomal vertebrates (represented by *Mus*). (See figure 11.6) This research stressed the similarities of the protostomes and the deuterostomes, showing that not only did the two groups have homologous Hox genes that were in the same order in their respective chromosomes, but also that the geographic order of expression was conserved. The expression pattern of Hox genes throughout the animal kingdom was so similar that Slack and colleagues[21] proposed that this constituted the fundamental basis of being an animal. (Interestingly, more recent gene expression maps—those presenting Hox gene expression in sponges[22]—have been used as evidence against this view.)

Moreover, further investigations have demonstrated that variations of this expression pattern can produce morphological changes. Changes in the Hox gene expression map in crustaceans were correlated with the changing number of maxillipeds, and changes in the Hox gene expression pattern in vertebrates correlated with changes in the number of cervical vertebrae.[23] In some cases, severe alterations all but eliminated the certain constellations of Hox gene expression, and these eliminated the cervical vertebrae as well as the lumbar vertebrae. Thus, in the limbless snakes, the Hox gene expression patterns make the thoracic—ribbed—vertebrae expanded at the expense of the Hox gene patterns allowing other vertebral types.[24] Similar changes in Hox genes change the structures of arthropods as well, distinguishing the shrimp from the lobster.[25]

Alterations of the Hoxd11 and Hoxd13 expression pattern have even been proposed to account for the formation of the autopod. Figure 11.7

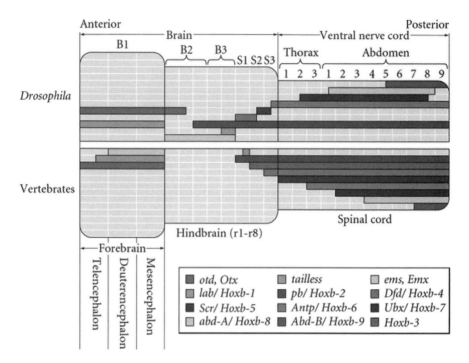

Figure 11.6
Expression of regulatory transcription factors in *Drosophila* and vertebrates. The *Drosophila* genes *ems*, *tll*, and *otd* are homologous to the vertebrate *Emx*, *Tll*, and *Otx* genes, respectively. The Hox genes expressed in *Drosophila* and in vertebrates have similar patterns in their respective hindbrains and neural cords.

depicts a Devonian fantasy of about 350 million years ago. Although there is no way of knowing the Hox gene expression of the lobe-finned fish (or how it might be fruitfully compared with a highly derived modern osteichthyan such as *Danio rerio*), the alterations in gene expression pattern between modern fish and modern tetrapods, and the importance of Hox genes in specifying limb parts, converged to make the different gene expression maps important evidence for a mechanism by which fish fins could be transformed into tetrapod limbs.[26]

Gene expression maps can also provide clues as to how genes used in one area of development can be recruited to another. For instance, the *Distal-less* gene, used to define the limb primordia of insects, became co-opted to produce the eyespots of butterfly wings,[27] and the *fgf10* gene, used in producing the tetrapod limb bud, may also be utilized for producing the carapacial ridge specific to turtles.[28]

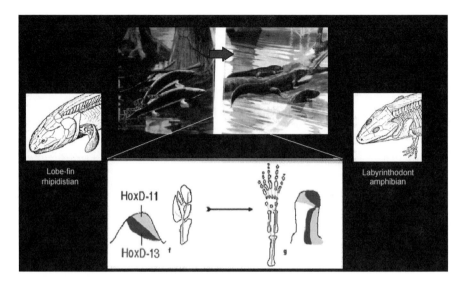

Figure 11.7
In the Devonian, labyrinthodont amphibian tetrapods emerged from an ancestor that was probably similar to lobe-fin rhipidistians. It has been proposed that the change in Hox gene expression patterns at the end of the fin created the autopod and enabled limbs to arise from the stalk of fins.

   Gene expression maps can also be used to suggest homologous relationships that are not obvious. For instance, regulatory regions from *Amphioxus Hox* genes expressed in the anterior neural tube were found to drive spatially localized expression of a reporter gene in vertebrate neural crest cells, in derivatives of neurogenic placodes, and in branchial arches, despite the fact that these cephalochordates lack both neural crest and neurogenic placodes.[29] This implies that there was already the expression pattern of "head" before there were morphologically cranial structures, and that certain regions of the cephalopod neural tube are homologus to vertebrate brains. Similar research has also homologized the cephalopod endostyle to the vertebrate thyroid gland. In a recent paper, Meinhardt[30] has used gene expression data to advance the case that the body of the diploblastic and radially symmetrical *Hydra* is homologous to the brain of triploblastic and bilaterally symmetric animals, and that the most anterior regions of the head form from the pedal (*Nk2.1*-expressing) domains of a *Hydra*-like ancestor. Using another cnidarian, *Nematostella*, Martindale and colleagues (2004) argue that the gene expression patterns indicate that this sea anemone has both bilateral symmetry and a mesodermal germ layer.

Lowe and colleagues[31] argue for homologous regions of the vertebrate and hemichordate nervous systems based on gene expression patterns. They use these proposed homologies to interpolate what the nervous system of the bilaterian ancestor of invertebrates and vertebrates may have looked like. Comparative gene expression mapping forms the evidence for Lowe's entire thesis.

Another use of gene expression maps has been to show deep homology. One of the other major principles emerging from evolutionary developmental biology is that analogous structures can be formed by homologous instructions. For instance, the dorsal neural tube of the deuterostomes and the ventral nerve cord of the protostomes appear to be constructed by the same set of gene and gene product interactions, suggesting that nature figured out how to make a nervous system only once. Similarly, there appear to be homologous genes and pathways to specify both insect and vertebrate limbs which have no homologies as walking structures.[32] In both types of limbs, gene expression maps show that the proximal-distal axis appears to be specified by *Fringe* (proximal) and *Distal-less* (distal); the dorsal-ventral axis, by *Wnt* (dorsal) and *Engrailed* (ventral); and the anterior-posterior axis, by *Hedgehog* expression (posterior).

NEW TECHNIQUES OF GENE EXPRESSION MAPPING

Computer-aided technology has been incredibly important in the gene-mapping community. Some of the greatest advances are in the area of three-dimensional expression-mapping. The first concerns three-dimensional reconstructions wherein the computer rapidly and "objectively" realigns embryonic sections so as to make them usable for three-dimensional reconstruction[33] (figure 11.8). This procedure combines the methodological advantages of whole-mount, in situ hybridization and the high-scale resolution of serial sectioning. After capturing phase-contrast and bright field views of serial sections, one can map in situ hybridization data through a series of algorithms. The subjective interactions of the processor have been eliminated.

Another three-dimensional aid is Geographic Information System (GIS) analysis. Epoxy resin casts of an organ (such as a tooth) are optically sectioned at 25–100 μm intervals, using a laser confocal microscope. High-resolution digital elevation models (DEMs) of the organ topology are produced from the image stacks, using the 3Dview version of the NIH-Image software. DEMs can be transferred to GIS software as well as interpreted by surface-rendering computer programs. All traditional mor-

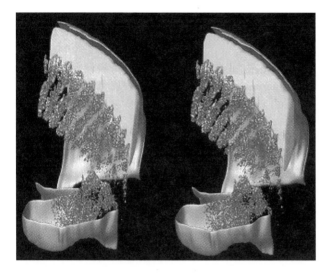

Figure 11.8
Fate mapping of myotome cells (purple) as they enter the limb bud. Computer reconstruction made from *myf-5* expression in the developing myotome cells. Crossing your eyes allows you to see a 3-dimensional view. Courtesy of J. Streicher and G. Müller.

phometrical measurements can be obtained from DEMs. However, the total shape data (i.e., DEMs) can be explored with GIS prior to the selection of appropriate measurements. Furthermore, since the DEMs can be stored as a database of organ morphologies, the three-dimensional morphology can be accessed remotely via the Web. Jernvall and his colleagues[34] have used this technology to measure the development of mammalian teeth. Moreover, they can place on this topography the various gene expression patterns of in situ hybridization.

The GIS technology was designed specifically for landscape topologies in real cartographic mapping, and it has been adopted by ecologists for their surveys. The method and analogy of geographic mapping to gene expression-mapping was made explicitly by Jernvall in his presentation at the SICB meeting in January, 2001 (figure 11.9). In these papers, Jernvall and colleagues have demonstrated that changes in gene expression in enamel knots (analyzed by in situ hybridization) cause the morphological differences between mouse and vole molars (as seen by the DEMs). Expression of *fgf4*, *Shh*, and *p21* prefigures changes in morphology, and the spatial distribution causes the subsequent location of cusps. This remarkable conclusion is predicated on extremely sophisticated gene expression and fate-mapping correlations.

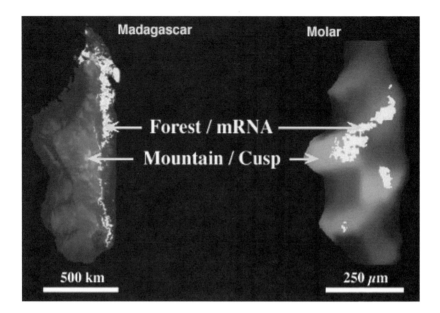

Figure 11.9
Explicit mapping analogy made by Jernvall in his GIS reconsructions of gene expression in the rodent molar. Courtesy of J. Jernvall.

CONCLUSION

The mapping program in embryology, initiated at the turn of the twentieth century, is an extremely active and important program in contemporary developmental biology. Both fate maps and gene expression maps summarize new data, and organize it in a way that relates it to other data and leads to new research. Mapping has been especially important in providing evidence for evolutionary developmental biology. It not only summarizes the evidence in the field but also serves the functions of relating the new data to classical data (thereby linking a new science to an established science) and of showing the importance of evo-devo to evolutionary biology. New procedures in monitoring gene expression patterns, localizing dyes in certain cells, and graphically representing these patterns have made mapping of paramount importance to developmental biology.

Developmental biologists have long been confronted with "mapping" problems and have solved them in creative ways. Fate maps compressed four-dimensional space onto two dimensions, and video animation is now allowing the temporal dimension to be shown in detail.

For evolutionary developmental biology, the current challenge is producing a five-dimensional representation: the four standard dimensions of space and time placed into the context of the paleontological temporal dimension.

NOTES

1. Monmonier, 1991.

2. See Galperin, 1998.

3. Vogt, 1929.

4. Vogt was an outdoorsman and was not averse to using geographic metaphors. In 1929, he speaks of Spemann's "conquest of an uncharted territory of knowledge" (Hamburger, 1988, p. 80). The fate maps resemble contemporary anthropological maps depicting the spread of Teutons or Magyars into new areas.

5. See Gilbert, 2000.

6. See Lane and Smith, 1999.

7. Sherwood and McClay, 2001.

8. Bauer et al., 1994.

9. Keller, 1976.

10. Hamburger, 1988.

11. Spemann and Mangold, 1924.

12. Spemann and Schotte, 1932.

13. Tung et al., 1962

14. Wray and Raff, 1990. More recently, Ferkowitz and Raff (2001) have used gene expression mapping to show the importance of Wnt5 expression in the transition from indirect- to direct-developing sea urchins.

15. Keller, 1976; Gont et al., 1993.

16. Vodicka and Gerhart, 1995.

17. Winklebauer and Schürfeld, 1999.

18. Riddle et al., 1993.

19. Kishimoto et al., 1997.

20. See Hirth and Reichert, 1999.

21. Slack et al., 1993.

22. Schierwater and Kuhn, 1998.

23. Gaunt, 1994; Burke et al., 1995.

24. Cohn and Tickle, 1999.

25. Averof and Patel, 1997.

26. Sordino et al., 1995; Shubin et al., 1997.

27. Brakefield et al., 1996.

28. Loredo et al., 2001.

29. Manzanares et al., 2000.

30. Meinhardt, 2002.

31. Lowe et al., 2003.

32. See Shubin et al., 1997.

33. See Streicher et al., 2000.

34. Jernvall et al., 2000.

References

Averof, Michael, and Patel, N. H. 1997. Crustacean appendage evolution associated with changes in *Hox* gene expression. *Nature* 388: 682–686.

Bauer, Daniel V., Huang, S., and Moody, S. A. 1994. The cleavage stage origin of Spemann's Organizer: Analysis of the movements of blastomere clones before and during gastrulation in *Xenopus*. *Development* 120: 1179–1189.

Brakefield, P. M., Gates, J., Keyes, D., Kesbeke, F., Wijngaarden, P. J., Monteiro, A., French, V., and Carroll, S. B. 1996. Development, plasticity, and evolution of butterfly eyespot patterns. *Nature* 384: 236–242.

Burke, Ann C., Nelson, C. E., Morgan, B. A., and Tabin, C. 1995. Hox genes and the evolution of vertebrate axial morphology. *Development* 121: 333–346.

Cohn, Michael J., and Tickle, C. 1999. Developmental basis of limblessness and axial patterning in snakes. *Nature* 399: 474–479.

Conklin, Edwin G. 1905. The organization and cell lineage of the ascidian egg. *Journal of the Academy of Natural Science, Philadelphia* 13: 1–119.

Ferkowicz, Michael J., and Raff, R. A. 2001. Wnt gene expression in sea urchin development: Heterochronies associated with the evolution of developmental mode. *Evolution and Development* 3: 24–33.

Galperin, Charles. 1998. From cell lineage to developmental genetics. *History and Philosophy of the Life Sciences* 20: 301–344.

Gaunt, Stephen J. 1994. Conservation in the Hox code during morphological evolution. *International Journal of Developmental Biology* 38: 549–552.

Gilbert, Scott F. 2000. *Developmental Biology*, 6th ed. Sunderland, Mass.: Sinauer.

Gont, Linda K., Steinbeisser, H., Blumberg, B., and De Robertis, E. M. 1993. Tail formation as a continuation of gastrulation: The multiple tail populations of the *Xenopus* tailbud derive from the late blastopore lip. *Development* 119: 991–1004.

Hamburger, Viktor. 1988. *The Heritage of Experimental Embryology: Hans Spemann and the Organizer.* Oxford: Oxford University Press.

Hirth, Frank, and Reichert, H. 1999. Conserved genetic programs in insect and mammalian brain development. *BioEssays* 21: 677–684.

Jernvall, Jukka, Keranen, S. V., and Thesleff, I. 2000. Evolutionary modification of development in mammalian teeth: Quantifying gene expression patterns and topography. *Proceedings of the National Academy of Sciences USA* 97: 14444–14448.

Keller, Ray E. 1976. Vital dye mapping of the gastrula and neurula of *Xenopus laevis*. II. Prospective areas and morphogenetic movements of the deep layer. *Developmental Biology* 51: 118–137.

Kishimoto, Yasayuki, Lee, K. H., Zon, L., Hammerschmidt, M., and Schulte-Merker, S. 1997. The molecular nature of zebrafish swirl: BMP2 function is essential during early dorsoventral patterning. *Development* 124: 4457–4466.

Lane, Mary C., and Smith, W. C. 1999. The origins of primitive blood in *Xenopus*: Implications for axial patterning. *Development* 126: 423–434.

Loredo, Grace A., Brukman, A., Harris, M. P., Kagle, D., LeClair, E. E., Gutman, R., Denney, E., Henkelman, E., Murray, B. P., Fallon, J. F., Tuan, R. S., and Gilbert, S. F. 2001. Development of an evolutionarily novel structure: Fibroblast growth factor expression in the carapacial ridge of turtle embryos. *Journal of Experimental Zoology/ Molecular and Developmental Evolution* 291: 274–281.

Lowe, Christopher, J., Wu, M., Salic, A., Evans, L., Lander, E., Stange-Thomann, N., Gruber, C. E., Gerhart, J., and Kirschner, M. 2003. Anteroposterior patterning in hemichordates and the origins of the chordate nervous system. *Cell* 113: 853–865.

Manzanares, Miguel, Wada, H., Itasaki, N., Trainor, P. A., Krumlauf, R., and Holland, P. W. 2000. Conservation and elaboration of Hox gene regulation during evolution of the vertebrate head. *Nature* 408: 854–857.

Martindale, Mark Q., Pang, K., and Finnerty, J. R. 2004. Investigating the origins of triploblasty: Mesodermal gene expression in a diploblastic animal, the sea anemone *Nematostella vectensis* (phylum, Cnidaria; class, Anthrozoa). *Development* 131: 2453–2474.

Meinhardt, Hans. 2002. The radially symmetric hydra and the evolution of the bilateral body plan: An old body became a young brain. *BioEssays* 24: 185–191.

Monmonier, Mark. 1991. *How to Lie with Maps.* Chicago: University of Chicago Press.

Riddle, Robert D., Johnson, R. L., Laufer, E., and Tabin, C. 1993. Sonic hedgehog mediates the polarizing activity of the ZPA. *Cell* 75: 1401–1416.

Schierwater, Bernd, and Kuhn, K. 1998. Homology of Hox genes and the zootype concept in early metazoan evolution. *Molecular and Phylogenetic Evolution* 9: 375–381.

Sherwood, David R., and McClay, D. R. 2001. LvNotch signaling plays a dual role in regulating the position of the ectoderm-endoderm boundary in the sea urchin embryo. *Development* 128: 2221–2232.

Shubin, Neil, Tabin, C., and Carroll, S. 1997. Fossils, genes, and the evolution of animal limbs. *Nature* 388: 639–648.

Slack, Jonathan M. W., Holland, P. W. H., and Graham, C. F. 1993. The zootype and the phylotypic stage. *Nature* 361: 490–492.

Sordino, Paolo, Van der Hoeven, F., and Duboule, D. 1995. Hox gene expression in teleost fins and the origin of vertebrate digits. *Nature* 375: 678–681.

Spemann, Hans, and Mangold, H. 1924. Induction of embryonic primordia by implantation of organizers from a different species. In B. H. Willier and J. M. Oppenheimer (eds.), 1974. *Foundations of Experimental Embryology*, pp. 144–184. New York: Hafner.

Spemann, Hans, and Schotté, O. 1932. Über xenoplatische Transplantation als Mittel zur Analyse der embryonalen Induction. *Naturwissenschaften* 20: 463–467.

Streicher, Johannes, Donat, M. A., Strauss, B., Sporle, R., Schughart, K., and Müller, G. B. 2000. Computer-based three-dimensional visualization of developmental gene expression. *Nature Genetics* 25: 147–152.

Tung, T. C., Wu, S. C., and Tung, Y. Y. F. 1962. The presumptive areas of the egg of *Amphioxus. Scientia Sinica* 11: 82–90.

Vodicka, Marie A., and Gerhart, J. C. 1995. Blastomere derivation and domains of gene expression in the Spemann Organizer of *Xenopus laevis. Development* 121: 3505–3518.

Vogt, Walther. 1929. Gestaltungsanalyse am Amphibienkeim mit örtlicher Vitalfärbung. II. Teil. Gastrulation und Mesodermbildung bei Urodelen und Anuren. *Wilhelm Roux's Archiv für Entwicklungsmechanik der Organismen* 120: 384–706.

Winklebauer, Rudolf, and Schürfeld, M. 1999. Vegetal rotation, a new gastrulation movement involved in the internalization of the mesoderm and endoderm in *Xenopus. Development* 126: 3703–3713.

Wray, Gregory A., and Raff, R. A. 1990. Novel origins of lineage founder cells in the direct-developing sea urchin *Heliocidaris erythrogramma. Developmental Biology* 141: 41–54.

———

12

TRACKING ORGANIC PROCESSES: REPRESENTATIONS AND
RESEARCH STYLES IN CLASSICAL EMBRYOLOGY AND
GENETICS

James Griesemer

THEMES

This chapter will explore two themes concerning scientific practices, illustrated by examples drawn from research on problems of heredity and development from the late nineteenth to the early twentieth century. Because these themes apply not only to the sciences that emerged but also to the history and philosophy of science, my argument about the nature of scientific practice has historiographic as well as philosophical implications.

First, scientists frequently follow a process in order to understand both its causal character and where it may lead. Radioactive tracers, fluorescent stains, genetic markers, and embryonic transplants all facilitate tracking processes and determining how physiological, molecular, and genetic outcomes result from known inputs. Indeed, one might argue that the notion of following a process unifies descriptions of science as theoretical representation, as systematic observation, and as technological intervention.[1] Following processes is a characteristic activity of science. Moreover, the concept of science as following a process cuts across many analytical distinctions commonly used to describe science (e.g., theory vs. observation, theory vs. experiment, hypothesis-testing vs. measurement, active manipulation vs. passive observation, scientific methods vs. scientific goals). A genetic marker "marks" the transmission process in breeding experiments, but it also may be the subject of causal investigation—the role played by the marker in development or disease, for example.

Second, scientific reports of process-following tend to be accompanied by representations that reflect commitments to follow processes in particular ways, foregrounding some aspects of phenomena and backgrounding others. Visual representation of genes as causes of genetic continuity and of somatic discontinuity, for example, focuses attention on continuities the eye can follow rather than on discontinuities that must be bridged in thought.[2] Moreover, how one follows a process constrains what one represents as followed. The representation in turn focuses attention on foregrounded elements as the significant and explanatory aspects of the process-as-followed. The result is constraint and guidance of how processes may be followed on other occasions, as well as what implications are (literally) drawn from reported work. In consequence, hereditary transmission, for example, is now hard to characterize other than by contrast with developmental expression—the causal dichotomy of the representation has become entrenched in "common sense."[3]

Finally, historians and philosophers interested in the split between embryology and genetics early in the twentieth century, after the rediscovery of Mendel's work, have tended to rely on narrative representations of the history of biology that foreground certain theories and experiments to the exclusion of others.[4] Narratives that put a single field of science in the limelight of attention at a given time constrain understanding of scientific change to occur in a linear sequence, as though genetics succeeded classical embryology rather than their branching into collateral lines of work from a common wellspring. It seems unlikely, given the social organization and realignment of scientific work in the twentieth century, that historical change would be adequately represented by a linear succession of disciplines.[5] I focus on the segmentation of lines of work early in the twentieth century that now, in evo-devo research, may be anastamosing in a process of intersection discussed by Gerson in this volume.

Since embryology is not well tracked in narratives that foreground the success of genetics after the split, succession cannot but fail to explain this episode of scientific change. We must instead follow the "bushy" divergences and reticulations of several sciences as they spawn new lines of work if we are to understand, and follow, science as a process. In my view, this means that we must follow not only the trails of changing theories and problem agendas, but also the tools, techniques, and methods for following processes which scientists deploy in their varying lines of work.

This following of scientists while they follow nature may be of some help toward assembling resources to describe and articulate the newly emerging field of evo-devo.[6] This field, comprising several intersecting lines of work, seeks to join genetic, developmental, and evolutionary research problems and programs that have been treated, quite successfully, as though they were separate throughout most of the twentieth century, but in reality they are more like the segments of a centipede: moving together with limited autonomy. It is indeed ironic that the result of this new synthesis may be the rediscovery of problems that are 100 years old.[7] Here, I follow in a line of analysts who question the origin myths of scientists looking back to founding scientific fathers to explain the roots of their own fields' successes and failures.[8] In this chapter I illustrate themes of process-following, foregrounding, and backgrounding by interpreting Mendel as a developmentalist in order to reveal the extent to which workers in the late nineteenth century sought a unified account of heredity, development, and evolution. My rhetorical aim is to cultivate a new perspective on our contemporary scientific landscape, a landscape in which genetics and developmental biology are considered to have very different origins, cultures of research practice, theoretical goals, and formalisms—in short, different research styles. But where previous interpreters of Mendel have related his developmentalism to nineteenth-century concerns with the stability of species, I seek to articulate a view of Mendel as an exemplar for those modern biologists who seek theoretical unification or intersection of domains segmented by earlier generations who took Mendelism to be the "wedge issue" of the day.

The bulk of the chapter is concerned to formulate a view of Mendel's activity that may help reorient our modern understanding of genetic research styles which have come to stand in contrast to embryological and morphological research styles. The broader goal is to provide insight into developmental research styles by indicating features shared with Mendel in following processes and drawing inferences about earlier stages in terms of later ones. The paper considers embryological research styles more with an eye toward formulating problems for further research than with presentation of firm conclusions of immediate use to the interpretation of evo-devo in the light of history.

ON FOLLOWING A PROCESS

Following processes is a key project for understanding causality in the world. Where, when, and how processes originate; what interactions happen to them along their way; and how they terminate is, in a word, what there is according to process ontologies. Regardless of the metaphysical standing of process ontologies, there is no doubt that scientists do follow processes, that this is an important and central activity in their work, and that they achieve causal understanding as a result of doing it.

The connection between following a process and causal understanding has long been explored by philosophers of science, though mainly by those concerned with the physical sciences. From the 1920s, Reichenbach used a "mark principle" to describe the causal relevance of factors in explanations of physical effects such as the propagation of light.

"A mark," Reichenbach wrote, "is the result of an intervention [in a process] by means of an irreversible process."[9] The mark principle helped Reichenbach articulate a technical, probabilistic criterion of causal relevance. His goal was to use the irreversibility of marking processes to infer the direction of time in marked causal processes.

My use of Reichenbach's mark principle here is not aimed at ontological problems as general as the direction of time, but rather is a tool with which to describe what biologists do when they follow processes. This is no mere analyst's category. Biologists, especially the embryologists I discuss below, often described their work as introducing and then following "marks" in order to establish the fate or prospective significance of marked parts of a dynamic process.[10] A shared assumption of genetic and embryological research styles is the notion that hybrid organic material (whether naturally or artificially produced) can serve both as the introduction of a mark into a process to facilitate tracking and as a causal intervention to see how the process might turn out differently than it otherwise would. That is to say, paying attention to particular aspects or properties in order to follow a process, marking interventions, and analyzing the causal character of a process in terms of counterfactual support may all be entwined in a single activity.

Reichenbach noted that marking interventions can be either "deliberately performed" or the product of "natural causes."[11] This distinction is relevant to the history of biology, where both experimental and observational means of marking organic processes are exploited, often in the same study, to achieve causal understanding of development in embryological and genetical research styles, as I will describe below.

Salmon elaborated Reichenbach's principle in a criterion of "mark transmission" to serve his analysis of causal processes:

> MT: Let P be a process that, in the absence of interactions with other processes, would remain uniform with respect to a characteristic Q, which it would manifest consistently over an interval that includes both of the space-time points A and B (A ≠ B). Then, a mark (consisting of a modification of Q into Q'), which has been introduced into process P by means of a single local interaction at point A, is transmitted to point B if P manifests the modification Q' at B and at all stages of the process between A and B without additional interventions.[12]

If a causal process from A to B can be thought of as the "development" of A, then this very abstract notion of causality contains the philosophical root problem of heredity/development. Heredity concerns the respect in which A stands in a certain causal relation to B, while development concerns the bringing about of B from A.[13] Both concern the very same causal process, both seek an account of causal process in terms of continuity and constancy or invariance, and both investigate it by tracking marks that identify it as a causal process.[14]

In natural histories of various sorts, biologists introduce or observe a local, irreversible mark in a process of interest by attending to a particular part of the process—for example, when ecologists follow an ecosystem by tracking a particular isotope of a circulating molecule such as CO_2 or N_2. In molecular biology, experiments often consist of introducing a mark and seeing where it ends up, such as in the Hershey-Chase experiment, which tracked the whereabouts of radioactively labeled phosphorus (P^{32}) and sulfur (S^{35}) in order to determine whether phosphorus-rich viral DNA or the sulfur-rich protein coat of a bacteriophage was the information-bearing infectious agent. These are only more systematic and disciplined versions of what casual observers do on a lazy summer afternoon when they track the flow of a river by watching the movement of a leaf floating downstream.

"Noteworthy observation" is facilitated by taking note of peculiarities that make certain parts of a process stand out and thus make tracking easier.[15] As Hans Spemann remarked in 1931, "You may well make a discovery without intending to do so, but not without noticing it."[16] Noteworthy attention in biological research often results from "mental marking," a kind of tracking that is not easily assigned to either of the traditional categories of passive observation and active experiment. It is nonmanipulative yet active work on the part of a tracker, and may serve experiment as well as observation.[17] Mental marking is an actual causal intersection connecting a natural process of interest and the scientist observing it, but it is ineffective as a means of causal intervention or control through the marking interaction.

Noticing a morphological feature (a structure, a pigment pattern, a cell in a particular location) of an embryonic region is an important type of mental marking in embryology. Noted morphological features can be tracked to where they end up several or many cell divisions or developmental stages later. One means of mental marking that emerged in late nineteenth-century embryology depended on microscope observations and camera lucida drawings: features noted in the microscope were marked on a drawing of the embryo and tracked through development via superimposed labels and arrows, resulting in a diagram that represented mark transmission/part transformation through the developmental process.

The feature noted in such cases is not itself an irreversible interaction between the observer and an embryo undergoing development but, rather, part of the natural process, and thus not a mark in Reichenbach's sense. Instead, the embryologist's noticing the feature and attending to it constitutes the marking interaction. Following the mark in attention, through continued observation and aided by techniques such as camera lucida drawing, constitutes tracking the marked process.

In cases closer to Reichenbach's discussion, biologists may physically change a property of a process "from Q to Q'." In a manipulative marking intervention, the experimenter focuses on a "target" of attention prior to the marking interaction and then introduces a mark that physically changes a property of the process in such a way that continuous mental attention is not required to track the process. This is a procedural benefit of the irreversibility of marks that Reichenbach required of causal processes. The mark can be tracked in intermittent "checkups" via subsequent observations of or interventions in the process to see if the process still carries the mark Q'. This operational notion of mark transmission thus also plays a theoretical role in identifying the process as causal. Theory and methodology are as intimately related as two sides of a coin.

In what follows, I explore the nature and representation of process-following in heredity/development with examples spanning the historical period of the split between diverging embryological lines of work and the new science of genetics. New styles for following processes, theories, and methods emerged around the turn of the twentieth century, and they illustrate the way in which representations produced as a process is followed provide reinforcing feedback that organizes attention into foreground and background concerns. Foregrounding and backgrounding of different aspects of the same biological process lead to different research styles, and since narratives tend to follow historical developments within styles, history often gives a false impression of disciplines that are separate because their theories describe different processes. The split between genetics and embryology early in the twentieth century, I maintain, occurred at the level of research styles, of theoretical commitments that facilitated the segmentation of a vision of a unified social world of biological research, not the discovery of separate realms of biological processes of heredity and development.[18]

REPRESENTING A PROCESS AS FOLLOWED

Gerson (this volume) describes research styles as "abstract commitments used to organize other, relatively concrete commitments. Styles typically appear as general philosophical or methodological positions (e.g., focusing on structural rather than functional considerations, or preferring the construction of formal models to the detailed description and analysis of particular cases). Any such pattern of commitments can serve as the basis of a subworld or intersection." Genetic and embryological research styles package commitments to follow processes according to particular sorts of marking interactions and tracking conventions together with commitments to represent processes in particular ways.[19] Attention-guiding feedback from scientific representations results from work to honor both sets of commitments. My central claim is that because geneticists and embryologists follow the same process, their research styles are constrained to be similar in certain "underlying" ways, despite the considerable divergences we associate with those disciplines. I explain the differences that historically emerged in the split between them in terms of how their diverging representational practices drove attention to different aspects of the one process and thus supported the development of distinct lines of work. I consider several kinds of examples drawn from lines of work traditionally classified as either genetics or embryology.

In the next section, I argue that Mendel's famous work *Experiments in Plant Hybridization* is clearly a work of developmental explanation, despite its championing by twentieth-century geneticists who came to view their social world and traditions of heredity research as separated from embryological research. Mendel's commitments to developmental explanation and representation are clear and undeniable from his report of what he followed, how he followed it, and how he represented what he followed.

The seeds of a new, "genetic" style of research also are clearly present in Mendel's work, however, as is suggested by historiographers who seek to place Mendel back in his nineteenth-century context rather than read him, as geneticists so easily do, as offering the first modern genetic theory of factor transmission.[20] My description of Mendel's research as following a developmental process is aimed to support, at the level of research styles, the broad view of historians that biologists in the second half of the nineteenth century sought a unified understanding of heredity, development, and evolution.[21]

I also illustrate how embryologists who tracked cell lineages in development, such as C. O. Whitman, E. B. Wilson, and E. G. Conklin, adopted representational styles that became increasingly abstract and eventually led to the emergence of the theoretical causal logic of the new, genetic style of explanations. In other words, the embryological origins of the gene theory can be detected in representational strategies and empirical methods as well as in the problem agendas, theories, and scientific pedigrees of embryologists. Since these strategies and methods were integral to embryological work, not only is the gene theory embryological in origin but genetic practice is "embryological" in origin as well.[22]

I argue, furthermore, that genetics still is "embryological" insofar as genetic research styles do not and cannot ignore or black-box embryological phenomena, although it has become standard to describe genetics as ignoring development in its abstract mapping of genotype to phenotype.[23] That is, at the level of research styles for tracking processes, the split between genetics and embryology is a matter of what is represented in the foreground versus the background of attention. The split cannot be manifested in a commitment to study one process to the exclusion of the other because heredity and development are only aspects of a single causal process. Studying one is ipso facto studying the other, just as studying cooking is ipso facto studying chemistry.

Genetics backgrounds embryological interests and problems in its methods and representational strategies for the sake of foregrounding problems and interests that are now recognized as genetic rather than embryological. But it does not follow from this that the embryological *process* is backgrounded by genetic *methods*. Thus, embryological concerns fill Mendel's many pages on how to prepare material for hybridization experiments. However, the foreground/background distinction is something recognizable only in hindsight, in the light of a *subsequently* preferred theoretical perspective or historical narrative that represents processes in terms of foregrounded problems. Moreover, as a consequence of the feedback that representations provide in the conduct of scientific work, it has become difficult to see that the phenomena, the methods, and the representations which appear throughout the history of genetic and embryological research are entangled, even though our focused attention makes them appear to be cleanly separated. The theoretical abstraction of heredity from development in modern scientific thought cannot provide a framework for understanding the history of heredity/development nor, to the extent that similar goals and strategies of unification or intersection are in play in current evo-devo research, can it provide a framework for understanding its scientific future. It was previously argued that modern distinctions (such as genotype/phenotype) are of little help in understanding the history or philosophy of the abstractions that led to them.[24]

Following a process scientifically requires a large measure of self-control and self-discipline. One aim of scientific representation is to guide a viewer's or reader's attention in ways corresponding to the discipline required of the researcher. If representations play dual roles of reporting scientific results to outside parties and as "working objects" at the laboratory bench or field site, then attention guided by particular sorts of representations can affect the ways in which commitments to follow processes are honored and understood.[25] The contexts of intervention and representation, like the contexts of discovery and justification they replace, cannot be separated, even for analytical purposes, if representations guide marking practices, and marking interventions are the basis of representations of processes followed.

MENDEL AS A DEVELOPMENTALIST

Mendel's achievement can be correctly interpreted as a theory of development as well as a foundation of the modern genetic theory and methodology of heredity.[26] I will argue that this was Mendel's intent and not merely the fancy of a revisionist historian or philosopher in order to illustrate three points. First, Mendel's project aimed to follow a biological process along the general lines described above. Second, Mendel considered this process to be developmental, and accordingly expressed his goal as pursuit of a theoretical understanding of a key aspect of development. Third, the representational strategies that Mendel devised as working tools to keep track of his process-following work and to communicate his theory led him to formulate several distinct notational conventions in his *Experiments*. These had the effect of focusing attention (foregrounding) on what we now take to be factor transmission and defocusing attention to the developmental aspects of the process, backgrounding them as a methodology for manipulating plants in pursuit of laws of transmission rather than as the target of theoretical investigation.[27]

My argument is not that we should disregard readings of Mendel as a (proto)geneticist in favor of some more "accurate" developmentalist reading of his work. My goal is instead to try to understand how the particularities of Mendel's experimental, theoretical, and representational practices contributed to his work's incorporation into a genetic conception of the process he was following rather than into the inclusive developmental one that I believe directed his concerns.

These points suggest that Mendel's representational strategies probably played an important role in shaping the attention of his readers, scientific followers, and historical interpreters to following the developmental process that interested Mendel. As a result, we no longer clearly see in his

writings that (1) Mendel's laws (of hereditary transmission) are not
Mendel's theory (of the development of hybrids), and (2) both heredity
and development are integral aspects of one process, each of which must
be attended to in order to track it.

What distinguishes foreground from background in Mendel's science,
as well as in subsequent interpretations, is the direction of the reader's
attention. The aspect foregrounded by Mendelians such as Correns and
Bateson—hereditary transmission of factors in development—is represented in laws of transformation that describe the characters which
Mendelians followed through experimentally generated processes.
Mendelian characters are marks, in the dual sense discussed above, that
Mendel noted and attended to as he followed and manipulated heredity/
development. The aspect of Mendel's work backgrounded by geneticists—
that there are two kinds of developmental constancy of characters which
pass through the hybrid offspring—is represented in Mendel's statement
of theory that leads to his laws and in statements describing the experimental practices necessary to systematically construct artificial processes of
hybridization and development that can be followed. For geneticists, what
Mendel calls his *theory* is merely a developmental means to lawful genetic
ends, while for the nineteenth-century unifier, the laws only codify and
support the developmental theory.

That Mendel's project aimed to follow a process is plain from the
opening of his *Experiments*, Mendel notes on the first page that the
observers who preceded him—Kölreuter, Gärtner, Herbert, Lecoq, and
Wichura—pursued experiments to "follow up [in] the developments of
the hybrids in their progeny" the "striking regularity with which the same
hybrid forms always reappeared whenever fertilisation took place between
the same species."[28] That is, the line of work to which Mendel contributed
is that of following the developmental process of hybrids into the progeny.
Mendel's achievement was to recognize that the statistical distribution of
offspring of particular kinds can reveal aspects of the developmental
processes occurring in their hybrid parents. This achievement stands
regardless of whether one reads the impact of his results as bearing on the
old problem of species stability, on new problems of hereditary transmission, or on the lawlikeness of hybrid development. The properties of interest throughout Mendel's work are kinds of developmental constancy of
characters.

As Olby and Sapp have noted, the concerns of these hybridists were
neither those of a modern geneticist nor exactly those of Mendel himself.
Mendel's work has accordingly been put to a variety of rhetorical pur-

poses.[29] Many of the hybridists' experiments, for example, were conducted to explore the possibility of a direct action of foreign pollen on characters such as seed, pod, and fruit color. Indeed, when Gärtner hybridized plants to explore this question, he did not grow the second generation from the hybrid seeds because this would have been irrelevant to the problem of the action of pollen on the seeds of the hybrids themselves.[30] Mendel's work, in contrast, focuses squarely on the problem of finding a "generally applicable law governing the formation and development of hybrids."[31] Following up the development of hybrids in their progeny is precisely a problem of tracking the consequences of the developmental process that took place in their hybrid parents. It is thus a problem of following a process set in motion and artificially controlled by hybridization experiments. The process is of a kind whose inputs had been carefully explored by Mendel's predecessors, but whose outputs were studied quite differently and to different ends by Mendel.

That Mendel aimed to follow a developmental process by constructing cross-hybrid plants from purebred lines is plain in his language. Attention to Mendel's linguistic usage clarifies his theoretical goals and their relations to the laws he inferred and tested on the basis of his experimental work. Terms with the root word "inherit" occur twice in English translations of Mendel's *Experiments*.[32] There are four occurrences of words with the root "transmit." In sharp contrast, words with the root "develop" occur fifty-eight times, often many times on a single page.[33] Of these fifty-eight occurrences, eight appear in the phrase "developmental series" (*Entwicklungsreihe*), which could be understood not as a term from biology but as a mathematical term to describe the "development" of the terms forming a combination series.[34]

The terms of a combination series describe different kinds of individual organisms that could appear among the progeny of a cross in terms of the combinations of kinds of factors they would receive from their parents. A + 2Aa + a and B + 2Bb + b are examples of developmental series.[35] They refer to three kinds in a progeny—A, Aa, and a—and to four (types of) individuals—A, Aa, aA, and a—that instantiate these three kinds. "AB + Ab + aB + ab + 2ABb + 2aBb + 2AaB + 2Aab + 4AaBb" is "indisputably a combination series in which the two expressions for the characters A and a, B and b are combined."[36]

The members of this series are the progeny organisms developed in subsequent generations from parents constructed in hybridization experiments. Their representation in a mathematical combination series later in *Experiments* describes a developmental series in both the mathematical sense of a progression of terms in a combinatorial expansion and the biological sense of a progression of offspring kinds in the developmental "expansion" of a progeny bred from the hybrids. I will comment below on this character notation in contrast to Mendel's other theoretical notations for gamete forms (pollen and egg cells), fertilizations, and results of fertilization.[37]

Most important, the order of terms in the combination and developmental series reveal something of Mendel's developmentalist thinking— a point to which I will return in discussing his four notations. The developmental variety that can be experimentally generated from hybrids was of central concern to Mendel because it posed a challenge to identification of underlying unity with constancy expressible in a law. Mendel's theoretical struggle was to relate the constancy of character *form* seen in the parental types appearing among a progeny and the constancy of *behavior* in the segregating character forms of hybrids.

Even discounting the meaning of "developmental series" as mathematical rather than biological, "develop" appears an order of magnitude more times than either "inherit" or "transmit." Moreover, Mendel's first mention of the concept of a developmental series makes it clear that he saw a connection between the mathematical expression of a combination series, which mathematically describes all possible combinations of characters, and the biological process of development of offspring of each combination or kind: "In order to discover the relations in which the hybrid forms stand towards each other and also towards their progenitors," Mendel writes, "it appears to be necessary that all members of the series developed in each successive generation should be, *without exception*, subjected to observation."[38] A developmental series thus represents a mathematical series of kinds of biological processes of development.

With this awareness of the relative frequencies of term usage, we can look afresh at Mendel's expression of his developmental concerns. Mendel writes abundantly about the development of plant characters: of buds opening before being "perfectly developed," of the withering of "certain parts of an otherwise quite normally developed flower" and "defective development of the keel," of seed shape and albumin developed immediately after artificial fertilization, of pods developed early or late, of seeds damaged by insects "during their development."[39] Thus, we know that his use of *Entwicklung* cannot always or routinely be interpreted as a vague or ambiguous term meaning "unfolding" or "evolution." These developmental concerns, tucked away in the "methods" section of Mendel's *Experiments*, are crucial to control in explicit protocols if Mendel is to be able to track the development of hybrids by means of crossing experiments. If plant parts do not develop in constant, controlled ways, no inferences can be made about the developmental process in the hybrids, through which Mendel tracks the characters, from their distribution in the progeny. That is, if development is not suitably well-behaved, character tracking will break down in the generations bred from the hybrids.

Moreover, when Mendel moves to talk about the inferences he will draw from the experimental results of generations bred from the hybrids, his developmental language does not slip into vague or metaphorical language about development: "The proportions in which the descendants of the hybrids develop and split up in the first and second generations" does not confound what we today would think of as separable processes of heredity and development. Rather, Mendel is writing about how hybrid organisms go through embryological development in such a way that developmental kinds of hybrids can be resolved, but only by tracking the characters into subsequent generations of progeny created by self-fertilization.

Mendel is clearly concerned to systematically analyze how his constructed hybrids develop, in the biological sense of the term. This of course does not imply that Mendel had any very precise embryological mechanism in mind for the development of the characters he investigated—how tall pea plants become tall, or how plants make violet-red flowers or yellow seeds. The developmental theory he espoused[40] is that constant progeny (i.e., progeny bearing parental characters) "can only be formed when the egg cells and the fertilising pollen are of like character, so that both are provided with the material for creating quite similar individuals, as is the case with the normal fertilisation of pure species."[41] But Mendel's language does signal that his interest was, as he repeatedly says it is, in the development of and from the hybrids. And it signals the ways in which his methodology of experimental hybridization was crafted to serve that goal well.

Mendel routinely refers to his theoretical goal as an understanding of the "development of hybrids," expressed in the form of a law. In the section "The Reproductive Cells of the Hybrids," Mendel lays out a hypothesis and an assumption, which I take to be his theory of the development of hybrids and which Bateson called "the essence of the Mendelian principles of heredity."[42] Mendel's theory is built on an induction from experience with hybridization: "So far as experience goes, we find it in every case confirmed that constant progeny can only be formed when the egg cells and the fertilising pollen are of like character."[43]

Mendel then goes on to formulate the hypothesis that extends his inductive generalization from hybrids of different character to hybrids of all types, and thus to offer a theoretical hypothesis to explain the development of hybrids in general:

> We must therefore regard it as certain that exactly similar factors must be at work also in the production of the constant forms in the hybrid plants. Since the various constant forms are produced in one plant, or even in one flower of a plant, the conclusion appears logical that in the ovaries of the hybrids there are formed as many sorts of egg cells, and in the anthers as many sorts of pollen cells, as there are possible constant combination forms, and that these egg and pollen cells agree in their internal composition with those of the separate forms.[44]

There can be no doubt that Mendel took this hypothesis to be his theory or that its target was to explain the development of hybrids, though there is one caveat addressed in an added assumption needed to render the theory testable (i.e., amenable to what Mendel called "experimental proof"), because he goes on to say: "In point of fact it is possible to demonstrate theoretically that this hypothesis would fully suffice to account for the development of the hybrids in the separate generations, if we might at the same time assume that the various kinds of egg and pollen cells were formed in the hybrids on the average in equal numbers."[45] A few lines later, after describing the form that these experimental proofs should take—what we now call backcross experiments—Mendel predicts the plant forms (character combinations) that must develop from the hybrids constructed, "if the above theory be correct."[46]

I claimed above that Mendel's theory is not Mendel's laws and that Mendel's theory is a theory of the development of hybrids. In exploring the relation between Mendel's theory and Mendel's laws, it is important to consider a frequently overlooked feature of scientific representations: they are often working objects, developed as bench or field tools for tracking phenomena and following processes, but subsequently pressed into service as tools for communicating results and interpretations.[47] Mendel's notation is often taken for granted as a tool for expressing his "laws" of segregation and independent assortment in hereditary transmission, while the interpretation of Mendel's theoretical goals remains controversial.[48] Mendel's notation looks antiquated to modern eyes because he represented kinds of characters—A, Aa, and a—rather than factor combinations or genotypes—AA, Aa, and aa—to express developmental and combination series.[49] Many historically sensitive expositions present Mendel's work in terms of a modern, un-Mendel-like genotype notation.[50] Mendel's own collection of notations tells an important story about his theoretical goals, gives clues for recognizing the developmental target of his explanatory laws of heredity, and, most important, reflects his shifting expository concerns in terms of what each notation foregrounds or backgrounds.

A change of notation within *Experiments* reflects a subtle shift of Mendel's attention from tracking processes of fertilization, hybridization, and development to explaining the development of hybrids in terms of the fertilization processes created by controlled breeding experiments. Mendel's first notation, "character development notation," foregrounded the problem agenda, methods, and empirical results that led to his explanation of the development of hybrids. The three subsequent notations highlight the inferred pathways of character transmission in the fertilization process that explain the patterns observed *while following the process of character development* in the hybrids. Once these germ cell, fertilization, and product-of-fertilization notations (collectively character-transmission-in-fertilization notation) are adopted and notational equivalence to the character development notation is established, it becomes feasible to formulate and attend to problems of character transmission as part of a new kind of theoretical enterprise of heredity research without speculating on the precise mechanisms and detailed behavior of development in hybrids.[51] In other words, the shift from character development to character transmission in fertilization notation facilitates a reversal of foreground and background commitments to lines of work that we recognize as alternately embryological or genetic. Instead of following developmental processes in the foreground with an implied background of characters transmitted in fertilization, the character transmission notation foregrounds patterns of transmission and backgrounds character development.

This is not to say that Mendel's work "really" concerns heredity rather than development, because his explanatory notation concerns transmission, nor to say that Mendel actually achieved this differentiation of lines of work. Rather, I claim that Mendel's notation made it easier to take for granted the developmental phenomena necessary to establish the laws "of heredity," and thus made it easier for followers to think of their work as distinct from developmental concerns. So easy, in fact, that the meaning of the character development notation has been recovered only with great analytical effort.[52]

The character development notation was motivated by Mendel's noticing the "double signification" of the dominant character in hybrids.[53] By following such characters into the second generation bred from the hybrids, Mendel's $3:1$ segregation result is resolved into a $2:1:1$ ratio.[54] The $2:1:1$ ratio is significant in two ways. First, if the dominant character (say, tall) is parental in character, then it is constant in the offspring and "it must pass unchanged to the whole of the offspring." But if the dominant character is hybrid in character, then it is not constant in the offspring. However, it is constant in a second sense: "it must maintain the same behaviour as in the first generation"; that is, the second generation bred from the hybrids (F_2 in modern terminology). So the first point is that the $2:1:1$ ratio reveals *two kinds* of developmental constancy. One kind is constancy of character *form* of the parental characters represented by T and t in the second and third terms, or $1:1$ component, of the ratio. The other kind is constancy in character *behavior* of the hybrids represented by Tt in the first term of the ratio.[55] $2:1:1$ relates the equal proportions of these two kinds of character constancy—$2:(1:1)$ or $2:2$— whereas the modern genotype notation, $1:2:1$, does not.

The second point is that the ratio is reported as $2:1:1$ to distinguish the two kinds of *character* constancy (hybrid constancy of behavior versus parental constancy of form), not two kinds of factors (T, t) or three kinds of factor combinations (TT, Tt, tt). The dominant character, tall, thus has two forms, Tt and T, where the first denotes the constant (segregating) behavior of the hybrid form, while the second denotes the constant form of development of the parental type in the progeny of hybrids. Reporting the proportions as $2:1:1$ allows the *characters* tall and short to be put into proportion at the same time that the two kinds of developmental constancy are put into proportion.

The double significance of the dominant character in a given generation is determined by tracing the process of development of hybrids forward into the progeny. The distribution of the progeny characters provides information about the developmental process (behavior) in the hybrid parent. Thus, although the tracking direction runs from ancestors to descendants, the inference and explanation of development of hybrids is from later steps to earlier ones. Moreover, the entire exercise is framed by prior knowledge from the hybridists' projects that the progeny of hybrids are variable and can be "transmuted" into parental types in virtue of the constant behavior of the hybrid characters. A similar pattern of tracking and explanation will be described for several kinds of embryological work.

At this point, Mendel introduces the developmental series A + 2Aa + a to show "the terms in the series for the progeny of the hybrids of two differentiating characters." The hybrid character appears in the middle of the series rather than on the left side (as in his reported $2:1:1$ ratio) because the point Mendel is preparing to make concerns not the distinction between the two kinds of constancy, but the mathematical behavior of the number of forms instantiating the two kinds of constancy. With continued self-fertilization the number of hybrid forms is reduced relative to those of parental forms at a ratio of $2^n-1:1$, where n is the number of generations. The purpose of the table on p. 14 of *Experiments* is to show that the constancy of behavior of the hybrid forms leads to a constant number of hybrids in each generation, while the numbers of parental types that are constant in form increases by 2^n-1 for each parental type, each generation.[56]

Mendel's character development notation reveals the core principle of his "law of development." Although there are two kinds of developmental constancy, there can be a single law of development to describe character constancy. Tracking the two kinds of constancy of dominant characters in the developmental series A + 2Aa + a explains the observation of $3:1$ ratios as *due to* $2:1:1$ ratios.[57]

Mendel goes on to consider multiple character hybridizations in which combination series express the (mathematical) products of developmental series. Thus, a hybridization of characters with developmental series A + 2Aa + a and B + 2Bb + b yields the combination series

AB + Ab + aB + ab + 2ABb + 2aBb + 2AaB + 2Aab + 4AaBb.

Note that in this series as well, the order of terms has developmental significance. The first four terms, Mendel points out, are constant in form in both characters. The next four terms are constant in form for one character and constant in behavior for the other. The last term is constant in behavior for both characters. The mathematical behavior of combination series reflects the developmental constancy of form and behavior of characters in hybridization experiments, which bring characters together by fertilization, exhibit constancy in development, and are parceled out in the progeny generation. Thus, although combination series expressions seem to us (moderns) to focus attention on factor transmission patterns that generate the ratios explained by hereditary laws of segregation and independent assortment, the order in which terms are presented reflects a basic developmental phenomenon that Mendel discovered through experimental control of hybrid development and a commitment to represent the phenomenon as developmental.

After formulating his theory of the development of hybrids in terms of the hypothesis that constant forms in the progeny result from the pairing of like with like, and the assumption that kinds of egg and pollen cells are formed in hybrids, on average, in equal numbers, Mendel describes a number of "experimental proofs" of the theory: confirmation of predictions of the ratios between hybrids and constant forms that must result in circumstances specified by experiments with controlled hybrid development. We need not go into details of the well-known backcross experiments, other than to observe that the predictions depend on the theoretical assumption that egg and pollen cells of all possible forms are developed in the hybrids in equal numbers. Thus the experimental proofs, like the theory itself, are developmental in character.

I mention the experimental proofs because they mark a transition between Mendel's attention to the developmental subjects leading to his theory, on the one hand, and explanation in terms of character transmission, on the other. Mendel makes the link between development and transmission the basis for claiming that the experimental proofs confirm his theory. That is, the developmental output of forms and progeny ratios from the hybrids, if the theory is true, is linked to the inputs to the hybrids via fertilization. The linkage comes in a sentence that begins with a statement about development and ends with a statement about fertilization: "the pea hybrids form egg and pollen cells which, in their constitution, represent in equal numbers all constant forms which result from the combination of characters united in fertilization."[58]

This linkage occasions a change from character development notation to a notation that foregrounds character transmission. Mendel had been following two processes in his experiments: (1) fertilization via crossing to produce hybrids and (2) development via the behavior of hybrids displaying dominant characters and their distribution in combination series among the progeny. If these processes are causally linked in a single process, then a notation designed to describe character development of hybrids ought to be formally equivalent to a notation suited to tracking characters in germ cells through fertilization. This is exactly what Mendel shows, noting that the simplest case of a developmental series for a pair of differentiating characters is "represented by the expression A+2Aa+a, in which A and a signify the forms with constant differentiating characters, and Aa the hybrid form of both." In considering the formation in development of four (kinds of) individuals from three classes of developmental constants (A, Aa, a), Mendel makes the notational link between development and fertilization by noting that the formation of the four individuals of three classes must involve "pollen and egg cells of the form A and a" taking part, "on the average equally in the fertilisation; hence each form twice,

The pollen cells   A + A + a + a

The egg cells   A + A + a + a

Figure 12.1
Germcell character notation.

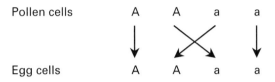

Pollen cells   A   A   a   a

Egg cells   A   A   a   a

Figure 12.2
Four kinds of fertilization processes.

since four individuals are formed."[59] The formulation of the germ cell character notation follows (figure 12.1): "There participate consequently in the fertilisation. . . ."

This notation concerns the inputs to fertilization, the process which is controlled by experimental hybridizations so as to yield the development of hybrids that can be followed into the progeny generations. Mendel elaborates on the fertilization process now that he has introduced a notation suitable for describing character transmission through germ cells which follows "the law of probability" (p. 25), extending the notation to represent the four kinds of fertilization processes that yield the possible kinds of individual hybrids (figure 12.2).

One further extension of the character transmission notation clarifies the results of the fertilization process to express the developmental kinds of hybrids that result from fertilization processes. Mendel writes (see figure 12.3), "The result of the fertilisation may be made clear by putting the signs for the conjoined egg and pollen cells in the form of fractions. . . . We then have . . ."

A   A   a   a

− + − + − + −

A   a   A   a

Figure 12.3
Mendel's character transmission notation.

A   A   a   a

− + − + − + − = A + 2Aa + a.

A   a   A   a

Figure 12.4
Notational Equivalent of Mendel's character development and character transmission
notations.

The preceding three figures together constitute a character-transmission-in-germ-cells notation, which is nevertheless different from the genotype notation of post-Mendelian genetics. Mendel's notation is designed to reveal the fertilization processes that bring together pollen and egg cells of like kind in which "consequently the product of their union must be constant, viz. A and a"—that is, the unions (fertilization processes) that bring about developmental constancy of form (parental characters, whether dominant or recessive). The other fertilizations (the second and third in the series) result in "a union of the two differentiating characters of the stocks, consequently the forms resulting from these fertilisations are identical with those of the hybrid from which they sprang. *There occurs accordingly a repeated hybridization.*"[60] Finally, Mendel establishes the notational equivalence of his character development and character transmission notations (figure 12.4): "We may write then. . . ."

"This represents the average result of the self-fertilisation of the hybrids when two differentiating characters are united in them."[61] The rest of the section describes more complex fertilization processes in which there are many more kinds of "participators" (pollen and egg cells of specified character).

Notational equivalence is the key to Mendel's whole theoretical argument. Fertilization processes, controlled in the setup of hybridization experiments, produce hybrids with two kinds of developmental constancy. Because of their mode of expression in combination series, these developmental series can be described in a single law of development deduced from the distribution of characters in the progeny of hybrids. Thus, the characters input to germ cells in fertilization by controlled hybridization experiments can explain, via the law of development of hybrids, not only the development of hybrids but also the distribution of characters in offspring. The character development notation focused attention on the inference backward from character distribution in the progeny to the development of the hybrid parents. The character transmission notation focuses attention on the inference forward from fertilization inputs, skipping across the development of hybrids, to the progeny distribution outputs. Thus, although the notations are formally equivalent theoretically, they foreground very different aspects of the process they both represent, and thus serve different research styles.

Perhaps I have belabored the obvious: Mendel inferred hybrid development from the distribution of characters in developmental series among the progeny. Transmission of characters, via the constitution of pollen and egg cells, explains developmental constancies of experimentally constructed hybrid parents in terms of the equivalence of their notations. But in doing so, Mendel thereby also created the possibility of using character combinations to predict character states of progeny without attention to the backward inference to the state of the hybrid parents that was required to construct his theory of development. The predictive power of character transmission notation foregrounds problems quite different from those rooted in Mendel's speculative theory of developmental processes in hybrids. That problem has become, in the hands and representations of geneticists, mere background to the discovery and use of the transmission notation suited to following characters' distributions across generations.

This exploration of Mendel's developmentalism illustrates not only the integral role of developmental thinking in the formulation of Mendel's theory and laws, but also the way in which his representational scheme linking processes of fertilization and development served a developmental goal at the same time it facilitated a shift of attention from the backward inference of hybrid development, from progeny distributions to the forward inference of character transmission.

As the founding father of genetics,[62] Mendel produced work that must count as part of genetics no matter how the field is construed. To discover that Mendel is a serious developmentalist no less than a protogeneticist does not in any way undermine the conceptual foundation of genetics, however. Rather, it reveals a deeper connection between developmental and genetic thought than is usually admitted.[63] In the next section, I briefly discuss another nineteenth-century founding father, August Weismann, whose doctrine of germinal continuity and somatic discontinuity girds the genetic perspective but was formed in a project of unification. In subsequent sections of this chapter, I consider embryological projects organized around following processes that also suggest how representational strategies can constrain and guide attention to foreground and background in ways that complement and contrast with the Mendelian focus on character transmission.

WEISMANN AS A GENETICIST

This anachronistic description emphasizes the point that nineteenth-century pursuits of a unified understanding of heredity/development[64] are open to various interpretations. Followers who articulate lines of work can shape styles and disciplines that diverge from the work of founders, whose representations provide wide scope for variously directing attention. The phenomenon results not so much from followers seeing the founder's work "as" some particular sort of thing, like seeing the movement of the sun at sunset "as" a motion of the Earth rather than of the sun.[65] It is more a matter of "seeing in"—seeing one possibility, pattern, or process "in" a representation rather than others "in" it. The focused attention that results in "seeing in" becomes the basis for abstracting to what is and can be seen "in focus," without excluding other views from potential awareness. Shifts of focus facilitate the emergence of new theoretical commitments that may change research styles.

Complementing Mendel, the developmentalist who became a founder of modern genetics, is August Weismann, the "geneticist" who pursued a theory of heredity that could explain development.[66] Weismann's work is separated from Mendel's by twenty years of significant developments in cytology that afforded a much more sophisticated and empirically grounded developmentalism on the cellular level. Weismann's explanation of development was rejected in his lifetime, while the fundamental implications of his unified theory for heredity were incorporated into the emerging science of genetics.

As I have argued elsewhere, Weismann held a symmetrical view of development as both cause and consequence of hereditary continuity.[67] Although his methods were not quantitative or experimental, like Mendel's, Weismann sought an explanation of hereditary continuity in terms of general, integrated principles of heredity/development in accord with the latest work in cytology, including his own work on the significance of polar bodies and reduction division in meiosis, as well as the differentiation of germinal from somatic cells.

Weismann's integrated account of differentiation as the result of separation of determinants in development, on the one hand, and hereditary continuity as the consequence of germ plasm sequestration inside the cells of the developing germ line, on the other hand, can be read as both developmental and hereditarian. Embryologists in the 1890s rejected Weismann's (and Roux's) "preformationist" theory of mosaic determinants of differentiation on account of the discovery of embryonic regulation by Driesch and others. In brief, the Roux-Weismann hypothesis became untenable, and Weismann's developmental determinism was broadly rejected.

However, those continuing to favor the nucleus as a privileged locus of developmental causation, such as E. B. Wilson, focused on germinal features of Weismann's symmetrical representations of development and on cellular processes of gametogenesis and fertilization in their own representations, breaking the symmetry between heredity and development in Weismann's theory, and making nuclear hereditary continuity the focus of theoretical attention.[68] Wilson's well-known and widely reproduced representations of Weismann's doctrine in his textbook, *The Cell in Development and Inheritance*, contributed to shifting attention toward heredity as a causal process separate from development.[69] Nuclear "monopoly" can thus be understood as foregrounding some aspects of a unified process of heredity/development in representations and backgrounding others.[70] This was a critical shift of theoretical attention by a key player whose textbook explicated and guided emerging lines of research. Wilson's representations formed the core of a new cytological perspective on the emerging discipline of genetics. The point, however, is that it was the shift of attention in Wilson's representation, not Weismann's, that served to differentiate genetics as a line of work separate from embryology.[71] While Weismann endeavored to track all the diverging cell lines of a developing body, Wilson's abstraction focused attention on the germinal cell line in a way that foregrounded a cytological interpretation of Mendelism.

FOLLOWING DEVELOPMENTAL FATE ACROSS CELL GENERATIONS

Several kinds of embryological work focused on the problem of differentiation, the development of organized heterogeneity out of the apparent homogeneity of the fertilized egg. The contested ground of when, where, and how in development differentiation takes place is of interest here for the reflected light it sheds on research styles. Commitment to study differentiation in particular ways is reflected in the choice of what to track in embryogenesis, how to track it, and, especially, when to begin and end tracking. The question of when to track differentiates not only lines of work in embryology but also embryological research from genetic research.[72] Thus, key aspects of the split between genetics and embryology trace not to following different biological processes of heredity and development, nor to the general structure of similar inferences from tracking the same process, but to tracking commitments that lead to different research styles.[73]

Various technical means were developed for following embryos through embryogenesis, the developmental process in which zygotes or eggs become adult organisms. Later in the twentieth century, development came to mean more than normal embryogenesis, including regeneration, dedifferentiation, cancer, and other phenomena.[74] My focus is not a full description or analysis of these varied kinds of embryological work, but rather a set of comparisons with Mendel's hybridization research style.

Methods for tracking differentiation and fate determination cut across traditional categories of observation, manipulation, and experimentation, but all involve procedures for following processes by means of marking interactions. The representational tools developed at the laboratory bench in each of several lines of work facilitated not only representations of the processes tracked but also a theoretical perspective on embryological work that foregrounded problems contrasting with those characterized as transmission genetics early in the twentieth century.

The process tracked by embryologists was literally the same one followed by Mendel: the development of hybrids. Embryologists also took variable or pure-breeding material from nature or from constructed stocks, and controlled conditions of both observation and development by means of a single set of marking interventions. However, embryological hybrids had origins and tracking significance different from genetic hybrids. Embryologists did not typically mark and control the development of the organisms they studied by introducing marks prior to fertilization in experimental breeding of a whole progeny, but rather by postfertilization marking of single embryos.

Thus, the points at which embryologists began tracking typically occurred later in the heredity/development process than those Mendel tracked. Nor did their inferences have the same temporal scope as Mendel's. Marks introduced later in a process produce a shorter causal "cone" of propagating tracks, and their introduction into single embryos rather than whole progenies also narrows the cone spatially. However, embryological inferences were constrained to run along paths similar to those of genetic inferences because they concerned the same process.

In most lines of embryological work, observation was focused and controlled by marking the embryological process, either with mental marks that served the production of hybrid representations or with physical marks that hybridized the developmental system under observation. Thus, instead of aiming at a theory of the development of hybrids, the embryologists aimed at a theory of (normal) development inferred from hybridizing marks.[75] In this inverted theoretical spectrum of problems and methods, the focus is on the path of development rather than on the hybrid character of what develops.

An important embryological problem of the late nineteenth century grew in response to Haeckel's view that differentiation of embryos into forms characteristic of particular branches of the phylogenetic tree begins with gastrulation, and that the ancestor of all life therefore resembles a gastrula.[76] Some, such as C. O. Whitman, doubted that organ differentiation is determined only at the stage where it is first observed morphologically.[77] The skeptics sought differentiation in earlier embryological stages, some by attempting to work at the cellular level to find its earliest manifestations, perhaps in the zygote or even the cytoplasm of the egg cell. These "cell lineage workers," already very familiar with the end results of development, shifted attention to early cleavage stages of blastulation.[78] They sought to identify the fate or prospective significance of cells that did not yet manifest the differentiated states of the kind of tissue or organ to be explained, whether epidermis or mesoderm, neural plate or lens, notochord or somite.

Cell lineage embryologists offered causal narratives in which the fate of a part or region becomes evident by tracking it through a sequence of stages leading to a differentiated outcome.[79] Thus, cells in embryogenesis were taken to have a double significance: a present significance at each point of observation in development and a prospective significance for future states.[80] A key representation of the results of such work came to be called a fate map, identifying cells or embryo regions in terms of fate or prospective significance rather than in terms of present significance.[81]

Although the history of this work is fascinating, I am concerned here with only one aspect: the laboratory bench representations used to track embryonic change so as to follow the process forward to differentiated outcomes.[82] Tracking work provides the basis for causal narrative accounts of prospective significance, which involves two shifts of attention: (1) from developmental outcome to some earlier stage of a central subject significant to the narrative from which to begin tracking, then (2) tracking the historical process forward in time, conceptually "back" to the future developmental outcome from which the narrative account began.[83]

At least three nonexclusive kinds of process-following in embryological work can be identified according to the kind of marking interaction used to track the process: (1) mental marking of embryos with corresponding physical marking of diagrams, (2) physical marking of embryos with artificial substances, and (3) physical marking of embryos by heterospecific tissue hybridization. These methods of following embryological differentiation led to a variety of visual representations that foregrounded phenomena, methods, and theories we now take to be embryological and that backgrounded phenomena, methods, and theories we now take to be genetic. In the following, I will focus on a few illustrative points.

FOLLOWING EMBRYOS BACK TO THE FUTURE

Mendel's methodology involved (1) identification of adult characters that bred true in pure-line preparations, (2) experimental crosses to produce hybrid organisms, (3) enumeration of the hybrid progeny by character type, (4) self-fertilization to breed more generations from the hybrids, and (5) inference of the developmental state of the hybrid parents from the statistical distribution of characters among their progeny. The result was a theory and a mathematical law of the development of hybrids that describe the constancies of form and behavior of characters in development. Additionally, Mendel offered experimental proofs confirming novel predicted progeny distributions from backcrosses, which he had not used to formulate his theory and law of development.

Embryologists of the late nineteenth and early twentieth centuries engaged in tracking styles that shared important elements with Mendel's hybridization work, but with distinct goals and details of method, technique, and experimental subjects. If the tracking styles of presumptively genetic and embryological research are similar, the disciplinary split must be explained by differences located elsewhere. Nineteenth-century representations of a unified biology of heredity/development such as Mendel's and Weismann's have been read as contributions to a theory of heredity by later workers who were able to focus on those hereditary aspects of the process that had been rendered easily abstracted from developmental aspects. F. R. Lillie claimed that the problem of the relations between genetics and development was not "visualized by Darwin and by Weismann, because, for each of them, the theory of development included the theory of heredity." A few lines later, Lillie notes that "Since Weismann, physiology of development and genetics have pursued separate and independent courses."[84] Just as abstractions of heredity from development foreground the problems of an emerging genetics research style, so abstractions of development from heredity drove the emergence of an embryological research style in the twentieth century. Tracking the development of research styles is not, by itself, enough to explain them, but it is an important part of the explanatory project to document the representational openness of nineteenth-century unifiers that facilitated the diversification of subsequent lines of research.[85]

TRACKING EMBRYO PARTS IN SEMI-DIAGRAMS

The descriptive cell lineage work of C. O. Whitman, E. B. Wilson, F. R. Lillie, E. G. Conklin, and others[86] presents important examples of mental marking and a mode of embryological inference aimed at interpreting the developmental significance of cells and embryonic regions.[87] In this work, scientists observed embryos developing under a microscope. A camera lucida device was fitted to the microscope in such a way that the observer could draw (trace) virtual images portraying the embryo in real time on a piece of paper adjacent to the specimen.[88] One goal of this work was to trace cell genealogy through embryogenesis as the cells divided, moved relative to each other, and became progressively obscured from view behind or within the growing mass of cells in blastulation, gastrulation, and beyond.

Whitman coined the term "semi-diagram" to describe the workbench representations he produced in the course of his work with this technique. Cell outlines depicting an embryo—an embryo portrait—were traced during observation. Labels for cells and embryo regions, and arrows indicating cell movements or cell genealogy, were then superimposed on the tracing to produce a representation that is semi-diagrammatic (i.e., both pictorial and symbolic) of the process tracked by means of minute shifts of visual attention between the developing specimen and the articulated drawing.[89]

Whitman mentally marked a specimen by focusing his attention during observation on noteworthy features such as position, relative size or shape, or pigmentation, of a cell of particular interest. At the same time, he physically marked his embryo portrait in ways that turned the picture into a semi-diagram of the process he was tracking (figure 12.5). Whitman's semi-diagrams thus exhibited representational features that facilitated foregrounding of either "hereditary" (cell genealogy) or "embryological" (cell fate, differentiation) aspects.

E. B. Wilson and many other regular visitors to the Marine Biological Laboratory at Woods Hole followed in Whitman's footsteps.[90] Wilson, in his own cell lineage work, made innovative representations of cell genealogy as he traced cell lines to later stages of development (figure 12.6). In these, he brought together the representational styles of Whitman and those of Theodor Boveri in Germany, with whom Wilson had worked.[91] Rather than depict only early embryo stages that could be fully pictured with the camera lucida technique, Wilson abstracted what he saw to produce a fully diagrammatic representation that was all symbol and no portrait. These genealogical diagrams showed only the pattern of diverging lineages, with labels to indicate some vestige of the spatial information contained in Whitman's semi-diagrams. Despite this difference, Whitman's and Wilson's techniques are similar: mentally mark the embryo in observation and physically mark a diagram to track a process of cell division leading from a determined state to a visible embryonic differentiation.

E. G. Conklin illustrates a different orientation to the problem of mentally marking embryos in order to track determined states through

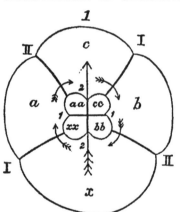

Semi-diagrammatic surface views of the egg of Clepsine in different stages of Cleavage.

Diag. 1. — The eight-cell stage, showing the relation of the embryonic axis to the first two cleavage-planes. The arrow, 2–2, shows the median plane of the embryo, and the four small arrows indicate the direction in which the four micromeres have rotated on the axis of the egg.

Figure 12.5
C. O. Whitman's "semi-diagram" (1887, diagram 1, p. 109). The camera lucida drawing of cell outlines constitutes a cell portrait, while the symbolic elements (letters and arrows) are diagrammatic. The combination of the two results in an image that is "semi-diagrammatic."

Figure 12.6 (*facing page*)
E. B. Wilson's "genealogical" cell-lineage diagram (1892, p. 382).

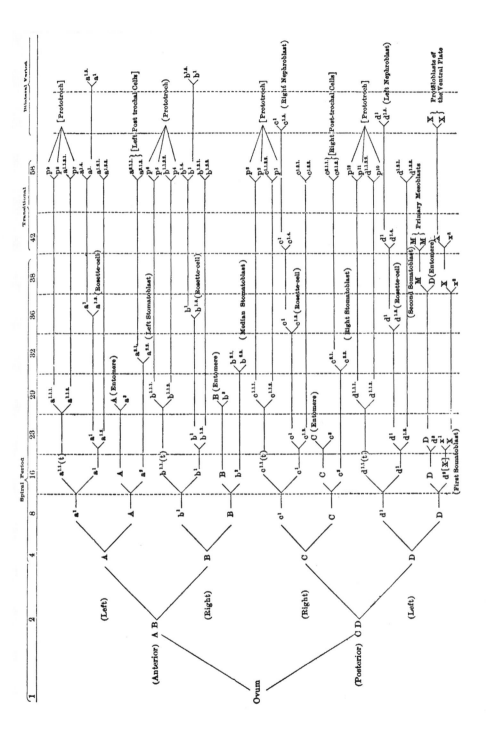

development.[92] Although Conklin, like other cell lineage workers, used camera lucida techniques and noteworthy observation to mentally mark cells or regions, in some cases he relied specifically on pigment markings of cells, which behaved as though the observer had introduced a persistent physical mark directly on the embryo. Other sorts of noteworthy features, such as cell position, size, or unusual shape (e.g., protruding lobes), tended to come and go during development, limiting their utility as marks. Conklin noticed that in the ascidian *Cynthia partita*, a yellow, crescent-shaped band of pigmented cells appeared at a certain point and could be followed through embryogenesis to the determination of mesodermal tissue (figure 12.7). Presence of yellow pigment in a cell at a later time meant membership in the cell lineage tracing back to the original mark.

Cell-lineage workers interpreted organ differentiation in terms of the consequences of prior determination within cell lines, as opposed to pre-deliniation of the actual differentiated states, at least for those species with what Conklin called "determinate cleavage" (see Conklin, 1905, p. 9). They aimed to avoid the commitments of earlier "preformationists," but at the same time focused their attention on the paths along which determined states lead to differentiated ones. As I will elaborate below, these paths of differentiation are the embryologist's version of a "hereditary" process abstracted and foregrounded from those aspects of the same process that concerned geneticists.

Here I want to make two points about research style. First, the pattern of tracking and inference parallels Mendel's work on the development of hybrids. A marking interaction is a kind of "hybridization" in the sense that it involves an intersection of two processes. In Mendel's work, an experimental cross brings together different characters and tracks them through progeny statistics. In cell-lineage work, a developing embryo and an articulating drawing are brought together by the minutely shifting, focused attention of an observing embryologist who mentally marks the distribution of cell lines and physically marks a distribution of symbols in diagrams. In both lines of work, the aim is inference about an earlier stage of the process on the basis of a distribution of progeny (organisms or cells) later on.

The second point concerns the dual role of the representations. In Mendel's demonstration of notational equivalence, the possibility of reinterpreting his work on the development of hybrid characters in terms of a factor theory of hereditary transmission was so strong that it took very careful analysis by historians to show that Mendel probably did not hold the factor theory with which he is credited.[93] Mendel's notational

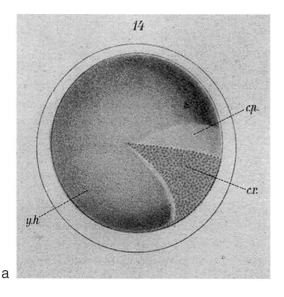

a

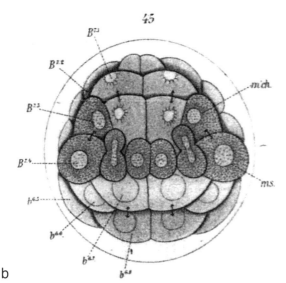

b

Figure 12.7
(a) E. G. Conklin's mental mark using yellow crescent in *Cynthia partita* (Conklin, 1905, pl. II, fig. 14). Conklin's captions for his figures 13 and 14 read: "Side views of egg, showing the formation of the crescent (cr.) from the yellow hemisphere; in all the figures the animal pole is above, the vegetal pole below. Above the yellow crescent is an area of clear protoplasm (c. p.)." In this figure, Conklin identifies a noteworthy feature that can be used as a mental mark to follow the yellow pigmented area through subsequent cell divisions. (b) A forty-four-cell stage embryo, "posterior view, showing separation of another mesenchyme cell from a muscle cell."

conventions facilitated not only his articulation of a theory of the development of hybrids but also a new, "genetic" interpretation of his developmental work by readers such as Bateson and Correns. Mendel's character notation was designed to serve his own theoretical purposes, not necessarily those of his followers. Mendel sought to focus his attention on the independent quantitative behavior of characters in hybrids in order to support his unified, law-based theory of character constancy. The hybrids were the key because, on the one hand, they exhibited a kind of constancy (of behavior) different from that of the parental characters (constancy of form) but, on the other hand, only the progeny generated from the hybrids provided information about character development in their parents. Thus, it was important to Mendel to find a single law of constancy in order to extend the theoretical interpretation of development of hybrids to the parental characters as well.

Thus, Mendel's representations functioned both as working objects of his own theoretical investigations and as facilitators, in abstraction, of a new set of research commitments. The role of his representations in his own theoretical project served developmentalist research commitments quite different from the genetics (and developmental biology) that followed. Similar conclusions apply to cell lineage work in embryology.

First, camera lucida drawings provided semi-diagrammatic, workbench means of attending to events occurring in the embryos under study. The "virtual" quality of camera lucida images allowed minute shifts of visual attention between specimen-watching and image-making so that each mark on the drawing provided moment-to-moment feedback to visual attention *to the embryo*. Second, because the image persists after the embryo has been tracked,[94] it could be manipulated via further symbolic annotation, and new diagrams such as Wilson's genealogical diagram could be drawn to abstract features of theoretical interest from the semi-diagrams.

As I have argued elsewhere, abstraction of genealogical form from cytoembryological content through the history of cell-lineage diagrams facilitated an identification of the cell-lineage workers' findings on fate determination in embryogenesis with Weismann's doctrine of germ plasm continuity and somatoplasm discontinuity.[95] Although "Weismannism" does not accurately reflect Weismann's views any more than "Mendelism" reflects Mendel's views, it is a clear example of how the working drawings of cell-lineage workers facilitated the theoretical abstraction of Weismannism and the conceptualization of Mendelism as the foundation for a modern causal theory of heredity. Thus, the Janus faces of scientific representations from both sides of the genetics/embryology divide facilitated the origin of genetics as a new line of work for a first generation of "embryologists" such as T. H. Morgan and W. E. Castle and as a distinct discipline for their students.[96]

Marking Embryos with Embryos

Thus far, I have focused on one line of classical descriptive cytoembryology. A similar tracking and reasoning pattern holds for manipulative experimental embryology. Walther Vogt, in a series of essays in the mid-1920s, pioneered manipulative physical marking of embryos with artificial mica chips and vital dyes.[97] (see figure 12.8.) These can be thought of as more controllable, artificial, interventionist versions of Conklin's mental-marking use of yellow-crescent pigment granules. The dyes could be accurately placed more or less at will on numerous, very small surface regions of embryos and tracked through many cell generations. This was a crucial step in extending embryological work on fate determination and differentiation to the embryos of vertebrates, which generally do not have the transparent embryonic cells of the invertebrate species studied by the descriptive cell-lineage workers.[98] While Vogt's artificial marking technique extended the research style of the descriptive cell-lineage workers, a more manipulative technique had emerged earlier that had both benefits and disadvantages compared with artificial marks.

Hans Spemann and others (e.g., Otto Mangold) engaged in transplantation experiments beginning early in the twentieth century.[99] The problem of interpreting qualitative differences of the developing cells of embryos led Spemann to his first interspecies transplant experiments in 1918 and 1919. In 1906, he had invented a technique using glass needles to conduct microsurgery on embryos.[100] He used it to transplant embryonic material between two species of newts in order to study which regions of the embryo contribute to the formation of the neural plate.

The key to the new technique's success as a marking procedure for Spemann was that one newt species, *Triton taeniatus*, has pigmented eggs, while another species, *Triton cristatus*, does not. When Spemann transplanted presumptive epidermis from *T. cristatus* into a hole cut with a glass needle into the presumptive medullary plate in *T. taeniatus* and transplanted the *taeniatus* presumptive medullary plate material into the hole made in *cristatus* presumptive epidermis, he thereby marked each embryo *in the same procedure* by which he made experimental manipulations to

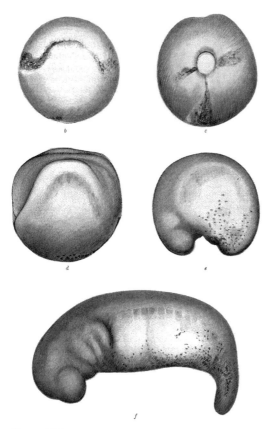

Figure 12.8
Walther Vogt (1925, fig. 14, p. 583). Vogt's physical marking technique using vital dyes. Different dyes, staining blue or red, could be traced through subsequent stages of development.

determine which kinds of presumptive tissue contributed to neural plate development. The *cristatus*-into-*taeniatus* transplant put white (unpigmented) tissue into a dark background while the complementing transplant put dark tissue into a white background (figure 12.9). The question was whether each tissue would differentiate according to its origin in epidermis or mesoderm or according to its transplanted location in the complementing kind of presumptive tissue. In these experiments, the point was not to study interspecies embryological hybrids, but only to take advantage of the marking effect of transplanted heterospecific material to explore whether regions were already determined to a particular fate or whether they were indifferent, and thus susceptible to induction.[101]

a

b

Figure 12.9
Hans Spemann and Hilde Mangold (1924, fig. 1, p. 16, and fig. 3, p. 17). Heteroplastic transplantation marking technique. Left panel: *cristatus* embryo at neurula stage with *taeniatus* implant (dark). Right panel: *taeniatus* embryo at neurula stage with *cristatus* implant (light).

Spemann's application of heterospecific transplantation in the organizer experiments of the early 1920s, conducted by his student Hilde Mangold, brought together several strands of thought and technique from the preceding decades of work.[102] The idea that the blastopore dorsal lip played a determining role in gastrulation was stimulated by Spemann's early experimental work in which he constricted the early blastula and noted that in cases where both constriction products received dorsal lip, each developed into a whole embryo, but where one received the dorsal lip and the other did not, the former developed into a small whole embryo while the latter developed into a *Baustück* of partially organized material. The idea of forming chimeric embryos grew out of experiments to explore the extent of cooperation among cells from different species.[103]

In the 1924 Spemann-Mangold paper, heterospecific transplants from *cristatus* into *taeniatus* of material from the region of the blastopore dorsal lip were conducted in order to track the progress of the lip material through gastrulation. The unpigmented *cristatus* cells appear in whole-embryo and histological sections as white cells against a background of dark cells, so structures to which the transplanted cells contribute (or wholly form) can easily be discerned. Reciprocal transplant experiments were not performed because the *cristatus* embryos rarely survived removal of the vitelline membrane during the transplant operation.[104]

The paper argues that while prior to gastrulation some regions contain cells as yet indifferent to their later fate, the dorsal lip of the blastopore is determined prior to gastrulation, so its movement inside the embryo leads it to play an inductive role. Spemann and Mangold called it an "organizer," and the region from which it was extracted, an "organization center." The transplanted dorsal lip influences its environment, but the nature of the effect depends on its precise location. Transplantation into the "normal zone of invagination" resulted in participation in normal gastrulation. Transplantation into an area of "indifferent" tissue resulted in autonomous invagination by the *cristatus* "organizer" and the development of a secondary embryo with varying degrees of differentiation and of integration with the primary embryo.[105]

Key conclusions of the paper are tentative. Although the authors are confident that the "organizer" plays an inductive role, they do not know the mechanism. They cannot distinguish with certainty between its playing the role of a mere trigger to normal gastrulation in indifferent tissue and of a determiner of the course of development subsequent to gastrulation according to a fate previously determined in the donor *cristatus* embryo. In the latter case, they write, "The organizer, by virtue of its intrinsic developmental tendencies, would essentially continue its development along the course which it had already started and it would supplement itself from the adjacent indifferent material" (p. 38). The integration of the transplanted material, they argue, rules out its complete self-determination in the new context. Thus, the paper stands as a crucial argument for a new epigenesis. Development proceeds neither in virtue of full predelineation by inherited factors or determinants, nor with fully flexible regulation at all stages. Indeed, prior to this concluding argument, Spemann and Mangold had argued that "Definitely directed inherent developmental tendency and capacity for regulation are not mutually exclusive."[106]

Here, determined developmental states with prospective significance play the same theoretical role as characters in Mendel's work on the development of hybrids. However, the causal scope of the hybridization was restricted because transplantations were made at the blastula stage (rather than at fertilization, as in Mendel's character hybridization work) and tracking ended when gastrulae began to deteriorate as a result of complications induced by the hybridization. Spemann and Mangold created hybrid cell genealogies in their experimental transplants, so in a way they were performing the same kind of manipulation as Mendel's experimental hybridizations.

Despite the similarities, however, only in the inverted theoretical spectrum of embryology is tracking attention focused on the *path* of developmental induction in the hybridized material rather than on its developmental *state*. The developmental path and the determining role of the transplanted dorsal lip material are tracked by means of the heterogeny of hybridized embryo parts (i.e., how the white and dark materials become distributed). Tracking is used to infer the developmental state of the transplanted material (whether determined by its origin or induced to a determined state later by its new environment). However, the point of establishing the earlier state is to explain a role in the pathway to differentiation in and beyond gastrulation so as to interpret developmental constancy of the process in terms of hybrid embryos rather than to interpret hybrid characters in terms of a law of constancy of form and behavior. As Conklin had noted, determined blastomeres "are constant in their manner of origin and development" (1905, p. 95).

Importantly, embryological inferences about developmental pathways were interpreted by Spemann and Mangold in terms, if not applications, that resemble descriptions of hereditary phenomena. They wrote about "differences within the organization center that could hardly have been *transmitted* to the induced embryonic primordium by stimulation of gastrulation alone."[107] The transmission of a determined state is no mere trigger, in other words; it requires epigenetic regulation by induction. Similar requirements were imagined for classical genes—"difference makers" transmitted from parent to offspring—to determine phenotypic differences by means of differential expression. Spemann and Mangold also wrote "that the possibility exists of a determining effect progressing from cell to cell . . . also during later developmental stages." "This conception," they continued, "of progressive determination leads of necessity back to the conception that there are points in the developing embryo from which determination emanates."[108] Their work with the experimental hybridization mark methodology goes beyond "the facts that were known earlier," which "sufficed only to establish the concept of a starting point for differentiation, but not to demonstrate the real existence of such centers. To obtain this evidence," they continued, "it is not enough to separate the region to be tested, which is believed to be such a center, from its potential field of activity. It must be brought into contact with other parts, normally foreign to it, on which it can demonstrate its capacities."[109]

These statements resemble Mendel's argument for his theory of the development of hybrids. The mental marking activity of observational, comparative embryology (like the mental marking of Mendel's hybridist predecessors) revealed deep constancies of embryological form but also overwhelming variation within phyla.[110] Only in the context of an experimental embryological hybrid, where parts are brought into interaction with parts "normally foreign to it," Spemann and Mangold argue, could the developmental capacities of a part be revealed. But unlike Mendel, Spemann and Mangold were interested in those capacities for what they reveal about the process of development, of "the main problem, i.e. the harmonious patterning subsequent to gastrulation."[111]

I do not claim that these embryological accounts of the process by which determined states are propagated "from cell to cell" constitute a theory of heredity in any current sense of the term. But they *are* elements of a theory of heredity in a sense appropriate to a turn-of-the-century unified theory of heredity/development. In order for transgeneration transmission of characters of the sort Mendel described to take place, within-generation transmission or progress of determined states must occur. Mendel's methodology did not permit cell-to-cell tracking, however, any more than Spemann's methodology allowed tracking the progeny distributions required to infer Mendelian hybrid states. Nevertheless, from both points of view, heredity and development are intertwined—in the new epigenesis envisioned in Spemann's organizer concept as much as in the new preformation envisioned in Mendel's development of hybrids.

Embryologists interested in the times and places of embryonic determination followed an inferential pattern similar to Mendel's: (1) identification of adult or other outcome states (sometimes the state of a neurula or gastrula rather than an adult); (2) marking in some way an early embryonic cell, tissue, or region; (3) tracking the marked parts through development to the outcome state; and then (4) inference of the fate or state of determination of the earlier marked part in terms of where, and how, the mark ended up in the later stage or adult. Few embryologists were interested in the sort of quantitative, mathematical laws that Mendel sought, but what is relevant to present concerns is the structure of the inferences in light of the marking interactions needed to track a process.

For both Mendel and the embryologists, work began with knowledge of an outcome state in terms of which to organize a tracking project from an early point in the process. The continuity of process tracked via marks from input to output permitted an inference of the role of the earlier stage, character, or part in determining, causing, or becoming the later stage, character, or part.

However, Mendel and the embryologists reached very different outcomes due to the particular tracking choices each made. Where Mendel used parental characters to mark development in its earliest stage (the zygote) by means of experimental hybridization, and then tracked the development of the hybrids by quantifying progeny distributions, embryologists marked early developmental stages and followed the marks continuously or continually through embryonic stages to an end stage of interest (often no farther than a gastrula or neurula). Although Mendel and the embryologists both engaged in tracking the very same process of heredity/development, where one draws attention to the relation between stages and skips over details of the process connecting them (Mendel), the other draws attention to the detailed transformations (embryologists), but within a narrowed temporal and spatial scope.

Heredity/Development Redux?

By the mid-1920s, it was clear to both geneticists and embryologists that their field of research had split, not only into separate lines of work but also into separate disciplines in distinct social worlds.[112] Those whose careers had taken them through the historical divergence, particularly leaders of the new disciplines of genetics and "physiology of development" (i.e., embryology on the way to developmental biology), had much to say about the relations between the fields as well as the prospects and desirability of "reunion" (as F. R. Lillie called it).

The present chapter serves a larger argument that the historical split between embryology and genetics in the early twentieth century—as well as the continuing conceptual difficulty of organizing genetic, developmental, and evolutionary theories into a coherent synthesis—are problems of meshing research styles and representational practices that cover the same terrain in different but intertwined ways. In my view, while discipline-forming rhetoric supported a parting of the ways between embryology and genetics in the early twentieth century, in a more fundamental sense the gene theory not only had an embryological origin, it never really left embryology at all in the broad and significant sense that includes Mendel's developmentalist project.[113]

It does not follow from the divergence of research styles and representational practices in genetics and embryology that nature is divided into separate processes of heredity and development. In this chapter, I have argued that we can detect the separation of fields in the representational practices of scientists following the one process of heredity/development. Putting heredity and development back together again can thus be thought of, in part, as a problem of conceptual reorientation—of change in theoretical perspective—to recognize that theories of heredity entail methodologies of development, and conversely. Moreover, in this chapter have I argued for a reorientation of thought about the theories: Mendel's theory was a theory of development, built on the same methodological and inferential structure of process-following as the work of embryologists. Putting Humpty-Dumpty together again may well be more a matter of instigating realignment of perspectives than of transferring tools or even problems between separated fields.[114]

There is a familiar joke about Thomas Hunt Morgan, winner of the Nobel prize for the theory of the gene in 1933, who began his career as a descriptive embryologist working on the phylogeny of sea spiders. Early in his career, Morgan turned to transmission genetics in *Drosophila*, but at the end of his career wrote a book titled *Embryology and Genetics*.[115] The joke is that the only synthesis Morgan achieved between embryology and genetics was the "and" in the book's title. Underlying the joke, and belying insiders' narratives, is the fact that Morgan persisted throughout his career in his belief that development and heredity were one subject, however asymmetrical his practical pronouncements might have been. In the year of his discovery of a white-eyed fruit fly and the beginnings of his acceptance of a gene theory of *both* hereditary transmission and trait development, Morgan wrote: "We have come to look upon the problem of heredity as identical with the problem of development."[116] His opposition to the Mendelian theory of character transmission was not that it mixed up hereditary transmission with developmental realization of hereditary potentials, but that it got the relation between them *wrong*.

In his 1926 Sedgwick Lecture (the same year he published *Experimental Embryology*), Morgan made it clear that he was *not* arguing that development was merely a subject to be ignored by transmission genetics for practical, pragmatic reasons. Rather, he urged that careless claims by *geneticists*, which gave the impression that genetics could explain everything about development, were just as lamentable as the belief by some embryologists that geneticists must be stupid to think their work could have a bearing on the important problems of development. His distinction of heredity as a phenomenon of (mainly nuclear) transmission, and development as a phenomenon of cytoplasmic change, acknowledges the divergent styles and methods of genetics and embryology; it does *not* entail a radical Weismannist separation of *causes* of transmission and expression, however much we might read it that way today.

Indeed, Morgan was a radical defender of the role of the cytoplasm. His purpose in 1926 was to clarify and negotiate shared understanding and cooperation among diverging specialties—a task of unification, not segregation. Geneticists, he argued, were working on a different aspect of the problem than embryologists were, not attempting to explain the phenomena of development genetically. He wrote:

> There has been some criticism of the theory that the genes are the exclusive factor in heredity, on the grounds that the cytoplasm can not be ignored in any complete theory of heredity. There is no need, I think, for misapprehension on this score. The confusion that is met with sometimes in the literature has resulted from a failure to keep apart the phenomenon of heredity, that deals with the transmission of hereditary units, and the phenomenon of embryonic development that takes place almost exclusively by changes in the cytoplasm.[117]

Morgan's separation of *phenomena* can be read either as a separation of fields and styles of research or as a proclamation of a separation in nature. To speak of "the phenomenon of heredity" as dealing with "the transmission of hereditary units" is to speak of the research commitments of genetics to follow transmission processes rather than cytoplasmic changes. And his suggestion, in the same paper, that embryological work must control genetic variability, or else the production of developmental variation (i.e., variation in the causal scope of embryological tracking work) by experimental control of environment will be confounded with genetic variation (i.e., variation in the tracking scope of genetic work), puts genetics in the cooperative service of embryology, not the other way around:

> It has long been known that the environment is one of the causes of variability in embryological development, even within the range of changes that are normal. It had not been so well appreciated, until genetics made the situation clear, that genetic elements may produce effects that superficially at least are often indistinguishable from those produced by the environment. Here, then, are two variables producing like results. Now, the most promising lead that we have at the present time *in the study of the development of the living organism* is to vary the environmental conditions—especially the temperature—in order to get data as to the nature of the processes that are taking place. Unless the other variable—the genetic constitution—is under control it is hopeless to try to get reliable data.[118]

Finally, the infamous joke book of 1934, *Embryology and Genetics*, formulates what Jan Sapp has called the "paradox of development": that genes must explain development if the theory of the gene is true, but how can they, since every cell of a differentiated multicellular body contains all the same genes?[119] Morgan stated the paradox this way: "At first sight it may seem paradoxical that a guinea pig that can develop areas of black hair should have white areas of hair if, as is the case, the cells of both areas carry all the genes."[120]

The point to notice is that this is a "paradox" about development that is to be faced by *geneticists*. It it a challenge to geneticists to understand development in order to put their genetic house in order, not a reductionist's hubris trying to show that problems of development are genetic problems. Indeed, Morgan's preferred view of the matter was that "The initial differences in the protoplasmic regions may be supposed to affect the activity of the genes. The genes will then in turn affect the protoplasm, which will start a new series of reciprocal reactions" (p. 134). This is to say that the problems of development will be solved not by looking to the genes, but to their activation by *protoplasm*.

Moreover, the critical commentaries of many prominent geneticists and embryologists throughout the period of the split (1910–1930), the rise to public prominence of genetics (1915–1930s), and the emergence of molecular biology (1940s–1970s) all show continuing concern to hold the problems of development and genetics together, despite divergent styles, concepts, theories, methods, and institutions. Thus, I favor mining genetics for insights, not in support of the reduction of development to developmental genetics but for the heuristic production of a unified perspective which recognizes that genetics entails an idealized and abstracted account of development and development entails an idealized and abstract account of heredity.[121] Rather than dismissing classical genetics as merely false in its black-boxing of development, and classical embryology as backwater in its ignoring of genes, I propose that we rethink classical genetics as a false model that may provide heuristic means to a truer theory of development.[122]

CONCLUSION: FOLLOWING SCIENCE IN THE LIMELIGHT

The argument of this chapter is historiographic as well as philosophical: there is a temptation to confuse the divergence of research styles in separating scientific social worlds with a historical progression of fields or lines of work that take the scientific limelight in turn. It is a reflection of the times that the historiography of twentieth-century biology has so frequently pursued "success narratives"—accounts of torch-passings from successful field to successful field as one "star" problem, method, model organism, theory, scientist, school of thought, line of work, or field takes the limelight from another. It is no doubt also a reflection of the times that the same historiography has taken the (rapid or retarded) rate of scientific progress as a primary explanatory goal.

These features are probably due in part to the uses to which scientists have put their own insider histories, and these have informed wider historical investigation. Standard histories of the origins of genetics, for example, "have explored the complex relations of genetics to embryology primarily in order to illuminate the emergence of the gene theory and genetics."[123] Fields in decline or "a period of depression," as the eminent embryologist Ross Harrison described his field in 1925, are ignored by such narratives during their low points with all the embarrassment felt in relation to a sick relative, until they finally die or rise again, as Harrison thought might have been happening in the "gold rush" to experimental embryology a decade later.[124]

Günter Stent, writing fifty years after Harrison, echoed his concerns about another root of modern embryology when he observed that despite "highly promising beginnings, the study of developmental cell lineage went into decline after the turn of this [the 20th] century. It remained a biological backwater for the next 50 years." Stent attributed the decline of cell-lineage work to the discovery of regulative and inductive phenomena during the "gold rush" which challenged the view that cell lineage could be a causal factor in cell differentiation.[125]

Stent's interest, as a scientist, in the revival of cell-lineage work in the 1980s was "accompanied by the introduction of analytical techniques more precise and far-reaching than those available to Whitman and other nineteenth-century pioneers."[126] Thus, attention to periods of decline tends to be limited to interest in pathologies of method and poverty of imagination as explanations for lack of success. How and why the cell-lineage style of work persisted through those fifty years of decline in the backwater, so as to make a comeback in the 1980s, is a question rarely asked by those whose historical interests center on success. As Maienschein argued, "The emergence of a program in genetics has served for scholars as a favorite example of productive scientific change, specifically of theory change, and research has focused on the apparently more productive program after the change *to* genetics."[127] If our historical question focuses instead on diversifying lines of work within segmenting social worlds, then the fate of cell-lineage work becomes a key problem in the genealogy of (one strand of) current efforts to reunify heredity and development.

Maienschein looked back to uncover the sources of this productive change in the commitments of earlier research programs in heredity/development. In this chapter, I have tried to articulate conceptual resources for changing theoretical perspectives with the hope that they will be helpful in tracing forward the continuation and development of research styles practiced throughout and following the split between embryology and genetics. Specifically, I have described ways in which the practices of fields in decline, such as descriptive embryology, may be maintained in the background, as methodological commitments, for the success of projects and problems foregrounded by rising fields such as genetics. While it may be true that the fortunes of classical genetics rose while those of descriptive embryology fell, the continued practice of scientific styles out of fashion requires historical investigation if we are to understand the emergence, problems and prospects, and historical continuity of hybrid or intersectional fields such as evo-devo at the end of the twentieth century. We need to know as much about embryology and the emergence of developmental biology from 1940 to 1980 as we do of its golden age from 1880 to 1940 in order to have anything serious to say about relations between embryology and genetics. But this is made more difficult by the hegemony of success narratives trained on the scientific limelight.

Conventional narratives of the history of biology have it that Darwinian evolutionary theory was in the limelight from the mid-nineteenth century until the phylogeny-reconstruction craze fueled by Haeckel's bio-genetic law ran into trouble. Embryology for its own sake took the torch late in the century with cell-lineage, fertilization, and experimental devel-opmental mechanics studies. The rediscovery of Mendelism and the emer-gence of the chromosome theory of heredity transformed the "new preformationism" into a respectable competitor to epigenesis in the early twentieth century, so that even staunch defenders of epigenesis such as T. H. Morgan were won over to a nuclear theory of the gene.[128] The rise of genetics, with its doable problems of hereditary-factor transmission,[129] displaced descriptive embryology and its nineteenth-century goal of a unified account of heredity, development, and evolution, and even eclipsed experimental embryology for a time. Experimental approaches such as Spemann and Mangold's organizer experiment brought experimental embryology back into vogue briefly, which earlier physiological and chem-ical embryologies had failed to do.[130]

However, the juggernaut of molecular biology at midcentury com-pleted the ascendancy of genetics, and with it the evolutionary synthesis of genetics and systematics, which left comparative morphology and devel-opment on the sidelines.[131] Only in the late twentieth century, with renewed attention to the reunification of evolution and development (which the skeptical embryologist Lillie had already discussed in 1927 as the problem of the "reunion" of genetics and embryology), was the hope rekindled that the severed and several concerns of evolutionary, genetic, and developmental biologists might share the limelight, spurred by the excitement of discoveries in developmental genetics and deep phylogeny reconstruction, as well as by a resurgence of interest in (and confusion of) epigenesis alongside the new molecular epigenetics.[132]

While such limelight narratives make instructive drama, they con-found understanding of social processes.[133] For the sake of constructing a narrative of field succession under a single spotlight of historical attention, interactions among contemporaneous, ramifying social worlds of researchers in different lines of work are submerged and made harder to investigate as continuous, going concerns. The conventional narrative frames the problems of various lines of embryological work mainly in rela-tion to the rise of genetics rather than in relation to each other. Whole fields, taken to be represented by a single line of work that is out of the limelight, are relegated to static background "context" or "infrastructure." They then merit scientific credit only as "materials and methods." Because of the fact that embryologists have been offstage and out of attention throughout much of the "age of genetics," it is hard to explain by means of a limelight narrative embryology's reemergence in such forms as developmental genetics, molecular developmental biology, evo-devo, and epigenetics. While Stent's 1988 notice of renewed interest in cell-lineage work was accurate, how is this line of work's historical *availability* for renewal to be explained by a genetics-centered narrative that had no need even to mention its existence for fifty years?[134]

My concern has been to describe research styles in ways that allow for the subdivision of social worlds of scientists rather than make the "naturalistic assumption" that scientists work in different worlds because they work on separate aspects of nature. When contemporaneous relations within and among fields are interpreted in terms of tracking and representing processes, we can do a better job of following the historical processes by which fields change. Likewise, when biologists understand the entwined nature of the biological processes they follow, they can do a better job of following them through the transformations that differentiate the research interests of particular fields and lines of work.

NOTES

1. Hacking, 1983.

2. Griesemer and Wimsatt, 1989.

3. Griesemer, 2000b.

4. See, e.g., Coleman, 1965; Churchill, 1974; Allen, 1975, 1985. Compose Gilbert, 1978, 1987, 2003a; Maienschein, 1987, 1991a, 1991b. For a provocative philosophical attempt to integrate foregrounded and backgrounded theories and concepts of heredity and development in evo-devo, see Winther (2001b).

5. See Gerson, 1998.

6. On following scientists, see Latour,1987, 1999.

7. Guralnick, 2002; Love and Raff, 2003.

8. E.g., Olby, 1979, 1997; Sapp, 1990.

9. Reichenbach, 1971, p. 198.

10. According to Conklin (1905, p. 93), the concept of fate traces to W. His (1874) and his concept of organ-forming germ regions of the egg protoplasm.

11. Conklin, 1905, p. 93.

12. Salmon, 1984, 148.

13. I describe the developmental process from A to B, but of course it is also characteristic of developmental thinking to view life as completing a cycle, or full circle, to return to an original condition.

14. There has been lively philosophical debate about Reichenbach's principle, in light of Salmon's elaboration of it, as to whether its invariance condition is too strong and whether its modal character raises more philosophical problems than the concept of causality it was meant to explain (e.g., Dowe, 1992; Salmon, 1994). But whatever the Reichenbach-Salmon theory's standing, scientists do engage in the enterprise their theory describes and do track the processes they follow by means of mark transmission to gain causal understanding.

15. On noteworthy observation, see Hacking, 1983, chap. 9.

16. Quoted in Sander and Faessler, 2001, p. 9.

17. See Bernard, 1865; Grinnell, 1919; Brandon, 1994 on nonmanipulative experiment.

18. This is not to say that the social world of biology *was* unified, but that a certain theoretical perspective took it to be. On the segmentation of the biological sciences in the late nineteenth century, see, e.g., Nyhart, 1995.

19. A commitment is an actual expenditure of resources, not merely the intention to spend them, and thus research commitments entail trade-offs: doing things one way entails that they are not done some other way (even if they could be on a future occasion, given enough resources, or could have been on the occasion in question, but were not). The commitment to *represent* in a particular way is what I call a "theoretical perspective" (Griesemer, 2000a).

20. Olby, 1979.

21. E.g., Allen, 1975, 1986; Maienschein, 1987.

22. On embryological origins of genetics, see Gilbert, 1978, 1987, 2003a; on representational strategies, see Griesemer and Wimsatt, 1989; Griesemer, 1994. Olby, 1997, argues that while Mendel was a developmentalist, his concerns were with developmental history (*Entwicklungsgeschichte*) and the debate over transmutation of species, not with embryological development (covered by the term "evolution"). While I agree with Olby that Mendel's predecessors in hybridization studies sought a "law-following developmental process of transmutation yielding new species" (Olby, 1997, sec. 4), I think Mendel's work stands between this older concern with species transmutability and a modern "embryological" concern with the nature of development, pursued by Mendel in inferences about the development of hybrid organisms. So perhaps it would be more accurate to say that genetic practice had a developmentalist origin, acknowledging that Mendel's practice did not arise from embryological practices of the first half of the nineteenth century (e.g., Von Baer's).

23. Churchill, 1974.

24. Maienschein, 1987; Griesemer and Wimsatt, 1989; Griesemer, 2000b, 2002a.

25. Daston and Galison, 1992; see also Gannett and Griesemer, 2003.

26. Griesemer, 2000b, 2002a. See also Gilbert, 1978, 1987, 2003a; Olby, 1979, 1997; Sapp, 1990.

27. It is not my view that Mendel actually held or understood his work to be about factor transmission. I agree with Olby (1979, 1997) that Mendel analyzed characters, not factors. However, I read Mendel's several notations for describing characters to suggest representations that facilitate foregrounding the factor transmission aspect, and thus that Mendel facilitated the reading his geneticist followers gave. Thus, I am interested in the other side of Olby's coin: just as there is a danger of whig interpretation, i.e. a story of progress toward the present, of Mendel's work through twentieth-century eyes, so there is a danger of assimilating him too closely with the goals, projects, theories, and commitments of his predecessors. Watershed figures do not resemble the water on either side of the divides they mark.

28. Mendel, 1965, p. 1. All quotations are from the 1965 edition based on the Royal Horticultural Society of London translation.

29. Olby, 1979, 1997; Sapp, 1990.

30. Olby, 1966, p. 43.

31. Mendel, 1965, p. 1.

32. I searched the Mangelsdorf print edition by hand and by a pdf download of the typeset Electronic Scholarly Publishing edition. The first is the Royal Horticultural Society of London translation. The second is a revised translation by Robert Blumberg for the MendelWeb project (http://www.netspace.org./MendelWeb/), further revised for ESP (http://www.esp.org/foundations/genetics/classical/gm-65.pdf).

33. There are also nine occurrences of the term "behavior" in the context of descriptions of behavior *in* hybrids or in hybrid unions that *produce* offspring of a given kind. I construe behavior *in* a hybrid to be an expression of a developmental property or process.

34. Olby, personal communication.

35. See Mendel, 1965, pp. 17, 19, 25, 30 for evidence that "developmental series" refers to expressions of this kind.

36. Ibid., p. 17.

37. Ibid., p. 25.

38. Ibid., p. 3; italics in original.

39. Ibid., pp. 3, 7, 9, 10, 11.

40. Also espoused by Naudin; see Morgan 1932, p. 262.

41. Mendel, 1965, p. 20.

42. See Bateson's footnote 1 to ibid., p. 21.

43. Ibid., p. 20.

44. Ibid., pp. 20–21.

45. Ibid., p. 21.

46. Ibid.

47. On working objects, see Daston and Galison, 1992. On diagrams as communication tools, see Griesemer and Wimsatt, 1989.

48. Some interpretations in the recent literature include support of evolution, antievolution, speciation by hybridization, laws of heredity, and development; see the discussion in Olby, 1979, 1997.

49. "Genotype" is a term introduced into English by Johannsen, 1911.

50. E.g., Bowler, 1989.

51. Mendel, 1965, pp. 25–26.

52. Olby, 1977.

53. Mendel, 1965, p. 11.

54. Modern genetics reports this ratio as $1:2:1$ to reflect the combinatorics of the developmental series of *genotypes*, AA+2Aa+aa. My point in emphasizing the order of terms in the developmental series of characters, A+Aa+a, is that Mendel, 1965, p. 13, is doing something developmentally interesting in reporting his result as $2:1:1$.

55. T, t represent characters, while Tt represents the hybrid form of the dominant character, not a distinct character. The different notations T, Tt for the same dominant character (tall) reflect the double significance of the dominant character in the development of hybrids.

56. Mendel, 1965, p. 14. Although it would be tempting to say that Mendel simply thought of characters as determined by pairs of factors and that his developmental series for characters (e.g., A+2Aa+a), are expansions of the binomial expression for germ cell combinations $(A+a)^2$, Olby, following Heimans, has given a compelling argument that Mendel could not have endorsed such an interpretation. Mendel, according to Olby (1979, p. 62; italics in original): "could conceive of *three* contrasted characters *which can exist together in the* F1 *hybrid*, but which are mutually exclusive in the germ cells. . . . There is, therefore, no escape from the conclusion that Mendel's conception of the character pair did not lead him to the conception of mutually *exclusive pairs of factors* also."

57. Ibid., p. 15.

58. Ibid., p. 25.

59. Ibid.

60. Ibid., p. 26; italics in original.

61. Ibid.

62. Sapp, 1990.

63. Griesemer, 2002a.

64. And evolution. Weismann, for instance, argued for a form of internal selection as a developmental process, leading to his being dubbed "neo-Darwinian" by Romanes. Although Weismann's strong internal selectionist view, like his mosaic account of development, was rejected, a variant of it made a comeback in the form of "developmental selection" in evolutionary work on clonal organisms (see Buss, 1987). While this chapter focuses on relations between heredity and development, limelight narratives of the modern evolutionary synthesis in the twentieth century present a problematic view of relations of both of these fields (and processes) to evolution. Love and Raff (2003), for example, argue that synthesis historiography has distorted the history of evo-devo relations in significant ways (e.g., by ignoring evolutionary morphology). For Weismann's larger commitments to a deeply intertwined account of variation, heredity, and development, see Winther (2001a).

65. Hanson, 1958.

66. A more accurate label for Weismann would be "Vererbungist." Churchill, 1987.

67. Griesemer and Wimsatt, 1989; Griesemer, 1994; 2000a, 2000b, 2000c, 2002a; Griesemer and Churchill, 2002.

68. This is not to say that Weismann had no hand in this way of thinking. His arguments against the inheritance of acquired characteristics in light of his doctrine of germinal continuity and somatic discontinuity (particularly in regard to his public debate with Herbert Spencer), and the expanding body of research implicating the nucleus and chromosomes in heredity, certainly contributed to Wilson's (and others') shift of attention.

69. Wilson, 1896, 1900, 1925. See also Allen, 1975; Churchill, 1974; Griesemer and Wimsatt, 1989; Maienschein, 1991b.

70. Sapp, 1987.

71. Although, as Griesemer and Wimsatt (1989) argue, Boveri had already begun to formulate the separation between problems of heredity and development in his representations of chromatin diminution in ascarids, which Boveri took to *support* Weismann's theory of development.

72. Grant Yamashita (2002) formulates the problem of germ-soma temporal differentiation as "When is a germ." Guralnick (2002) discusses temporal aspects of cell-lineage work, in particular its "teleological" character, in explaining prospective significance in terms of later developmental outcomes.

73. Clearly this cannot be the whole story, and I do not intend to explain every difference between these fields in terms of tracking activity. Rather, I think that the practices and representations associated with tracking and following provide key indicators of lines of divergence, including conceptual and theoretical developments as well as differences of problem agenda and methodology.

74. See, e.g., Berrill, 1971.

75. If one takes Mendel to have been investigating the older hybridist problem of the stability of Linnean species (see Olby, 1979, 1997), then his project is even closer to that of the embryologists I discuss because, presumably, his aim would have been to use his account of the development of hybrids to make inferences about how normal development prevents species instability.

76. See, e.g., Weindling, 1991; Richards, 1992.

77. Whitman, 1878, 1887. Others working at the same time as Haeckel and Whitman also sought differentiation earlier than gastrulation, e.g., Lankester and His. See Whitman, 1878; Maienschein, 1978, 1991a; Guralnick, 2002.

78. Maienschein, 1978, 1991a, 1991b; Stent, 1988; Galperin, 1998; Guralnick, 2002.

79. Griesemer, 2002b.

80. While the embryologists identified a double temporal significance of embryonic cells—for the adult state as well as for the immediate developmental context—Mendel had identified a double temporal significance in the future potential of the dominant character—to produce tall offspring in the first generation that are constant in hybrid behavior and tall offspring in subsequent generations bred from these hybrids that are constant in form.

81. Gilbert, 2003b.

82. See Guralnick (2002) for an argument that cell lineage work declined after 1907 because there was no point continuing to document the same pattern of development at the phylum level while failing to provide adequate tools for exploring the large amounts of variation in developmental pattern at lower taxonomic levels.

83. Griesemer, 2002b; on the notion of a central subject, see Hull, 1975.

84. Lillie, 1927, p. 361.

85. For an important attempt to explain these changes as part of a larger process of the "rationalization" of research, see Gerson, 1998. Gould (1983) described a "hardening" of the modern evolutionary synthesis that reflected a "narrowing" of perspective and attention to natural selection in evolutionary theory which is similar to the narrowings that result in subsequent lines of work on genetic or embryologic aspects of heredity/development. For Gould's account of a variety of alternative perspectives that are made possible by the representational openness of early evolutionary theory, see Gould, 2002.

86. Whitman, 1878, 1887; Wilson, 1892; Lillie, 1895, 1899; Conklin, 1897, 1905. See also Maienschein, 1978, 1991a; Stent, 1988; Galperin, 1998; Guralnick, 2002.

87. Griesemer, 2002b.

88. See Hammond and Austin (1987) on the history and technology of camera lucida microscopy.

89. Whitman, 1887.

90. Wilson, 1892; see Maienschein, 1978, 1991a.

91. Maienschein, 1991b, p. 228.

92. Conklin, 1905.

93. See Olby, 1979, 1997.

94. Of course, the embryo could be fixed and sectioned and further images made, or the embryo could be turned into an image. See Sander and Faessler, 2001.

95. Griesemer and Wimsatt, 1989; Griesemer, 1994, 2000a, 2002b.

96. To take only one almost random example (i.e. within easy reach in my office), Castle's textbook, *Genetics and Eugenics* (1916; 3rd ed., 1927), displays Wilson's diagram of "Weismann's doctrine" on p. 56 in part II, "The Historical Development of Genetics." Part III, "The Essential Facts of Genetics," begins with "Mendel's Law of Heredity." This order of presentation, using Weismannism as a conceptual guide to the understanding of Mendel's work as foundational for modern biology, was common in American biology textbooks from about 1910 to 1933.

97. See, e.g., Vogt, 1925; fig. 8. For further discussion see, e.g., Sander and Faessler, 2001.

98. Brauckmann and Gilbert, 2004.

99. Willier and Oppenheimer, 1974, p. 145.

100. Hamburger, 1988, p. 20.

101. See Willier and Oppenheimer, 1974; Hamburger, 1988; and Sandler and Faessler, 2001 for further details of Spemann and Mangold's experimental techniques and the history of the organizer experiment.

102. Spemann and Mangold, 1924; Sandler and Faessler, 2001. All quotations from Spemann and Mangold are from the Hamburger translation reprinted in Willier and Oppenheimer, 1974, and in 2001 in the *International Journal of Developmental Biology* 45(1):13–38. Page references are to the 2001 reprint.

103. Sandler and Faessler, 2001, pp. 4–6.

104. Spemann and Mangold, 1924, p. 29.

105. Ibid., p. 37.

106. Ibid., p. 35.

107. Ibid.; emphasis added.

108. Ibid., pp. 35, 36.

109. Ibid., p. 37.

110. Guralnick, 2002.

111. Spemann and Mangold, 1924, p. 35.

112. Compare Morgan, 1926; Spemann, 1926; Lillie, 1927; Brachet, 1927; Conklin, 1929; Harrison, 1937.

113. Gilbert, 1978; see also Griesemer, 2000b, 2002a.

114. Cf. Love and Raff, 2003.

115. Morgan, 1934.

116. Morgan, 1910, p. 449.

117. Morgan, 1926, p. 490.

118. Ibid., p. 492; italics added.

119. Sapp, 1991.

120. Morgan, 1934, p. 134, quoted in Griesemer, 2000b, p. 271.

121. Griesemer, 2000b.

122. Wimsatt, 1987.

123. Maienschein, 1987, p. 79.

124. Harrison, 1937, p. 370.

125. Stent, 1988, p. 225.

126. Ibid., p. 226.

127. Maienschein, 1987, p. 80.

128. On the new preformationism, see Gould, 1977; on Morgan, see, e.g., Gilbert, 1978, 1987, 2003a.

129. Kohler, 1994.

130. Spemann, 1927; cf. Gilbert, 2003a.

131. Mayr and Provine, 1980.

132. Newman and Müller, 2000.

133. Kuhn, 1970.

134. Perhaps "rehabilitation" would be more apt than "renewal." If cell-lineage work fell out of favor because it was (1) descriptive in an age of experimentalism, (2) holistic and genealogical in an age of rising nuclear causal-analytic determinism (by appealing to the ancestral cell line as a cause in development), and (3) focused on comparative morphology in an age of rising formalism and quantification, then these "faults" would somehow have to be excused and explained away to justify renewed interest.

REFERENCES

Allen, G. E. 1975. *Life Science in the Twentieth Century.* New York: Cambridge University Press.

――――. 1978. *Thomas Hunt Morgan: The Man and His Science.* Princeton, N.J.: Princeton University Press.

――――. 1986. "T. H. Morgan and the Split Between Embryology and Genetics, 1910–1935." In T. J. Horder, J. A. Witkowski, and C. C. Wylie (eds.), *A History of Embryology,* pp. 113–146. New York: Cambridge University Press.

Bernard, C. 1865. An Introduction to the Study of Experimental Medicine. H. C. Green (translator), 1965, New York: Dover Publications.

Berrill, N. J. 1971. *Developmental Biology.* New York: McGraw-Hill.

Bowler, P. J. 1989. *The Mendelian Revolution: The Emergence of Hereditarian Concepts in Modern Science and Society.* Baltimore: Johns Hopkins University Press.

Brachet, A. 1927. "The Localization of Development Factors." *Quarterly Review of Biology* 2(2): 204–229.

Brandon, R. N. 1994. "Theory and Experiment in Evolutionary Biology." *Synthese* 99: 59–73.

Brauckmann, S., and S. F. Gilbert. 2004. "Sucking in the Gut: A Brief History of Early Studies on gastrulation." In C. D. Stern (ed.), *Gastrulation: From Cells to Embryo.* Cold Spring Harbor, N.Y.: Cold Spring Harbor Laboratory Press.

Buss, L. W. 1987. *The Evolution of Individuality.* Princeton, N.J.: Princeton University Press.

Castle, W. E. 1927. *Genetics and Eugenics: A Text-book for Students of Biology and a Reference Book for Animal and Plant Breeders,* 3rd ed. Cambridge, Mass.: Harvard University Press.

Churchill, F. B. 1974. "William Johannsen and the Genotype Concept." *Journal of the History of Biology* 7: 5–30.

――――. 1987. "From Heredity Theory to *Vererbung*: The Transmission Problem, 1850–1915." *Isis* 78: 337–364.

Coleman, W. 1965. "Cell, Nucleus, and Inheritance: An Historical Study." *Proceedings of the American Philosophical Society* 109: 124–158.

Conklin, E. G. 1897. "The Embryology of *Crepidula,* a Contribution to the Cell Lineage and Early Development of Some Marine Gastropods." *Journal of Morphology* 13: 1–266.

――――. 1905. "Organization and Cell-lineage of the Ascidian Egg." *Proceedings of the Academy of Natural Sciences (Philadelphia)* 13: 1–119.

————. 1929. "Problems of Development." *American Naturalist* 63(684): 5–36.

Daston, L., and P. Galison. 1992. "The Image of Objectivity." *Representations* 40: 81–128.

Dowe, P. 1992. "Wesley Salmon's Process Theory of Causality and the Conserved Quantity Theory." *Philosophy of Science* 59(2): 195–216.

Galperin, C. 1998. "From Cell Lineage to Developmental Genetics." *History and Philosophy of the Life Sciences* 20: 301–344.

Gannett, Lisa, and James Griesemer. 2003. "The ABO Blood Groups: Mapping the History and Geography of Genes in *Homo sapiens*." In Hans-Jörg Rheinberger and Jean-Paul Gaudillière (eds.), *Classical Genetic Research and Its Legacy: The Mapping Cultures of Twentieth-Century Genetics*, pp. 119–172. New York: Routledge.

Gerson, Elihu M. 1998. "The American System of Research: Evolutionary Biology, 1890–1950." Ph.D. thesis, University of Chicago.

Gilbert, S. 1978. "The Embryological Origins of the Gene Theory." *Journal of the History of Biology* 11(2): 307–351.

Gilbert, S. F. 1987. "In Friendly Disagreement: Wilson, Morgan, and the Embryological Origins of the Gene Theory." *American Zoologist* 27: 797–806.

————. 2003a. "The Embryological Origins of the Gene Theory" (revised and adapted). http://www.devbio.com/article.php?ch=4&id=23. First posted March 7, 2003; last edit May 20, 2003.

————. 2003b. "The Organization and Cell-Lineage of the Ascidian Egg." http://www.devbio.com/article.php?ch=1&id=3. First posted March 4, 2003; last edit March 17, 2003.

Gould, S. J. 1977. *Ontogeny and Phylogeny.* Cambridge, Mass.: Belknap Press of Harvard University Press.

————. 1983. "The Hardening of the Modern Synthesis." In M. Grene (ed.), *Dimensions of Darwinism*, pp. 71–93. Cambridge: Cambridge University Press.

————. 2002. *The Structure of Evolutionary Theory.* Cambridge, Mass.: Belknap Press of Harvard University Press.

Griesemer, J. R. 1994. "Tools for Talking: Human Nature, Weismannism and the Interpretation of Genetic Information." In Carl Cranor (ed.), *Are Genes Us? The Social Consequences of the New Genetics*, pp. 69–88. New Brunswick, N.J.: Rutgers University Press.

————. 2000a. "Development, Culture and the Units of Inheritance." *Philosophy of Science (Proceedings)* 67: S348–S368.

————. 2000b. "Reproduction and the Reduction of Genetics." In P. Beurton, R. Falk, and H.-J. Rheinberger (eds.), *The Concept of the Gene in Development and Evolution: Historical and Epistemological Perspectives*, pp. 240–285. Cambridge: Cambridge University Press.

————. 2000c. "The Units of Evolutionary Transition." *Selection* 1: 67–80.

————. 2002a. "Limits of Reproduction: A Reductionistic Research Strategy in Evolutionary Biology." In M. H.V. van Regenmortel and D. Hull (eds.), *Promises and Limits of Reductionism in the Biomedical Sciences*, pp. 211–231. Hoboken, N.J.: Wiley.

————. 2002b. "Space <=> Time: Temporality and Attention in Iconographies of the Living." In H. Schmidgen (ed.), *Experimental Arcades: The Materiality of Time Relations in Life Sciences, Art, and Technology (1830–1930)*, preprint 226, pp. 45–57. Berlin: Max Plank Institut für Wissenschaftsgeschichte.

Griesemer, J., and F. B. Churchill. 2002. "Weismann, August Friedrich Leopold." In Mark Pagel (ed.), *Encyclopedia of Evolution*, vol. 2, pp. 1149–1151. New York: Oxford University Press.

Griesemer, J. R., and W. C. Wimsatt. 1989. "Picturing Weismannism: A Case Study of Conceptual Evolution." In M. Ruse (ed.), *What the Philosophy of Biology Is: Essays Dedicated to David Hull*, pp. 75–137. Dordrecht: Kluwer Academic.

Grinnell, J. 1919. "The English Sparrow Has Arrived in Death Valley: An Experiment in Nature." *The American Naturalist* 53: 468–472.

Guralnick, R. 2002. "A Recapitulation of the Rise and Fall of the Cell Lineage Research Program: The Evolutionary-Developmental Relationship of Cleavage to Homology, Body Plans and Life History." *Journal of the History of Biology* 35: 537–567.

Hacking, I. 1983. *Representing and Intervening*. Cambridge: Cambridge University Press.

Hamburger, V. 1988. *The Heritage of Experimental Embryology: Hans Spemann and the Organizer*. New York: Oxford University Press.

Hammond, J., and J. Austin. 1987. *The Camera Lucida in Art and Science*. Bristol, U.K.: Adam Hilger.

Hanson, N. R. 1958. *Patterns of Discovery: An Inquiry into the Conceptual Foundations of Science*. Cambridge: Cambridge University Press.

Harrison, R. J. 1937. "Embryology and Its Relations." *Science* 85(2207): 369–374.

His, W. 1874. *Unsere Körperform und das physiologische Problem ihrer Entstehung: Briefe an einen befreundeten Naturforscher*. Leipzig: F.C.W. Vogel.

Hull, D. L. 1975. "Central Subjects and Historical Narratives." *History and Theory* 14: 253–274.

Johannsen, W. 1911. "The Genotype Conception of Heredity." *American Naturalist* 45: 129–159.

Kohler, R. J. 1994. *Lords of the Fly: Drosophila Genetics and the Experimental Life*. Chicago: University of Chicago Press.

Kuhn, T. S. 1970. *The Structure of Scientific Revolutions*, 2nd ed. Chicago: University of Chicago Press.

Latour, B. 1987. *Science in Action.* Cambridge, Mass.: Harvard University Press.

———. 1999. *Pandora's Hope: Essays on the Reality of Science Studies.* Cambridge, Mass.: Harvard University Press.

Lillie, F. R. 1895. "The Embryology of the Unionidae. A Study in Cell-Lineage." *Journal of Morphology* 10(1): 1–100.

———. 1899. "Adaptation in Cleavage." *Biological Lectures from the Marine Biological Laboratory, Woods Hole, Mass.*, third Lecture, pp. 43–67. Boston: Athenaeum Press.

———. 1927. "The Gene and the Ontogenetic Process." *Science* 66: 361–368.

Love, A. C., and R. A. Raff. 2003. "Knowing Your Ancestors: Themes in the History of evo-devo." *Evolution and Development* 5(4): 327–330.

Maienschein, J. 1978. "Cell Lineage, Ancestral Reminiscence, and the Biogenetic Law." *Journal of the History of Biology* 11(1): 129–158.

———. 1987. "Heredity/Development in the United States, Circa 1900." *History and Philosophy of the Life Sciences* 9: 79–93.

———. 1991a. *Transforming Traditions in American Biology, 1880–1915.* Baltimore: Johns Hopkins University Press.

———. 1991b. "From Presentation to Representation in E. B. Wilson's *The Cell.*" *Biology and Philosophy* 6(2): 227–254.

———. (ed.). 1986. *Defining Biology: Lectures from the 1890s.* Cambridge, Mass.: Harvard University Press.

Mayr, E., and W. Provine (eds.). 1980. *The Evolutionary Synthesis: Perspectives on the Unification of Biology.* Cambridge, Mass.: Harvard University Press.

Mendel, G. 1965. *Experiments in Plant Hybridisation*, trans. Royal Horticultural Society of London. Cambridge, Mass.: Harvard University Press.

Morgan, T. H. 1910. "Chromosomes and Heredity." *The American Naturalist* 44: 449–496.

———. 1926. "Genetics and the Physiology of Development." *American Naturalist* 60(671): 489–515.

———. 1932. "The Rise of Genetics." *Science* 76: 261–267.

———. 1934. *Embryology and Genetics.* New York: Columbia University Press.

Newman, S. A., and G. B. Müller. 2000. "Epigenetic Mechanisms of Character Origination." *Journal of Experimental Zoology (Molecular and Developmental Evolution)* 288: 304–317.

Nyhart, L. K. 1995. *Biology Takes Form: Animal Morphology and the German Universities, 1800–1900.* Chicago: University of Chicago Press.

Olby, R. 1966. *Origins of Mendelism.* New York: Schocken Books.

————. 1979. "Mendel No Mendelian?" *History of Science* 17: 53–72.

————. 1997. "Mendel, Mendelism and Genetics." MendelWeb, Edition 97.1 (Feb. 22, 1997). http://www.mendelweb.org/MWolby.intro.html.

Reichenbach, H. 1991. *The Direction of Time*, ed. Maria Reichenbach. Berkeley: University of California Press.

Richards, R. J. 1992. *The Meaning of Evolution: The Morphological Construction and Ideological Reconstruction of Darwin's Theory.* Chicago: University of Chicago Press.

Salmon, W. 1984. *Scientific Explanation and the Causal Structure of the World.* Princeton, N.J.: Princeton University Press.

————. 1994. "Causality Without Counterfactuals." *Philosophy of Science* 61: 297–312.

Sander, K., and P. Faessler. 2001. "Introducing the Spemann–Mangold Organizer: Experiments and Insights That Generated a Key Concept in Developmental Biology." *International Journal of Developmental Biololgy* 45: 1–11.

Sapp, J. 1987. *Beyond the Gene: Cytoplasmic Inheritance and the Struggle for Authority in Genetics.* New York: Oxford University Press.

————. 1990. "The Nine Lives of Gregor Mendel." In H. E. Le Grand (ed.), *Experimental Inquiries*, pp. 137–166. Dordrecht: Kluwer Academic.

Spemann, H. 1926. "Croonian Lecture: Organizers in Animal Development." *Proceedings of the Royal Society of London* B102(716): 177–187.

Spemann, H., and H. Mangold. 1924. "Über Induktion von Embryonalanlagen durch Implantation artfremder Organisatoren." *Archiv für mikroskopische Anatomie und Entwicklungsmechanik* 100: 599–638. Reprinted as "Induction of Embryonic Primordia by Implantation of Organizers from a Different Species," trans. V. Hamburger. *International Journal of Developmental Biology* 45(2001): 13–38.

Stent, G. 1988. "Cell Lineage and Segmentation in Development." In M. Markus, S. C. Müller, and G. Nicolis (eds.), *From Chemical to Biological Organization*, pp. 225–234. New York: Springer-Verlag.

Vogt, W. 1925. "Gestaltungsanalyse am Amphibienkeim mit örtlicher Vitalfärbung. Vorwort über Wege und Ziele. I. Teil. Methodik und Wirkungsweise der örtlichen Vitalfärbung mit Agar als Farbträger." *Wilhelm Roux's Archiv für Entwicklungsmechanik der Organismen* 106: 542–610.

Weindling, P. 1991. *Darwinism and Social Darwinism in Imperial Germany: The Contribution of the Cell Biologist, Oscar Hertwig, 1849–1922.* Stuttgart: G. Fischer.

Whitman, C. O. 1878. "The Embryology of *Clepsine*." *Quarterly Journal of Microscopic Science* 13: 215–315.

————. 1887. "A Contribution to the History of the Germ-Layers in *Clepsine*." *Journal of Morphology* 1(1): 105–182.

Willier, B. H., and J. M. Oppenheimer (eds.). 1974. *Foundations of Experimental Embryology*, 2nd ed. New York: Hafner Press.

Wilson, E. B. 1892. "The Cell-Lineage of *Nereis*. A Contribution to the Cytogeny of the Annelid Body." *Journal of Morphology* 6: 361–480.

———. 1896. *The Cell in Development and Inheritance*, 2nd ed. London: Macmillan.

———. 1925. *The Cell in Development and Heredity.* New York: Macmillan.

Wimsatt, W. C. 1987. "False Models as Means to Truer Theories." In M. Nitecki and A. Hoffman (eds.), *Neutral Models in Biology*, pp. 23–55. London: Oxford University Press.

Winther, R. G. 2001a. "August Weismann on Germ-Plasm Variation." *Journal of the History of Biology* 34: 517–555.

———. 2001b. "Varieties of Modules: Kinds, Levels, Origins, and Behaviors." *Journal of Experimental Zoology (Molecular and Developmental Evolution)* 291: 116–129.

Yamashita, G. 2002. "When Is a Germ? Or, Germ, Soma, and the Construction of Time in Embryology." In H. Schmidgen (ed.), *Experimental Arcades: The Materiality of Time Relations in Life Sciences, Art, and Technology (1830–1930)*, preprint 226, pp. 189–203. Berlin: Max Plank Institut für Wissenschaftsgeschichte.

13

THE JUNCTURE OF EVOLUTIONARY AND DEVELOPMENTAL
BIOLOGY

Elihu M. Gerson

In their collection of papers on the history of evolution research, Mayr and Provine[1] make clear that the Modern Synthesis which has formed the core of evolutionary research since the 1940s did not include any significant participation by students of development. The history of developmental research in the same period shows an equal lack of interest with evolutionary concerns on the part of developmental biologists.[2] The last generation, however, has seen the extraordinary success of a new intersection between development and evolution research. This connection, tentative and fragile in the late 1970s, has grown to become one of the most productive and intense areas of research in the life sciences. This chapter addresses several questions about that connection: How did the separation of the two lines of research come about? What happened to bring them back together? How can we think about this kind of intersection analytically, in a way useful to scientists as well as to science studies scholars?

My approach to these questions is analytic, synoptic, and organizational. Rather than attempting a detailed narrative of events in many specialties over many decades, I focus on the long-term pattern of relationships among lines of research in comparative biology. My primary concern here is to sketch a way of understanding the growth and decay and regrowth of connections among lines of research. I focus on lines of work because they are the hinge of events: simultaneously part of both the intellectual work and its relations with the rest of society.

Changes in the large-scale pattern of institutions since the early twentieth century have facilitated some ways of doing research and hindered others. One result of this process was the separation of evolutionary and developmental research early in the twentieth century, followed by their reconnection and intersection in the late twentieth century. This chapter sketches some (only some) of the more salient institutional trends in this process; I have provided a more detailed analysis of the American case elsewhere.[3] In particular, I shall focus on four factors. First, research at the end of the nineteenth century ran up against the limits in the capacities of the available research technology, making it difficult or impossible to address important questions. Second, the organization of many institutions (including science and academia) was strongly rationalized (i.e., arranged with economic efficiency in mind) in the first half of the twentieth century. This process facilitated some aspects of research and retarded others. One side effect of this process was to encourage a split between research on development and research on evolution. Third, some styles of research (i.e., strategies for making large-scale decisions) became more prevalent while others declined. The effect was to reinforce the rationalizing trend and accelerate the division between development and evolution research. Finally, certain changes in the concepts used by scientists in both fields facilitated the intersection of the two fields late in the twentieth century.

HISTORICAL BACKGROUND: EVOLUTION AND DEVELOPMENT TOGETHER AT FIRST

The point of nineteenth-century natural history was not merely ever-finer description, but the analysis of connections among the parts of nature. This was accomplished by comparing across patterns of different kinds (morphological, developmental, geographical, paleontological), and inferring classifications and relationships among classifications. Darwin's hypothesis of descent with modification was the most important of these relationships because it tied together so many different kinds of evidence. The analysis of descent relationships (phylogeny) was a major part of the research stimulated by Darwin's theory. Darwin's argument relied heavily on geographic, morphological, and paleontological evidence, but the connection with developmental concerns was quickly drawn. Especially in Germany, the work of Müller, Haeckel, and their followers (adapting an established tradition of morphological work) provided a close tie between comparative embryology and the study of phylogeny.[4] Similarity among developmental patterns was especially important because early stages of development gave a great deal of information about relationships which was otherwise unobtainable. The idea that developmental sequence reflects the phylogenetic history of the species (the "biogenetic law") was the centerpiece of this work. A similar program of research grew up in England.[5]

Through the late nineteenth century, the focus of morphological attention shifted from major organ systems to single organs to cells to parts of cells. At the same time, the focus of taxonomic attention shifted from classes and orders to species, subspecies, and varieties. Similarly, the focus of attention in stratigraphy (hence, paleontology) shifted from eras to periods to ages. A similar pattern thus characterized work in all the kinds of phenomena studied by naturalists. Their focus was increasingly on detecting ever smaller differences and interpreting their significance.

At the end of the nineteenth century, this mode of comparative research ran into difficulties. In order to make progress, scientists had to look earlier and earlier in development, specify more profound morphological characters, deal with smaller and smaller taxonomic groups, and look earlier and earlier in the rocks for paleontological specimens. Each succeeding round of research became more difficult to manage technically, and correspondingly more expensive to organize and manage. The progress of research on inheritance, for example, indicated that morphological features were controlled at the subcellular level. By the end of the century, these difficulties had reached a crisis. Limits of technology made it almost impossible to detect very small differences. Microscopy, for example, was inadequate to examine the microstructure of events surrounding fertilization and the earliest stages of growth. Fossils rarely preserved the fine morphological detail needed to make fine taxonomic distinctions. Moreover, the chain of homologous inferences from one classification to another became very long, and hence there was great room for argument over interpretations. In consequence, the strategy seemed to fail just where it was used to address the most general and important questions, such as the relationships among the major groups of arthropods and the phyletic origin of the vertebrates.[6] As a result of these difficulties, many scientists began to focus on other problems. William Bateson, for example, simply threw up his hands and decided to study variation.[7]

SEGMENTING EVOLUTIONARY CONTINUITY AND DEVELOPMENTAL CONTINUITY

The prevalent view of evolution in the late nineteenth century was that evolution converts similarity and continuity into difference and diversity via adaptation. In this view, adaptation was about the relationships among parts of nature. This included the fit among parts of the body vis-à-vis each other, and vis-à-vis the environment. It also included the fit of species vis-à-vis each other via behavior and community structure. This view put morphological and developmental concerns at the heart of evolutionary thought, and made significant room for ecological and behavioral considerations as well. For example, Haeckel's vision (as captured in the biogenetic law) focused on similarity among different kinds of continuity—phylogeny and ontogeny were together part of a larger process. This construed ontogeny as a consequence or epiphenomenon of phylogeny. There was thus an intimate connection between the development of individuals and the succession of generations that rested on material continuity.

In this view, inheritance (i.e., heredity and development together) was about the preservation (re-production) of similarity across generations. Lineage and genealogy (succession of generations) were the chief categories for thinking about what lay between individual organisms and species. Evolutionary morphology was thus crucially about part-whole relationships and how they change. For many, even inheritance was about part-whole relations (i.e., genealogy), even after the rediscovery of Mendel in 1900. Alternative approaches to development in the late nineteenth century began to focus on mechanical explanations of development.[8] This approach emphasized development rather than phylogeny, and the search for efficient and material causes rather than formal and final ones. This difference in emphasis had the effect of starting a wedge between studies of development and studies of evolution.

In the twentieth century this split widened in important ways. In the nineteenth century, Weismann's distinction between germ plasm and somatoplasm fostered a conceptual separation between generations, on the one hand ("sequestration of germ plasm"), and ontogeny, on the other.[9] This relatively mild distinction was picked up in the United States and converted into "Weismannism," which black-boxed the life cycle as irrelevant to transmission.[10] The effect was to create a much deeper division between development and transmission as problems of research.

At the same time, statistical approaches became increasingly important. Their approach was oriented toward the distribution of variation among individuals rather than continuity or differences among species. This work laid the foundations of modern multivariate statistics.[11] In 1909, Wilhelm Johannsen extended these ideas by distinguishing between variation among individuals and variation among "lines." This notion was elaborated further by J. A. Harris, R. A. Fisher, and Sewall Wright. In 1922, Fisher extended this line of thinking once again by removing the need to consider lineage explicitly. Instead, variation could be analyzed in a population without respect to genealogy. With this result, geneticists were no longer obliged to know or care how an individual fitted into a lineage.[12]

This work in population genetics enabled a reconceptualization of evolution. The older view was that evolution was conversion of continuity into diversity via adaptation of parts. In the new view, evolution became the conversion of differences among instances of a single kind (individual organisms) into differences among multiple kinds (species). This is the conventional view of current biology. Synchronic populations rather than diachronic lineages became the chief category for thinking about what lay between individuals and species.

Fisher's statistical stroke of 1922 was paralleled by a simultaneous second one in evolutionary theory. In the view of evolutionary morphology, individual life cycles were the product of a species' growth; the succession of generations gave rise to more individuals. In the same year Walter Garstang turned this upside down by defining phylogeny as a sequence of life courses.[13] The history of a clade thus became a simple aggregation of individual lives. There was no longer a casual connection between phylogeny and ontogeny. The work of Fisher and Garstang meant that evolution and development could be studied in virtually complete independence of one another.

Evo-Devo: Sources of Change in the Late Twentieth Century

The conventional history of biology has focused on the critical role of populational concepts in the development of modern evolutionary theory, and that is correct. It is a cliché that development was excluded from the Modern Synthesis.[14] But the changes that gave rise to the Modern Synthesis were also a shift from concern with part-whole relations (centered on adaptation and morphology) to instance-kind relations (centered on sorting and segregating). This distinction is crucial.

Construing something in terms of part-whole or compositional relations means looking at its organization as a system of parts which work together. For example, my finger is part of my hand, which is part of my arm, which is part of my body. The focus is on the ways in which different parts are connected and interact. Comparisons of different arms (or forelimbs) is in the service of understanding differences in function.

By contrast, construing something in terms of instance-kind or aggregative relations means looking at it as an example of a class of things. Rather than seeing fingers, hands, and arms as connected things, this view sees them as objects with properties such as size, shape, and color. Comparison of different forelimbs is in the service of understanding differences in the variation of properties, and hence of differences among kinds.

Neither of these approaches is superior to the other; they are simply different (and complementary) ways of framing the objects of research. But these differences have profound implications. As the shift in frame occurred early in the twentieth century, nature metaphorically became not Paley's craftsman making watches one at a time, but Chaplain's factory making things from interchangeable parts in huge batches. The Modern Synthesis entrenched and solidified this shift of concerns, and hence the division of research labor. In doing so, it was completing a process which had been underway for half a century.

In the last part of the twentieth century, the study of evolution and the study of development rejoined enthusiastically. The result, commonly called "Evo-Devo," has become one of the most vigorous parts of biology.[15] Many of the larger problems which motivate research in the new area are clearly descended from (or are the same as) the problems that motivated researchers in the early twentieth century. Yet there are important differences as well. These are not simply a function of the more specialized knowledge and advanced techniques used by today's researchers. Rather, they stem from changes in the organization of research problems and specialties over a long period. The most striking thing about Evo-Devo is that it consists of a system of effective alliances among just those lines of research that which were excluded from the web of events that led to the Modern Synthesis. It was, after all, not just development that was excluded from the Modern Synthesis, but all the lines of work that did not easily follow the shift in stylistic emphasis from part-whole relations to instance-kind relations. These included morphology, problems of community structure in ecology, phylogeny, and even studies of gene function ("physiological genetics"), not to mention behavior and comparative physiology.[16]

In the first generation after the Modern Synthesis, these areas of research were the weaklings of comparative biology in terms of available attention and resources. Their intersection and alliance in the last decades of the twentieth century became one of the most active and promising parts of comparative biology. This reversal of fortune came about because the institutional context of research changed in ways that opened possibilities which had been discouraged for a long time. The changes were complex, and the influences they had are more so. Here, I will focus on four kinds of recent change that have interacted to foster the intersection of evolutionary and developmental biology: (1) developments in research technology, (2) the rationalization of work both in the larger society and in lines of research, (3) changes in the prevalence of certain styles of research, and (4) certain changes in the conceptual organization of evolutionary research. There were many other important factors, but these four allow us to see the outlines of the story most clearly.

Developments in Research Technology

I said above that natural history ran into serious technical difficulties at the end of the nineteenth century. These problems made it difficult to follow a program of comparative research that retained the close ties between development biology and evolutionary biology. The last part of the twentieth century, in contrast, was quite striking for its development of many techniques and procedures which provide fresh traction on the traditional problems of comparative biology in general, and on the problems of Evo-Devo research in particular. These new techniques and procedures have dramatically sharpened the questions that comparative biologists can ask effectively. The most important of these research technologies can be summed up in three broad, mutually supportive, and overlapping groups: molecular technology, computing technology, and visualization technology.

By "molecular technology," I mean the techniques of molecular biology applied to problems of Evo-Devo research. The development of molecular biology was one of the outstanding success stories of twentieth-century science.[17] The primary significance of this work for Evo-Devo does not lie in the development of molecular explanations of transmission or even in gene function, for all that these are extremely important. Rather, the significance of molecular biology is that it provides a family of tools and techniques for observing and manipulating the earliest stages of development and, hence, the most profound aspects of morphology.[18]

The study of developmental genetics in turn provides the basis for important ties with evolutionary theory.[19] Here, studies of gene expression and function merge imperceptibly into the study of development proper. The associations among the early, the old, the small, and the profound hold as much interest now as a century ago, for all that the interpretive context is very different. Evo-Devo researchers can, by these using the molecular techniques, fruitfully concern themselves with questions which were impossibly difficult only recently.

Computing technology has come to play an important role in all parts of research. Molecular genetics, in particular, has made extensive use of computational or information-theoretic metaphors,[20] but these uses are not my concern here. Instead, I am concerned with the use of computing (whether as hardware, software, or both) as a part of the procedures used to conduct research. This includes collecting, processing, visualizing, storing, organizing, and analyzing data; modeling and simulating biological processes; and communicating the results of these efforts. These uses of computing now pervade every aspect of the research process. Since there are so many applications, it is difficult to single out any one of them as especially significant. The important result of large-scale computing use is that the effective power of comparative approaches has been vastly extended, and this is turn means that many of the problems which could not be addressed effectively at the end of the nineteenth century were being addressed at the end of the twentieth.

Consider the development of computer-based museum specimen catalogs as an example.[21] In recent years there has been a concerted effort by the world's museums to automate their specimen catalogs and make them available to scholars over the Internet. This seems a mundane and obvious administrative chore, but its implications reach quite far. Information storage and retrieval technology is crucial to comparative research. A century ago, index cards and standardized note-taking procedures were the high-tech cutting edge of museum practice, as they were in industry.[22] The development of specimen databases is a straightforward extension of traditional practice. It is significant because collection databases and the Internet enable a vast increase in the amount of comparative work that can be done, a substantial increase in the thoroughness of its coverage, and a dramatic drop in the costs of doing it. In this context, it is important to keep in mind that a significant number of discoveries are made in the course of reexamining specimens already in museum collections.

Another example is provided by the debates over classification methods since the 1960's.[23] One important consequence of these debates has been a marked increase in the rigor of phylogeny and classification studies. This increase would have been far more difficult to establish as a practical part of research without computing technology to deal with the data analysis requirements.

Visualization techniques have also played an important role in advancing Evo-Devo. Procedures for representing evolutionary and developmental processes have always been important.[24] In recent years, the number and variety of techniques available has grown rapidly. Many, but not all, of these new techniques are computer-based; almost all are computer-aided.

Improvements in microscopy, for example, have made a substantial difference to almost all lines of biological research. In the early twentieth century, light microscopy had reached its effective limits. The development of electron microscopy at the time of World War II opened up many new possibilities.[25] Since then, several new forms of microscopy have been developed. The most important of these for biological research purposes is scanning electron microscopy, which enables detailed morphological study across several levels of scale which were unreachable before.

X-ray technology has also proved to be an important tool for visualization.[26] One particularly interesting application of this technology is the use of X-ray tomography (CAT scanning) to construct computerized pictures of skeletal elements. Rowe, for example, used an industrial CAT scanner to map the skull of an ancient didelphid. The resulting database allows the structure of the skull to be displayed on a computer monitor. The image can be manipulated in many ways, so that the most detailed analysis in three dimensions is possible without destroying the specimen. And, of course, multiple scans can be juxtaposed and compared, either visually on screen or as data sets for computerized analysis.[27] Building on this work, Rowe has led the development of a substantial research facility devoted to CAT scanning that is available to the scientific community.[28]

Many other examples are possible, but the point is clear: an abundance of new technologies for working with and comparing specimens means that many of the classical nineteenth-century problems of evolutionary biology, intimately connected to developmental questions, can be fruitfully addressed once again.

Rationalization of Work
In the first two-thirds of the twentieth century there was widespread emphasis on rationalization—doing more with the same resources, or the same with fewer resources. Across many institutions, concern focused on making activities more efficient and lower in unit cost. This in turn meant an emphasis on homogeneous raw materials and an elaborate division of labor designed to ensure high volumes of production. In particular, rationalization meant eliminating contingencies and connections among things wherever possible, and emphasizing homogeneity, removal of properties held to be irrelevant, independence of parts, and simplicity of relations.

A critical nineteenth-century innovation in this respect was the invention of the "American System" of manufacture.[29] Craft-oriented production, like Paley's watchmaker, focused on the hand-fitting of variable parts. By contrast, in the American System all parts of a given type are made to such close tolerances that any one of a type can substitute for any other in the production process. This meant that parts could be treated as if they were equivalent instances of a kind, with assurance of their correct articulation with other parts of different types. This in turn enabled much lower labor costs, since objects could be assembled by workers with less skill, who did not have to hand-fit variable parts together. It also enabled a more fine-grained division of labor and higher volumes of production. In the early twentieth century, this approach was extended from the manufacture of objects to the manufacturing process itself. Each step became a matter of repeating simple, standardized actions in exactly the same way. The classic expression of this approach is the work of F. W. Taylor.[30]

The organization of research, as with every other industry, was rationalized in the twentieth century. The industrialization of academic research went furthest and fastest in the United States, but it also took place in other industrializing countries. One way in which the rationalization process affected research was to encourage the segmentation of research specialties, which increasingly focused on narrower and narrower problems. Rationalizing research lines were particularly concerned to eliminate ("black box") problems of adaptation, complexity, interaction, and dependence. The development of statistical theory, intimately involved with problems of genetics and evolution, treated organisms as abstract data points with few properties.[31]

Some lines of research fell in with this way of doing things very easily—notably genetics, especially population genetics. Others found it very difficult to do so, because the structure of their problems (and, hence, their intellectual coherence as lines of work) made it very difficult to shift. Research in behavior as an evolutionary phenomenon, for example, suffered greatly because behavior was conceptualized as an adaptive mechanism "fitting" individual organisms and species to their environments.[32] Research on community structure suffered for the same reason, and ecology failed to thrive until it was reconceptualized as concerned with statistical distributions of species.[33] Research on morphology and on development, inseparable from the question of parts and their relationships in space and time, was almost completely separated from work on evolution.

More generally, work on multifaceted problems was discouraged by this pattern of specialization. Since evolution and development were each complex, multisided problems, their intersection was especially problematic. Indeed, there was a clear two-part pattern to the development of new specialties in biology. First, some specialties (e.g., genetics) concentrated on problems and approaches which could be rationalized relatively easily, while other specialties (notably, development) focused on problems and approaches which were difficult to rationalize. Second, the specialties with difficult-to-rationalize problems were less successful in creating new breakthroughs, and attracting funds and attention. But they rationalized nonetheless, if more slowly and painfully, as did the organizations which housed and financed them.

The Modern Synthesis at midcentury, then, was an alliance among specialties which had easily rationalizable problems (or, more precisely, whose problems seemed easily rationalizable at the time). These included population genetics and cytogenetics, a systematics redefined to focus on counting characters, and a paleontology focused on a statistical and analytic treatment of clades, not description of taxa.[34] The most difficult problems of evolutionary theory (e.g., speciation, the nature and role of adaptation) were treated by replacing interaction among parts and among organisms with simple quantitative differences in reproductive success. This is analogous to using sales volume as a measure of engineering quality.

Stylistic Change

Every approach to research requires scientists to make many specific commitments about the ways they allocate resources. Each hypothesis, for example, implies certain kinds of data collection and analysis, and each technique requires certain materials and tools. The decisions to work in one way rather than another in turn are guided by styles of research. Styles are abstract commitments used to organize other, relatively concrete, commitments.[35] Styles typically appear as general philosophical or methodological positions (e.g., focusing on structural rather than functional considerations, or preferring the construction of formal models to the detailed description and analysis of particular cases).[36] Any such pattern of commitments can serve to frame a line of research problems. Stylistic conflicts within a line of research are routine.

One important dimension of style was the commitment to partitioning versus integrative approaches. Some kinds of research focus on dividing phenomena into constituent parts or instances, in order to understand things in terms of their constituents and the relationships among them. This is the partitioning style. In contrast, the integrative style concentrates on setting phenomena in context, understanding them in terms of their relations with the world around them. In order to find out why a chicken crosses the road, a partitioning scientist might undertake neurosurgery, while an integrative scientist might design experiments with different roads, different chickens, and different traffic conditions.

The rationalization process so characteristic of twentieth-century science (and other institutions) was preeminently characterized by the partitioning style. By the middle of the century, it was becoming clear that certain classes of problem were not responding to these rationalizing styles of work organization. These problems are those that are irredeemably concerned with relations among elements or interaction among parts and processes. In the manufacturing industries, these issues showed up as problems of optimization, resource allocation, and coordination. These problems appeared almost everywhere, even in accounting.[37] Several managerial and engineering disciplines emerged to deal with these problems in the middle years of the century, typically under the rubric of "systems."[38]

The same emphasis on systems and issues of coordination which characterized other institutions appeared in biology research as well—especially in development and ecology. But here an interesting contrast appeared. In many uses of the systems approach, the language of systems was a way of extending the rationalizing trend in the partitioning style. The problem was to overcome the intricacies imposed by contingency and complex dependencies through the use of more sophisticated analysis. This view was especially strong in the emergent sister disciplines of operations research and industrial engineering. It was brought into biology by Robert MacArthur, who applied these approaches, learned in his wartime work, to ecological problems under the influence of G. E. Hutchinson.[39]

In contrast, other biologists, notably Von Bertalanffy and Waddington, saw the systems approach as a way of transcending the limits of the partitioning approach.[40] For them (and they were notably students of development and morphology), the systems approach was a way to conceptualize and deal with the complex interrelations of system and environment. Here, the focus was on ways of representing the effects of context. Adherents of these two ways of thinking about systems often talked past one another, but rarely got into explicit debates over the meaning of "system"; rather, they were invariably focused on overcoming the limitations of the partitioning approach, either by reforming it or by substituting some kind of integrative view. By the 1970s, it was clear that systems approaches were not, in general, very effective ways of organizing work either in research or in other institutions. Some applications of industrial engineering ideas in the 1960s failed loudly enough to bring the entire systems approach into question.[41]

Progress in molecular genetics was showing that simple partitioning strategies were not going to provide adequate explanations of gene expression.[42] Various of critics of the Modern Synthesis raised objections to it from a more interpretive perspective.[43] In many areas of biological research touching on evolutionary problems, scientists were beginning to reformulate their work in terms more sympathetic to integrative approaches.[44] Increasing openness to integrative approaches facilitated the emergence of an explicit Evo-Devo intersection in the late 1970s and early 1980s.

Another important stylistic shift in the late twentieth century was the reopening of concern with parts and wholes, in contrast to instances and kinds. The Modern Synthesis was firmly rooted in the instance-kind approach. The approach of population genetics and the widespread adoption of the populational approach to systematics and paleontology firmly entrenched the idea that species were sets of instances. Indeed, some trends in evolutionary research in the 1960s and 1970s seemed to build upon this approach and entrench it further. Williams's attack on group selection, for example, emphasized the instance-kind approach. The emergence of sociobiology in the 1960s extended the approach to the analysis of behavior as well.[45]

But efforts to work in the instance-kind style were also running into difficulties by the 1960s. There were questions about the adequacy of the approach, even as the central alliance of the Modern Synthesis was showing strain.[46] Questions about adaptation were reappearing in morphology.[47] There was increased concern with symbiosis, the problem area least amenable to an instance-kind approach.[48] Developmental biology had never given up its part-whole orientation to problems of morphogenesis. A major review of the field published in 1955 shows a thoroughgoing commitment to construing the developmental process as one of part-whole relations.[49] Doubts about the adequacy of the instance-kind style appeared even in population genetics, the heartland of the approach.[50] In short, the instance-kind style proved inadequate to deal with problems involving adaptation, communication, coordination, articulation, and cooperation. At the same time, these problems were becoming of increasing interest and importance. As evolutionary research in the 1960s and 1970s began to consider approaches that did not rely on commitment to the instance-kind style, the possibility of a stylistic rapprochement between evolutionary and developmental approaches emerged.

Conceptual Change

The concepts that scientists use to organize their work change all the time. Some of these changes are important, but are not closely connected to the basic issues of the field. From an evolutionist's point of view, for example, the many striking changes in the concept of "gene" that took place over the twentieth century do not make for substantive differences in the ways that species and speciation are conceptualized. Other conceptual changes, by contrast, enable important shifts in the way lines of research interact, formulate their projects, and interpret their results.

The beginning of a rapprochement between development and evolution researchers in the late 1970s was facilitated by conceptual changes which reduced the distance between the two lines of work. Two of the more important conceptual shifts in this period were the idea of species as individuals and the notion of heterochrony.

The Modern Synthesis saw individual organisms as members of populations. It thus conceptualized the relationship of individuals to their species as one of instances and kinds. In 1974, Ghiselin suggested that species are individuals rather than kinds.[51] Rather than independent instances of an abstract class, Ghiselin saw species as concrete individuals characterized by historical existence (i.e., birth and death) and relationships among members which provided coherence to the whole. In particular, species are made up of breeding relationships among member organisms, and hence of lineages. In this view, the relationship of a particular organism to its species is one of part to whole, rather than one of instance to kind. The idea was endorsed by the philosopher David Hull in 1976.[52]

This idea provided an important conceptual basis for a rapprochement between researchers committed to different approaches in evolution and development research.[53] It is far easier to think about integrated approaches to evolution and development if one does not have to reconcile two incompatible ways of thinking. This increase in conceptual homogeneity makes it easier to formulate relatively specific research problems. Viewing species as individuals thus was an important, if unobvious, facilitating step which made the connection between evolutionary and developmental biology easier as the idea gained broad acceptance.

The Modern Synthesis in Britain was not quite so thoroughgoing in its rationalization of problems and concepts as it was in the United States. There was some room for consideration of development in the British Synthesis; the morphologist Gavin De Beer published studies of some issues in the comparative study of development.[54] De Beer was extending the work of Garstang on the relationship between development and evolution.[55] Accepting Garstang's idea of evolution as a sequence of life cycles, De Beer reframed many of the questions which had concerned Haeckel and his followers. In particular, he concentrated on heterochrony, the idea that some species retained or lost characters associated with particular phases of the life cycle. This work fell on stony ground; for a generation after World War II, it was little more than a sideline which generated little new research.

Gould published a major study of heterochrony which did much to revive concern with the relationship between evolution and development.[56] He was concerned to make the notion of heterochrony respectable, and in order to do this, he had to remove the stigma of Haeckelian recapitulation from it. He did this by systematically reviewing the history of the recapitulation idea and its vicissitudes, taking great care to show that the modern notion of heterochrony was quite compatible with "Darwinian" (i.e., Modern Synthesis) thought.

Gould also modified the idea of heterochrony in important ways. For De Beer, heterochrony was (paradoxically) a static notion: he saw different kinds of character pattern, with "early" and "late" features of development appearing in different (but presumably related) species. This view is structural, not processual. Gould's innovation was to think of heterochrony as a matter of different rates of development among different parts of the body across different species. Some of the organs of some species, that is, aged faster (or slower) than others, or halted their development sooner or later in the life cycle.

This way of looking at heterochrony was important because it cuts in two different directions. On the one hand, it naturally provokes questions about the mechanisms that control differential rates of growth in morphogenesis, and this naturally leads to questions about the genetic control of growth. On the other hand, if leads to questions about relatively large morphological differences among clades, and to thoughts of evolutionary innovation and radiation. These are questions of great interest to paleontologists and others interested in large-scale evolution.[57] Heterochrony is thus a "bridge" concept which spans the distinctions among size or scale of phenomena which became an important part of evolutionary biology after the Modern Synthesis. The study of heterochrony has developed into an important part of Evo-Devo.[58] More generally, it provides a conceptual framework in which paleontologists, developmental biologists, and morphologists can talk about parts and their relations at a large (i.e., early, old, profound) scale, just as they did before World War I. And it does so, moreover, in a way which does not necessarily bring contumely from geneticists and systematists.

CONCLUSIONS

All these technical and conceptual changes have placed the relationship between evolutionary and developmental biology on a new basis. Rather than two separate worlds, the two lines of research have become a major cluster of intersections in which many problems and research efforts are simultaneously part of both evolutionary and developmental biology.

The changes in the intellectual content of research during the twentieth century were tied to, and reflected, corresponding changes in the larger society. The pattern of influence is complex, but the main outline is clear. For the first two-thirds of the twentieth century, rationalization was a driving force across institutions and levels of organizational scale. Lines of research which could be easily rationalized had a considerable advantage in the quest for every kind of financial and organizational support. At one level, organizations which adopted "businesslike" administrative procedures could more easily attract funds and public support.[59] At another level, research programs whose projects were of a size and cost convenient for the efficient production of Ph.D.s and undergraduate laboratory classes (e.g., fruit fly genetics) were at a comparative advantage by comparison with programs that were administratively inconvenient (e.g., biogeography).

There were many other contingencies, but at every point, some programs were better off, better able to compete, than others. Some programs (such as Mendelian genetics based on breeding experiments) enjoyed this relative advantage quite often, and these flourished. Others (such as community ecology) struggled. Overwhelmingly, the relatively successful projects were those which fit comfortably into a rationalizing world.

The same relative advantages occurred with stylistic commitments as well. Some styles were commensurate with the prevailing trend toward rationalization (e.g., partitioning, instance-kind), and others were not (interpretive, part-whole). Approaches which adopted the commensurable styles had advantages over those which did not, and were thus relatively successful.

There were exceptions to the trend, in research as in other parts of the institutional structure.[60] Some lines of research (notably, vertebrate paleontology) were able to establish supporting alliances outside the rationalizing system of alliances among disciplines, sponsors, and universities. But in general, programs stylistically, intellectually, and organizationally amenable to rationalization enjoyed a steady incremental accumulation of advantage.

This process reached its peak with the late Modern Synthesis. In the late 1960s, the limits of the rationalizing approach had been reached, and attention started to turn once again to problems which had been bypassed in the previous generation. By the end of the 1970s, there was a clear counter-trend under way, in the larger society as well as in biology. This emphasized coordination, adaptation, the articulation of parts, and holism. In this context, the intersection of evolutionary and developmental biology in the late 1970s was a natural step in a world that was quite ready for it. The cumulative advantages that flowed to rationalizing work in the first part of the century began flowing instead to efforts that emphasized inter-disciplinary collaborations and a fascination with levels, networks, and complexity.

Evo-Devo has now built up substantial institutional and intellectual momentum. There are journals, conferences, and all of the institutional apparatus of a flourishing line of research. Where is the juncture going? Will it be successful? Will there be a coherent discipline of Evo-Devo? Is a new and more general theory emerging? I don't know. Biologists will organize and settle their research problems well enough, and they have little to gain from the kibitzing of an ill-trained amateur such as myself. My concern as a sociologist is not with the answers scientists come up with, but with understanding how and why they come up with them, the ways in which they work, and the arrangements which facilitate and con-strain that work.

Yet I can't help but note that the successful development of Evo-Devo does not depend on the emergence of a single successful perspec-tive or theory which will then guide research. We often have disciplines without such coherent perspectives; indeed, I think they are the typical case. Rather than common answers, the field will be defined by a joint (and competitive) attack on common problems. And if it's anything like all the other disciplines, there will be many debates and differences in viewpoint. This is ordinary metabolism, not pathology.

ACKNOWLEDGMENTS

I am grateful to James R. Griesemer, Jan Sapp, William C. Wimsatt, and Rasmus G. Winther for many helpful discussions over the years, and to Manfred Laubichler and Jane Maienschein for helpful comments on earlier drafts. I am grateful to M. Sue Gerson for continuing support.

NOTES

1. 1980; see especially Hamburger, 1980.

2. Mayr, 1991; Sapp, 1987, 2003.

3. Gerson, 1998.

4. Müller, 1862; Haeckel, 1866. See also Russell, 1916; Nyhart, 1995.

5. Bowler, 1989, 1996; Gee, 1996.

6. Bowler, 1996; Gee, 1996; Nyhart, 2002.

7. Coleman, 1970.

8. Churchill, 1973; Gould, 1977.

9. Churchill, 1987; Winther, 2001.

10. Griesemer and Wimsatt, 1989; Maienschein, 1991; Allen, 1986.

11. MacKenzie, 1981; Kim, 1994; Porter, 1986; Gigrenzer et al., 1989.

12. Johannsen, 1909, 1911; Harris, 1911a, 1911b; Fisher, 1918, 1922; Sewall Wright, 1918. Provine (1986) is the best single account.

13. Garstang, 1922.

14. For populational concepts, see, e.g., Mayr, 1982; Provine, 1986. For the exclusion of development, see Churchill, 1980; Hamburger, 1980.

15. Some major works and recent collections on the intersection include Akam et al., 1994; Arthur, 1997; Bonner, 1983; Goodwin et al., 1983; Gould, 1977; Hall, 1992; Hall and Wake, 1999; Jackson et al., 2001; Møller and Swaddle, 1997; Raff, 1996; Raff and Kaufman, 1983; Wagner, 2001, West-Eberhard, 2003.

16. Ghiselin, 1980; Bowler, 1996; Goldschmidt, 1938.

17. Olby, 1974; Judson, 1979; Kay, 1993, 2000.

18. This point is complemented by the theoretical and philosophical work of Riedl (1978) on "burden" and Wimsatt (e.g., 1986, 2001) on "generative entrenchment."

19. For development genetics, see, e.g., Gehring, 1998. For ties to evolutionary theory, see Raff and Kaufman, 1991; Raff, 1996; Schank and Wimsatt, 1986.

20. Oyama, 1985; Griesemer, 2005.

21. This example is based on unpublished work in collaboration with James R. Griesemer.

22. Goode, 1991; Grinnell, 1910; Yates, 1989.

23. Hull, 1988.

24. E.g., Hopwood, 1999.

25. Rasmussen, 1997.

26. Kevles, 1997.

27. Rowe, 1996.

28. The facility can be seen at http://www.ctlab.geo.utexas.edu/. The associated digital morphology library at http://www.digimorph.org provides impressive examples of the technical trends discussed here.

29. Smith, 1977; Hounshell, 1984; Hoke, 1990.

30. Taylor, 1911.

31. Gigerenzer et al., 1989.

32. Research that treated behavior as instance-kind relations flourished in the neighboring discipline of psychology. O'Donnell, 1985; Acree, 1978.

33. McIntosh,1975; Hagen, 1992; Kohler, 2002.

34. For paleontology, see Cain, 1990; Laporte, 2000.

35. Gerson, 1998.

36. Harwood's analysis (1993) of styles of research coincides well with my view.

37. E.g., Johnson and Kaplan, 1987.

38. Heims, 1991; Hughes and Hughes, 2000; Hammond, 2003.

39. Cody and Diamond, 1975.

40. Von Bertalanffy, 1933, 1952; Waddington, 1940, 1968–1972.

41. Hoos, 1972; Berlinski, 1976.

42. E.g., Stent, 1981.

43. See especially Gould and Lewontin, 1979. But also see Ho and Saunders, 1984; Goodwin and Saunders, 1992.

44. E.g., Alberch et al., 1979; Gould, 1977; Levins, 1966; Lewontin, 1974.

45. Williams, 1962; Hamilton, 1996.

46. Grene, 1958, 1961; Rudwick, 1964; Mayr, 1959.

47. E.g., Bock and Von Wahlert, 1965; Love, 2003, and this volume.

48. Sagan, 1967.

49. Willier et al., 1955.

50. Lewontin, 1974.

51. Ghiselin, 1974.

52. Hull, 1976.

53. E.g., Gould, 2002.

54. De Beer, 1930, 1940a, 1940b.

55. Garstang, 1922.

56. Gould, 1977.

57. E.g., Alberch et al., 1979.

58. McKinney, 1988; McKinney and McNamara, 1991; McNamara, 1997; Zelditch, 2001.

59. See, e.g., Barrow, 1990.

60. Berk, 1994; Scranton, 1997; Shenhav, 1999; and Perrow, 2002 discuss the exceptions to the trend in the manufacturing sector.

REFERENCES

Abbott, Andrew. 1995. "Things of Boundaries." *Social Research* 62: 857–882.

Acree, Michael C. 1978. "Theories of Statistical Inference in Psychological Research: A Historico-Critical Study." Ph.D. dissertation, Clark University.

Akam, Michael, P. Holland, P. Ingham, and G. Wray (Eds.). 1994. *The Evolution of Developmental Mechanisms*. Cambridge: Company of Biologists.

Alberch, Per, Stephen J. Gould, George F. Oster, and David B. Wake. 1979. "Size and Shape in Ontogeny and Phylogeny." *Paleobiology* 5: 296–317.

Allen, Garland. 1986. "T. H. Morgan and the Split between Embryology and Genetics, 1910–1935." In T. J. Horder, J. A. Witkowski, and C. C. Wylie (eds.), *A History of Embryology*, pp. 113–146. Cambridge: Cambridge University Press.

Amundson, Ronald. 1994. "Two Concepts of Constraint: Adaptationism and the Challenge from Developmental Biology." *Philosophy of Science* 61: 556–578.

———. 2001. "Adaptation and Development: On the Lack of Common Ground." In S. H. Orzack and E. Sober (eds.), *Adaptationism and Optimality*, pp. 303–334. New York: Cambridge University Press.

Arthur, Wallace. 1997. *The Origin of Animal Body Plans: A Study in Evolutionary Developmental Biology*. New York: Cambridge University Press.

———. 2000. "Intraspecific Variation in Developmental Characters: The Origin of Evolutionary Novelties." *American Zoologist* 40: 811–818.

Baldwin, James M. 1902. *Development and Evolution*. New York: Macmillan.

Barrow, Clyde W. 1990. *Universities and the Capitalist State: Corporate Liberalism and the Reconstruction of American Higher Education, 1894–1928*. Madison: University of Wisconsin Press.

Becker, Howard S. 1960. "Notes on the Concept of Commitment." *American Journal of Sociology* 66: 32–40.

————. 1982. *Art Worlds.* Berkeley: University of California Press.

Berk, Gerald. 1994. *Alternative Tracks: The Constitution of American Industrial Order, 1865–1917.* Baltimore: Johns Hopkins University Press.

Berlinski, David. 1976. *On Systems Analysis: An Essay Concerning the Limitations of Some Mathematical Methods in the Social, Political, and Biological Sciences.* Cambridge, Mass.: MIT Press.

Bock, Walter J., and G. von Wahlert. 1965. "Adaptation and the Form-Function Complex." *Evolution* 19: 269–299.

Bonner, John T. (Ed.). 1983. *Evolution and Development.* Berlin: Springer-Verlag.

Bowler, Peter J. 1983. *The Eclipse of Darwinism: Anti-Darwinian Evolution Theories in the Decades Around 1900.* Baltimore: Johns Hopkins University Press.

————. 1989. "Development and Adaptation: Evolutionary Concepts in British Morphology, 1870–1914." *British Journal of the History of Science* 22: 283–297.

————. 1996. *Life's Splendid Drama: Evolutionary Biology and the Reconstruction of Life's Ancestry, 1860–1940.* Chicago: University of Chicago Press.

Cain, Joseph A. 1990. "George Gaylord Simpson's 'History of the Section of Vertebrate Paleontology in the Paleontological Society.'" *Journal of Vertebrate Paleontology* 10: 40–48.

Cartwright, Nancy. 1983. *How the Laws of Physics Lie.* New York: Oxford University Press.

————. 1989. *Nature's Capacities and Their Measurement.* New York: Oxford University Press.

Chandler, Alfred D. 1990. *Scale and Scope: The Dynamics of Industrial Capitalism.* Cambridge, Mass.: Belknap Press of Harvard University Press.

Churchill, Frederick B. 1973. "Chabry, Roux, and the Experimental Method in 19th Century Embryology." In Ronald N. Giere and Richard S. Westfall (eds.), *Foundations of Scientific Method: The Nineteenth Century,* pp. 161–205. Bloomington: Indiana University Press.

————. 1980. "The Modern Evolutionary Synthesis and the Biogenetic Law." In E. Mayr and William B. Provine (eds.), *The Evolutionary Synthesis: Perspectives on the Unification of Biology,* pp. 112–122. Cambridge, Mass.: Harvard University Press.

————. 1987. "From Heredity Theory to *Vererbung*: The Transmission Problem, 1850–1915." *Isis* 78: 337–364.

Clements, Frederick. 1907. *Plant Physiology and Ecology.* New York: Holt.

Cody, Martin L., and J. Diamond. 1975. "Introduction." In Martin L. Cody and Jared Diamond (eds.), *Ecology and Evolution of Communities,* pp. 1–12. Cambridge, Mass.: Belknap Press of Harvard University Press.

Coleman, W. 1970. "Bateson and Chromosomes: Conservative Thought in Science." *Centaurus* 15: 228–314.

De Beer, Gavin R. 1930. *Embryology and Evolution*. Oxford: Clarendon Press.

————. 1940a. "Embryology and Taxonomy." In Julian S. Huxley (ed.), *The New Systematics*, pp. 365–394. Oxford: Clarendon Press.

————. 1940b. *Embryos and Ancestors*. Oxford: Clarendon Press.

Fisher, Ronald A. 1918. "The Correlations Between Relatives on the Supposition of Mendelian Inheritance." *Transactions of the Royal Society of Edinburgh* 52: 399–433.

————. 1922. "On the Dominance Ratio." *Proceedings of the Royal Society of Edinburgh* 42: 321–341.

Galison, Peter. 1997. *Image and Logic: A Material Culture of Microphysics*. Chicago: University of Chicago Press.

Garstang, Walter. 1922. "The Theory of Recapitulation: A Critical Re-statement of the Biogenetic Law." *Zoological Journal of the Linnaean Society* 35: 81–101.

Gaskins, Richard H. 1992. *Burdens of Proof in Modern Discourse*. New Haven, Conn.: Yale University Press.

Gee, Henry. 1996. *Before the Backbone: Views on the Origin of the Vertebrates*. London: Chapman & Hall.

Gehring, W. J. 1998. *Master Control Genes in Development and Evolution: The Homeobox Story*. New Haven: Yale University Press.

Gerson, Elihu. M. 1998. "The American System of Research: Evolutionary Biology, 1890–1950." Ph.D. dissertation, University of Chicago.

Ghiselin, Michael T. 1974. "A Radical Solution to the Species Problem." *Systematic Zoology* 23: 536–544.

————. 1980. "The Failure of Morphology to Assimilate Darwinism." In E. Mayr and William B. Provine (eds.), *The Evolutionary Synthesis: Perspectives on the Unification of Biology*, pp. 180–192. Cambridge, Mass.: Harvard University Press.

Gigerenzer, Gerd, Swijtink, Zeno E., Porter, Theodore M., Daston, Lorraine J., Beatty, John, and Kruger, Lorenz. 1989. *The Empire of Chance: How Probability Changed Science and Everyday Life*. New York: Cambridge University Press.

Gilbert, Scott F. 1978. "The Embryological Origins of the Gene Theory." *Journal of the History of Biology* 11: 307–351.

————. 1988. "Cellular Politics: Goldschmidt, Just, Waddington and the Attempt to Reconcile Embryology and Genetics, 1938–1940." In R. Rainger, Keith R. Benson, and J. Maienschein (eds.), *The American Development of Biology*. Philadelphia: University of Pennsylvania Press.

————. 2000. "Genes Classical and Genes Developmental: The Different Uses of Genes in Evolutionary Syntheses." In P. Beurton, R. Falk, and H.-J. Rheinberger (eds.), *The Concept of the Gene in Development and Evolution: Historical and Epistemological Perspectives*, pp. 159–177. New York: Cambridge University Press.

————. 2001. *Developmental Biology*, 6th. ed. Sunderland, Mass.: Sinauer.

Goldschmidt, Richard B. 1938. *Physiological Genetics*. New York: McGraw-Hill.

Goode, George B. 1991. *The Origins of Natural Science in America: The Essays of George Brown Goode*, ed. Sally G. Kohlstedt. Washington, D.C.: Smithsonian Institution Press.

Goodwin, Brian C., N. Holder, and C. C. Wylie (Eds.). 1983. *Development and Evolution*. Cambridge: Cambridge University Press.

Goodwin, Brian C., and Peter T. Saunders (Eds.). 1992. *Theoretical Biology: Epigenetic and Evolutionary Order from Complex Systems*. Baltimore: Johns Hopkins University Press.

Gould, Stephen J. 1977. *Ontogeny and Phylogeny*. Cambridge, Mass.: Belknap Press of Harvard University Press.

————. 2002. *The Structure of Evolutionary Theory*. Cambridge, Mass.: Belknap Press of Harvard University Press.

Gould, Stephen J., and Richard C. Lewontin. 1979. "The Spandrels of San Marco and the Panglossian Paradigm: A Critique of the Adaptationist Program." *Proceedings of the Royal Society of London* B205: 581–598.

Grene, Marjorie. 1958. "Two Evolutionary Theories." *British Journal for the Philosophy of Science* 9: 110–127, 185–193.

————. 1961. "Statistics and Selection." *British Journal for the Philosophy of Science* 12: 25–42. Reprinted in Grene's *The Understanding of Nature*, pp. 154–171. Boston: D. Reidel, 1974.

Griesemer, James R. 2005. "The Informational Gene and the Substantial Body: On the Generalization of Evolutionary Theory by Abstraction." In Martin R. Jones and Nanay Cartwright (eds.) *Idealization XII*, pp. 59–115. Amsterdam: Rodopi.

Griesemer, James R., and William C. Wimsatt. 1989. "Picturing Weismannism: A Case Study of Conceptual Evolution." In M. Ruse (ed.), *What the Philosophy of Biology Is: Essays Dedicated to David Hull*, pp. 75–137. Boston: Kluwer Academic.

Grinnell, Joseph. 1910. "The Uses and Methods of a Research Museum." *Popular Science Monthly* 77: 163–169.

Haeckel, Ernst. 1866. *Generelle Morphologie der Organismen*, 2 vols. Berlin, Reimer.

Hagen, Joel B. 1992. *An Entangled Bank: The Origins of Ecosystem Ecology*. New Brunswick, N.J.: Rutgers University Press.

Hall, Brian K. 1992. *Evolutionary Developmental Biology*. London: Chapman & Hall.

Hall, Brian K., and M. H. Wake (Eds.). 1999. *The Origin and Evolution of Larval Forms.* San Diego: Academic Press.

Hamburger, Viktor. 1980. "Embryology and the Modern Synthesis in Evolutionary Theory." In Ernst Mayr and William B. Provine (eds.), *The Evolutionary Synthesis: Perspectives on the Unification of Biology*, pp. 97–112. Cambridge, Mass.: Harvard University Press.

Hamilton, William D. 1996. *Narrow Roads of Gene Land: The Collected Papers of W. D. Hamilton, vol. 1, Evolution of Social Behavior.* New York: W. H. Freeman.

Hammond, Debora R. 2003. *The Science of Synthesis: Exploring the Social Implications of General Systems Theory.* Boulder: University Press of Colorado.

Harris, J. Arthur. 1911a. "The Biometric Proof of the Pure Lines Theory." *American Naturalist* 45: 346–364.

———. 1911b. "The Distribution of Pure Line Means." *American Naturalist* 45: 686–700.

Harwood, Jonathan. 1993. *Styles of Scientific Thought: The German Genetics Community, 1900–1933.* Chicago: University of Chicago Press.

Heims, Steve J. 1991. *The Cybernetics Group.* Cambridge, Mass.: MIT Press.

Ho, Mae-Wan, and Peter T. Saunders (Eds.). 1984. *Beyond Neo-Darwinism: An Introduction to the New Evolutionary Paradigm.* London: Academic Press.

Hoke, Donald R. 1990. *Ingenious Yankees: The Rise of the American System of Manufactures in the Private Sector.* New York: Columbia University Press.

Hoos, Ida. 1972. *Systems Analysis in Public Policy: A Critique.* Berkeley: University of California Press.

Hopwood, Nick. 1999. " 'Giving Body' to Embryos: Modeling, Mechanism, and Microtome in Late Nineteenth-Century Anatomy." *Isis* 90: 462–496.

Hounshell, David A. 1984. *From the American System to Mass Production, 1800–1932: The Development of Manufacturing Technology in the United States.* Baltimore: Johns Hopkins University Press.

Hughes, Agatha C., and Thomas P. Hughes (Eds.). 2000. *Systems, Experts, and Computers: The Systems Approach in Management and Engineering, World War II and After.* Cambridge, Mass.: MIT Press.

Hull, David L. 1976. "Are Species Really Individuals?" *Systematic Zoology* 25: 174–191.

———. 1988. *Science as a Process: An Evolutionary Account of the Social and Conceptual Development of Science.* Chicago: University of Chicago Press.

Jackson, Jeremy B. C., Scott Lidgard, and Frank K. McKinney (Eds.). 2001. *Evolutionary Patterns: Growth, Form, and Tempo in the Fossil Record. In Honor of Allan Cheetham.* Chicago: University of Chicago Press.

Johannsen, Wilhelm. 1909. *Elemente der exacten Erblichleitslehre.* Jena: Gustav Fischer.

Johannsen, W. 1911. "The Genotype Conception of Heredity." *American Naturalist* 45: 129–159.

Johnson, H. Thomas, and Robert S. Kaplan. 1987. *Relevance Lost: The Rise and Fall of Management Accounting.* Boston: Harvard Business School Press.

Judson, Horace F. 1979. *The Eighth Day of Creation.* New York: Simon and Schuster.

Kay, Lily E. 1993. *The Molecular Vision of Life: Caltech, the Rockefeller Foundation, and the Rise of the New Biology.* New York: Oxford University Press.

———. 2000. *Who Wrote the Book of Life? A History of the Genetic Code.* Stanford, Calif.: Stanford University Press.

Kevles, Bettyann H. 1997. *Naked to the Bone: Medical Imaging in the Twentieth Century.* Reading, Mass.: Addison-Wesley.

Kim, Kyung-Man. 1994. *Explaining Scientific Consensus: The Case of Mendelian Genetics.* New York: Guilford Press.

Kohler, Robert E. 2002. *Landscapes and Labscapes: Exploring the Lab-Field Border in Biology.* Chicago: University of Chicago Press.

Laporte, Léo. 2000. *George Gaylord Simpson: Paleontologist and Evolutionist.* New York: Columbia University Press.

Laubichler, Manfred D. 2000. "Homology in Development and the Development of the Homology Concept." *American Zoologist* 40: 777–788.

Levins, Richard. 1966. "The Strategy of Model-building in Population Biology." *American Scientist* 54: 421–431.

Lewontin, Richard C. 1974. *The Genetic Basis of Evolutionary Change.* New York: Columbia University Press.

Love, Alan C. 2003. "Evolutionary Morphology, Innovation, and the Synthesis of Evolutionary and Developmental Biology." *Biology & Philosophy* 18: 309–345.

Mackenzie, Donald A. 1981. *Statistics in Britain, 1865–1930: The Social Construction of Scientific Knowledge.* Edinburgh: Edinburgh University Press.

Maienschein, Jane. 1987. "Heredity/Development in the United States, Circa 1900." *History and Philosophy of the Life Sciences* 9: 79–93.

———. 1991. *Transforming Traditions in American Biology, 1880–1915.* Baltimore: Johns Hopkins University Press.

Mayr, Ernst. 1959. "Where Are We?" *Cold Spring Harbor Symposia in Quantitative Biology* 24: 1–14.

Mayr, E. 1982. *The Growth of Biological Thought: Diversity, Evolution, and Inheritance.* Cambridge, Mass.: Harvard University Press.

Mayr, Ernst. 1991. "An Overview of Current Evolutionary Biology." In L. Warren and H. Koprowski (eds.), *New Perspectives on Evolution*, pp. 1–14. New York: Wiley-Liss.

Mayr, Ernst, and William B. Provine (Eds.). 1980. *The Evolutionary Synthesis: Perspectives on the Unification of Biology.* Cambridge, Mass.: Harvard University Press.

McIntosh, Robert P. 1975. "H. A. Gleason—'Individualistic Ecologist' 1882–1975: His Contribution to Ecological Theory." *Bulletin of the Torrey Botanical Club* 102: 253–273.

McKinney, Michael L. 1988. *Heterochrony in Evolution: A Multidisciplinary Approach.* New York: Plenum Press.

McKinney, Michael L., and Kenneth J. McNamara. 1991. *Heterochrony: The Evolution of Ontogeny.* New York: Plenum Press.

McNamara, Kenneth J. 1997. *Shapes of Time: The Evolution of Growth and Development.* Baltimore: Johns Hopkins University Press.

Møller, Anders P., and John P. Swaddle. 1997. *Asymmetry, Developmental Stability, and Evolution.* New York: Oxford University Press.

Müller, Fritz. 1862. *Für Darwin.* Translated by W. S. Dallas as *Facts and Arguments for Darwin.* London: John Murray, 1869.

Nyhart, Lynn K. 1995. *Biology Takes Form: Animal Morphology and the German Universities, 1800–1900.* Chicago: University of Chicago Press.

———. 2002. "Learning from History: Morphology's Challenges in Germany ca. 1900." *Journal of Morphology* 253: 2–14.

O'Donnell, J. M. 1985. *The Origins of Behaviorism: American Psychology, 1870–1920.* New York: New York University Press.

Olby, Robert C. 1974. *The Path to the Double Helix.* Seattle: University of Washington Press.

Oyama, Susan. 1985. *The Ontogeny of Information: Developmental Systems and Evolution.* New York: Cambridge University Press.

Perrow, Charles. 2002. *Organizing America: Wealth, Power, and the Origins of Corporate Capitalism.* Princeton, N.J.: Princeton University Press.

Porter, Theodore M. 1986. *The Rise of Statistical Thinking: 1820–1900.* Princeton, N.J.: Princeton University Press.

Provine, W. B. 1986. *Sewall Wright and Evolutionary Biology.* Chicago: University of Chicago Press.

Raff, Rudolf A. 1996. *The Shape of Life: Genes, Development, and the Evolution of Animal Form.* Chicago: University of Chicago Press.

Raff, Rudolf A., and Thomas C. Kaufman. 1991. *Embryos, Genes, and Evolution: The Developmental-Genetic Basis of Evolutionary Change.* 2nd ed., Bloomington: Indiana University Press, 1991.

Raff, Rudolf A., and G. A. Wray. 1989. "Heterochrony: Developmental Mechanisms and Evolutionary Results." *Journal of Evolutionary Biology* 2: 409–434.

Rasmussen, Nicolas. 1997. *Picture Control: The Electron Micrscope and the Transformation of Biology in America, 1940–1960.* Stanford, Calif.: Stanford University Press.

Riedl, Rupert. 1978. *Order in Living Organisms: A Systems Analysis of Evolution.* New York: Wiley.

Rowe, Timothy. 1996. "Coevolution of the Mammalian Middle Ear and Neocortex." *Science* 273: 651–654.

Rudwick, Martin J. S. 1964. "The Inference of Function from Structure in Fossils." *British Journal for the Philosophy of Science* 15: 24–40.

Russell, Edward S. 1916. *Form and Function: A Contribution to the History of Animal Morphology.* London: John Murray.

Sagan, Lynn. 1967. "On the Origin of Mitosing Cells." *Journal of Theoretical Biology* 14: 225–275.

Sapp, Jan. 1987. *Beyond the Gene: Cytoplasmic Inheritance and the Struggle for Authority in Genetics.* New York: Oxford University Press.

———. 2003. *Genesis: The Evolution of Biology.* New York: Oxford University Press.

Sarkar, Sahotra. 1999. "From the *Reaktionsnorm* to the Adaptive Norm: The Norm of Reaction, 1909–1960." *Biology & Philosophy* 14: 235–252.

Schank, Jeffrey C., and William C. Wimsatt. 1986. "Generative Entrenchment and Evolution." *PSA86* 2: 33–60.

Scranton, Philip. 1997. *Endless Novelty: Specialty Production and American Industrialization, 1865–1925.* Princeton, N.J.: Princeton University Press.

Shenhav, Yehouda A. 1999. *Manufacturing Rationality: The Engineering Foundations of the Managerial Revolution.* New York: Oxford University Press.

Smith, Merritt R. 1977. *Harper's Ferry Armory and the New Technology: The Challenge of Change.* Ithaca, N.Y.: Cornell University Press.

Stent, Gunther S. 1981. "Strengths and Weaknesses of the Genetic Approach to the Development of the Nervous System." *Annual Review of Neuroscience* 4: 163–194.

Strauss, Anselm L. 1978. "A Social Worlds Perspective." *Studies in Symbolic Interaction* 1: 119–128.

———. 1991. *Creating Sociological Awareness: Collective Images and Symbolic Representations.* New Brunswick, N.J.: Transaction Books.

Taylor, Frederick W. 1911. *The Principles of Scientific Management*. New York: Harper & Brothers.

Von Bertalanffy, Ludwig. 1933. *Modern Theories of Development: An Introduction to Theoretical Biology*, trans. J. H. Woodger. Oxford: Oxford University Press.

————. 1952. *Problems of Life: An Evaluation of Modern Biological Thought*. New York: Wiley.

Waddington, Conrad H. 1940. *Organisers & Genes*. Cambridge: Cambridge University Press.

————. (Ed.). 1968–1972. *Towards a Theoretical Biology*, 4 vols. Edinburgh: Edinburgh University Press.

Wagner, Günter P. 1996. "Homologues, Natural Kinds, and the Evolution of Modularity." *American Zoologist* 36: 36–43.

————. (Ed.). 2001. *The Character Concept in Evolutionary Biology*. San Diego: Academic Press.

Wagner, Günter P., C.-H. Chiu, and Manfred Laubichler. 2000. "Developmental Evolution as a Mechanistic Science: The Inference from Developmental Mechanisms to Evolutionary Processes." *American Zoologist* 40: 819–831.

Wake, David B., Mabee, Paula M., Hanken, James, and Wagner, Günter P. 1991. "Development and Evolution—The Emergence of a New Field." In Elizabeth C. Dudley (ed.), *The Unity of Evolutionary Biology: Proceedings of the Fourth International Congress of Systematic and Evolutionary Biology*, vol. 1, pp. 582–588. Portland, Ore.: Dioscorides Press.

West-Eberhard, Mary J. 2003. *Developmental Plasticity and Evolution*. New York: Oxford University Press.

Williams, George C. 1966. *Adaptation and Natural Selection: A Critique of Some Current Evolutionary Thought*. Princeton, N.J.: Princeton University Press.

Willier, Benjamin H., Paul A. Weiss, and Viktor Hamburger (Eds.). 1955. *Analysis of Development*. Philadelphia: Saunders.

Wimsatt, William C. 1974. "Complexity and Organization." In Kenneth F. Schaffner and R. S. Cohen (eds.), *PSA 1972*, pp. 67–86. Dordrecht: D Reidel.

————. 1980. "Reductionist Research Strategies and Their Biases in the Units of Selection Controversy." In T. Nickles (ed.), *Scientific Discovery: Case Studies*, pp. 213–259. Dordrecht: Reidel.

————. 1986a. "Forms of Aggregativity." In A. Donagan, Anthony N. Perovich, and Michael V. Wedin (eds.), *Human Nature and Natural Knowledge: Essays Presented to Marjorie Grene*, pp. 259–291. Boston: Reidel.

———. 1986b. "Developmental Constraints, Generative Entrenchment, and the Innate-Acquired Distinction." In P. W. Bechtel (ed.), *Integrating Scientific Disciplines*, pp. 185–208. Dordrecht: Martinus Nijhoff.

———. 1987. "False Models as Means to Truer Theories." In Matthew H. Nitecki and Antoni Hoffman (eds.), *Neutral Models in Biology*, pp. 23–55. Chicago: University of Chicago Press.

———. 1997. "Aggregativity: Reductive Heuristics for Finding Emergence." *Philosophy of Science* 64: S372–S384.

———. 2000. "Emergence as Non-aggregativity and the Biases of Reductionisms." *Foundations of Science* 5: 269–297.

———. 2001. "Generative Entrenchment and the Developmental Systems Approach to Evolutionary Processes." In S. Oyama, Paul E. Griffiths, and R. D. Gray (eds.), *Cycles of Contingency: Developmental Systems and Evolution*, pp. 219–238. Cambridge, Mass.: MIT Press.

Winther, R. G. 2001. "August Weismann on Germ-plasm Variation." *Journal of the History of Biology* 34: 517–555.

Wolpert, Lewis. 1994. "The Evolutionary Origin of Development: Cycles, Patterning, Privilege and Continuity." In M. Akam, P. Holland, P. Ingham, and G. Wray (eds.), *The Evolution of Developmental Mechanisms*, pp. 79–84. Cambridge: Company of Biologists.

———. 1999. "From Egg to Adult to Larva." *Evolution and Development* 1: 3–4.

Wright, Sewall. 1918. "On the Nature of Size Factors." *Genetics* 3: 367–374.

Yates, JoAnne A. 1989. *Control Through Communication: The Rise of System in American Management*. Baltimore: Johns Hopkins University Press.

Zelditch, Miriam L. (Ed.). 2001. *Beyond Heterochrony: The Evolution of Development*. New York: Wiley-Liss.

III

REFLECTIONS

14

TAPPING MANY SOURCES: THE ADVENTITIOUS ROOTS OF
EVO-DEVO IN THE NINETEENTH CENTURY
Brian K. Hall

> The fact is, there is no such thing as a science of embryology; it is not
> even a definite branch of a science . . . and the results of the study of
> development can be given with full clearness and in an intelligible
> manner only when formulated as parts of the general doctrine of the
> science under which they fall.[1]

The title of the workshop, "From Embryology to Evo-Devo," quite obviously is not meant to be taken literally. No one knows better than the
organizers—Jane Maienschein and Manfred Laubichler—that embryology
did not *become* evo-devo. Evo-devo is not the late twentieth- and early
twenty-first-century version of nineteenth-century embryology. Embryology became developmental biology. There is, however, no doubt that the
roots of evo-devo can be traced to the evolutionary embryology of the
last four decades of the nineteenth century.[2] Indeed, in the 1880s, evolutionary embryologists such as Ray Lankester argued, as in the epigraph,
that embryology had no existence outside evolution. The issues that con-
sumed evolutionary embryologists then—comparative embryology, how
embryos provide evidence for classification of and phylogenetic relationships among organisms, whether embryos provide evidence of evolutionary history—remain substantive issues for evo-devo today.

But evo-devo goes well beyond nineteenth-century embryology and
is far more integrative. For a start we have genetics and molecular biology,
both of which are integral elements of evo-devo, while life history theory
and phenotypic plasticity provide, respectively, the ecological and ontogenetic elements of evo-devo;[3] we also now have a deeper knowledge of
variation and the sources of variation, and accept natural selection as the
major mechanism sorting that variation. (Hallgrimsson and Hall, 2005).
Heredity and variation—the foundational elements of Darwin's theory of
descent with modification—found ready acceptance after publication of
The Origin of Species in 1859. Although natural selection was accepted only
slowly, some prominent zoologists (Balfour, Haeckel, Lankester, Gegenbaur;
pioneers of evolutionary embryology), accepted natural selection and,

indeed, saw that natural selection operated on embryonic and larval stages as readily as, if not more readily than, on adults. In addition, we now know of other types of selection, including stabilizing selection, so that when confronted with limitations on morphological variation—constraint, *Baupläne*, conserved stages—whether in extant organisms or in fossilized remains, we can explain the maintenance, if not the origin, of such bounded morphology (Hall and Olson, 2003; Hall, 2004).

We have paleontology, which provides the perspective of deep time, and the much deeper understanding of the origins and diversification of groups provided by the fossil record.[4] Placing little faith in fossils—the fossil record was too sparse and incomplete, and the gaps far too plentiful—nineteenth-century zoologists relied on embryos, and while reluctant to insert links into chains of fossil organisms, they were quite prepared to insert stages into gaps in the embryonic record if those insertions would reconstruct the "true" phylogeny of the group. Indeed, evolutionary embryologists went to far-flung corners of the globe in search of embryonic stages of animals such as platypus and lungfish, which were considered missing links between reptiles and mammals and fish and tetrapods, respectively (box 14.1; Hall, 1999b, 2001).

Even this is an inadequate representation of evo-devo, which is much more than evolutionary embryology *plus* these other fields. Evo-devo seeks to integrate these fields into a comprehensive understanding of how development and developmental change are causally related to evolution and to evolutionary change. In what follows, I provide a short evaluation of the themes and elements outlined above, an evaluation that should be read in conjunction with those by Gerd Müller and Gunter Wagner in this volume. Although we did not plan it, the three of us summarize the workshop within historical, contemporary, and future frameworks, respectively. An unusual and most effective aspect of the workshop was the inclusion of three individuals involved day by day in the execution of evo-devo. As it happens, this was a first, previous Dibner workshops having dealt almost exclusively with fields established by individuals no longer living. Workshop participants were able to discuss and evaluate a science/discipline as it unfolds, an opportunity reflected in the productive, even exciting discussion and cross-fertilization of ideas, knowledge, backgrounds, and approaches. I also provide some case studies to illustrate the roles of institutions, technology and, en passant, of interdisciplinary interactions in the origin, rise, and fall of evolutionary embryology.

Box 14.1

Embryos as Missing Links

The late nineteenth-century agenda of evolutionary embryology was implemented by the electors and holders of the Balfour studentship established at Cambridge University after the untimely death of Francis Balfour in 1882. The recipient of the first studentship, William Caldwell, traveled to Australia and determined the mode of reproduction and embryonic development of the platypus, finally enabling its systematic position to be determined. The Balfour studentship, so ably used by William Caldwell,[1] was soon recognized as "the Zoological blue ribbon of Cambridge."[2]

Mounting expeditions to collect new zoological specimens, often from remote and dangerous regions, preoccupied some of the best zoologists in the 1880s and 1890s. Almost without exception their research took them to far-flung parts of the globe: Australia, South America, Africa, the Near East, and the Far East. Almost without exception these expeditions were long, arduous, often unsuccessful, and in John Samuel Budgett's case, fatal.[3] Almost without exception they involved a search for primitive, ancestral, or archetypal animals thought likely to shed light on relationships between major groups of animals or on the ancestry of the vertebrates. Hence, the searches for *Ornithorhynchus* and the Australian lungfish *Neoceratodus* by William Caldwell, for *Balanoglossus* by William Bateson, for *Peripatus* and *Nautilus* by Arthur Willey, and for *Polypterus* by John Budgett. Obtaining knowledge of embryonic development was the central aim all these expeditions. Themes that emerged from this research included (1) the importance, indeed the necessity, of viewing animals in their natural states; (2) the importance of defining and delimiting life history stages, (3) the importance of adaptation to local conditions, and (4) the necessity of separating such adaptation from the primitive features characteristic of the group to which the species belonged. Thus, by the 1890s, evolutionary embryology had expanded into something much more like the evo-devo of today than it is normally given credit for.

Notes

1. Hall, 1999b.
2. Shipley, 1907, pp. 29–30.
3. Hall, 2001.

Embryology and Evolution

Karl von Baer launched comparative embryology within the framework of the classification of organisms. It was more than a happy coincidence that his four embryological types coincided with the four *embranchements* erected by Cuvier. Subsequent discovery of a notochord in both *Amphioxus* (figure 14.1) and ascidians by Alexander Kowalevsky, and in *Balanoglossus* by William Bateson, allowed new groups of animals to be recognized (chordates as one example) and, perhaps more important, provided a novel approach to the origins of the vertebrates and their relationship to the invertebrate phyla (box 14.2). The later search for missing links in embryos of exotic animals was part of the same search for relationships and phylogenetic origins (box 14.1).

In 1878, Thomas Huxley wrote an essay titled "Evolution in Biology".[5] Even casual inspection shows that the bulk of this work is about embryology and its history, and not about evolutionary change through time. In part, this is because of the twofold use of the term "evolution" for individual and phylogenetic development, a duality discussed by Huxley in his essay. To write a history of embryology in 1878 under the title "Evolution in Biology" was perfectly appropriate. Evolution (from the Latin *evolutio*, "unrolling") had been used for embryonic development since the eighteenth century.[6] By October 1, 1880, in an address at the opening of the Mason Science College in Birmingham, Huxley was using

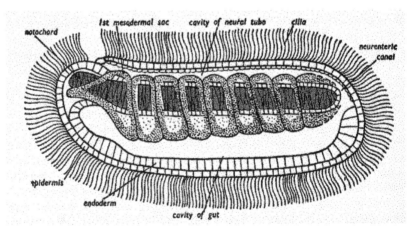

Figure 14.1
A larval *Amphioxus*, drawn in lateral view with anterior to the left. The notochord is shaded. Modified from Murray, 1958.

Box 14.2
Discoveries in Embryos and Classification

Seven years after the publication of *The Origin of Species*, Alexander Kowalevsky demonstrated that the ascidian tadpole developed in a manner similar not only to that known for amphibians but also to the pattern Kowalevsky had described in *Amphioxus*.[1] Ascidians had been grouped with the mollusks, but Kowalevsky's description of a notochord, gill slits, and neural folds that invaginated from folds of the ectoderm—that is, an *Amphioxus*-like, vertebrate pattern of embryogenesis—dramatically demonstrated the affinity between ascidian tadpoles and vertebrates. Leading workers such as Darwin, Haeckel, and Gegenbaur accepted Kowalevsky's findings and conclusions, and so the view that vertebrates arose from an ascidian tadpole-like ancestor became the favored theory of the origin of the vertebrates.[2] Comparative embryology had morphed into evolutionary embryology.

Such studies also enabled new groups to be recognized and new relationships to be proposed. In 1870, Karl Gegenbaur designated a new taxonomic group, the Enteropneusta, for the balanoglossid worms studied by Bateson. Four years later Haeckel established the phylum Chordata. Seven years later, Lankester recognized the Urochordata (tunicates and their allies) and Cephalochordata (*Amphioxus* and its allies). Along with the Vertebrata, these classes constituted the phylum Chordata.[3]

Notes

1. Kowalevsky, 1866, 1871.
2. Ghiselin, 1994; Maienschein, 1994.
3. See Bowler (1996) for detailed discussion and references.

"development" to mean embryonic development: "When a biologist meets with an anomaly, he instinctively turns to the study of development to clear it up."[7] Terminology notwithstanding, the message remained: Huxley reprinted his 1878 essay fifteen years later with only one alteration (and that in a footnote), which acknowledged that he had not done justice to Erasmus Darwin in the first printing.[8] Huxley felt no need to modify his strongly embryological slant on the place of evolution in biology. Embryology was central, and so Huxley provided a strong endorsement of Haeckel's gastraea theory: "Each [animal of higher organization], starting from the condition of a simple enucleated cell, becomes a cell-aggregate; and this passes through a condition which represents the gastrula stage, before taking on the features distinctive of the group to which it belongs. "Stated in this form, the 'gastraea theory'" appears to the present writer to be one of [the] most important and well founded of recent generalisations. So far as individual plants and animals are concerned, therefore, evolution is not a speculation but a fact; and it takes place by epigenesis."[9] Bowler's comment on "the subordination of embryology to phylogenetic research" aptly describes evolutionary embryology of the 1860s to the 1880s.

EMBRYOLOGY DESERTS EVOLUTION BUT RETURNS THREE SCORE AND TEN YEARS LATER

By the 1880s, however, interest in embryology as phylogeny was beginning to wane as individuals such as Gregory Bateson, Reginald Punnett, and others became frustrated with the multitude of subjective scenarios that could be constructed from sequences of embryos, on the one hand, and the need to "invent" stages in embryogenesis to make the developmental sequence fit the desired evolutionary sequence, on the other (Hall, 1992, 2000). As one example, Balfour in 1881 invented the "proto-gnathostomata" as one of six hypothetical groups in his phylogeny of the chordates. Proto-gnathostomes had (or were imagined to have) branchial bars, one of which—the mandibular bar—was converted into the jaw skeleton.[10]

Then came Wilhelm Roux's famous experiment in which he killed one of the cells of a two-cell-stage frog blastula to test how cells became different (differentiated) as development progresses. The ability to do such embryonic physiology provided a major impetus for evaluating embryos for their own sake in order to uncover embryological mechanisms, rather than studying embryos solely for their evolutionary content. Despite (perhaps because of) Hans Driesch's experiment with echinoderm embryos, which produced a result opposite to Roux's experiment— isolated blastomeres from sea urchins produced complete embryos; isolated blastomeres from frog embryos did not—experimental embryology became the focus of study and flourished, paradoxically, in part because of the construction of a marine station in Italy for the express purpose of testing Darwin's evolutionary theory.

Experimental embryology took hold, flourished, and was transformed into the developmental biology of the twentieth and twenty-first centuries. Evolutionary embryology as developmental biology contributes to evo-devo by providing comparative embryology, embryos as characters in phylogenetic analyses and reconstructions, and by continuing to provide information pertinent to the origin and relationships of the metazoans. A major adjunct to these studies has been the rise of molecular biology, especially the recognition of the conservation of key "developmental" ("master, regulative, or signaling") genes involved in the specification of such fundamental animal features as an anterior-posterior axis with head opposing tail, bilateral symmetry, and repetitive (segmental) organization of body parts or regions. Questions such as the conservation of embryonic stages and whether early metazoans were direct or indirect developers (i.e., whether they possessed a larval stage) are now addressed by evo-devoists in comparative studies using development, molecular, life history, phylogenetic, and paleontological approaches. This is not Haeckel's "ontogeny recapitulates phylogeny," but Garstang's "ontogeny creates phylogeny" (Garstang, 1929; Hall, 2000, 2003).

Similar developmental genes (bone morphogenetic proteins, fibroblast growth factors) provide the upstream signals that activate developmental pathways in features as diverse as induction of mesoderm, limb buds, kidneys, teeth, and skeleton. The challenge now is to identify the downstream cascades and networks activated or repressed by these signaling genes and to understand the role of gene regulation in phenotypic change.

Furthermore, we now appreciate that developmental mechanisms with important roles in evolutionary change exist beyond the level of the genes and that developmental properties emerge as development progresses. So, cell-to-cell and tissue-to-tissue interactions, interactions between environmental or ecological factors and ontogenetic stages, and even interactions between predator and prey, influence ontogenetic and phylogenetic transformations (Hall, 1999a, 2003b; Hall et al., 2004).

The rise of phylogenetics and cladistic analysis, along with the myriad molecular characters now available, has transformed nineteenth-century classification and systematics into a robust discipline that identifies relationships and provides hypotheses to be tested. With a robust phylogeny in hand, the direction of evolutionary change in developmental characters, stages, and processes can be determined.

Ecological factors are increasingly taken into account through analysis of the developmental and evolutionary bases of altered life history strategies as diverse as phenotypic plasticity, seasonal polymorphisms, and the loss of larval stages (Schlichting and Pigliucci, 1998).

INSTITUTIONS, INSTRUMENTS, AND INTERDISCIPLINARY INTERACTIONS

The individuals drawn to evolutionary embryology based their studies on the comparative embryology initiated by Von Baer, on Darwin's theory of descent with modification, and on Haeckel's theory of the recapitulation of phylogeny by ontogeny. They could not, however, have undertaken their studies without the required instrumentation (technology) and institutional support, both of which provided interdisciplinary interactions that furthered evolutionary embryology. I address the roles of institutions, instruments, and interdisciplinary interactions by analyzing the development of the Naples Zoological Station and the invention of an automatic microtome, which allowed rapid serial sectioning of embryos. We take the latter for granted. It was, however, a major technological advance that enhanced the speed of research in the same way that automated DNA sequencers enhanced and advanced research.

There is more than a passing parallel between the conditions that allowed evolutionary embryology to thrive in the 1860s and 1870s and those which led to evo-devo a century later. Epigenetics and emergent properties provided new theoretical underpinnings, molecular genetics and phylogenetics provided new techniques, and a number of institutions provided world stages equivalent to that in Naples from 1872 on. The rapid successes achieved at Naples led to new journals, national conferences and international congresses, and exchange of ideas on an international scale. Evo-devo has achieved the same status in a world that is much more globalized than it was a century ago.

Stazione Zoologica Napoli

The Stazione Zoologica Napoli, perhaps the most important institution to foster the development of evolutionary embryology, opened on the shores of the Bay of Naples early in 1874. The station was established with the express intention of providing a facility where individuals or groups of scientists from around the world could test Darwin's theory; it is the model for conducting international research and fostering intellectual excitement, advance, and interaction. If was the brainchild, passion, and lifelong accomplishment of Felix Anton Dohrn (1840–1909), known as Anton. As its founder and first director, Dohrn provided the drive, financing, scientific vision, and organization that made the station such a success.[11]

A successful and financially rewarding sugar refinery gave Anton's father, Carl August Dohrn, the independence to pursue his real interests, which included a famous entomological collection, editing *Entomologische Zeitung* (1843–1887), extensive correspondence with scientists all over the world, translating Spanish dramas, and the passionate pursuit of music.[12] Anton Dohrn received his zoological training at the University of Jena, where the twenty-eight-year-old Ernst Haeckel introduced him to the writings of Charles Darwin. Dohrn, like Haeckel, was one of the first to succumb to the new doctrine of evolution. Dohrn, as his first assistant, was present at Haeckel's first defense of Darwin's theory at an 1863 meeting of German scientists and doctors at Stettin, on the occasion of the twenty-fifty anniversary of the Stettin Entomological Society. Verification of Darwin's theory became his life's work and the rationale for establishing the marine station. As far as Dohrn was concerned, zoology should concentrate on Darwinism (natural selection and the struggle for existence) and Haeckelian recapitulation; Dohrn referred to himself as "the Statesman of Darwinism" who needed "to create, to organize, to develop."[13]

In 1868 Dohrn began work in Messina, Sicily, on the comparative morphology and embryology of marine animals. Johannes Müller's work at this and other Mediterranean sites had demonstrated the value of using marine organisms to elucidate general biological principles. So rich was the Mediterranean fauna that Haeckel discovered over 140 species of radiolarians during the 1859–1860 field season alone.[14] Darwin's theory inspired Dohrn to devote his attention to the evolution of the arthropods and to a search for the ancestry of the vertebrates among the annelids. An extensive correspondence on these issues with Darwin and Huxley began in late 1867. Dohrn, however, was frustrated by the difficulties associated with attempting to carry out his research, including finding the animals and keeping them alive in totally inadequate laboratory facilities consisting of a rudimentary, portable aquarium ("laboratory").[15]

From such beginnings and frustrations Dohrn conceived the idea of creating centers (zoological stations) where zoologists of any nation could gather and work, and where the public could view the wonders of marine life (Dohrn, 1872c). Dohrn was just the man to pull off such a feat. His administrative and organizational expertise were evident; he was offered (and refused) the directorship of the Hamburg Zoological Gardens, and was dissuaded by Huxley from applying for the directorship of the Zoological Museum in Calcutta.[16] Another reason for Huxley's advice was Dohrn's financial independence and the role Huxley felt Dohrn could play in developing a zoological station in Europe:

> If it were necessary for you to win your own bread, one's advice might be modified. Under such circumstances one must do things which are not entirely desirable. But for you who are your own master and have a career before you, to bind yourself down to work six hours a day at things you do not care about and which others could do just as well, while you are neglecting the things which you do care for, and which others could not do so well, would, I think, be amazingly unwise.

Liberavi animam! don't tell my Indian friends I have dissuaded you, but on my conscience I could give no other advice.

Huxley encouraged Dohrn to use his name in whatever way would aid development of the station.[17] Late in December 1869 Dohrn wrote to Darwin to outline his plans for a station. A letter to friends in Berlin, written a month later, radiates joyful, enthusiastic exuberance and optimism from every line:

> I am going to establish in Naples a large aquarium for the public . . . on two floors, with both large and small aquaria, sea animals will be on show to the public, and a further floor will be reserved for the scientists. I am going to lead this thing—once it is going smoothly it will take up very little time. . . . All can be done very cheaply. No loss of dying animals. Hurrah, it's a marvellous idea.
>
> I have already calculated that for 120 visitors daily for 9 months of the year I can have profits, running, and everything. And how many more will come! The splendour of the animals. The boredom of the people! And in rainy weather! you must congratulate me, the idea is ready money, freedom, independence, and a nice home for my dear friends in Naples![18]

Dohrn had more in mind than a mere research station.

The British Association for the Advancement of Science (BAAS) took an active role in the plans. In September 1870 Dohrn attended its Liverpool meeting, where the establishment of a station was endorsed and where Huxley introduced him to Charles Darwin. The BAAS set up a committee to promote the foundation of zoological stations in various parts of the world, and received frequent reports of progress from Dohrn, who served as secretary to the committee.[19] The committee reported in August 1871 that construction was to begin in October, with completion by January 1873. Given these good tidings, the committee urged establishment of a zoological station in the British Islands, reminding the BAAS of its support for dredging explorations. Indeed, after Darwin's death there was a movement to establish a memorial fund to found a marine station in Darwin's name.[20]

By October 1871 Dohrn, all his books and equipment, Ray Lankester, and a few other friends were installed at the Palazzo Torlonia in Mergellina to plan the station. Ever the practical man, Dohrn decided to build his station in Naples rather than Messina, which he favored initially. Naples, with a population of over half a million people, attracted 30,000 tourists a year. Dohrn was sure these tourists would pay to view his proposed public aquarium, and thus underwrite the costs of the research facility. The Neapolitan city fathers were very suspicious of the bearded foreigner who wanted to place in the Royal Park, one of the most beautiful parks in the center of the city, a large building of uncertain usage—theater, restaurant, hotel, palazzo, bordello? After much cajoling of the reluctant city authorities, construction of a station in stucco and marble began under Dohrn's supervision in the early spring of 1872 (see Dohrn, 1872c for his evaluation of need for zoological stations).

His initial plans were much less grandiose than the complex that was built, although it was impressive (figure 14.2). In his letter to Darwin in December 1869, Dohrn talked of "a little house of perhaps four rooms, an Aquarium connected with the sea and the house,—the Aquarium of perhaps 60 feet in Cubus, where one might have streaming water,—a boat for dredging work, dredges, nets, ropes. . . ."[21] Three years later the plans included 7,000 square feet (650 square meters) of floor space, aquarium tanks, twenty-four laboratories, a library of 25,000 volumes, and living accommodations. Stocking the library was greatly facilitated by support from Huxley, who promised "all my books, past, present and to come for

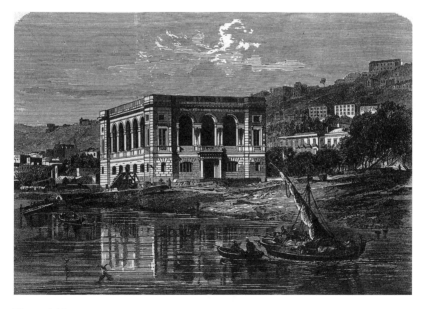

Figure 14.2
Stazione Zoologica, 1873. Reproduced with permission from an original lithograph in the Historical Archives, Stazione Zoologica Anton Dohrn di Napoli.

the Aquarium," adding: "The best part about them is that they will not take up much room. Ask for Owen's by all means; 'Fas est etiam ab hoste doceri.'"[22] The BAAS also provided its publications, and in 1873 the Royal Society provided a set of the *Philosophical Transactions*.

Construction of the station took all Dohrn's energies and talents, and most of his private wealth; two thirds of the costs came from Dohrn and his father. Crises in the municipal government and the outbreak of the Franco-Prussian War added unexpected setbacks. In February 1872 Dohrn was sufficiently concerned over the financial prospects that he wrote to Huxley that he was contemplating withdrawing from the project. Huxley's response was twofold. He promised to write to Darwin, and he consulted with Michael Foster, who told him that an appeal should be made to Frank Balfour (figure 14.3), still an undergraduate. Balfour, Darwin, Huxley, and Foster individually and collectively did much to encourage the development of the Stazione Zoologica. As well as playing a vital role in the financing and successful development of the station,

Figure 14.3
A bust of Francis (Frank) Maitland Balfour, by the sculptor Adolf von Hildebrand—who executed the busts of Darwin and Von Baer for Stazione Zoologica—commissioned by Balfour's family after Frank's death. It is now in the Balfour Library of the Department of Zoology, Cambridge University. This copy was donated by J. Willis Clark, the university registrar, and Darwin's son George. The connection between Balfour, Darwin, Von Baer, Cambridge, Naples, theory, instrumentation, and implementation could not be more appropriate. Photographed by the author.

Frank Balfour was one member of the first pair of British zoologists to work there (Dew-Smith was the other). In a letter written on February 13, 1874, to Anton Dohrn, Darwin commented: "You will find him [Balfour] a most amiable young man, with much ability." In the same letter Darwin expressed the earnest desire that his third son, Francis (also known as Frank), should spend some months in Naples "to learn the art of observing marine animals."[23] Dohrn and Balfour exchanged almost 100 letters over the ten years between 1872 and 1882, precisely the type of interaction Dohrn hoped would flow from establishing the station.

Darwin saw the value of the station, and was enthusiastic and supportive from the outset. He gave £120 in 1874—a subscription of £100 for himself, and £10 each for his sons George and Francis—and a further £100 in 1880. The latter was an amount equivalent to the first Bressa Prize of 12,000 lire, awarded to Darwin by the Royal Academy of Sciences of Turin in 1879. Darwin wrote to Dohrn on February 15, 1880:

> Perhaps you saw in the papers that the Turin Society honoured me to an extraordinary degree by awarding me the *Bressa* Prize. Now it occurred to me that if your station wanted some piece of apparatus, of about the value of £100, I should very much like to be allowed to pay for it. Will you be so kind as to keep this in mind, & if any want should occur to you, I w^d send you a cheque at any time.—With all good wishes for your own success & happiness & for the prosperity of the Station.[24]

To cover escalating construction costs, Dohrn instituted a system of "working tables." Research space would be made available to governments, universities, or scientific organizations for an annual sum. Each institution would have its own "table" for research. Michael Foster and Thomas Huxley urged Cambridge University to acquire two research tables, which Cambridge did in December 1875, committing £75 per year for five years and increasing the amount to £100 in 1883. Between 1876 and 1931 the BAAS provided £4,640 toward leasing tables at Naples.[25] Each "table" was, in effect, a condensed laboratory with chemical reagents, anatomical and microscopical tools and apparatus, and drawing materials. Along with rental of a research table came a daily supply of fresh animals, use of the library, and access to the laboratory "servants" (figure 14.4), who were instructed to ask after the needs of the guests every quarter of an hour![26]

This inspired scheme of research tables, playing the pretensions, aspirations, and reputations of governments and institutions against one

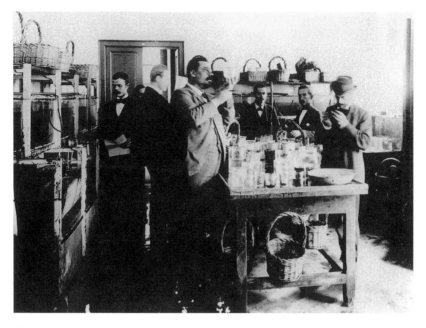

Figure 14.4
The sorting room at Stazione Zoologica circa 1890, with scientists and servants at work. Reproduced with permission from an original photograph in the Historical Archives, Stazione Zoologica Anton Dohrn di Napoli.

another, was a resounding success. In addition to admission fees to the public aquarium and fees for the research tables, the station did well financially from the sale of preserved marine organisms and, en passant, contributed to the development of methods of fixation, section cutting, staining, and optical microscopy that were exported to Britain and Continental countries.[27]

Embedding and sectioning small and delicate specimens involved the laborious handling and processing of single sections. Specimens embedded in cocoa butter were cut, and each section was mounted on a separate slide. Even as late as 1880, Arthur Shipley, a student of Balfour's at Trinity College, Cambridge, recorded that it took the afternoons of half a term to section a small *Amphioxus*. Shipley commented drolly that "each section had to be mounted on a separate slide—when really we would have been better employed on the river or on the football-field."[28] Improvement in the quality of microscopic observations that could be made from such sections was one tangible outcome. So, too, was the invention of an automatic microtome that would produce serial sections.

The Cambridge Instrument Company and the Automatic Microtome

Developments in instrumentation played a substantial role in the rise of evolutionary embryology. Microscopes were of high quality, but no adequate means were available to slice an organism into sections, especially organisms as small as embryos. The few instrument makers were located outside universities and were fully occupied with industrial design and manufacture. The physiologist Albert Dew-Smith[29] and Horace Darwin—fifth and youngest son of Charles—both undergraduates at Cambridge, played instrumental roles in correcting this lack.

The Cambridge Scientific Instrument Company Horace Darwin entered Trinity College with Frank Balfour in 1871, two years after Dew-Smith.[30] After graduating, Darwin spent three years apprenticed to an engineering firm. Returning to Cambridge in 1877, he established himself as a consulting engineer. The need for instrumentation for research quickly became evident to him. In the spring of 1878 James Stuart, Professor of Mechanism and Applied Mechanics, began to make equipment for members of the university, partly to fill the need, partly to cover the salaries of the two mechanics/instrumentmakers/instructors employed in a workshop that in time became the university engineering laboratories. The chief mechanic was Robert Fulcher. During 1878 he and Dew-Smith came to an arrangement to use the machinery in the workshop to construct equipment on a commercial basis, Fulcher providing the expertise and Dew-Smith the money (in part from an auction at Sotherby's of some of his books, engravings, and drawings that yielded £3080.3s). Fulcher and Dew-Smith also published the *Journal of Physiology* for Michael Foster.

When it became clear in 1879 that Dew-Smith and Fulcher wanted to find their own premises, James Stuart sought out Horace Darwin, who had been receiving commissions from Dew-Smith, and proposed a partnership. Darwin turned it down. Although Fulcher designed a sliding microtome in 1879–1880, he lacked the expertise in mechanical engineering needed to supply the needs of their clients. An acrimonious breakup of the partnership between Dew-Smith and Fulcher took place on December 8, 1880. By December 28, Horace Darwin could write to a client that as of January 1, 1881, "The business now known as Robert Fulcher will be carried on, provided no unforeseen circumstances occur, as The Cambridge Instrument Company under the management of the Proprietors, who will be A. Dew-Smith and myself."[31] Within the Darwin family, the new company became known as "Horace's Shop."

Links between the instrument company, Michael Foster, and Cambridge physiology continued to be forged. Horace Darwin became a leader in the design of scientific instruments for all types of work and the dominant force in the company. In designing and marketing a microtome that could cut ribbons of serial sections the company played a crucial role in the development of natural science at Cambridge—indeed, throughout the world, as the needs of evolutionary embryology and the technological skills of a man receptive to those needs came together.

The Automatic Microtome   Two Cambridge undergraduates, William Caldwell and Richard Threlfall, discovered that consecutive histological sections could be cut so as to form a continuous ribbon. This simple finding ushered in serial sectioning, the development of the automatic microtome, a renaissance in microscopical anatomy, and materially advanced evolutionary embryology.[32] In his entry on embryology in the eleventh edition of the *Encyclopaedia Britannica*, Adam Sedgwick emphasized the importance of Caldwell's contribution:

> In methods, while great improvements have been made in the processes of hardening and staining embryos, the principal advance has been the introduction in 1883 by W. H. Caldwell in his work on the development of *Phoronis* of the method of making tape-worm like strings of sections as a result of which the process of mounting in order all the sections obtained from an embryo was much facilitated and the use of an automatic microtome rendered possible.[33]

Microtomes existed well before Caldwell and Threlfall developed their automatic model. John Hill in the 1760s designed a cutting engine to cut thin slices of wood. By the first half of the nineteenth century, various forms of section cutters had been designed, usually based on a vertical brass tube to hold the specimen beneath a glass disk with a central hole. A micrometer screw advanced the specimen, and individual thin sections were cut with a cutthroat barber's razor drawn across the glass disk. By 1840 these instruments were known as microtomes. The Cambridge Scientific Instrument Company was the agent for such a microtome manufactured by Zeiss in Germany. A version of the Rivet microtome (the first sliding microtome where the specimen was advanced along a sliding plane, made in wood by Rivet in 1868), modified by Francis Balfour, was being sold by the Cambridge Scientific Instrument Company early in 1881. The description in the list of prices was "Section Cutter: Balfour's modification of Rivet's instrument. It is so arranged that each division corresponds to .1 mm and the Vernier attached permits sections to be cut to .01 mm in thickness for each division. Suitable flat knives are supplied with the Instrument. £6.6s.0d."[34] One of the first letters (January 10, 1881) written on behalf of the new company was to W. H. Caldwell of Trinity College, concerning his requirements for the Zeiss microtome, for which the new company was now the agent.

Caldwell was in Naples in May 1882, investigating the embryology of *Phoronis* and brachiopods, when he discovered that paraffin sections cut on the Zeiss microtome adhered to one another across the blade to form a ribbon if the temperature was just right. Back in Cambridge, Caldwell communicated this finding to Threlfall. One or the other, it is unclear which, saw that this "made the construction of a section-cutting machine a possibility."[35] By mid-July, Threlfall had produced drawings for a machine with a fixed knife and moving specimen; all previous microtomes worked on the principle of holding the specimen still and moving the knife. Ten or eleven months later, a machine constructed at Threlfall's expense by the Cambridge Scientific Instrument Company underwent tests at the Cavendish Laboratories. By the autumn of 1883, Threlfall was experimenting with the most suitable paraffin for embedding specimens to obtain ribbons.

Caldwell and Adam Sedgwick (who had replaced Balfour at Trinity College) tested the prototype microtome. It succeeded beyond expectation. The instrument, driven by a water motor, was used in the morphology laboratory of Trinity College for the rest of the century. The only restoration work done on this original instrument was to fit a new belt to carry the ribbon of sections. The original design specified that the belt be made from the strings of a Cambridge B.A. gown. So, too, was the replacement.[36] With a belt almost a meter long, the original Threlfall-Caldwell microtome resided until quite recently, rather sadly and ill-displayed, in a dusty glass case in the corridor outside the Balfour Library in the Zoology Department at Cambridge.

Caldwell was a guest at the Physiological Society's meeting of July 21, 1883, and gave a presentation on "the mode of using a new continuous mechanical automatic microtome."[37] The first notification in print of the invention and utility of the microtome was in two entries added to the second edition of Foster and Balfour's *Elements of Embryology*. The editors, Adam Sedgwick and Walter Heape, wrote:

> The microtome we are most accustomed to is a "sliding microtome" made by Jung of Heidelberg; it gives excellent results. Recently however

Messrs. Caldwell and Threlfall have designed an automatic microtome which has been used with success at the Cambridge Morphological Laboratory and promises to effect a great saving of time and trouble in cutting sections.

Since writing the account of section-cutting on p. 434, we have obtained more experience as to the practical working of Messrs. Caldwell and Threlfall's microtome there mentioned. We find that it cuts more accurately and better than any other microtome with which we are acquainted, and can confidently recommend it to investigators and teachers with large classes.[38]

Threlfall and Caldwell had an amicable relationship; witness Threlfall's rationale for not filing for patent rights, a rationale that reflects an intriguing mixture of naïveté, high principles, and sharing of credit with Caldwell: "I am sometimes asked why I did not patent the machine. The answer is that, as it was a machine for the furtherance of scientific research and had no commercial application, both Caldwell and I considered that to take out a patent would have been as improper as it would be for a physician to patent a medical discovery."[39] How times have changed!

THE BUBBLE BURSTS

Neapler Entwicklungsmechaniker and Genetics

In 1874, a high point in evolutionary embryology, with recapitulation in the air and a new marine station on the ground, Wilhelm His published *Unserer Körperform*, his mechanical analysis of embryos as sheets and tubes that bend, pinch, and fold, an approach that paved the way for a more physiological approach to embryos and to embryology, and, six years later, to the establishment of *Entwicklungsmechanik*, a totally new approach to the physiology of embryos. According to His and his followers, the laws of physics and chemistry, not speculations of hypothetical ancestors, would provide the understanding of embryology being sought by so many. The proverb recounted by Walter Garstang in the closing paragraph of his 1929 presidential address to Section D of the BAAS might well have served as the motto of the experimental embryologists: "There is an old German proverb which needs to be hung over the mantelpiece of those of us who have a bent for speculation: *Behaupten ist nicht beweisen* (to claim is not to prove).[40]

As the seeds of evolutionary embryology prospered in Naples in the 1870s and early 1880s, so did the seeds of the *Entwicklungsmechanik* that overtook and supplanted evolutionary embryology in the later 1980s. Pioneers in experimental work on embryos such as Thomas Hunt Morgan, E. B. Wilson, Theodor Boveri, Kurt Herbst, and Hans Driesch initiated their studies at Naples; they were known throughout Europe as *Neapler Entwicklungsmechaniker*. The station Dohrn had built to test Darwin's theories was now host to an embryology that ignored evolution. Embryologists were now experimentalists. However, many who a decade before would have gone into evolutionary embryology now flowed to the new field of genetics. These geneticists trained at the same institutions as their forebears, spent time in Naples, and often began by investigating embryos—often from exotic locales—before turning their backs on embryos.[41] A case in point is Reginald Punnett.

After graduating in zoology from Cambridge in 1898, Punnett spent six months at Stazione Zoologica, followed by a stint in Carl Gegenbaur's laboratory in Heidelberg, apprenticeships that prepared him for comparative and evolutionary embryology. Moving then to Saint Andrews University in Scotland, Punnett began studies on the embryology of nemerteans, studies that were the subject of about a dozen of papers (e.g., Punnett, 1900a, 1900b, 1900c, 1901, 1902), including three on animals collected from the Torres Strait and Singapore. All is standard evolutionary embryology so far.

Punnett became dissatisfied, however, with the contributions morphology and embryology were making to evolutionary studies. His conversion to genetics came in 1901 while he was convalescing from an appendectomy, an operation made possible by the financial security provided by a Caius College fellowship (he was then back in Cambridge). Punnett decided to test with mice the idea that sex ratios were influenced by diet, and he wrote to William Bateson, who was conducting experiments on Mendelian inheritance. Punnett collaborated with Bateson from 1904 to 1910, Bateson himself having deserted evolutionary embryology after his spectacular discovery of the notochord (stomochord) in *Balanoglossus* (box 14.3). As did Bateson, Punnett played a major role in establishing genetics as a separate and distinct science; the Punnett square is used to determine the outcome of genetic crosses by high school and university students alike.[42]

Successes came quickly in the new field of genetics, as they did in molecular genetics seventy years later. Punnett and Bateson discovered the Mendelian basis of sex determination, sex linkage, and autosomal genetic linkage.[43] Evelyn Hutchinson, one of no more than half a dozen students in Punnett's genetics course, attended the lecture at which Punnett announced demurely that he had finished all the calculations for the sweet

Box 14.3
Betrayed by *Balanoglossus*

The following letter concerning his *Balanoglossus* work was written by William
Bateson to his mother (November 22, 1886) from Kazalinsk, a desolate location
immediately northwest of the Aral Sea in Russian (West) Turkistan, where
Bateson was seeking evidence for environmental influences on morphological
diversity in fauna in the lakes in that region:

Entre-nous, the *Balanoglossus* business was a very easy victory, and wasn't much
work at all. The thing did itself. Of course, the "*kudos*" turned up most substantial trumps, but the thing isn't valuable really. Five years hence no one will
think anything of that kind of work, which will be very properly despised. It
hasn't any bearing whatever on the things we want to know. It came to me at
a lucky moment and was sold at the top of the market; presently steam will be
introduced into Biology and wooden ships of this class won't sell well.[1]

Bateson's position on embryological approaches was also made very clear
in the preface to *Materials for the Study of Variation, Treated with Especial Regard to
Discontinuity in the Origin of Species,* his massive accumulation of examples of discontinuous variation in the origin of species, published in 1894. Although the
preface was written on December 29, 1893, many of the arguments against
embryology are taken verbatim from his application for the position of deputy
to the Linacre professor of human and comparative anatomy at Oxford three
years earlier:

Some years ago it was my fortune to be engaged in an investigation of the
anatomy and development of *Balanoglossus.* At the close of that investigation it
became necessary to analyze the meaning of the facts obtained, and especially
to shew their bearing upon those questions of relationship and descent which
modern morphology has attempted to answer. To this task I set myself as I best
might, using the common methods of morphological argument and interpretation, and working all the facts into a scheme which should be as consistent as
I could make it.

But the value of this and of all such schemes, by which each form is duly
ushered to its place, rests wholly on the hypothesis that the methods of argument are sound. Over it all hung the suspicion that they were not sound. This
suspicion seemed at that time so strong that in preface to what I had to say
I felt obliged to refer to it, and to state explicitly that the analysis was undertaken in pursuance of the current methods of morphological criticism, and
without prejudging the question of possible or even probable error in those
methods. . .

Were we all agreed in our assumptions and as to the canons of interpretation, there might be some excuse, but we are not agreed. Out of the same
facts of anatomy and development men of equal ability and repute have brought
the most opposite conclusions. To take for instance the question of the ancestry of the Chordata, the problem on which I was myself engaged, even if we
neglect fanciful suggestions, there remain two wholly incompatible views as to

Box 14.3 (continued)

the lines of Vertebrate descent, each well supported and upheld by many. From the same facts opposite conclusions are drawn. Facts of the same kind will take us no further. The issue turns not on the facts but on the assumptions. Surely we can do better than this. Need we waste more effort in these vain and sophistical disputes?

If facts of the old kind will not help, let us see facts of a new kind. That the time has come for some new departure most naturalists are now I believe beginning to recognize. For the reasons set forth in the introduction I suggest that for this new start the Study of Variation offers the best chance.[2]

Some 105 years of debate over homology of the "notochord" of *Balanoglossus* and the notochord of chordates has changed the picture.[3] Hemichordates are now accorded the status of a separate prechordate phylum allied to the echinoderms. The notochord is actually a stomochord—a buccal pouch or buccal diverticulum projecting from the mouth cavity. Development of stomochord and notochord are very different, the stomochord arising from gut endoderm and the notochord, from dorsal mesoderm. *Balanoglossus* has gone full circle from echinoderm to a fourth class of chordates and back to an association with the echinoderms, acquiring status as the phylum Hemichordata on the way.

Notes

1. Beatrice Bateson, 1928, pp. 19–20.
2. William Bateson, 1894, pp. v–vi.
3. A common theme that is as crucial for evo-devo now as it was for descriptive and evolutionary embryology a century ago, is homology. Whether features are morphological, behavioral, molecular, gene networks, developmental sequences, or developmental stages, how do we know that we are comparing equivalent features? Surprisingly, homology was mentioned only once in the workshop. Yet, it is fundamental and has been so for millennia. Witness Aristotle's classifications of organisms on the basis of shared features or correlations of features: deer with antlers lack tusks; deer with tusks lack antlers, animals with flippers are related, and so forth.

pea and that, indeed, there were as many linkage groups as haploid chromosomes. Thus did Punnett first demonstrate that the chromosome theory applied to plants.[44]

Punnett founded poultry genetics and was a major force in the introduction of genetics to commercial breeders of livestock as well as to the general public. Two books—*Mendelism* (1905) and *Heredity in Poultry* (1923)—were important instruments in his quest to educate others in the new science of genetics. For thirty-five years Punnett put his stamp on the newborn, then the infant, then the robust adult science of genetics. The temptations of genes over embryos, then as now, were powerful and hard to resist.

There is a long history of the separation of genetics (especially population genetics) from embryology, and the consequent exclusion of embryology from the Modern Synthesis. The incorporation of molecular genetics and developmental biology in evo-devo is one of the major arguments for regarding evo-devo as much more than the twenty-first-century offspring of embryology. Its roots spread much more widely and adventitiously. It was not from embryology to evo-devo, but rather a synthesis and integration of genetics, embryology, morphology, paleontology, ecology, and evolution that produced evo-devo.

NOTES

1. Lankester, 1878, p. 113.

2. The discipline in which zoologists turned from the morphology of adult animals to comparative embryology is referred to as evolutionary morphology or evolutionary embryology. Hall, 1992, 1999a, 2000; Nyhart, 1995; Bowler, 1996.

3. Schlichting and Pigliucci (1998), Carroll et al. (2001), Wilkins (2002), and Hall et al. (2003) provide introductions to and analyses of these aspects of evo-devo.

4. Conway Morris, 1994; Erwin and Wing, 2000; Hall, 2002.

5. Reprinted in T. H. Huxley, 1893.

6. See Bowler (1989) and Hall (1992) for discussions of the changing use of the term "evolution."

7. Quoted in T. H. Huxley, 1986, p. 46.

8. T. H. Huxley, 1893, p. 209 and footnote, p. 226.

9. Ibid., p. 202; Bowler, 1988, p. 1.

10. As late as 1933, Hans Gadow waxed enthusiastic over another hypothetical group, the "Proto-Gano-Dipnoi." Following Balfour's embryological researches, "it was taken

for granted that these, to the exclusion of all other fishes, represented the ancestral stage of all Gnathostomata" (Gadow, 1933, p. 66, n. 1). Evelyn Hutchinson speaks of Gadow as a marvelous relic of the nineteenth century, a man who began his course of lectures by writing John 1:1 in transliterated Greek on the blackboard, and who once followed Lutheran tradition by nailing to the door of the zoology laboratory a set of theses on the homology of the ear ossicles (Hutchinson, 1979, pp. 88–89).

11. A mountain of material has been published on Anton Dohrn and Stazione Zoologica Napoli. Accessible short histories are the overviews by Oppenheimer (1980), Müller (1996), and Fantini (1998/1999, 2000). Groeben and Müller (1975), whose background material is very helpful (esp. pp. 11–21), cite more than 100 sources; see also Partsch (1980) and Groeben (1982). Perhaps the most comprehensive biographies of Dohrn are by Kühn (1950) and Theodor Heuss (1991).

12. Carl Dohrn collected European folk songs that were otherwise only part of the oral tradition of their native countries, performed whole operas to his own accompaniment, and was a close friend of Felix Mendelssohn-Bartholdy, who was Anton's godfather. Anton's inheritance of his father's love and encyclopedic knowledge of music cemented two long-lasting relationships in Naples, one with William Lloyd, the English engineer who constructed the exhibition aquariums at the Naples station, the other with the cell and developmental biologist E. B. Wilson, one of the first Americans to undertake research at the station.

William Lloyd, who also constructed the Crystal Palace Aquarium in London, and aquariums in Hamburg and Berlin, did not take to Dohrn on first acquaintance. One morning Dohrn was ascending the steps of the station, whistling a tune. Lloyd, who was coming down, asked whether he realized that the tune was from Mendelssohn's Violin Concerto. When Dohrn acknowledged that he did, and that Mendelssohn was his godfather, Lloyd immediately took to him, and they became fast friends. A similar incident occurred with E. B. Wilson. As the two were sailing past the island of Procida, Dohrn began to hum the opening theme from Schumann's A Major String Quartet. Much to Dohrn's surprise (after all, his companion was an American!), Wilson immediately continued the melody. This musical bond blossomed into a friendship that extended to Dohrn's son Reinhard, and Wilson's staunch support for the station in the 1920s (Groeben and Müller, 1975, pp. 41, 87, 90).

13. Dohrn, 1872a; Groeben, 1985, pp. 4, 15.

14. Groeben, 1985, p. 7.

15. While at Millport, on the isle of Cumbrae in Scotland, in the summer of 1867 and again in 1868, after the annual meeting of the BAAS, Dohrn collaborated with the amateur naturalist David Robertson, in whose house he was a guest, on designing and constructing this portable aquarium (Groeben, 1984, 1985). Robertson founded the Millport Biological Station in 1885.

16. Huxley thought the Indian climate not good for anyone with a tendency to dysentery—"I doubt very much if you would stand it for six months."—and that the time spent "naming and arranging specimens" would leave no time, and the Indian climate no energy, for anything else. This and the following quotation are from a letter

from Huxley to Dohrn, January 15, 1868, quoted from L. Huxley (1900), vol. 1, p. 303.

17. Huxley and Dohrn's relationship extended to succeeding generations. Huxley's grandson Julian was a lifelong friend of Anton's son Reinhard, who succeeded his father as director. When support for the station waned after World War I, Julian Huxley, like his grandfather before him, communicated with the president of the BAAS to seek his and that institution's continued support for the station and for Reinhard Dohrn (Waters and Van Helden, 1992, pp. 84–85). A recent graduate, Julian Huxley launched his research career with an important study on the relationship between embryology and evolution as expressed during regeneration in the sponge *Sycon*; the research was carried out at Naples in the summer of 1909. Huxley devoted chapter 5 of the first volume of his autobiography *Memories* to Naples and his time there (Huxley, 1970).

18. Cited in Groeben and Müller, 1975, p. 11.

19. Other committee members were Thomas Huxley; Ray Lankester; Wyville Thomson, professor of natural history at Edinburgh; and P. L. Sclater, secretary of the Zoological Society (Howarth, 1931, p. 196).

20. Dohrn, 1872b; Cunningham, 1882.

21. Groeben, 1982, pp. 25–26.

22. Letter from Huxley to Dohrn, August 23, 1872; quoted from L. Huxley (1900), vol. 1, p. 376.

23. Quoted from Groeben, 1982, p. 54.

24. Quoted in F. Darwin, 1958, p. 310.

25. Geison, 1978, p. 126; Groeben, 1982, p. 105, n. 122; Tanner, 1917, p. 288; Howarth, 1931, p. 280.

26. Ever the cautious host, Dohrn sent potential users a letter advising them to drink good red wine, avoid beer (especially in the summer), drink only Serino water, take fruit in moderation, and completely avoid the oysters collected near the harbor. Should the latter injunction not be followed and "indigestion" occur, visitors were advised "to keep to a strict diet (egg-soup, grilled meat, red wine and abdominal bandage), to stop work for some days and to make an excursion to Capri, Sorrento, Amalfi or to Vesuvius (Hotel Eremo). [As a last resort] It may become necessary to seek out a local doctor." (Groeben and Müller, 1975, pp. 42–43)."

27. The successful establishment of the Naples Zoological Station was important in promoting the spread of zoological or marine stations around the world; twenty-three more had opened by the turn of the nineteenth century. Naples was not the first marine station to be established—Concarneau in Brittany, on the Bay of Biscay, opened in 1859, and Banyuls-sur-Mer in France, in 1863—but Naples was the first to be so successful on such a grand and diversified scale as both public aquarium and research institution. It was the model for stations opened in Sydney, Australia, in 1878; Woods

Hole, Massachusetts, in 1888; Plymouth, England, a year later; and Helgoland in Germany in 1892. Personal connections were important in these developments. The first American to work at Naples, from November 1881 to May 1882, was Charles Otis Whitman, who married the second American to work at Naples, Emily Nunn, and was the first director of the marine station at Woods Hole.

28. Shipley, 1924, p. 161.

29. A student of Michael Foster's at Trinity College, Dew-Smith undertook some of the earliest studies on response of mollusk hearts to electric current (Dew-Smith, 1874; Dew-Smith and Foster, 1875, 1876). Dew-Smith and Balfour were the first two British scientists to undertake research at Stazione Zoologica (February–June 1874).

30. Born Albert George Dew, he entered Trinity College in 1869 and a year later inherited a substantial fortune that changed his life. He changed his name to Dew-Smith, decorated the walls of his rooms with Burne Jones prints, began to collect rare prints and precious stones, and did little further work as an undergraduate. Robert Louis Stevenson, an occasional visitor to Trinity College, is said to have modeled the character Allwater, in *The Ebb Tide*, on Dew-Smith. By 1876, having published five papers, Dew-Smith abandoned physiological research and devoted his time to helping Michael Foster administer and finance the development of physiology at Cambridge, providing the funds for equipment and, from 1875 on, additional monies over the £50 provided by Trinity College to pay the salary of Foster's demonstrator, J. N. Langley, who succeeded Foster as professor of physiology. Treated as if he were a member of Trinity College, he also donated substantial sums to the *Journal of Physiology*, founded by Foster in 1878 (Cattermole and Wolfe, 1987, p. 9, for the character in *The Ebb Tide*; Geison 1978, pp. 182, 187, for further details).

31. Letter of December 28, 1880, from Horace Darwin to Professor Poynting, newly appointed Professor of Physics at Mason College in Birmingham. Quoted from Cattermole and Wolfe (1987, p. 20), which is the best source for information on Dew-Smith and on the Cambridge Scientific Instrument Company. Dew-Smith turned his attention to lithography and to photography, from which he derived much artistic pleasure. Indeed, he became one of the best photographers of his day, especially known for photographs of his contemporaries; his photographic portraits of Frank and Gerald Balfour, Michael Foster, Francis Galton, and others are preserved in the Trinity College Library.

32. Hughes, 1959, p. 16; Bracegirdle, 1978, pp. 81, 263–264; Cattermole and Wolfe, 1987, pp. 166–178.

33. Sedgwick, 1910, p. 328.

34. Handwritten list of products sent by Dew-Smith to Professor J. D. Munsen in the United States on February 17, 1881. Quoted from Cattermole and Wolfe, 1987, p. 172.

35. Hughes, 1959, pp. 16, 357.

36. Cattermole and Wolfe, 1987, pp. 175–176.

37. Sharpey-Schäfer, 1927, p. 71

38. Foster and Balfour, 1883, pp. 434, 471. The first printed description of the micro-tome, an unsigned editorial in the 1885 volume (volume 5) of the *Journal of the Royal Microscopical Society*, was written by Caldwell. Few photographs of the microtome were published; it appears in Caro's history of the dye industry (1892) and in Threlfall (1930), the latter reproduced as plate 42 in Bracegirdle (1978). The Cambridge Scientific Instrument Company built twelve microtomes between 1882 and 1884. Because the Caldwell-Threlfall microtome was expensive to produce—it sold for £31.10s—Horace Darwin invented a simpler microtome in 1885. The "Darwin Rocker" was so much less expensive to produce (£5.5s), and so robust and effective, that the Cambridge Rocking Microtome, the "Cambridge Rocker," dominated the market. The Rocker delivered 100 sections a minute as a ribbon, and twice that number if motor-driven. The manual model was the instrument on which I was trained to cut serial sections by my Ph.D. supervisor, P. D. F. Murray, a Cambridge man. It is also the microtome I purchased when establishing my laboratory in Canada in 1968, eighty-three years after the first one appeared. It is in use as I type (word-process).

39. Threlfall, 1930, p. 361.

40. Garstang, 1929, p. 98.

41. Close links between institutions in the United Kingdom and continental Europe fostered the rapid advance of evolutionary embryology, including attracting students from outside Europe. Because many of these students moved into other fields, the influence of evolutionary embryology spread rapidly and into unexpected areas. Henry Fairfield Osborn and William B. Scott, subsequently major players in vertebrate paleontology in the United States, traveled to England to study at London and Cambridge: Scott in 1877, Osborn a year later. Scott studied comparative anatomy with Huxley in London and embryology with Balfour in Cambridge before he moved to Heidelberg, where he completed a Ph.D. on the development of *Petromyzon* with Carl Gegenbaur. Scott and Osborn both then joined the Princeton faculty. Scott remained for fifty years; Osborn moved to Columbia after ten years and then to the American Museum of Natural History. Thus the distinctive approach to vertebrate paleontology in the United States owes much to the training in evolutionary embryology received by Scott and Osborn.

42. See Bateson et al. (1905, 1906) for the initial collaborative work between Bateson, Saunders, and Punnett; see Crew (1967) for biographical details on Punnett.

43. See Sturtevant (1965, chap. 4) for an accessible discussion, and Roberts (1929) for a more in-depth analysis.

44. Hutchinson, 1979, p. 100.

REFERENCES

Balfour, Francis Maitland. 1881. *A Treatise on Comparative Embryology*, vol. 2. London: Macmillan.

Bateson, Beatrice. 1928. *William Bateson, F.R.S.: Naturalist. His Essays & Addresses with a of His Life*. Cambridge: Cambridge University Press.

Bateson, William. 1894. *Materials for the Study of Variation, Treated with Especial Regard to Discontinuity in the Origin of Species*. London: Macmillan. Reprinted Baltimore: Johns Hopkins University Press, 1992.

Bateson, W., Saunders, E. R., and Punnett, Reginald Crundall. 1905. "Experimental Studies in the Physiology of Heredity." *Reports to the Evolution Committee of the Royal Society of London* 2: 1–131.

———. 1906. "Experimental Studies in the Physiology of Heredity." *Reports to the Evolution Committee of the Royal Society of London* 3: 1–53.

Bowler, Peter J. 1988. *The Non-Darwinian Revolution: Reinterpreting a Historical Myth*. Baltimore: Johns Hopkins University Press.

———. 1989. *Evolution: The History of an Idea*, Rev. ed. Berkeley: University of California Press.

———. 1996. *Life's Splendid Drama: Evolutionary Biology and the Reconstruction of Life's Ancestry 1860–1940*. Chicago: University of Chicago Press.

Bracegirdle, B. 1978. *A History of Micro Technique*. Ithaca, N.Y.: Cornell University Press.

Caldwell, William H. 1885. "Automatic Microtome." *Journal of the Microscopical Society* 5: 150–153.

Caro, H. 1892. "Ueber die Entwickelung der Theerfarben-Industrie." *Berlin Deutsch Chemisches Geseltschaft* 25: 955–1105.

Carroll, Sean B., Grenier, Jennifer K., and Weatherbee, Scott D. 2001. *From DNA to Diversity: Molecular Genetics and the Evolution of Animal Design*. Malden, Mass.: Blackwell Science.

Cattermole, M. J. G., and Wolfe, A. F. 1987. *Horace Darwin's Shop: A History of the Cambridge Scientific Instrument Company, 1878 to 1968*. Boston: Adam Hilger.

Conway Morris, Simon. 1994. "Why Molecular Biology Needs Palaeontology." *Development* 1994 (suppl.): 1–13.

Crew, Francis A. E. 1967. "Reginald Crundall Punnett." *Biographical Memoirs of Fellows of the Royal Society of London* 13: 323–326.

Cunningham, J. T. 1882. "The Darwin Memorial." *Nature* 26: 124.

Darwin, Charles Robert. 1859. *The Origin of Species by Means of Natural Selection*. London: John Murray.

Darwin, Francis (Ed.). 1958. *The Autobiography of Charles Darwin and Selected Letters*. New York: Dover. An unabridged reproduction of *Charles Darwin: His Life Told in an Autobiographical Chapter and in a Selected Series of His Published Letters*. New York: Appleton, 1893.

Dew-Smith, Albert. 1874. "On Double Nerve Stimulation." *Journal of Anatomy and Physiology* 8: 74–82.

Dew-Smith, Albert, and Foster, Michael. 1875. "On the Behaviour of the Hearts of Mollusks Under the Influence of Electric Currents." *Proceedings of the Royal Society of London* 23: 318–343.

————. 1876. "The Effects of the Constant Current on the Heart." *Journal of Anatomy and Physiology* 10: 735–771.

Dohrn, Anton. 1872a. "Der gegenwärtige Stand der Zoologie und die Gründung zoologische Stationen." *Preuss. Jahrbücher* 30: 137–161.

————. 1872b. "Report of the Committee, Consisting of Dr. Anton Dohrn, Professor Rolleston, and Mr. P. L. Sclater, Appointed for the Purpose of Promoting the Foundation of Zoological Stations in Different Parts of the World." *Report of the British Association for the Advancement of Science* 1872: 192.

————. 1872c. "The Foundations of Zoological Stations." *Nature* 5: 277–280.

Erwin, Douglas H., and Wing, Scott L. (Eds.). 2000. *Deep Time: Paleobiology's Perspective, suppl. to Paleobiology* 26 (4). Lawrence, Kan.: Paleontological Society.

Fantini, B. 1998/1999. "The History of the Stazione Zoologica Anton Dohrn: An Outline." *Stazione Zoologica Anton Dohrn Activity Report* 1998/1999: 71–109.

————. 2000. "The Stazione Zoologica Anton Dohrn and the History of Embryology." *International Journal of Developmental Biology* 44: 523–535.

Foster, Michael, and Balfour, Francis Maitland. 1883. *The Elements of Embryology*, 2nd ed. A. Sedgwick and W. Heape (eds.). London: Macmillan.

Gadow, Hans. 1933. *The Evolution of the Vertebral Column: A Contribution to the Study of Vertebrate Phylogeny*, J. F. Gaskell and H. L. H. H. Green (eds.). Cambridge: Cambridge University Press.

Garstang, Walter. 1929. "The Origin and Evolution of Larval Forms." *Report of the 96th Meeting of the British Association for the Advancement of Science*: 77–98.

Geison, Gerald L. 1978. *Michael Foster and the Cambridge School of Physiology: The Scientific Enterprise in Late Victorian Society.* Princeton, N.J.: Princeton University Press.

Ghiselin, M. T. 1994. "The Origin of Vertebrates and the Principle of Succession of Functions. Genealogical Sketches by Anton Dohrn 1875." *History and Philosophy of the Life Sciences* 16: 3–94. Translated from the German, with introduction and bibliography.

Groeben, Christiane. 1984. "The Naples Zoological Station and Woods Hole." *Oceanus* 27: 60–69.

————. 1985. "Anton Dohrn—The Statesman of Darwin." *Biological Bulletin* 168 (suppl.): 4–25.

———— (Ed.). 1982. *Charles Darwin, 1809–1882, Anton Dohrn, 1840–1909: Correspondence.* Naples: Macchiaroli.

Groeben, Christiane, and Müller, I. 1975. *The Naples Zoological Station at the Time of Anton Dohrn*, Richard Ivell and Christl Ivell (trans.). Naples Italy.

Hall, Brian Keith. 1992. *Evolutionary Developmental Biology.* London: Chapman & Hall.

————. 1999a. *Evolutionary Developmental Biology*, 2nd ed. Dordrecht: Kluwer Academic.

————. 1999b. "The Paradoxical Platypus." *BioScience* 49: 211–218.

————. 2000. "Balfour, Garstang and de Beer: The First century of Evolutionary Embryology." *American Zoologist* 40: 718–728.

————. 2001. "John Samuel Budgett (1872–1904): In pursuit of *Polypterus.*" *BioScience* 51: 399–407.

————. 2002. "Palaeontology and Evolutionary Developmental Biology: A Science of the 19th and 21st Centuries." *Palaeontology* 45: 647–669.

————. 2003. "Evo-Devo: Evolutionary Developmental Mechanisms." *International Journal of Developmental Biology* (special issue on evolutionary developmental biology).

————. 2004. "Evolution as the Control of Development by Ecology." In *Environment, Development and Evolution: Towards a Synthesis*, Brian Keith Hall, Roy Pearson, and Gerhard B. Müller (eds.), pp. ix–xxiii. Cambridge, Mass.: MIT Press.

————. 2007. " 'Spandrels': Metaphor for Morphological Residue or Entrée into Evolutionary Developmental Mechanisms?" In *The Spandrels of San Marco 25 Years Later*, Denis Walsh (ed). Oxford: Oxford University Press.

Hall, Brian Keith, and Olson, Wendy (eds.). 2003. *Keywords and Concepts in Evolutionary Developmental Biology.* Cambridge, Mass.: Harvard University Press.

Hall, Brian Keith, Pearson, Roy, and Müller, Gerhard B. (eds.). 2004. *Environment, Development, and Evolution: Towards a Synthesis.* Cambridge, Mass.: MIT Press.

Hallgrimsson, Benedikt, and Hall, Brian Keith (eds.). 2005. *Variation.* San Diego: Academic Press.

Heuss, T. 1991. *Anton Dohrn: A Life for Science.* Berlin: Springer-Verlag.

His, Wilhelm. 1874. *Unserer Körperform und das physiologische Problem ihrer Entstehung.* Leipzig: Engelmann.

Howarth, O. J. R. 1931. *The British Association for the Advancement of Science: A Retrospect 1831–1931*, 2nd ed. London: BAAS.

Hughes, A. 1959. *A History of Cytology.* London: Abelard-Schuman.

Hutchinson, G. E. 1979. *The Kindly Fruits of the Earth: Recollections of an Embryo Ecologist.* New Haven, Com.: Yale University Press.

Huxley, Julian Sorrell. 1970–1973. *Memories*, 2 vols. New York: Harper & Row.

Huxley, Leonard. 1900. *Life and Letters of Thomas Henry Huxley*, 2 vols. London: Macmillan.

Huxley, Thomas Henry. 1893. *Darwiniana*, vol. 2 of his *Collected Essays*. London: Macmillan.

―――. 1986. *Selections from the Essays*, A. Castell (ed.), Arlington Heights, Ill.: Harlan Davidson.

Kowalevsky, Alexander O. 1866. "Entwickelungsgeschichte der einfachen Ascidien." *Mémoirs of the Academy of Science St. Petersbourg* 10 (Series 7): 1–19.

―――. 1871. "Weitere Studien über die Entwicklung der einfachen Ascidien." *Archiv Für Mikroskopishche Anatomic* 7: 101–130.

Kühn, A. 1950. "Anton Dohrn und die Zoologie seiner Zeit." Publications of the Naples Zoological station (Supplement), 191 pp.

Lankester, Edward Ray. 1878. "Balfour on Elasmobranch Fishes." *Nature* 18: 113–115.

Maienschein, Jane. 1994. "'It's a Long Way from *Amphioxus*': Anton Dohrn and Late Nineteenth Century Debates About Vertebrate Origins." *History and Philosophy of the Life Sciences* 16: 465–578.

Müller, I. 1996. "The Impact of the Zoological Station in Naples on Developmental Physiology." *International Journal of Developmental Biology* 40: 103–111.

Murray, Patrick Desmond Fitzgerald. 1958. *Biology: An Introduction to Medical and Other Studies*. London: Macmillan.

Nyhart, Lynn. 1995. *Biology Takes Form: Animal Morphology and the German Universities, 1800–1900*. Chicago: University of Chicago Press.

Oppenheimer, Jane Marion. 1980. "Some Historical Backgrounds for the Establishment of the Stazione Zoologica at Naples." In *Oceanography: The Past*, M. Sears and D. Merriman (eds.), pp. 179–187. New York: Springer.

Partsch, K. J. 1980. *Die Zoologische Station in Neapel: Modell internationaler Wissenschaftszusammenarbeit*. Göttingen: Vandenhoek & Ruprecht.

Punnett, Reginald Crundall. 1900a. "On the Formation of the Pelvic Plexus, with Especial Reference to the Nervus Collector, in the Genus *Mustelus*." *Philosophical Transactions of the Royal Society of London* B192: 331–351.

―――. 1900b. "Note on a Hermaphrodite Frog." *Annals and Magazine of Natural History* 6: 179–180.

―――. 1900c. "On some Nemerteans from Torres Straits." *Proceedings of the Zoological Society of London* 1: 825–831.

―――. 1901. "On a Collection of Nemerteans from Singapore." *Quarterly Journal of Microscopical Sciences* 44: 111–139.

———. 1902. "Observations on Some Nemerteans from Singapore." *Proceedings of the Royal Society of Edinburgh* 23: 91–92.

———. 1905. *Mendelism.* London: Macmillan.

———. 1923. *Heredity in Poultry.* London: Macmillan.

Roberts, H. F. 1929. *Plant Hybridization Before Mendel.* Princeton, N.J.: Princeton University Press.

Schlichting, Carl D, and Pigliucci, Massimo. 1998. *Phenotypic Evolution: A Reaction Norm Perspective.* Sunderland, Mass.: Sinauer.

Sedgwick, Adam. 1910. "Embryology." In *Encyclopaedia Britannica*, 11th ed., vol. 9, pp. 314–329.

Sharpey-Schäfer, Edward A. 1927. *History of the Physiological Society During Its First Fifty Years, 1876–1926.* London: Cambridge University Press.

Shipley, Arthur E. 1907. "Biographical Sketch." In *The Work of John Samuel Budgett, Balfour Student of the University of Cambridge*, John Graham Kerr (ed.), pp. 1–55. Cambridge: Cambridge University Press.

———. 1924. *Cambridge Cameos.* London: Jonathan Cape.

Sturtevant, Alfred H. 1965. *A History of Genetics.* New York: Harper & Row.

Tanner, J. R. (ed.). 1917. *The Historical Register of the University of Cambridge: Being a Supplement to the Calendar, with a Record of University Offices, Honours and Distinctions to the Year 1910.* Cambridge: Cambridge University Press.

Threlfall, Richard. 1930. "The Origin of the Automatic Microtome." *Biological Reviews and Biological Proceedings of the Cambridge Philosophical Society* 5: 357–361.

Waters, C. K., and Van Helden, A. (eds.). 1992. *Julian Huxley: Biologist and Statesman of Science.* Houston, Tex.: Rice University Press.

Wilkins, Adam S. 2002. *The Evolution of Developmental Pathways.* Sunderland, Mass.: Sinauer.

15

Six Memos for Evo-Devo

Gerd B. Müller

At the beginning of the twenty-first century the causal analysis of organismal form has taken on a new label: Evo-Devo. This shorthand for evolutionary developmental biology designates a new research discipline that arose from the fusion of two earlier programs, the analyses of proximate and ultimate causation of organismal diversity—development and evolution. Beyond uniting these core subjects, Evo-Devo is in the process of becoming a conceptual hub for an even larger integration of research areas in organismal biology, including genetics, ecology, paleontology, behavior, cognition, and other fields. Clearly the new discipline is still in a process of consolidation, searching for unifying principles, but already a number of institutional attributes foreshadow its future importance in the biosciences. New journals specialize in Evo-Devo issues, monographs and textbooks appear, scientific meetings take up its subjects, specific courses are taught, research departments and positions are established, science foundations implement funding programs, professional societies and divisions form, and, most notably, historical and metatheoretical treatments arise, such as the present volume.

Although Evo-Devo thus bears all the attributes of becoming a distinct discipline, and although the consolidation of the field has taken place in recent times and under the close observation of both its empirical actors and its theoretical analysts, surprisingly different accounts of the origins, the contents, the aims, and the significance of the nascent discipline are being told. Some even question whether the field exists at all. This heterogenic reception of Evo-Devo is largely due to a neglect of the conceptual motivations that propel and distinguish the new discipline. These do not consist of the simple attempt to combine two formerly independent research agendas into one, but are primarily rooted in the limitations of each individual discipline with regard to the causal explanation of its principal subject: the form and structure of organisms. Developmental biology alone had no handle on the factors that cause embryonic mechanisms to change organismal structure over time, and evolutionary biology was missing the rules of developmental transformation that relate genotype to phenotype. Evo-Devo claims to provide an extended, more inclusive explanatory framework than either developmental or evolutionary theory alone. This claim will be met only if it is demonstrated that Evo-Devo is able to solve biological problems that are not solved by the traditional disciplines or one of their subdomains, such as population genetics or developmental genetics. Evo-Devo must also prove its heuristic potential: its capacity to induce new scientific questions, methods, and strategies that translate into genuine research projects.

This chapter briefly summarizes six areas that characterize the distinctivenes of the Evo-Devo discipline: (1) the historical roots of its fundamental ideas, (2) its conceptual innovations, (3) its specific research agendas, (4) the methodological consequences, (5) its explanatory capacity, and (6) its integrative effect on other disciplines. These six areas, briefly sketched out, I regard as essential to keep in mind in all metatheoretical interpretations of the Evo-Devo discipline. Hence I call them memos.

HISTORY

It has been argued alternatively, and simultaneously, that Evo-Devo is the modern continuation of the recapitulation program, that it is rooted in comparative embryology and systematics, or that it "erupted out of the discovery of the homeobox."[1] However, although the origin of Evo-Devo is related to all these issues, and has multiple conceptual roots (see part II and contributions in this volume), the fundamental motive force underlying its formation was not the reactivation of earlier research goals, nor a single technical discovery, but the increasing awareness of important explanatory deficits of the leading paradigm of evolutionary biology: adaptation. Criticism of adaptationism and of the powers of natural selection is old, but in the late 1970s and early 1980s, concern accumulated about the inability of neo-Darwinian theory to account for a number of characteristic phenomena of phenotypic evolution. These included the biases in the variation of morphological traits,[2] rapid changes of form evident from the fossil record,[3] the origin of nonadaptive traits,[4] apparent dissociations between genetic and phenotypic evolution,[5] and the origination of higher-level morphological organization such as homology, body plans, and novelty[6]—to name but some of the major open questions.[7]

It became increasingly obvious that these explanatory deficits of neo-Darwinism were due to its treatment of development as a "black box" and the consequent absence of the generative rules that relate between genotype and phenotype.[8] Consequently, spawned primarily by Gould's, Bonner's, and Raff and Kaufman's books,[9] and several initiating conferences (Dahlem, 1981; Sussex, 1982; Plzen, 1984; Columbia, 1985; Woods Hole, 1985; Vienna, 1986; Dijon, 1988; College Park, 1990), a research movement arose that resulted in empirical studies explicitly addressing the relations between developmental and evolutionary mechanisms.[10] Such studies used classical techniques of comparative and experimental embryology at first, and later, increasingly, the methodologies of molecular biology. At the same time theoretical treatments began to concentrate on the relationship between development and evolution.[11] This new agenda, early on, was termed "ontophyletics,"[12] and later, alternatively, "evolutionary embryology,"[13] "evolutionary developmental biology,"[14] or "developmental evolution."[15] First monographs and edited volumes that brought the many diverse works which had a bearing on these issues together did not use any of those labels.[16] But subsequent textbooks marked the consolidation of the field by using the title Evolutionary Developmental Biology,[17] now a widely accepted label that is frequently replaced by the shorthand Evo-Devo. A recent debate on whether evo-devo and devo-evo have the same agenda is addressed in the section "Agenda," below.

These early constituting initiatives by evolutionary theorists and experimental embryologists were soon met by a complementary movement in developmental genetics, induced by the discovery of unexpected similarities in the gene regulatory apparatuses among distantly related species.[18] This approach gained increasing momentum through the advances in molecular methodology, which provided a new basis for the comparative analysis of developmental processes at the genetic level and of their roles in the embryonic foundations of animal body plans.[19] The bulk of empirical work in Evo-Devo shifted toward this kind of analysis,[20] but historically it springs from a secondary, technique-driven root in Evo-Devo, not from its original conceptual motives, a fact that is frequently misrepresented.

It is true that an intellectual continuity of modern Evo-Devo with earlier conceptualizations of the relationship between development and evolution exists. This historical thread of ideas goes back to von Baer and Haeckel, Garstang and Balfour, and earlier to Serres, Meckel, Geoffroy, and even as far as Aristotle. More immediate forerunners in the twentieth century include Baldwin, Berill, de Beer, Devillers, Whyte, Przibram, Schmalhausen, Severtzoff, Waddington, and the earlier writings by Bonner and other workers directly involved in the present synthesis. In all these individual attempts, morphological phenomena represented the explanandum. This underlines that Evo-Devo's predominant conceptual roots are in comparative morphology and morphogenesis (see also chapters by Laubichler, Love, and Maienschein in this volume), representing the core of an organismal research program that was neither triggered by, nor limited to, the genetic principles of developmental evolution. But although certain theoretical foundations of Evo-Devo reach deep into the history of biological thought, no unified and institutionalized research program had existed before, especially not in the sense of a mechanistic, multilevel analysis of the causal interactions between development and evolution. Hence it is appropriate to locate the beginnings of Evo-Devo as a mechanistic science in the early 1980s.

CONCEPTS

For Evo-Devo to legitimately constitute an independent discipline, it must be conceptually distinct and must have different explanatory capacities than other agendas in either evolutionary or developmental biology have. This has been pointed out many times,[21] but recent claims to the effect that Evo-Devo "is focused on the developmental genetic machinery that lies behind embryological phenotypes"[22] and similar statements by leading developmental geneticists[23] seem to indicate that present-day Evo-Devo cannot be distinguished from comparative developmental genetics and also contains no alternative or additional concepts with regard to the prevailing evolutionary paradigm. Although many good arguments were put forward asserting that additional factors of evolution should be considered besides the standard neo-Darwinian repertoire, the proposed factors are mostly defined as mutation- and selection-based changes, such as in the case of "developmental reprogramming."[24] Hence they do not go beyond the received theory. It seems that both theoretical and historiographical analyses have not been sufficiently rigorous in identifying the conceptual innovations contained in Evo-Devo. Is there a new conceptual framework emerging, and if so, what are its constitutive characteristics and how does it differ from previous endeavors?

Most characteristically, and in sharp contrast to the traditional concepts, Evo-Devo does not limit itself to the analysis of phenotypic variation and adaptation, but explicitly addresses the generative mechanisms underlying the evolution of organismal form. These generative mechanisms encompass, but are not restricted to, the genetic circuitry involved in individual development and transgenerational inheritance. Going well beyond, the multilevel approach of Evo-Devo includes the physical properties of biological materials, the self-organizational capacities of cells and tissues, the dynamics of epigenetic interaction among developmental modules, the role of geometry and form in developmental processes, the influence of external and environmental parameters, and all other factors that impinge on the developmental generation of form—whether or not their role in evolution is associated with concomitant changes of the gene regulatory apparatus.

Second, and again distinct from existing paradigms, Evo-Devo explores the influences that the above-named properties of lineage-specific developmental systems have on the evolution of phenotypes. The principal ones of these effects concern the limitations and biases of phenotypic variation, the appearance of morphological innovations, and the generation of homologous character complexes and body plans. Herein lies a primary conceptual distinction of evolutionary developmental biology: that it not only concentrates on what is being modified and maintained by natural selection but also includes how new characters originate, how certain design motifs are established, and what generative rules apply to the evolutionary realization of biological form.

Hence the Evo-Devo framework consists of two interconnected parts, evolution's influence on development and development's influence on evolution (figure 15.1). Because one set of processes affects the other in a reciprocal manner, this constitutes a genuinely dialectical and systemic research agenda in which it is necessary to continuously redefine the standpoint of analysis. This has led to some confusion, and a discussion about whether evo-devo and devo-evo are the same.[25] Clearly, the two subagendas are not the same, because different questions are addressed in each domain, but I argue that only the systemic combination of both agendas makes Evo-Devo a distinct discipline. This view is not supported by those who fear that unscientific notions are imported into Evo-Devo by embracing the devo-evo agenda.[26]

Today, a large number of individual concepts are subsumed under the general rubric of Evo-Devo. Some of these were formulated well before the Evo-Devo era, and are now revived and elaborated on the basis of new kinds of evidence. Others were spurred by the new focus on the development-evolution interface and the necessity to define principles of its mechanistic realization. Since it is not possible to discuss all approaches within the limited space of this chapter, I provide a summary list of major concepts (table 15.1) grouped by the main themes defined in figure 15.1, except that the environment-development interaction is represented as a separate category. The list is not exhaustive, and includes only concepts

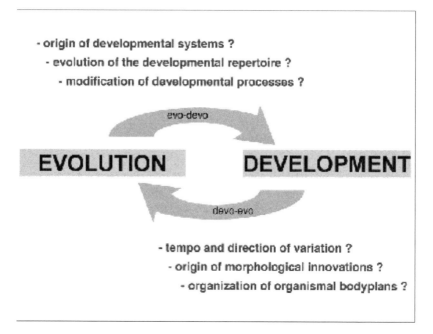

Figure 15.1
The dualistic structure of Evo-Devo. The two Subagendas, evo-devo and devo-evo, address different kinds of questions.

relating to the mechanistic interactions between developmental and evolutionary processes. Concepts concerning taxonomic or phylogenetic relationships (such as recapitulation) are not included. From this overview it is evident that a large number of very diverse approaches constitute the Evo-Devo discipline. None of these per se can claim to cover the entire field, but all have contributed to its present unification. Hence a systematic analysis of the items of table 15.1, as commenced by the present volume, will contribute much toward an understanding of the history and the conceptual roots of Evo-Devo.

In summary, Evo-Devo presents itself as an integrative, profoundly organismal program addressing the causal and reciprocal interrelations between development and evolution at multiple scales and multiple levels of analysis. It unites concepts from developmental biology and evolutionary biology, from genetics and epigenesis, from molecular and organismal analysis, from historical and mechanistic biology, with the aim of providing a more comprehensive explanation of organismal form. Evolutionary developmental genetics is one branch of this program, clearly a very

Table 15.1
EvoDevo concepts grouped by main topics addressed

EvoDevo Concepts	Selected Key References
Origin of Developmental Systems	
Life cycle evolution	Bonner, 1974
Cell lineage competition	Buss, 1987
Generic forms	Newman, 1992, 1994
Evolution of Developmental Repertoire	
Duplication of regulatory genes	McGinnis & Krumlauf, 1992; Holland, 1999
Evolving regulatory networks	Davidson et al., 1995; Wilkins, 2003; Wray & Lowe, 2000
Dissociation, Recruitment, Cooption	Duboule, 1998; Keys et al., 1999; True & Carroll, 2002
Modularity	Wagner, 1996; Raff, 1996; Callebaut & Rasskin-Gutman, 2005
Evolutionary Modification of Developmental Processes	
Heterochrony	Haeckel, 1866; de Beer, 1930; McKinney, 1991
Morphoregulation	Edelman, 1986, 1988
Ontogenetic repatterning	Wake & Roth, 1989
Dissociability	Needham, 1933; Raff, 1996
Environment-Development Interaction	
Reaction norms	Wolterek, 1909; Schlichting & Pigliucci, 1998
Dauermodifikation	Jollos, 1939
Polyphenism	Gilbert, 2001; West-Eberhard, 2003
Plasticity	Wilson, 1894; Baldwin, 1902; West-Eberhard, 2003
Life history theory	Stearns, 1992
Baldwin effect	Baldwin, 1896
Phenotypic Variation	
Ontogenetic buffers	Katz, 1981
Developmental constraints	Alberch, 1982; Maynard Smith et al., 1985
Developmental drive	Arthur, 2001
Evolvability	Wagner & Altenberg, 1996; Kirschner & Gerhart, 1998
Phenotypic Innovation	
Caenogenesis	Haeckel, 1866
Developmental side effects	Müller, 1990
Neophenogenesis	Johnston & Gottlieb, 1990; Gottlieb, 1992
Epigenetic causation	Newman & Müller, 2000
Developmental exaptation	Chipman, 2001
Environmental induction	West-Eberhard, 2003

Table 15.1 (continued)

Genetic And Epigenetic Fixation	
Internal selection	Whyte, 1965
Autonomization	Schmalhausen, 1942; Müller, 2003
Canalization	Waddington, 1942; Wagner et al., 1997; Wilkins, 2003
Burden	Riedl, 1978
Generative entrenchment	Wimsatt, 1986
Assimilation	Waddington, 1956; Wilkins, 2003; Nanjundiah, 2003
Hierarchization	Riedl, 1978; Buss, 1987; Salazar-Ciudad et al., 2001

Note: The cited references are a selection only, referring mostly to authors who founded a concept or provided major new contributions to it.

substantial branch, but it is not synonymous with Evo-Devo (as suggested by some). The conceptual focus of Evo-Devo is on the problem of phenotypic evolution. This point is of key importance, because Evo-Devo will remain viable as an independent research discipline only if it is able to create a consensus on its fundamental aims. Otherwise, it is likely to become cannibalized by the evolutionary genetics program and may have to wait for yet another resurrection.

AGENDA

From the conceptual frame we should distinguish the empirical agendas contained in Evo-Devo as defined by specific research questions. These always belong to either one of the two constitutive subdomains of the Evo-Devo inquiry (figure 15.1). The first is "How do the evolutionary mechanisms affect the processes of individual development?" and the second is "How do the properties of developmental systems affect the processes of evolution?" In line with the distinctions discussed above, research questions belonging to the first domain could collectively be called evo-devo questions, whereas those of the second domain would be devo-evo questions. This is not a terminology I particularly encourage; rather, I use it in order to help sort out some of the confusions about the "proper" Evo-Devo agenda.

In each of the two major domains several specific research questions can be distinguished (figure 15.1). Typical questions of the evo-devo kind concern the origins of developmental systems, the evolution of the developmental repertoire, and the evolutionary modification of developmental processes. Substantial progress has been achieved in the two latter areas, in particular with regard to evolutionary modifications of developmental timing, heterochrony,[27] and the evolution of gene regulatory mechanisms and gene networks. Descriptive and comparative investigations of regulatory gene expression associated with developmental processes have expanded explosively, and a number of special journal issues and books concentrate on these topics.[28] However, although highly informative in its own right, the genetic part of the agenda by itself will not be able to fulfill the Evo-Devo promise, as pointed out above.[29]

The second kind of questions, the devo-evo type, explores the effects of developmental properties on evolutionary change (figure 15.1). This includes questions concerning the tempo and direction of phenotypic evolution, in particular the roles of developmental constraint,[30] as well as the effects on the establishment of phenotypic organization, such as homology and body plans,[31] and the part of development in the origin of morphological novelty.[32] Novelty and innovation are presently receiving much attention, with several model cases that are taken to molecular levels of analysis such as avian feathers,[33] butterfly eyespot patterns,[34] and cephalopod body plans.[35] Phenotypic innovation may prove to be one of the central topics in which a developmental approach is indispensable for an evolutionary explanation. The works that concentrate on these areas indicate that it is not so much mutational events and new genes that are the origin of complex novel structures, but rather developmental reorganization and the co-optation of established regulatory pathways into new developmental functions. In addition, the self-organizational properties of cell and tissue masses, as well as the topology and shape of embryonic structure, contribute significantly to the formation of a basic repertoire of morphological modules that are later elaborated by variation and selection, and become routinized by genetic overdetermination.[36]

Recently these agendas have been expanded toward including the environment in which all of the Evo-Devo processes take place. It is increasingly recognized that the environment is a major causal factor acting on developmental and evolutionary processes, primarily via developmental plasticity. The extended agenda, which could be called Eco-EvoDevo, was the topic of several recent symposia (e.g., Altenberg, 2001; Anaheim, 2002) and has been described in a number of articles and books.[37] Both the devo-evo mechanisms mentioned above and the eco-devo mechanisms implied by the environmental interaction with developmental processes bring us back to the evo-devo questions, in the sense that developmentally or environmentally induced change must be captured and integrated by the evolving genetic circuitry in order to become heritable, a kind of process that is still much debated.[38]

METHOD

Methodologically, Evo-Devo is characterized not by unique tools but by the specific design of the research projects and the influence it exerts on the development of new strategies to obtain the relevant data. The primary methodological basis of Evo-Devo is the comparative method.[39] Evo-Devo has greatly promoted the extension of the comparative method from morphological description to molecular and experimental analyses. In particular, the comparative analysis of the embryonic expression patterns of regulatory genes in a variety of taxa has produced a wealth of new data.[40] At present, molecular comparison at the genetic level is beginning to reach out to nonmodel organisms, highlighting the recruitment of ancestral regulatory modules for new developmental functions in morphological evolution.[41]

Experimental approaches, too, are employed within the comparative framework and have conformed to Evo-Devo purposes, with experimental designs specifically set up to answer evolutionary questions. One example is the comparison of gene expression in mutant embryos with wild type and ancestral expression patterns (but see the caveats by Janies and DeSalle).[42] The results of these and other kinds of experiments are now not considered only for their immediate developmental meaning, but are interpreted before a much more explicit phylogenetic background. Indicative of this trend is the increasing use of taxonomic schemata in developmental biology publications. A number of experimental Evo-Devo domains have arisen, among them analyses of developmental timing, thresholds, modularity, and plasticity.[43] Evo-Devo objectives have also influenced the choice of organisms, thus expanding the variety of studied species, in contrast to traditional developmental genetics, which, by necessity, concentrates on standard model organisms.

Evo-Devo also prompts the development of new analytical tools. One example is the computational analysis of developmental gene expression and its evolutionary alteration. Since developmental evolution has been shown to reside less in the establishment of new control genes than in the modification of the dynamics of gene, cell, and tissue interactions, the precise topology, timing, and quantity of gene expression as related to changes of cell behavior and tissue properties becomes a target issue. This requires tools for the proper representation and quantification of three-dimensional (and eventually four-dimensional) gene expression in developing embryos in order to determine the exact differences in spatiotemporal gene activation that are associated with phenotypic variation and innovation. A number of such tools have been developed,[44] and new bioinformatic techniques for the analysis of such data are being explored[45] with the aim of understanding the topological evolution of spatial gene expression patterns. This does not automatically mean that it will soon become a lot easier to draw evolutionary inferences from such data, because of the many problems associated with the quantitative analysis of gene expression patterns and their actual roles in development and evolution.[46]

Successful attempts have also been made at modeling the behavior and consequences of evolving developmental networks.[47] Among other results, such models indicate that there is an evolutionary tendency of emergent developmental processes to be replaced by hierarchically organized processes, pointing out that greater flexibility may have existed in primitive developmental systems, whereas the routinized and genetically entrenched ontogenies of highly evolved species (today's model organisms) may constitute a derived condition.[48] These findings throw new light on the earlier concepts of canalization and assimilation.[49] Although many of the models used in Evo-Devo remain abstract, there is also biomorphic modeling of concrete developmental systems that illustrates how the differential activation of genes and gene products may affect the morphogenetic outcome of variation and innovation events.[50] These new techniques are characteristic consequences of the Evo-Devo agenda, and have sharpened the questions that biologists can ask effectively in their analysis of embryonic development. The rapid progress in molecular technology, computing technology, and visualization technology greatly assists the development of modeling approaches to Evo-Devo.[51]

Explanation

Evo-Devo extends neo-Darwinian explanation. Whereas the latter concentrates on the quantitative variation and fixation of traits in populations, the range of natural phenomena that can be addressed by Evo-Devo includes the qualitative phenotypic change and the origination of traits. Foremost among these are questions concerning higher levels of phenotypic organization, such as the establishment of standardized building units (homology), the arrangement of such units in lineage-specific sets (body plans), the directionality of change imposed by the developmental systems (constraint), the generation of new structural elements (novelty), the repeated generation of similar forms in independent taxa (homoplasy), and the redeployment of integrated sets of organization (modularity). In addition, Evo-Devo aims at explaining how development itself evolves and how the control of developmental processes is mutually effected by genetic, epigenetic, and environmental factors.

The significant commonality of the phenomena listed above is that they all require the inclusion of a generative component for their explanation (i.e., they pertain to the problem of origination, a problem that should be regarded as distinct from variation,[52] and hence is not explicitly addressed by the neo-Darwinian theory). This is particularly evident in phenotypic innovation (morphological novelty), such as the origin of avian feathers, chelonian carapaces, cephalopod arms, and insect wings. Developmental factors were implicated early on in the discussion of the mechanisms involved in novelty generation,[53] and it is in this domain that the Evo-Devo approach contributes more to a causal explanation than does classical population genetics, for which innovation is out of reach. This is because in these kinds of situations the transmission genetic approach lacks projectability,[54] whereas the Evo-Devo concept can account for emergent properties that underlie origination at the phenotypic level.

For the first time in the history of evolutionary theory, the knowledge of the generative capacities and limitations of developmental systems permits us to predict what is likely to phenotypically arise in evolution (in contrast to population genetics, which can be predictive only about the variation and fixation of given traits). This generative predictability becomes particularly evident when Evo-Devo leads to modeling. Recent examples show that different aspects of phenotypic evolution, such as the shape, variation, and complexity of morphological characters, can be predicted by computational approaches that simulate differential gene expression patterns and their morphogenetic consequences.[55] These and other mathematical approaches[56] indicate that only a limited number of phenotypic solutions can be attained evolutionarily, even in the presence of ample genetic variation, because of the generative characteristics of a specific developmental system.

In the domains sketched out above, Evo-Devo can be said to possess greater explanatory force[57] than neo-Darwinism. This does not invalidate the neo-Darwinian framework, which has greater explanatory force in other, population-related domains. Although the explanatory project of Evo-Devo differs from that of population genetics, a number of unification strategies can be conceived by which to reconcile the two approaches.[58] But in whichever way such a reconciliation takes place, it will inevitably lead to an expansion of the theoretical framework of neo-Darwinism in what might be called the developmental synthesis or the Evo-Devo synthesis.

Integration

Although the label "Evo-Devo" seems to indicate merely a fusion of two fields of scientific inquiry (i.e., evolutionary and developmental biology), the emerging new research paradigm has a much larger integrative capacity. At the core are the conceptual Evo-Devo framework and the associated empirical research agendas addressing genetic and epigenetic mechanisms of development in a comparative way. But strong connections exist with the morphological research traditions.[59] In fact, Evo-Devo is readdressing some of the classical problems of morphology from a new, mechanistic and causal perspective, and thus forms links with several morphological disciplines, such as comparative and functional anatomy, paleontology, and comparative embryology. Many of these problems are related to large-scale evolutionary events studied by paleontologists. Important historical questions about the origin of multicellular organisms, the early generation of all major body plans, and the tempo and patterns of phenotypic change can now be addressed using Evo-Devo tools. This integration of the analysis of historical patterns with that of extant developmental systems not only expands our understanding of phenotypic and developmental evolution but also establishes strong connections with systematics.[60]

A second area, besides morphology, in which Evo-Devo spurs integrative progress is at the intersection of ecology, physiology, and life history theory. The realization that environmental influences have important effects on physiological processes, and hence on development and on reproductive strategies, is not a new one, but work on this subject receives a refreshing boost through the consideration of the evolutionary consequences of such interactions. This includes the study of physiological and metabolic processes that mediate interactions between environmental signals and the plasticity of development, reemphasizing the concepts of reaction norm, polyphenism, and assimilation.[61] Life history theory offers a formal framework for integrating such data into evolutionary theory.[62] A number of recent symposia and books are devoted to these "Eco-Evo-Devo" themes.[63] Because of the present threats to undisturbed environments within which organisms develop and evolve, this aspect of Evo-Devo assumes societal and political importance, as indicated by suggestions that an ecological developmental biology must become a critical part of normative developmental biology if we are to base agricultural and industrial policies on scientifically accurate data.[64]

Yet another strong integrative effect of Evo-Devo thinking can be noticed in the behavioral sciences, including neurobiology, cognition research, and psychology. Behavior in particular has its own tradition of linking development to evolution,[65] a tradition that meets with modern evolutionary developmental neurobiology,[66] and receives increasing attention in cognition biology.[67] In these approaches the structure and dynamics of developmental systems become a primary determinant of evolution. It cannot be overemphasized that these approaches go beyond the standard evolutionary psychology, which is intellectually constrained by its steadfast adherence to the adaptive paradigm. The integrative developments in these areas are promoted by the theoretical framework of evolutionary developmental biology—"Evo-Devo meets the mind"[68]—and are beginning to include the study of evolution of language and communication.[69]

As a final point, it should be noted that Evo-Devo has formed stronger ties than many other biological disciplines with various fields of philosophy. On one hand, this is due to the appeal of a nonreductionistic science whose results pertain to a number of long-standing philosophical issues, such as mind-body, nature-nurture, emergence, natural kinds, epistemology, and ethics. On the other hand, with its focus on the causal mechanisms of evolutionary change, Evo-Devo contributes new information to the great debates within evolutionary biology, indicating a shift toward a more synthetic and systems-oriented period of scientific inquiry, a topic of much interest for philosophers of science. And it harbors the possibility of a new synthesis, a developmental synthesis, in the major theoretical framework of the biosciences: evolutionary theory. This would contribute to the naturalization of a number of philosophical subjects.[70] However, there is a mutual relationship. Not only does Evo-Devo inform philosophy, but the philosophical considerations of such crucial subjects as the concept of the gene, the units of selection, and the emergence of complexity inform Evo-Devo and help define its theoretical structure. This is a unique aspect, rarely seen in the biosciences, and can only mean that Evo-Devo is on the right track.

Acknowledgments

This essay is dedicated to Italo Calvino. I thank Sharon Minsuk and Manfred Laubichler for their critique and comments on earlier drafts.

NOTES

1. Arthur, 2002.

2. Alberch, 1982; Maynard Smith et al., 1985.

3. Gould and Eldredge, 1977; Gould, 1983.

4. Gould and Lewontin, 1979.

5. De Beer, 1971.

6. Ibid.; Roth, 1984.

7. More detail in Reid, 1985; see also Müller and Newman, 2003b; Müller, 2005.

8. Hamburger, 1980.

9. Gould, 1977; Bonner, 1982; Raff and Kaufman, 1983.

10. E.g., Katz et al., 1981; Katz, 1983; Alberch and Gale, 1983, 1985; Raff et al., 1984; Müller, 1986, 1989.

11. E.g., Alberch, 1980, 1982a; Goodwin, 1982; Fink, 1982; Hall, 1983; Katz, 1983; Kluge, 1985; Wake, 1986; Wagner, 1986.

12. Katz, 1981, 1983.

13. Müller, 1991.

14. Hall, 1992; Wake, 1996.

15. Wray, 1994; Wagner et al., 2000.

16. Raff and Kaufman, 1983; Goodwin et al., 1983; Arthur, 1984.

17. Hall, 1992, 1998.

18. McGinnis et al., 1984.

19. E.g., Akam, 1989; Holland, 1992, 1996.

20. See, e.g., Carroll et al., 2001; Davidson, 2001.

21. E.g., Wagner, 2000; 2001.

22. Arthur, 2002.

23. Davidson, 2001; Tautz, 2002.

24. Arthur, 2000.

25. Hall, 2000; Gilbert, 2003.

26. Hall, 2000.

27. E.g., McKinney, 1991; McNamara, 1997.

28. Special journal issues include *Developmental Genetics* 15 (1994); *Development* (suppl., 1994); Proceedings of the National Academy of Sciences USA 97 (2000); *Journal of Anatomy* 1999 (2001). Among the books are Carroll et al., 2001; Davidson, 2001; Wilkins, 2001.

29. See also Wagner, 2000, 2001.

30. Alberch, 1982; Maynard Smith et al., 1985.

31. Wagner, 1989; Laubichler, 2000.

32. Müller, 1990; Müller and Wagner, 1991; Newman and Müller, 2000; Müller and Newman, 2005.

33. Prum and Brush, 2002; Harris et al., 2002.

34. Keys et al., 1999; Nijhout, 2001.

35. Lee et al., 2003.

36. Newman and Müller, 2000.

37. Gilbert, 2001; Gilbert and Bolker, 2003; Hall et al., 2004; West-Eberhard, 2003.

38. Wilkins, 2003; Nanjundiah, 2003.

39. Müller, 1991; Atkinson, 1992.

40. Carroll et al., 2001; Davidson, 2001.

41. E.g., Lee et al., 2003.

42. But see the caveats by Janies and DeSalle, 1999.

43. See Müller, 1991; Von Dassow and Munro, 1999; West-Eberhard, 2003.

44. Streicher et al., 2000; Weninger and Mohun, 2002; Sharpe et al., 2002; Weninger et al., 2006.

45. Costa et al., 2004, 2005.

46. Janies and DeSalle, 1999.

47. E.g., Salazar-Ciudad et al., 2001; Szathmary, 2001; Kaneko, 2003.

48. Newman and Müller, 2000.

49. Waddington, 1942, 1956.

50. Jernvall, 2000; Jernvall et al., 2000.

51 See Gerson, this volume.

52. Müller and Newman, 2003a, 2005.

53. Müller, 1990; Müller and Wagner, 1991.

54. Wagner, 2000.

55. Jernvall, 2000; Jernvall et al., 2000.

56. E.g., Rasskin-Gutman, 2003.

57. Amundson, 1989; Sterelny, 2000.

58. See Griesemer, this volume.

59. See Laubichler, Love, and Maienschein, this volume.

60. Rieppel, 1993.

61. Schlichting and Pigliucci, 1998; Gilbert, 2001; Sarkar, 2003; West-Eberhard, 2003.

62. Stearns, 1992.

63. Gilbert and Bolker, 2003; Hall et al., 2004; West-Eberhard, 2003.

64. Gilbert, 2001.

65. Gottlieb, 1987, 1992.

66. Bock and Cardew, 2000.

67. Heyes, 2003.

68. Griffiths, 2006, in press.

69. Oller and Griebel, 2004.

70. Callebaut, 1993.

REFERENCES

Akam, M. 1989. "Hox and HOM: Homologous Gene Clusters in Insects and Vertebrates." *Cell* 57: 347–349.

Alberch, P. 1980. "Ontogenesis and Morphological Diversification." *American Zoologist* 20: 653–667.

———. 1982. "Developmental Constraints in Evolutionary Processes." In J. T. Bonner (ed.), *Evolution and Development*, pp. 313–332. Berlin: Springer-Verlag.

Alberch, P., and E. A. Gale. 1983. "Size Dependence During the Development of the Amphibian Foot: Colchicine-Induced Digital Loss and Reduction." *Journal of Embryology and Experimental Morphology* 76: 177–197.

———. 1985. "A Developmental Analysis of an Evolutionary Trend: Digital Reduction in Amphibians." *Evolution* 39: 8–23.

Amundson, R. 1989. "The Trials and Tribulations of Selectionist Explanations." In K. Hahlweg and C. A. Hooker (eds.), *Issues in Evolutionary Epistemology*, pp. 413–432. Albany: State University of New York Press.

Arthur, W. 1984. *Mechanisms of Morphological Evolution*. Chichester, U.K.: Wiley.

————. 2000. "The Concept of Developmental Reprogramming and the Quest for an Inclusive Theory of Evolutionary Mechanisms." *Evolution & Development* 2: 49–57.

————. 2001. "Developmental Drive: An Important Determinant of the Direction of Phenotypic Evolution." *Evolution & Development* 3: 271–278.

————. 2002. "The Emerging Conceptual Framework of Evolutionary Developmental Biology." *Nature* 415: 757–764.

Atkinson, J. W. 1992. "Conceptual Issues in the Reunion of Development and Evolution." *Synthese* 91: 93–110.

Averof, M., and S. M. Cohen. 1997. "Evolutionary Origin of Insect Wings from Ancestral Gills." *Nature* 385: 627–630.

Baldwin, J. M. 1896. "A New Factor in Evolution." *American Naturalist* 30: 89–101.

————. 1902. *Development and Evolution*. New York: Macmillan.

Bock, G. R., and G. Cardew (Eds.). 2000. *Evolutionary Developmental Biology of the Cerebral Cortex*. New York: Wiley.

Bonner, J. T. 1974. *On Development: The Biology of Form*. Cambridge, Mass.: Harvard University Press.

————. (Ed.). 1982. *Evolution and Development*, vol. 22. Berlin: Springer-Verlag.

Buss, L. W. 1987. *The Evolution of Individuality*. Princeton, N.J.: Princeton University Press.

Callebaut, W. (Ed.). 1993. *Taking the Naturalistic Turn: Or, How Real Philosophy of Science Is Done*. Chicago: University of Chicago Press.

Callebaut, W., and D. Rasskin-Gutman (Eds.). 2005. *Modularity: Understanding the Development and Evolution of Natural Complex Systems*. Cambridge, Mass.: MIT Press.

Carroll, S. B. 2000. "Endless Forms: The Evolution of Gene Regulation and Morphological Diversity." *Cell* 101: 577–580.

Carroll, S. B., J. K. Grenier, and S. D. Weatherbee. 2001. *From DNA to Diversity: Molecular Genetics and the Evolution of Animal Design*. Malden, Mall.: Blackwell Science; 2nd ed: 2005.

Chipman, A. D. 2001. "Developmental Exaptation and Evolutionary Change." *Evolution & Development* 3: 299–301.

Costa, L. da F., M. S. Barbosa, E. T. Manoel, J. Streicher, and G. B. Müller. 2004. "Mathematical Characterization of Three-Dimensional Gene Expression Patterns." *Bioinformatics* 20: 1653–1662.

Costa, L. da F., B. A. N. Travençolo, A. Azeredo, M. E. Beletti, G. B. Müller, D. Rasskin-Gutman, G. Sternik, M. Ibañes, and J. C. Izpisúa Belmonte. 2005. Field Approach to

Three-Dimensional Gene Expression Pattern Characterization. *Applied Physics Letters* 86: 143901–143903.

Davidson, E. H. 2001. *Genomic Regulatory Systems: Development and Evolution.* San Diego: Academic Press.

Davidson, E. H., K. Peterson, and R. A. Cameron. 1995. "Origin of the Adult Bilaterian Body Plans: Evolution of Developmental Regulatory Mechanisms." *Science* 270: 1319–1325.

De Beer, G. R. 1930. *Embryology and Evolution.* Oxford: Clarendon Press.

————. 1971. *Homology: An Unsolved Problem.* London: Oxford University Press.

Duboule, D. 1998. "Vertebrate Hox Gene Regulation: Clustering and/or Colinearity?" *Current Opinions in Genetics and Development* 8: 514–518.

Edelman, G. M. 1986. "Evolution and Morphogenesis: The Regulator Hypothesis." In J. P. Gustafson, G. L. Stebbins, and F. J. Ayala (eds.), *Genetics, Development, and Evolution*, pp. 1–27. New York: Plenum Press.

————. 1988. *Topobiology: An Introduction to Molecular Embryology.* New York: Basic Books.

Eldredge, N., and S. J. Gould. 1972. "Punctuated Equilibria: An Alternative to Phyletic Gradualism." In T. J. M. Schopf (ed.), *Models in Paleobiology*, pp. 82–115. San Francisco: Freeman, Cooper.

Fink, W. L. 1982. "The Conceptual Relationship Between Ontogeny and Phylogeny." *Paleobiology* 8: 254–264.

Gilbert, S. F. 2001. "Ecological Developmental Biology: Developmental Biology Meets the Real World." *Developmental Biology* 233: 1–32.

————. 2003. "Evo-Devo, Devo-Evo, and Devgen-Popgen." *Biology and Philosophy* 18: 347–352.

Gilbert, S. F., and J. A. Bolker. 2003. "Ecological Developmental Biology: Preface to the Symposium." *Evolution & Development* 5: 3–8.

Goodwin, B. C. 1982. "Development and Evolution." *Journal of Theoretical Biology* 97: 43–55.

Goodwin, B. C., N. Holder, and C. C. Wylie (Eds.). 1983. *Development and Evolution.* Cambridge: Cambridge University Press.

Gottlieb, G. 1987. "The Developmental Basis of Evolutionary Change." *Journal of Comparative Psychology* 101: 262–271.

————. 1992. *Individual Development and Evolution: The Genesis of Novel Behavior.* Oxford: Oxford University Press.

Gould, S. J. 1977. *Ontogeny and Phylogeny.* Cambridge, Mass.: Belknap Press of Harvard University Press.

————. 1983. "Punctuated Equilibrium and the Fossil Record." *Science* 219: 438–440.

Gould, S. J., and N. Eldredge. 1977. "Punctuated Equilibria: The Tempo and Mode of Evolution Reconsidered." *Paleobiology* 3: 115–151.

Gould, S. J., and R. C. Lewontin. 1979. "The Spandrels of San Marco and the Panglossian Paradigm: A Critique of the Adaptationist Programme." *Proceedings of the Royal Society of London,* B205: 581–598.

Griffiths, P. E. 2007. "Evo-Devo meets the mind: Towards a developmental evolutionary psychology." In R. Sansom and R. Brandon (eds.), *Integrating Evolution and Development,* in press. Cambridge, Mass.: MIT Press.

Haeckel, E. 1866. *Generelle Morphologie der Organismen,* 2 vols. Berlin: Georg Reimer.

Hall, B. K. 1983. "Epigenetic Control in Development and Evolution." In B. C. Goodwin, N. Holder, and C. C. Wylie (eds.), *Development and Evolution,* pp. 353–379. Cambridge: Cambridge University Press.

————. 1992. *Evolutionary Developmental Biology.* London: Chapman & Hall.

————. 1998. *Evolutionary Developmental Biology,* 2nd ed. Dordrecht: Kluwer.

Hall, B. K. 2000. "Evo-Devo or Devo-Evo—Does It Matter?" *Evolution & Development* 2: 177–178.

————., and W. M. Olson (Eds.). 2003. *Keywords and Concepts in Evolutionary Developmental Biology.* Cambridge, Mass.: Harvard University Press.

Hall, B. K., R. Pearson, and G. B. Müller (Eds.). 2004. *Environment, Development, and Evolution: Towards a Synthesis.* Cambridge, Mass.: MIT Press.

Hamburger, V. 1980. "Embryology and the Modern Synthesis in Evolutionary Theory." In E. Mayr and W. B. Provine (eds.), *The Evolutionary Synthesis: Perspectives on the Unification of Biology,* pp. 96–112. Cambridge, Mass.: Harvard University Press.

Harris, M. P., J. F. Fallon, and R. O. Prum. 2002. "Shh-Bmp2 Signaling Module and the Evolutionary Origin and Diversification of Feathers." *Journal of Experimental Zoology/Molecular and Developmental Evolution* 294: 160–176.

Heyes, C. 2003. "Four Routes of Cognitive Evolution." *Psychological Review* 110: 713–727.

Holland, L. Z., P. W. Holland, and N. D. Holland. 1996. "Revealing Homologies Between Body Parts of Distantly Related Animals by *in situ* Hybridization to Developmental Genes: *Amphioxus* Versus Vertebrates." In J. D. Ferraris and S. R. Palumbi (eds.), *Molecular Zoology,* pp. 267–295. New York: Wiley-Liss.

Holland, P. 1992. "Homeobox Genes in Vertebrate Evolution." *BioEssays* 14: 267–273.

Holland, P. W. 1999. "Gene Duplication: Past, Present, and Future." *Seminars in Cell and Developmental Biology* 10: 541–547.

Janies, D., and R. DeSalle. 1999. "Development, Evolution, and Corroboration." *Anatomical Record* 257: 6–14.

Jernvall, J. 2000. "Linking Development with Generation of Novelty in Mammalian Teeth." *Proceedings of the National Academy of Sciences USA* 97: 2641–2645.

Jernvall, J., S. V. Keranen, and I. Thesleff. 2000. "Evolutionary Modification of Development in Mammalian Teeth: Quantifying Gene Expression Patterns and Topography." *Proceedings of the National Academy of Sciences USA* 97: 14444–14448.

Johnston, T. D., and G. Gottlieb. 1990. "Neophenogenesis: A Developmental Theory of Phenotypic Evolution." *Journal of Theoretical Biology* 147: 471–495.

Jollos, V. 1939. *Grundbegriffe der Vererbungslehre, insbesondere Mutation, Dauermodifikation, Modifikation.* Berlin: Gebrüder Borntraeger.

Kaneko, K. 2003. "Organization Through Intra-inter Dynamics." In G. B. Müller and S. A. Newman (eds.), *Origination of Organismal Form*, pp. 195–220. Cambridge, Mass.: MIT Press.

Katz, M. J. 1983. "Ontophyletics: Studying Evolution Beyond the Genome." *Perspectives in Biology and Medicine* 26: 323–333.

Katz, M. J., R. J. Lasek, and I. R. Kaiserman-Abramof. 1981. "Ontophyletics of the Nervous System: Eyeless Mutants Illustrate How Ontogenetic Buffer Mechanisms Channel Evolution." *Proceedings of the National Academy of Sciences USA* 78: 397–401.

Keys, D. N., D. L. Lewis, J. E. Selegue, B. J. Pearson, L. V. Goodrich, R. L. Johnson, et al. 1999. "Recruitment of a Hedgehog Regulatory Circuit in Butterfly Eyespot Evolution." *Science* 283: 532–534.

Kirschner, M., and J. Gerhart. 1998. "Evolvability." *Proceedings of the National Academy of Sciences USA* 95: 8420–8427.

Kluge, A. G., and R. E. Strauss. 1985. "Ontogeny and Systematics." *Annual Reviews of Ecology and Systematics* 16: 247–268.

Laubichler, M. D. 2000. "Homology in Development and the Development of the Homology Concept." *American Zoologist* 40: 777–788.

Lee, P. N., P. Callaerts, H. G. de Couet, and M. Q. Martindale. 2003. "Cephalopod Hox Genes and the Origin of Morphological Novelties." *Nature* 424: 1061–1065.

Maynard Smith, J., R. Burian, S. Kauffman, P. Alberch, J. Campbell, B. Goodwinet, et al. 1985. "Developmental Constraints and Evolution." *Quarterly Review of Biology* 60: 265–287.

McGinnis, W., R. L. Garber, J. Wirz, A. Juroiwa, and W. J. Gehring. 1984. "A Homologous Protein-Coding Sequence in *Drosophila* Homeotic Genes and Its Conservation in Other Metazoans. *Cell* 37: 403–408.

McGinnis, W., and R. Krumlauf. 1992. "Homeobox Genes and Axial Patterning." *Cell* 68: 283–302.

McKinney, M. L., and K. J. McNamara. 1991. *Heterochrony: The Evolution of Ontogeny.* New York: Plenum Press.

McNamara, K. J. 1997. *Shapes of Time: The Evolution of Growth and Development.* Baltimore: Johns Hopkins University Press.

Müller, G. B. 1986. "Effects of Skeletal Change on Muscle Pattern Formation." *Bibliotheca Anatomica* 29: 91–108.

———. 1989. "Ancestral Patterns in Bird Limb Development: A New Look at Hampé's Experiment." *Journal of Evolutionary Biology* 2: 31–47.

———. 1990. "Developmental Mechanisms at the Origin of Morphological Novelty: A Side-Effect Hypothesis." In M. H. Nitecki (ed.), *Evolutionary Innovations,* pp. 99–130. Chicago: University of Chicago Press.

———. 1991. "Experimental Strategies in Evolutionary Embryology." *American Zoologist* 31: 605–615.

———. 2003. "Homology: The Evolution of Morphological Organization". In G. B. Müller and S. A. Newman (eds.), *Origination of Organismal Form,* pp. 51–69. Cambridge, Mass.: MIT Press.

———. 2005. "Evolutionary Developmental Biology." In F. Wuketits and F. J. Ayala (eds.), *Handbook of Evolution,* vol. 2, pp. 87–115. Weinhein: Wiley-VCH.

Müller, G. B., and S. A. Newman. 2003a. "Origination of Organismal Form: The Forgotten Cause in Evolutionary Theory." In G. B. Müller and S. A. Newman (eds.), *Origination of Organismal Form,* pp. 3–10. Cambridge, Mass.: MIT Press.

———. (Eds.). 2003b. *Origination of Organismal Form: Beyond the Gene in Developmental and Evolutionary Biology.* Cambridge, Mass.: MIT Press.

———. 2005. "The Innovation Triad: An Evo-Devo Agenda." *Journal of Experimental Zoology/Molecular and Developmental Evolution* 304: 487–503.

Müller, G. B., and G. P. Wagner. 1991. "Novelty in Evolution: Restructuring the Concept." *Annual Review of Ecology and Systematics* 22: 229–256.

Nanjundiah, V. 2003. "Phenotypic Plasticity and Evolution by Genetic Assimilation." In G. B. Müller and S. A. Newman (eds.), *Origination of Organismal Form,* pp. 245–263. Cambridge, Mass.: MIT Press.

Needham, J. 1933. "On the Dissociability of the Fundamental Processes in Ontogenesis." *Biological Review* 8: 180–223.

Newman S. A. 1992. "Generic Physical Mechanisms of Morphogenesis and Pattern Formation as Determinants in the Evolution of Multicellular Organization." In J. B. Mittenthal and A. B. Baskin (eds.), *Principles of Organization in Organisms,* pp. 241–267. Reading, Mass.: Addison-Wesley.

———. 1994. "Generic Physical Mechanisms of Tissue Morphogenesis: A Common Basis for Development and Evolution." *Journal of Evolutionary Biology* 7: 467–488.

Newman, S. A., and G. B. Müller. 2000. "Epigenetic Mechanisms of Character Origination." *Journal of Experimental Zoology/Molecular and Developmental Evolution* 288: 304–317.

Nijhout, H. F. 2001. "Elements of Butterfly Wing Patterns." *Journal of Experimental Zoology/Molecular and Developmental Evolution* 291: 213–225.

Oller, D. K., and U. Griebel (Eds.). 2004. *The Evolution of Communication Systems: A Comparative Approach.* Cambridge, Mass.: MIT Press.

Prum, R. O. 1999. "Development and Evolutionary Origin of Feathers." *Journal of Experimental Zoology/Molecular and Developmental Evolution* 285: 291–306.

Prum, R. O., and A. H. Brush. 2002. "The Evolutionary Origin and Diversification of Feathers." *Quarterly Review of Biology* 77: 261–295.

Raff, R. 1996. *The Shape of Life.* Chicago: University of Chicago Press.

Raff, R. A., J. A. Anstrom, C. J. Huffman, D. S. Leaf, J.-H. Loo, R. M. Showman, and D. E. Wells. 1984. "Origin of a Gene Regulatory Mechanism in the Evolution of Echinoderms." *Nature* 310: 312–314.

Raff, R. A., and T. C. Kaufman. 1983. *Embryos, Genes, and Evolution: The Developmental-Genetic Basis of Evolutionary Charge.* New York: Macmillan.

Rasskin-Gutman, D. 2003. "Boundary Constraints for the Emergence of Form." In G. B. Müller and S. A. Newman (eds.), *Origination of Organismal Form*, pp. 305–322. Cambridge, Mass.: MIT Press.

Reid, R. G. B. 1985. *Evolutionary Theory: The Unfinished Synthesis.* Ithaca, N.Y.: Cornell University Press.

Riedl, R. 1978. *Order in Living Organisms: A Systems Analysis of Evolution*, trans. R. P. S. Jefferies. Chichester, U.K.: Wiley.

Rieppel, O. 1993. "The Conceptual Relationship of Ontogeny, Phylogeny, and Classification: The Taxic Approach." *Evolutionary Biology* 27: 1–32.

Roth, V. L. 1984. "On Homology." *Biological Journal of the Linnean Society* 22: 13–29.

Salazar-Ciudad, I., S. A. Newman, and R. V. Sole. 2001. "Phenotypic and Dynamical Transitions in Model Genetic Networks. I. Emergence of Patterns and Genotype-Phenotype Relationships." *Evolution & Development* 3: 84–94.

Sarkar, S. 2003. "Generalized Norms of Reaction for Ecological Developmental Biology." *Evolution & Development* 5: 106–115.

Schlichting, C., and M. Pigliucci. 1998. *Phenotypic Evolution: A Reaction Norm Perspective.* Sunderland, Mass.: Sinauer.

Schmalhausen, I. I. 1949. *Factors of Evolution. The Theory of Stabilizing Selection.* Philadelphia: Blakiston Company (University of Chicago Press edition 1986).

Sharpe, J., U. Ahlgren, P. Perry, B. Hill, A. Ross, J. Hecksher-Sorensen, et al. 2002. "Optical Projection Tomography as a Tool for 3D Microscopy and Gene Expression Studies." *Science* 296: 541–545.

Stearns, S. C. 1992. *The Evolution of Life Histories*. New York: Oxford University Press.

Sterelny, K. 2000. "Development, Evolution, and Adaptation." *Philosophy of Science* 67(suppl.): 369–387.

Streicher, J., M. A. Donat, B. Strauss, R. Spörle, K. Schughart, and G. B. Müller. 2000. "Computer Based Three-Dimensional Visualization of Developmental Gene Expression." *Nature Genetics* 25: 147–152.

Szathmary, E. 2001. "Evolution: Developmental Circuits Rewired." *Nature* 411: 143–145.

Tautz, D. 2002. "Evo-Devo—Evolution von Entwicklungsprozessen." *Laborjournal* 5: 18–21.

True, J. R., and S. B. Carroll. 2002. "Gene Co-option in Physiological and Morphological Evolution." *Annual Review of Cell and Developmental Biology* 18: 53–80.

Von Dassow, G., and E. Munro. 1999. "Modularity in Animal Development and Evolution: Elements of a Conceptual Framework for Evo-Devo." *Journal of Experimental Zoology/Molecular and Developmental Evolution* 285: 307–325.

Waddington, C. H. 1942. "Canalization of Development and the Inheritance of Acquired Characters." *Nature* 150: 563–565.

———. 1956. "Genetic Assimilation." *Advances in Genetics* 10: 257–290.

Wagner, G. P. 1986. "The Systems Approach: An Interface Between Development and Population Genetic Aspects of Evolution." In D. M. Raup and D. Jablonski (eds.), *Patterns and Processes in the History of Life*, pp. 149–165. Berlin: Springer-Verlag.

———. 1989. "The Biological Homology Concept." *Annual Review of Ecology and Systematics* 20: 51–69.

———. 1996. "Homologues, Natural Kinds, and the Evolution of Modularity." *American Zoologist* 36: 36–43.

———. 2000. "What Is the Promise of Developmental Evolution? Part I: Why Is Developmental Biology Necessary to Explain Evolutionary Innovations?" *Journal of Experimental Zoology/Molecular and Developmental Evolution* 288: 95–98.

———. 2001. "What Is the Promise of Developmental Evolution? Part II: A Causal Explanation of Evolutionary Innovations May Be Impossible." *Journal of Experimental Zoology/Molecular and Developmental Evolution* 291: 305–309.

Wagner, G. P., and L. Altenberg. 1996. "Complex Adaptations and the Evolution of Evolvability." *Evolution* 50: 967–976.

Wagner, G. P., G. Booth, and H. Homayoun-Chaichian. 1997. "A Population Genetic Theory of Canalization." *Evolution* 51: 329–347.

Wagner, G. P., C. Chiu, and M. Laubichler. 2000. "Developmental Evolution as a Mechanistic Science: The Inference from Developmental Mechanisms to Evolutionary Processes." *American Zoologist* 40: 819–831.

Wake, D. B. 1986. "Phylogenetic Implications of Ontogenetic Data." In J. Chaline and B. Laurin (eds.), *Ontogenesis and Evolution*, pp. 489–501. Dijon: CNRS DARWININA.

———. 1996. "Evolutionary Developmental Biology—Prospects for an Evolutionary Synthesis at the Developmental Level." *Memoirs of the California Academy of Sciences* 20: 97–107.

Wake, D. B., P. Mabee, J. Hanken, and G. Wagner. 1991. "Development and Evolution: The Emergence of a New Field." In E. C. Dudley (ed.), *The Unity of Evolutionary Biology*, pp. 582–588. Portland, Ore.: Dioscorides Press.

Wake, D. B., and G. Roth. 1989. "The Linkage Between Ontogeny and Phylogeny in the Evolution of Complex Systems." In D. B. Wake and G. Roth (eds.), *Complex Organismal Functions: Integration and Evolution in Vertebrates*, pp. 361–377. New York: Wiley.

Weninger, W. J., and T. Mohun. 2002. "Phenotyping Transgenic Embryos: A Rapid 3D Screening Method Based on Episcopic Fluorescence Image Capturing." *Nature Genetics* 30: 59–65.

Weninger W. J., S. H. Geyer, T. J. Mohun, D. Rasskin-Gutman, T. Matsui, I. Ribeiro, L. da F. Costa, J. C. Izpisúa-Belmonte, and G. B. Müller. 2006. "High Resolution Episcopic Microscopy: Rapid 3D-analysis of Gene Expression and Tissue Architecture." *Anatomy and Embryology* 211(3): 213–221.

West-Eberhard, M. J. 2003. *Developmental Plasticity and Evolution*. Oxford: Oxford University Press.

Whyte, L. L. 1965. *Internal Factors in Evolution*. New York: George Braziller.

Wilkins, A. 2001. *The Evolution of Developmental Pathways*. Sunderland, Mass.: Sinauer.

———. 2003. "Canalization and Genetic Assimilation." In B. K. Hall and W. M. Olson (eds.), *Keywords and Concepts in Evolutionary Developmental Biology*, pp. 23–30. Cambridge, Mass.: Harvard University Press.

Wilson, E. B. 1894. "The Embryological Criterion of Homology." *Biological Lectures of the Marine Biological Laboratory of Wood's Hole* 1894: 101–124.

Wimsatt, W. C. 1986. "Developmental Constraints, Generative Entrenchment, and the Innate-Acquired Distinction." In W. Bechtel (ed.), *Integrating Scientific Disciplines*, pp. 185–208. Dordrecht: Martinus Nijhoff.

Wolterek, R. 1909. "Weitere experimentelle Untersuchungen über Artveränderung, speziell über das Wesen quantitativer Artunterschiede bei Daphniden." *Verhandlungen der Deutschen Zoologischen Gesellschaft* 1909: 110–172.

Wray, G. A. 1994. "Developmental Evolution: New Paradigms and Paradoxes." Developmental Genetics 15: 1–6.

Wray, G. A., and C. J. Lowe. 2000. "Developmental Regulatory Genes and Echinoderm Evolution." *Systematic Biology* 49: 151–174.

16

The Current State and the Future of
Developmental Evolution
Günter P. Wagner

Birds and Ornithologists Together

When I told a friend about my experience at the Dibner workshop on the history of evolutionary developmental biology, and how much I enjoyed the interaction with card-carrying historians of biology, he waved his hand and said: "What do birds have to do at a meeting of ornithologists?" Well, maybe not much, but if birds and ornithologists would speak the same language, they might want to attend the same meetings. As a biologist at the meeting, I was obviously one of the birds, and I felt that it gave me the opportunity to raise my state of existence above that of a beast by the opportunity to reflect on the state of my science in a non-disciplinary setting. While I cannot say whether the ornithologists (i.e., historians) at our meeting got anything out of my presence, I want to offer an insider's view of the most recent developments in the field that is not yet documented as recorded history. Below is a sketch of my current view about the state of the field.

Evolutionary Developmental Biology: A Throwback to the Geoffroy-Cuvier Debate?

In a paper on the history of molecular biology, Gunter Stent wrote that the origin of a discipline follows a simple pattern of successional stages: first a romantic phase, then a phase of dogmatism, and finally the mature "academic" phase. I think this schema very nicely describes the origin of evolutionary developmental biology, except that I would call the second phase an "enthusiastic" phase.[1] The romantic stage connects to the prehistory of devo-evo reaching back to the second half of the nineteenth century, and culminated in the Field Museum conference on macro-evolution in 1980 and the Dahlem conference on development and evolution in 1981. The latter two conferences expressed a reaction among evolutionary biologist to the then dominant neo-Darwinian theory of evolution that focused on population genetic theory and virtually extinguished organismal biology as part of evolutionary biology. Developmental evolution, though, was not the only reaction to this development; phenotype-based life history theory was another.[2] These conferences also clearly show that devo-evo as a scientific movement was well articulated many years before molecular developmental biology joined the act. This happened only after the discovery of homeobox genes in vertebrates in 1987, which made it clear that animals with very different body plans express very similar genes.[3]

This discovery led the "enthusiastic" phase of developmental evolution, which most closely resembled early days of comparative anatomy, as exemplified in the Geoffroy-Cuvier debate. It was a time when the new tools of comparative developmental genetics led to exaggerated claims of having "solved" the problems of homology by recording the expression of a few genes. It was a time when comparative biology needed to be reinvented by a discipline, developmental biology, that saw no use of it until the early 1990s. Most exemplary is the resuscitation of the idea that a deuterostome animal is an upside-down protostome animal after it was discovered that dorsoventral polarity is determined by the same genes in both proto- and deuterostome animals, but in inverse orientation. I think this fact was an important discovery, but the problem does not end there.[4] (figure 16.1).

Today evolutionary developmental biology has the institutional hallmarks of a successful scientific discipline. There are many faculty positions offered at the best universities, three international journals are dedicated to the subject, an increasing number of textbooks are available for those uninitiated to the field, and a special National Science Foundation panel has been established to support research in the area.[5] So it seems safe to assume that developmental evolution has entered its academic phase, in which a common language is established and problems are discussed rationally on the basis of empirical evidence. History, however, shows that enthusiasm and institutional successes do not guarantee the continued scientific leadership of a discipline.[6] Ultimately the success of a discipline depends on its ability to deliver on the promise the practitioners saw when the field was introduced. A minimal requirement for delivering on such a promise is a common understanding among the practitioners of a discipline on what types of data count as evidence and how such evidence distinguishes between correct and incorrect hypotheses. While developmental evolution (devo-evo) has made great progress over recent years, as witnessed in textbooks, the standards of evidence are handled quite variably in different parts

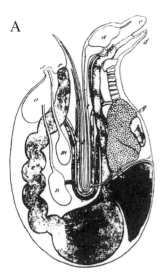
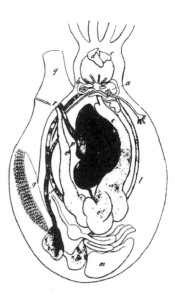

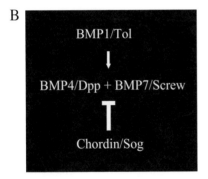

Figure 16.1

The theory of dorsoventral inversion. (A) In the early days of comparative anatomy, extreme forms of comparison were proposed, including a homology of vertebrate and invertebrate central nervous systems (CNS). In protostome invertebrates the CNS is located ventrally, while in vertebrates it is found in a dorsal position. The figure shows a duck folded on its back to make the arrangement of organs roughly comparable with that of a cephalopod. This figure was meant to ridicule attempts to force comparisons among animals with different body plans. (B) The theory of vertebrate/protostome CNS homology gained renewed currency after it was discovered that the nervous system of both kinds of animals is determined by a cascade of homologous genes (deRobertis and Sasai, 1996), suggesting a dorsoventral inversion event in the stem lineage of vertebrates. The figure shows the functional interactions among some of the genes involved in nervous system determination. A dorsoventral inversion might well have happened in evolution, but in this chapter I use this as an example of bold hypothesizing without any follow up, which was characteristic of evolutionary developmental biology during the mid-1990s. It is argued that evolutionary developmental biology has to develop a methodology which makes these hypotheses directly testable. Reproduced from Appel, *The Geoffroy-Cuvier Debate* (1987) by permission of Oxford University Press.

of the field. This is not surprising, given the number of independent disciplines contributing to devo-evo, each with its own research tradition and standards of evidence. The ultimate success of devo-evo, however, will depend on a synthesis of these disparate methodological standards.

A historical example that tells a cautionary tale for devo-evo is that of evolutionary morphology. In the second half of the nineteenth century, evolutionary morphology was a fast-growing, successful discipline similar to devo-evo today. It was a synthesis of comparative anatomy and morphology with evolutionary Darwinian thinking. The most prominent leaders of evolutionary morphology were the invertebrate biologist Ernst Haeckel (1834–1919) and the vertebrate biologist Carl Gegenbaur (1826–1903). Indeed, some of the same questions formulated by those researchers, more than 100 years ago, have reemerged in the new devo-evo synthesis; for example, the origin of vertebrates and the fin-limb transition, among others.[7] By the end of the nineteenth century, however, evolutionary morphology had lost its ability to recruit a critical mass of young scientists, and thus lost much of its momentum. Lynn Nyhart has argued that the reason for the loss of evolutionary morphology's prominent status was, in part, the inability of its practitioners to agree on standards of evidence to evaluate the veracity of evolutionary scenarios.[8] Specifically, there were no methods to evaluate the primacy of anatomical versus developmental data for any particular evolutionary hypothesis, leaving the researcher to rely on personal bias. Examples are the controversies over the origin of paired appendages in vertebrates and the role of segmentation of the vertebrate head.[9] This led to heated personal debates which ultimately convinced the younger generation that the field was not very promising.

THE STATUS OF THE DISCIPLINE

Devo-evo is starting from a stronger position than evolutionary morphology. Nowadays the study of phylogenetic relationships and character evolution are somewhat separated problems. Evolutionary morphology had the enormous task of resolving both simultaneously while including hypotheses regarding the process of evolution. In contrast, the objective of devo-evo is to explain the developmental basis for evolutionary changes of the phenotype, not to reconstruct phylogenetic relationships or patterns of character transformation.

The reconstruction of phylogenetic relationships is currently handled by a diverse array of morphological and molecular methods.[10] Patterns of character change across phylogenetic hypotheses are studied by explicit comparative methods.[11] These advances eliminate many of the difficulties of nineteenth-century evolutionary morphology by providing explicit phylogenetic frameworks and patterns of character change. In spite of these advances, however, the field still lacks effective ways to decide which devo-evo scenario actually supports the developmental mechanisms responsible for an evolutionary character transformation. Which data and inference methods should be used to establish whether a particular developmental change was instrumental in effecting the character transformation? The debate about the role of Hox genes in developmental evolution is a symptom of this problem.[12] In a nutshell that does not contain all the ramifications of this debate, the question is whether the major phenotypic effects of homeotic mutations are the genetic basis of major evolutionary transformations. Are the developmental mechanisms that lead to dramatic phenotype effects, such as the transformation of an antenna into a leg, also an explanation of major evolutionary transformations? Can we conclude from the developmental genetic evidence that the origin of new body parts and body part identity was due to a small or even a single mutation (macromutation)?[13] Or are the genetics of homeotic mutations the result of secondary modifications of the developmental process after the character arose? The problem boils down to establishing plesiomorphic developmental traits as well as distinguishing between developmental changes that have been instrumental in evolution and those which were merely secondary modifications of developmental pathways.[14]

The biggest intellectual danger of any evolutionary research is the temptation to find satisfaction in ingenious "just so" stories.[15] Devo-evo, as the youngest member of the evolutionary sciences, is in particular danger of falling into this trap, as other branches of evolutionary biology did in the past.[16] I agree with Gibson and Frohlich that the way out of this corner has to come from a combination of experimental and hypothesis-driven research programs. But what kinds of hypotheses are critical in evaluating these scenarios, and what is the inference method underlying the testing of hypotheses? There is no generally accepted methodology for the science of devo-evo to address these questions.

CAN A MICROEVOLUTIONARY APPROACH BE THE ANSWER?

One area in which devo-evo has become more rigorous is dealing with microevolutionary problems. In recent years the power of micro-evolutionary approaches to developmental evolution have been clearly demonstrated. Examples are the work on the role of Ubx variation in morphological evolution among closely related *Drosophila* species, the evolution of sexually dimorphic pigmentation patterns in *Drosophila*, and the QTL approach to skeletal evolution among closely related populations of sticklebacks.[17] The advantage of this approach is that it has to bridge much smaller evolutionary distances and, in many cases, can work with a much higher information density than the examples discussed above (i.e., many closely related species with similar developmental trajectories and genomes). While this approach is likely to continue to provide major insights into the mechanisms of developmental evolution, it is also limited in scope. One of the main sources of intellectual excitement in devo-evo is the prospect of understanding major evolutionary transformations.[18] If developmental evolution were to focus exclusively on microevolutionary processes, the field would abandon that major objective.[19] In other words, even a very successful microevolutionary approach to developmental evolution would not fulfill the expectations that have been raised: bridging the gap between evolutionary genetics and macroevolutionary pattern. A more rigorous comparative approach will have to complement microevolutionary work on developmental evolution.

Can There Be a Satisfactory Answer to Macroevolutionary Questions?

Regardless of how desirable a contribution of devo-evo to the study of major evolutionary transformations would be, the acid test will always be whether we can muster sufficient analytical power to gain scientific knowledge on the subject. While it is much too early to say whether this challenge can be answered successfully, I think that the most likely avenue to success may be found in a synthesis in which the guiding hand of comparative anatomy and paleontology determines the agenda of a collaboration between molecular evolution and mechanistic molecular biology. While many of my colleagues will agree about the need for more mechanistic molecular biology in devo-evo, the role of morphology may be less obvious. The contribution of rigorous comparative anatomy and paleontology is essential because they define the problem domain of this research effort.[20] What, exactly, is the pattern of evolutionary change we want to explain? What are the units of evolutionary change at the phenotypic level? What is the functional context of the evolutionary transformations? Neither developmental biology nor evolutionary genetics is prepared to answer these questions by themselves. But well-founded hypotheses on these issues are essential to ground the genetic work in organismal reality.

Another challenge to this enterprise consists in covering vast evolutionary distances and at the same time establishing conceptual continuity between macroevolutionary patterns and known evolutionary and molecular mechanisms. I am referring here to an idea, conceptual continuity, that was coined to explain what molecular cell biology achieved during the second half of the twentieth century. It is not, as commonly thought, the "reduction" of biology to chemistry. Instead, the real beauty of molecular cell biology is that it is able to unite previously disconnected fields, such as transmission genetics, cytogenetics, histology, biochemistry, and physiology, into a unified narrative. Most important, the unified narrative is not based on a single overarching theory, as is sometimes the case in physics, but on a detailed explication of, say, how the hydrolysis of ATP leads to the shortening of a muscle cell, generating force at the organ level.[21]

We think devo-evo is still in a stage where the contributing disciplines are in conceptual *discontinuity*. We understand, to some extent, the patterns of character transformation, the mechanisms of population genetics, the molecular processes that lead to genetic variation, and how gene expression contributes to the development of morphological characters. That is all great progress, but what we do much less well is connect these pieces. Given a character transformation (say, the fin-limb transition), and given that we know something about limb development, how do we know which developmental changes were responsible for the morphological changes?[22] How do we know whether there were natural selective forces acting on the developmental process during the morphological change? How do we know the extent of pleiotropic genetic effects that may have contributed to character conservation or had incidental effects?[23]

Molecular evolution has proven to be a powerful, but not foolproof, method for deep phylogenetic inferences as well as for detecting traces of past evolutionary processes, such as natural selection and population bottlenecks. In other areas of biology, a fruitful interaction between organismal biology, molecular evolution, and experimental molecular biology has been established. One example is the work of Mathews and collaborators on the evolution of phytochrome A in early angiosperms, showing the action of natural selection on the light-sensitive domains of this molecule and thus confirming paleoecological evidence about the biological context of the origin of angiosperms.[24] Here we witness the confluence of mechanistic knowledge about the function of a molecule, its role in the life of the plants, the paleoecological boundary conditions for the origin of angiosperms, and the population genetic processes actually causing evolutionary change. This is a nice example of conceptual continuity in evolutionary biology.

Ideally, a satisfactory methodology in devo-evo should also be able to show the action of natural selection on certain developmental genes in connection with the origin of a derived character, as well as demonstrate that these genetic changes are mechanistically responsible for the derived phenotype. Furthermore, we need to understand the organismal context that set the stage, in an enabling or constraining way, for the molecular changes to have evolutionary significance. What is missing, then, is a proof of the mechanistic efficacy of the identified molecular changes. This ultimately may prove to be the biggest challenge.

How Well Does the Experimental Method Apply to Devo-Evo?

In recent years considerable progress has been made in understanding the developmental basis of morphological evolution.[25] Research programs have predominantly been set in a framework of finding associations between developmental mechanisms that are involved in the development of a derived trait but are absent in species that represent the ancestral character.[26] Examples include new regulatory links between genes that are necessary for the development of the derived character but are absent in the ancestral state. For instance, the male-specific pigmentation of abdominal segments A5 and A6 in *Drosophila melanogaster* is derived within the close relatives of *melanogaster*.[27] Its evolution depends on the origin of two regulatory inputs from *AbdB* and *dsx* onto *bab* that are present in the *melanogaster* group but absent in other drosophilids. Another example is the origin of eyespot patterns in nymphalid butterflies, which has been linked to the evolution of a repressive input on *Cubitus Interruptus* by *Engrailed*.[28]

Another type of link between development and evolution is that of associating differences in the expression levels of developmental genes with differences in morphological traits. For instance, there is evidence that the location of body regions in amniotes is associated with shifts in the axial expression of Hox genes.[29] Differences between the bristle patterns on the mesothoracic femur of *Drosophila melanogaster* and *Drosophila simulans* can be associated with variation of *Ubx* expression.[30] There is also an association between the expression patterns of *Hoxa-11* in the forelimb and hind limb of *Xenopus levis* and the character of the mesopodial (wrist/ankle) bones.[31] But as fascinating as these findings are, most of them are still just associations.

Of course, the ultimate goal would be to establish a causal role for these developmental differences. By demonstrating a causal role, I mean the demonstration that the developmental genetic differences associated with a derived character state are sufficient to produce the derived character state.[32] The test would require introducing the genetic difference into the genome of an organism that represents the ancestral character state and observing whether this has the predicted effect of creating the derived character state. This of course requires some kind of genetic manipulation, and encouraging results have been published which show that in favorable circumstances this is experimentally tractable. For example, the insertion of the zebra fish enhancer of *Hoxd-11* into the mouse *Hoxd-11* genomic domain results in an anterior shift in the position of the sacrum similar to the more anterior position of the pelvic elements in zebra fish.[33]

The difference between bristle patterns in *Drosophila melanogaster* and *Drosophila simulans* cited above was traced to noncoding differences in the respective *Ubx* gene by analyzing F1 hybrids with one *Ubx* null allele.[34] In ascidians the derived absence of a larval tail in *Mogula oculata* was linked to a downregulation of the gene *Manx* compared with the levels of *Manx* expression in species with tails.[35]

These and other successes raise the expectation that a complete causal understanding of morphological differences is attainable. Here I want to discuss a recent mathematical model from the theory of so-called combinatory landscapes which may imply that this promise could be unattainable for evolutionary innovations.

The result I want to discuss is based on a new approach to modeling phenotypic evolution. The approach originated from work on the evolution of RNA secondary structure and is an extension of the pioneering work of Wright on adaptive landscapes.[36] The idea is to consider evolutionary changes as movements in an abstract space in which each point represents a different genotype. The neighborhood relationships among the genotypes reflect the ease with which one genotype is transformed into another. Genotypes closer together in this "genotype space" can more easily be transformed into each other by mutations or recombination, while it takes many steps to transform a genotype into a "distant" genotype. This is familiar to everyone working with DNA or protein sequences, where each sequence can be seen as a point connected to all the other sequences that differ by a single base pair or amino acid substitution (i.e., a "sequence space"). The sequence space represents all possible evolutionary transitions that are possible by point mutations. This idea a has counterpart in computer science and other disciplines in the concept of a configuration space.[37]

To model phenotypic evolution, we have to go a step further. One can also think of phenotypes as points in an abstract phenotype space (or morphospace).[38] In this case, the neighborhood relationships are defined by the underlying genetic changes that transform one heritable phenotype into another. Hence, the phenotype space is defined by the mapping from the underlying genotype space to the corresponding phenotypes, the so-called genotype-phenotype map. One essential observation, at least for macromolecules where phenotypes are molecular secondary or tertiary structures, is that the mapping from sequences (genotypes) to shapes (phenotypes) is from very many sequences to few shapes.[39] In other words, there are many more sequences than there are molecular shapes. This is already reflected in the degeneracy of the genetic code for proteins, but the degeneracy here is much more pronounced than that in the amino acid code. A generic feature of these highly degenerate mappings is that they show extensive nets of neutral (i.e., phenotypically equivalent) genotypes. Very often many of these equivalent genotypes are connected by single nucleotide substitutions. These connected sets of neutral genotypes are called neutral networks.[40] In this model, different phenotypes are represented by different neutral networks.

Using this picture of phenotypic evolution, one can consider a change in the phenotype as a transition from the neutral network representing the ancestral phenotype to the neutral network of the derived phenotype. Whether one phenotype can be transformed directly into another depends on the neighborhood relationships of their respective neutral networks in genotype space. Only phenotypes with neighboring networks can be transformed into each other in a single step. In addition, the relative ease with which one phenotype, A, can become another, B, compared with C, depends on how long the "border" between the respective neutral networks is. If there are many more genotypes in neutral network A that are poised to become B compared with the number of genotypes that are close to C, then the probability of a transition A → B is greater than the probability of A → C (figure 16.2).[41] Asymmetric transition probabilities can also be explained by this model. In figure 16.2 the probability of A → B is higher than the probability of B → A because a larger fraction of states of A are poised to become B than there are of B to become A.

This model of phenotypic evolutionary dynamics can be represented as topological relationships. This observation led Stadler and collaborators

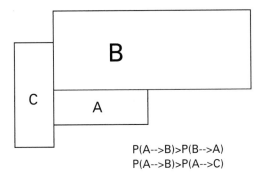

$$P(A-->B)>P(B-->A)$$
$$P(A-->B)>P(A-->C)$$

Figure 16.2
The rectangles in this figure represent neutral networks of genotypes representing phenotypes A, B, and C. In this model the relative lengths of borders between neutral networks determine the likelihood of a mutation transforming one phenotype into another. For instance, there are many more genotypes in neutral network A that are poised to become B than there are genotypes that are close to C. Hence, in this model, the probability of an A → B transition is greater than the probability of an A → C transition. Asymmetric transition probabilities between phenotypic states can also be explained by this model. The probability of A → B is higher than the probability of B → A because a larger fraction of states in A are poised to become B than there are in B to become A.

to propose a topological approach to the theory of phenotypic evolution.[42] Two extreme situations are conceivable. A character state B is easily reachable from character state A if the neutral network of B is accessible from each point on the neutral network of A (figure 16.3). Examples of this have been found in computational models of RNA secondary structure and have been called "shadows."[43] This is most likely a model for most examples of continuous quantitative variation. On the other hand, if the derived character state is accessible from only a small fraction of the ancestral neutral network, then the transition is predicted to be rare. A transition will be possible only if the population occupies a position on the neutral network close to the transition point (figure 16.3). The smaller the fraction of "crossing points" on the ancestral neutral net, the smaller the probability that the population will find itself poised for the transition. Furthermore, natural selection can do nothing to make the transition more likely. The positions within a neutral net are, well, neutral—that is, identical from the standpoint of natural selection, and thus natural selection can do nothing to prod the population to occupy a position at a crossing point.[44] Hence, one reason for certain phenotypic transitions in evolution being rare might be exactly that: a particular mutation needs to happen

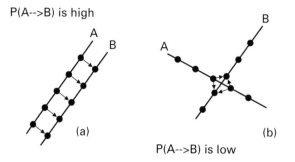

P(A-->B) is high

(a)

(b)

P(A-->B) is low

Figure 16.3

Two extreme cases of accessibility of phenotype B from phenotype A. In this figure the neutral network of genotypes representing the phenotypes A and B are drawn as straight lines, and individual genotypes as black dots. Mutations leading from neutral network A to B are indicated by small arrows. (a) The transition from A to B is possible from every genotype in A. The transition probability is high, and there are no restrictions on accessibility of B from A. This model may be representative for genetic variation of quantitative characters with ample genetic variation. (b) In this case the transition from A to B is possible from only a small fraction of genotypes in A, the ones close to the "crossing point" between A and B. Transitions between A and B are expected to be rare, not only because there are few genotypes that can make the transition by a single mutation, but also because selection cannot push the ancestral population toward the contact point between A and B. An evolutionary transition can happen only if the ancestral population happens to be close to a contact point with B by accident (i.e., by genetic drift).

in a genome that is poised for this transition. If the genetic background is not at a crossing point, that mutation will have no effect or a different effect. In a recent computational study of RNA secondary structure evolution, Peter Schuster and his collaborators found a negative correlation between the phenotypic effect of a mutation and its frequency.[45]

The problem with analyzing this mode of evolution is not that there is no causal connection between genetic and phenotypic evolution; in fact, there is. But it might be impossible to demonstrate this connection experimentally. This is because experiments can be done only with the species alive now, and there is no guarantee that any recent species has a genotype poised to replicate the evolutionary transition in question (figure 16.4). If we are lucky, we will be able to identify an outgroup species that represents the ancestral phenotypic character state to some level of approximation. But of course there is no reason to think that the population will be residing at the same position on the neutral net as the ancestral population that actually made the transition. More likely, the outgroup species

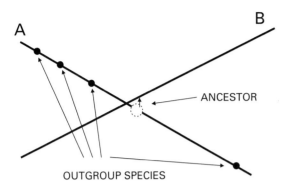

Figure 16.4

The problem of demonstrating which particular genetic change caused a rare evolutionary transition between A and B. The figure represents the same situation as in figure 16.3b. The transition must have happened in an ancestral population with a genotype that was close enough to the crossing point between A and B to allow a mutation to cause this transition. But recent outgroup species that represent the ancestral phenotype state A are not likely to have the genotype that would allow a small genetic manipulation (represented here by small arrows) to create the phenotype B. The reason is that the rarity of the transition in this model is due to the assumption that there are many more genotypes in the neutral network of A far from the contact point with B than there are genotypes close to the contact point. Hence, it is very unlikely that a living species has a genotype close to the contact point with B, and it will be impossible to experimentally demonstrate exactly which genetic change caused the transition to B in evolution.

will be located on some random point on the neutral network rather than exactly at the crossing point (figure 16.4). If we then perform a transgenic experiment and introduce the genetic change that was responsible for the origin of the derived character, there may be no effect or an effect different from the one predicted. This will happen even if we find the "correct" genetic difference; we simply will not be able to demonstrate the effect because we do not know the genetic background that made the transition possible in the first place. This is a methodological problem, but it is a real one that cannot be overcome by technical advances. It is a problem we must face if the above model of phenotypic evolution applies to at least a significant subset of phenotypic transformations. It goes to the heart of what is knowable in developmental evolution.

The above reasoning may apply specifically to rare phenotypic transitions if the rarity of the transition is due to developmental constraints. Of course, if a transition does not happen because there is no selection in favor of the derived character state, the reasoning does not apply. In a paper on the concept of evolutionary innovation, Gerd Müller and I argued that innovations represent such a class of transitions. We stated that it is characteristic of evolutionary innovations that they are realized in spite of, and by overcoming, ancestral constraints.[46] One can even argue that this suggestion can be used to conceptually distinguish between adaptation and innovation. The distinction is worth maintaining only if it implies a different research program for studying innovations than for the study of adaptations: to explain innovations, one needs to identify the specific developmental changes that overcame the ancestral developmental constraints on the ancestral character state.[47]

In a nutshell, I am arguing that if innovations are in fact caused by overcoming ancestral constraints, and if we can model this process as a rare evolutionary event (in the sense of figures 16.3b and 16.4), then it might be impossible to experimentally demonstrate exactly which genetic changes caused the evolutionary innovation.

The limitation of what can be known in developmental evolution suggested by figure 16.4 may be severe, but first we must ascertain whether the model applies to any real evolutionary situation. Maybe realized evolutionary transitions always take the "path of least resistance" and all of them represent "topologically likely" mutational transitions.[48] The molecular nature of these mutations and their effect on the phenotype may then be experimentally provable with any, or at least some, outgroup species by means of appropriate genetic manipulations. Whether this is possible is an empirical question that can be answered only by trying to identify molecular differences that could be responsible for the evolutionary phenotypic transformation. There has been some success in showing causal efficiency of certain molecular genetic differences. The question, then, is whether the success is limited to certain types of transformations (e.g., loss of characters or quantitative and meristic variations), or whether there are examples where explanations for true novelties can be tested experimentally.[49] Demonstrating causality of a particular genetic difference in creating a new phenotypic character would falsify the model for rare transformations based on neighborhood relationships between neutral nets.

The closest we have gotten to understanding the genetic basis of an unquestionable innovation is the work of Keys and collaborators on the origin of eyespot patterns.[50] It would be extremely important to know whether causing a loss of ci repression by en is sufficient to cause the development of something resembling an eyespot in a nonnymphalid butterfly.

QUO VADIS DEVO-EVO?

Certainly a plurality of methods will have to be deployed to meet the challenge of developmental evolution, in particular if we do not short-change the expectations associated with the field. These expectations, I would argue, center around a deeper understanding of the mechanisms of major evolutionary transitions with their corollaries, an understanding of body plans and the evolutionary novelties. Exactly how this synthesis will be achieved is unpredictable, but if we take our clues from molecular biology, a possible answer might seem plausible. As mentioned above, the intellectual contribution of molecular biology can be seen as a continuous narrative that starts at the molecular and atomic levels and ends in an explanation of cellular and organismal properties.[51] Its main achievement is to allow conceptual continuity across several levels of organization and to connect historically independent disciplines such as genetics and biochemistry and cytology. This situation can be mistaken for "reducing" biology to chemistry, but this is possibly a mistake.[52] But putting aside this philosophical issue, I would envision that a successful field of developmental evolution will take on a similar shape: a conceptually continuous narrative that connects the molecular processes which create genetic variation, the developmental processes which lead to phenotypes, their function, and the population genetic processes which ultimately effect evolutionary change. Then a synthesis of developmental and evolutionary biology obviously will have been achieved.

ACKNOWLEDGEMENTS

The author thanks the organizers of the Dibner workshop on the history of evolutionary developmental biology for the opportunity to participate in this event, and Manfred Laubichler for comments and suggestions on this chapter.

NOTES

1. Stent, 1968; Wagner et al., 2000.

2. Stearns, 1984.

3. Hart et al., 1987; Schughart et al., 1987.

4. Arendt and Nübler-Jung, 1994; DeRobertis and Sasai, 1996.

5. Carroll et al., 2001; Davidson, 2001; Wilkins, 2002; And the Journals *Evolution and Development*, *Genes, Development and Evolution*, and *JEZ-Part B: Molecular and Developmental Evolution*.

6. Nyhart, 1995.

7. Hall, 1992; Laubichler and Maienschein, 2003; Raff, 1996.

8. Nyhart, 1995, 2002.

9. Kuratani, 2003; Mitgutsch, 2003; Nyhart, 1995.

10. Hillis et al., 1996.

11. Felsenstein, 2004.

12. Budd, 1999; Gellon and Mcginnis, 1998; Robert, 2001; Schwartz, 1999.

13. Schwartz, 1999.

14. Budd, 1999; Müller and Newman, 1999; Newman and Müller, 2001; Stern, 2000; Wagner et al., 2000.

15. Gould and Lewontin, 1978.

16. Chiu and Hamrick, 2002; Frohlich, 2003; Gibson, 1999.

17. Gibson and Hogness, 1996; Gompel and Carroll, 2003; Stern, 1998; Sucena et al., 2003; Kopp et al., 2000; Peichel et al., 2001.

18. Hall, 1992; Raff, 1996.

19. Ganfornina and Sanchez, 1999.

20. Larsson and Wagner, 2002.

21. Herrmann, 1998.

22. Coates, 2003; Shubin, 1995; Wagner and Chiu, 2001.

23. E.g., Galis and Metz, 2001.

24. Mathews et al., 2003.

25. Carroll et al., 2001; Chiu and Hamrick, 2002.

26. Wagner et al., 2000.

27. Kopp et al., 2000.

28. Keys et al., 1999.

29. Belting et al., 1998; Burke et al., 1995; Gerard et al., 1997.

30. Stern, 1998.

31. Blanco et al., 1998.

32. Wagner et al., 2000.

33. Gerard et al., 1997.

34. Stern, 1998.

35. Swalla and Jeffery, 1996.

36. Cupal et al., 2000; Fontana and Schuster, 1998; Schuster et al., 1994; B. M. Stadler et al., 2001; Provine, 1986.

37. P. F. Stadler, 1996.

38. Eble, 2000.

39. Schuster et al., 1994.

40. Ibid.

41. Fontana and Schuster, 1998.

42. B. M. Stadler et al., 2001.

43. Fontana and Schuster, 1998.

44. This is not completely true because the probability of which part of the network is occupied is not entirely random. If the mutation rate is high enough, the population is more likely to be located on places with a high density of neutral states (Nimwegen et al., 1999). Nevertheless, this effect does not make the transition any more likely.

45. Schuster et al., in preparation.

46. Müller and Wagner, 1991. See also Eberhard (2001) and Wagner and Müller (2002) for a recent controversy on that subject.

47. Prum, 1999.

48. Schluter, 1996.

49. Müller and Wagner, 1991.

50. Keys et al., 1999.

51. Herrmann, 1998.

52. Laubichler and Wagner, 2001; Rosenberg, 1997.

REFERENCES

Arendt, D., and K. Nübler-Jung. 1994. Inversion of dorso-ventral axis? *Nature* 371: 26.

Belting, H. G., C. S. Shashikant, and F. H. Ruddle. 1998. Modification of expression and cis-regulation of Hoxc8 in the evolution of divergent axial morphologies. *Proceedings of the National Academy of Sciences USA* 95: 2355–2360.

Blanco, M. J., B. Y. Misof, and G. P. Wagner. 1998. Heterochronic differences of Hoxa-11 expression in *Xenopus* fore- and hind limb development: Evidence for a lower limb identity of the anuran ankle bones. *Development, Genes and Evolution* 208: 175–187.

Budd, G. E. 1999. Does the evolution in body patterning genes drive morphological change—or vice versa? *BioEssays* 21: 326–332.

Burke, A. C., C. E. Nelson, B. A. Morgan, and C. Tabin. 1995. Hox genes and the evolution of vertebrate axial morphology. *Development* 121: 333–346.

Carroll, S. B., J. K. Grenier, and S. D. Weatherbee. 2001. *From DNA to Diversity*. Malden, Mass.: Blackwell Science.

Chiu, C.-H., and M. W. Hamrick. 2002. Evolution and development of the primate limb skeleton. *Evolutionary Anthropology* 11: 94–107.

Coates, M. I. 2003. The evolution of paired fins. *Theory in Biosciences.* 122: 266–287.

Cupal, J., S. Kopp, and P. F. Stadler. 2000. RNA shape space toplogy. *Artificial Life* 6: 3–23.

Davidson, E. 2001. *Genomic Regulatory Systems: Development and Evolution*. San Diego: Academic Press.

DeRobertis, E. M., and Y. A. Sasai. 1996. A common plan for dorso-ventral patterning in Bilateria. *Nature* 380: 37–40.

Eberhard, W. G. 2001. Multiple origins of a major novelty: Moveable abdominal lobes in male sepsid flies (Diptera: [S]epsidae), and the question of developmental constraints. *Evolution and Development* 3: 206–222.

Eble, G. 2000. Theoretical morphology: State of the art. *Paleobiology* 26: 498–506.

Felsenstein, J. 2004. *Inferring Phylogenies.* Sunderland, Mass.: Sinauer.

Fontana, W., and P. Schuster. 1998. Continuity in evolution: On the nature of transitions. *Science* 280: 1451–1455.

Frohlich, M. W. 2003. An evolutionary scenario for the origin of flowers. *Nature Review Genetics* 4: 559–566.

Galis, F., and J. A. Metz. 2001. Testing the vunerability of the phylotypic stage: On modularity and evolutionary conservation. *Journal of Experimental Zoology, Part B: Molecular and Developmental Evolution* 291: 195–204.

Ganfornina, M. D., and D. Sanchez. 1999. Generation of evolutionary novelty by function shift. *Bioessays* 21: 432–439.

Gellon, G., and W. Mcginnis. 1998. Shaping animal body plans in development and evolution by modulation of Hox expression patterns. *BioEssays* 20: 116–125.

Gerard, M., J. Zakany, and D. Duboule. 1997. Interspecies exchange of a Hoxd enhancer in vivo induces premature transcription and anterior shift of the sacrum. *Developmental Biology* 124: 3493–3500.

Gibson, G. 1999. Developmental evolution: Going beyond the: "just so." *Current Biology* 9: 942–945.

Gibson, G., and D. S. Hogness. 1996. Effect of polymorphism in the *Drosophila* regulatory gene Ultrabithorax on homeotic stability. *Science* 271: 200–203.

Gompel, N., and S. B. Carroll. 2003. Genetic mechanisms and constraints governing the evolution of correlated traits in drosophilid flies. *Nature* 424: 931–935.

Gould, S. J., and R. Lewontin. 1978. The spandrels of San Marco and the Panglossian paradigm: A critique of the adaptationist programme. *Proceedings of the Royal Society of London* 205: 581–598.

Hall, B. K. 1992. *Evolutionary Developmental Biology*. New York: Chapman & Hall.

Hart, C. P., L. D. Bogarad, A. Fainsod, and F. H. Ruddle. 1987. Polypurine/polypyrimidine sequence elements of the murine homeobox loci Hox-1, -2, and -3. *Nucleic Acid Research* 15: 5495.

Herrmann, H. 1998. *From Biology to Sociopolitics: Conceptual Continuity in Complex Systems*. New Haven, Conn.: Yale University Press.

Hillis, D. M., C. Moritz, and B. K. Mable (Eds.). 1996. *Molecular Systematics*. Sunderland, Mass.: Sinauer.

Keys, D. N., D. L. Lewis, J. E. Selegue, B. J. Pearson, L. V. Goodrich, et al. 1999. Recruitment of a hedgehog regulatory circuit in butterfly eyespot evolution. *Science* 283: 532–534.

Kopp, A., I. Duncan, and S. B. Carroll. 2000. Genetic control and evolution of sexually dimorphic characters in *Drosophila*. *Nature* 408: 553–559.

Kuratani, S. 2003. Evolutionary developmental biology and vertebrate head segmentation: A perspective from developmental constraint. *Theory in Biosciences* 122: 230–251.

Larsson, C. E., and G. P. Wager. 2002. The pentadactyl ground state of the avian wing. *Journal of Experimental Zoology, Part B: Molecular and Developmental Evolution* 294: 146–151.

Laubichler, M. D., and J. Maienschein. 2003. Ontogeny, anatomy, and the problem of homology: Carl Gegenbaur and the American tradition of cell lineage studies. *Theory Biosciences* 122: 194–203.

Laubichler, M. D., and G. P. Wagner. 2001. How molecular is molecular developmental biology? A reply to Alex Rosenberg's Reductionism Redux: Computing the embryo. *Biology and Philosophy* 16: 53–68.

Mathews, S., J. G. Burleigh, and M. J. Donoghue. 2003. Adaptive evolution in the photosensory domain of phytochrome A in early angiosperms. *Molecular Biology and Evolution* 20: 1087–1097.

Mitgutsch, C. 2003. On Carl Gegenbaur's theory of head metamerism and the selection of taxa for comparisons. *Theory Biosciences* 122: 204–229.

Müller, G. B., and S. A. Newman. 1999. Generation, integration, autonomy: Three steps in the evolution of homology. In G. R. Bock and G. Cardew (eds.), *Homology*, pp. 65–79. New York: Wiley.

Müller, G. B., and G. P. Wagner. 1991. Novelty in evolution: Restructuring the concept. *Annual Review of Ecology and Systematics* 22: 229–256.

Newman, S. A., and G. B. Müller. 2001. Epigenetic mechanisms of character origination. In G. P. Wagner (ed.), *The Character Concept in Evolutionary Biology*. San Diego: Academic Press.

Nyhart, L. K. 1995. *Biology Takes Form: Animal Mophology and the German Universities, 1800–1900*. Chicago: University of Chicago Press.

————. 2002. Learning from history: Morphology's challenges in Germany ca. 1900. *Journal of Morphology* 252: 2–14.

Peichel, C. L., K. S. Nereng, K. A. Ohgi, B. L. Cole, P. F. Colosimo, et al. 2001. The genetic architecture of divergence between threespine stickleback species. *Nature* 414: 901–905.

Provine, W. B. 1986. *Sewall Wright and Evolutionary Biology*. Chicago: University of Chicago Press.

Prum, R. O. 1999. Development and evolutionary origin of feathers. *Journal of Experimental Zoology, Part B: Molecular and Developmental Evolution* 285: 291–306.

Raff, R. 1996. *The Shape of Life: Genes, Development, and the Evolution of Animal Form*. Chicago: University of Chicago Press.

Robert, J. S. 2001. Interpreting the homeobox: Metaphors of gene action and activation in development and evolution. *Evolution and Development* 3: 287–295.

Rosenberg, A. 1997. Reductionism redux: Computing the embryo. *Biology and Philosophy* 12: 445–470.

Schluter, D. 1996. Adaptive radiation along genetic lines of least resistance. *Evolution* 50: 1766–1774.

Schughart, K., C. Kappen, and F. H. Ruddle. 1987. Mammalian homeobox-containing genes: Genomic organization, structure, expression and evolution. *British Journal of Cancer* 58: 9–13.

Schuster, P., W. Fontana, P. F. Stadler, and I. Hofacker. 1994. From sequences to shapes and back: A case study in RNA secondary structure. *Proceedings of the Royal Society of London* B255: 279–284.

Schwartz, J. H. 1999. *Sudden Origins: Fossils, Genes, and the Emergence of Species*. New York: Wiley.

Shubin, N. 1995. The evolution of paired fins and the origin of tetrapod limbs. *Evolutionary Biology* 28: 39–86.

Stadler, B. M. R., P. F. Stadler, G. P. Wagner, and W. Fontana. 2001. The topology of the possible: Formal spaces underlying patterns of evolutionary change. *Journal of Theoretical Biology* 213: 291–279.

Stadler, P. F. 1996. Landscapes and their correlation functions. *Journal of Mathematical Chemistry* 20: 1–45.

Stearns, S. C. 1984. Models in evolutionary ecology. In K. Wohrmann and V. Loeschke (eds.), *Population Biology and Evolution*. Berlin: Springer-Verlag.

Stent, G. 1968. That was the molecular biology that was. *Science* 160: 390–395.

Stern, D. L. 1998. A role of Ultrabithorax in morphological differences between *Drosophila* species. *Nature* 396: 463–466.

———. 2000. Evolutionary developmental biology and the problem of variation. *Evolution* 54: 1079–1091.

Sucena, E., I. Delon, I. Jones, F. Payre, and D. L. Stern. 2003. Regulatory evolution of shavenbaby/ovo underlies multiple cases of morphological parallelism. *Nature* 424: 935–938.

Swalla, B. J., and W. R. Jeffery. 1996. Requirement of the Manx gene for expression of chordate features in a tailless ascidian larva. *Science* 274: 1205–1208.

van Nimwegen, E., J. P. Crutchfield, and M. Huynen 1999. Neutral evolution of mutational robustness. *Proceedings of the National Academy of Science U.S.A.* 96: 9716–9720.

Wagner, G. P., and C.-H. Chiu. 2001. The tetrapod limb: A hypothesis on its origin. *Journal of Experimental Zoology, Part B: Molecular and Developmental Evolution* 291: 226–240.

Wagner, G. P., C.-H. Chiu, and M. Laubichler. 2000. Developmental evolution as a mechanistic science: The inference from developmental mechanisms to evolutionary processes. *American Zoologist* 40: 819–831.

Wagner, G. P., and G. B. Müller. 2002. Evolutionary innovations overcome ancestral constraints: A re-examination of character evolution in male sepsid flies (Diptera: Sepsidae). *Evolution and Development* 4: 1–6.

Wilkins, A. S. 2002. *The Evolution of Developmental Pathways*. Sunderland, Mass.: Sinauer.

About the Authors

Garland E. Allen is Professor of Biology at Washington University in St. Louis, where he has been teaching both biology and history of science since 1967. His major research interests include the history of genetics, evolution and embryology, and their interrelationships in the post-Darwinian era through the mid-twentieth century, as well as the biological and social history of eugenics in the United States. He is the author of *Life Sciences in the Twentieth Century* (Cambridge, 1978), *Thomas Hunt Morgan, the Man and His Science* (Princeton, 1978), *Biology: Scientific Process and Social Issues* (Wiley, 2001), several co-edited volumes, and four editions of the textbooks, *Matter, Energy and Life* and *The Study of Biology* (Addison Wesley/Benjamin Cummings, 1966, 1972, 1976 and 1982) all with Jeffrey J. W. Baker.

Frederick B. Churchill is Professor Emeritus of History and Philosophy of Science at Indiana University. His research focuses on the interface of evolutionary, developmental and heredity theories in nineteenth- and early twentieth-century life sciences. He is co-editor (with Helmut Risler) *of August Weismann, Selected Letters and Documents*, 2 Vols. (1999). Over the years he has written on Wilhelm Johannsen, Rudolf Virchow, Wilhelm Roux, Oskar Hertwig, Julian S. Huxley, and recently on Alfred Kinsey. Currently he is engaged in writing a biography of Weismann.

Elihu M. Gerson is Director of the Tremont Research Institute in San Francisco. He received degrees in sociology from Queens College and the University of Chicago. He studies the social organization of distributed technical work, especially the history and organization of comparative biology. He also wonders about the methodological problems associated with this research.

Scott F. Gilbert is the Howard A. Schneiderman Professor of Biology at Swarthmore College, where he teaches embryology and evolutionary developmental biology. He is the author of the textbook, *Developmental Biology*, the co-author of *Bioethics and the New Embryology*, and the co-editor of *Embryology: Constructing the Organism*. His research concerns the developmental genetic changes involved in how the turtle forms its shell.

James Griesemer is Professor and Chair of Philosophy and a faculty member in the Center for Population Biology and the Science and Technology Studies Program, University of California, Davis. He works on the history, philosophy, and social studies of biology. Past studies include the material culture of modeling in natural history museums and visual representation in genetics, embryology, and evolutionary biology. His recent work focuses on the analysis of reproduction, development, and inheritance in the evolutionary process. He is editor, (with Eörs Szathmáry) of Tibor Gánti's *The Principles of Life* (Oxford University Press, 2003).

Brian K. Hall, University Research Professor and George M. Campbell Professor of Biology at Dalhousie University, began his career as an experimental embryologist, studying the development of the embryonic vertebrate skeleton with an evolutionary approach. A synthesis of evolution and development, *Evolutionary Development Biology*, was published in 1992 (2nd ed. 1998), followed by *Keywords and Concepts in Evolutionary Developmental Biology* (2003), co-edited with Wendy Olson.

Manfred D. Laubichler is Assistant Professor of Biology in the School of Life Sciences and Affiliated Assistant Philosophy at Arizona State University. His research interests include the conceptual and theoretical foundations of evolutionary developmental biology, the conceptual basis of social complexity, quantitative and population genetics and the history of theoretical and developmental biology. He is an associate editor of the *Journal of Experimental Zoology Part B: Molecular and Developmental Evolution* and a founding associate editor of the journal *Biological Theory*.

Alan Love is Assistant Professor in the Department of Philosophy at the University of Minnesota. His research concentrates on epistemological issues in the biological sciences, especially as they arise in the disciplinary synthesis of evolutionary developmental biology. Other philosophical interests include the nature of scientific explanation, conceptual change in the recent history of biology, and the role of historical investigation in philosophy of science.

Jane Maienschein is Regents' Professor and Parents Association Professor at Arizona State University, where she also directs the Center for Biology and Society. She specializes in history and philosophy of biology and the way biology, bioethics, and biopolicy play out in society. Focusing on embryology, genetics, and cell biology, Maienschein combines detailed analysis of the epistemological standards, theories, laboratory practices, and experimental approaches with study of people, institutions, and changing social, political, and legal context in which science thrives. She loves teaching and has won the History of Science Society's Joseph Hazen Education Award and Arizona State University's distinguished faculty awards. Her three books and ten co-edited books include most recently *Whose View of Life? Embryos, Cloning, and Stem Cells.*

Gerd B. Müller is Professor of Zoology at the University of Vienna, Austria, where he heads the Department of Theoretical Biology. He is also Chairman of the Konrad Lorenz Institute for Evolution and Cognition Research in Altenberg, Austria. His primary scientific interest is in evolutionary developmental biology, with a focus on the relationship between genetic and epigenetic determinants of morphological organization. This work includes the development of three-dimensional computational techniques for the analysis of gene activity patterns in morphogenetic processes.

Stuart A. Newman is Professor of Cell Biology and Anatomy at New York Medical College, Valhalla, New York. He works in the fields of cell differentiation, theory of biochemical networks and cell pattern formation, protein folding and assembly, and mechanisms of morphological evolution. Newman has been an INSERM Fellow at the Pasteur Institute, Paris, a Fogarty Senior International Fellow at Monash University, Australia, a visiting scientist at the University of Paris-Sud, the Indian Institute of Science in Bangalore, and the University of Tokyo. He is co-author (with Gabor Forgacs) of *Biological Physics of the Developing Embryo* (Cambridge, 2005).

Marsha Richmond is Associate Professor of Science and Technology, Department of Interdisciplinary Studies, Wayne State University, Detroit, Michigan. She formerly served as an Editor of the correspondence of Charles Darwin. Her research focuses on heredity, genetics, evolution, and cell theory in the late nineteenth and early twentieth centuries, and on the entry of women into academic biology. She is currently finalizing a book manuscript on Richard Goldschmidt and sex determination and beginning work on a monograph on women in the early history of genetics, 1900–1935.

Günter P. Wagner is Alison Richard Professor of Ecology and Evolutionary Biology and Chair of the Department of Ecology and

Evolutionary Biology at Yale University. He is a MacArthur Fellow, a Fellow of the American Association for the Advancement of Science, and a Corresponding Member of the Austrian Academy of Sciences in Vienna. He is the editor-in-chief of the *Journal of Experimental Zoology*. Among his numerous publications are the edited volumes *The Character Concept in Evolutionary Biology* (Academic Press, 2001) and *Modularity in Development and Evolution* (University of Chicago Press, 2004 with Gerhard Schlosser.)

William Wimsatt is Professor of Philosophy, and is a member of the Committee on Evolutionary Biology and the Committee on Conceptual and Historical Studies of Science at the University of Chicago. He works on methodological problems arising in the analysis of complex systems, especially in evolutionary biology but also in various of the human and physical sciences. He has also written on the history of classical genetics, on the nature of scientific visualization, and the problems of scientific change. Currently he is working especially on the role of development in evolution and on problems in the construction of adequate theories of cultural evolution.

John P. Wourms is Professor of Biological Sciences at Clemson University. His research, which is supported by NSF and NOAA, focuses on the evolution of development; evolution of viviparity and maternal-embryonic relationships in fishes; and the history of embryology. He has carried out extensive field studies viz. East Africa, Australia, Cruise 5 of the *Te Vega* across the Indian Ocean, and deep submersible studies in the Monterey Canyon. He serves on the editorial boards of the *Journal of Morphology* and *Environmental Biology of Fishes*.

INDEX

Note: Figures, tables, and boxes are indicated by italics